AESS Interdisciplinary Environmental Studies and Sciences Series

More information about this series at http://www.springer.com/series/13637

Krystyna Stave · Goraw Goshu
Shimelis Aynalem
Editors

Social and Ecological System Dynamics

Characteristics, Trends, and Integration in the Lake Tana Basin, Ethiopia

AESS
Association for Environmental Studies and Sciences

Springer

Editors
Krystyna Stave
School of Public Policy and Leadership
University of Nevada Las Vegas
Las Vegas, NV
USA

Goraw Goshu
Blue Nile Water Institute and College of Agriculture and Environmental Sciences
Bahir Dar University
Bahir Dar
Ethiopia

Shimelis Aynalem
Blue Nile Water Institute and College of Agriculture and Environmental Sciences
Bahir Dar University
Bahir Dar
Ethiopia

ISSN 2509-9787 ISSN 2509-9795 (electronic)
AESS Interdisciplinary Environmental Studies and Sciences Series
ISBN 978-3-319-45753-6 ISBN 978-3-319-45755-0 (eBook)
DOI 10.1007/978-3-319-45755-0

Library of Congress Control Number: 2017932415

Printed on acid-free paper

This Springer imprint is published by Springer Nature
The registered company is Springer International Publishing AG
The registered company address is: Gewerbestrasse 11, 6330 Cham, Switzerland

Preface

The Lake Tana Basin, the catchment area of Ethiopia's largest freshwater lake, is the source of the Blue Nile River, the main tributary of the world's longest transboundary river. This rich social-ecological system is home to more than three million people and diverse ecological resources. Water, fish, and forest resources, agricultural and other products are used in the region and beyond. Human activities in the basin affect the characteristics of local ecosystem resources, and, in turn, affect the quality and sediment load of the water that flows downstream from the basin. Within the basin, a growing population puts pressure on agricultural land and other resources and generates a flow of people from rural to urban areas. In addition to food, urban growth increases demand for construction resources like eucalyptus poles and energy resources such as charcoal from rural lands. People, animals, and goods flow into the city from the rural surroundings and back out again daily. The links between urban and rural areas, upstream and downstream ecosystems, and socioeconomic and ecological system components in the region are complex and dynamic, making it difficult to see how changes in one part feed through the system to change other components and, therefore, to develop policies for sustainable management.

This book is the result of conversations among the three editors in late 2013 about these dynamic connections and policy challenges. Krys Stave, a system dynamicist studying social-ecological systems, had recently arrived at Bahir Dar University as a Fulbright Scholar. Goraw Goshu, a water quality and environmental expert, had been setting the course for the recently established Blue Nile Water Institute as its Director. Shimelis Aynalem, who was teaching and conducting research on the region's birds, fauna, and wetlands, had recently published a book on the Birds of Lake Tana. Krys talked about her difficulties finding published research about the region and her conversations with researchers in the region about their work and concerns. Goraw described the regional conferences the BNWI had held to identify issues, the research being conducted by Institute scientists, and the repository of M.Sc. and Ph.D. theses, conference papers, manuscripts, research reports, and policy briefs the Institute was building. One of the Institute's projects was to organize these resources into a digital library, but until that happened, access to the information was very limited. Shimelis related stories from his field work about unique and little known features of the Lake Tana ecosystem. It became clear that, while there was a lot of good work being done locally, it was getting little recognition from policy makers at all levels and from other researchers outside the region. In addition, local researchers felt their work and concerns were not being fully considered by decision-makers and decision-makers felt researchers were not giving them the information they needed to answer specific applied questions. There was a need for greater integration across disciplines and between policy-makers and researchers.

The book project was thus created to make the work of local researchers visible to other researchers and decision-makers at all levels and promote regional integration. It has four objectives: compile the work of researchers in the region about its socioeconomic and ecological conditions, make local research available in an easily accessible format to external researchers and others wanting to learn about the region, assess the state of knowledge and identify research needs, and develop a systems framework to integrate what is known and provide a platform for further collaborative work.

We recruited local researchers, decision-makers, and practitioners mainly from Bahir Dar University and resource management organizations in the region to contribute chapters about their areas of expertise. We asked that each chapter provide an overview of the chapter topic, a detailed baseline characterization of what is known, a summary of research that has been done on the topic to date, and a discussion of research questions and issues for further study. A total of 49 authors contributed to the 35 chapters. The book, organized in five Parts, describes basic characteristics of the Lake Tana basin, their dynamic interactions, sustainability issues, and a systems framework for collaborative and integrative management. The first Part sets the stage, with an introduction to the region's challenges, a theoretical systems perspective to organize the details to follow, and a historical overview. The next three parts contain the heart of the book, covering ecosystem characteristics, social system characteristics, and management institutions and innovations. The final part synthesizes the work and looks forward. It includes a systems map developed as the basis for further cross-disciplinary collaboration and concludes with a summary of research needs and potential projects. The book is not comprehensive. For example, it does not touch on public health. Rather, we sought to provide a foundation and framework for addressing the question of how to sustainably manage resources in the face of social and ecological pressures. The key theme that ties the chapters together is this: human and environmental system components are linked in diverse and reciprocal ways, and the linkages govern the way the system responds to disturbances. Understanding these connections and dynamics is critical for managing the system for sustainability.

The book has two main audiences: policy-makers, national and international researchers, development practitioners, non-governmental organizations, and others seeking basic scientific knowledge and an integrated understanding of change in this particular region, and those facing similar challenges in other places. For the first, this book compiles baseline characteristics and trends in an easily accessible form and builds a base for collaborative problem-solving. For the second, it is an example and framework for understanding and communicating the science in this complex system in a way that supports policy-making.

An edited volume like this is the collective work of many people. First and foremost, it would not exist without the efforts of the exceptional chapter authors. We thank them for their contributions, timely responses to reviews, and patience while we assembled the work as a whole. In addition to the people who contributed chapters, a number of others offered encouragement, material and administrative support, and excellent suggestions. Dr. Baylie Damtie, President of Bahir Dar University, was instrumental in championing the book and urging researchers to participate in the project. The Office of the Vice President for Research and Community Services of Bahir Dar University, the Blue Nile Water Institute and the College of Agriculture and Environmental Sciences provided the environment for researchers to participate in the book. We thank them for their invaluable support.

To ensure the quality of the work, a number of people reviewed chapters and provided general comments on the topics and direction of the book. They include: Jonathan Patz (Global Health Institute, University of Wisconsin, Madison, Wisconsin, USA), Zeleke Mekuriaw (International Livestock Research Institute, Bahir Dar, Ethiopia), Ian Francis (ESRO Aberdeen and Royal Society of the Protection of Birds, UK), David Cabot (Environmental Consultancy Services, Carrigskeewaun, Carrowniskey, Westport, Co. Mayo, Ireland), Koos Vijverberg (Netherlands Institute of Ecology, Wageningen, the Netherlands) and Masresha Fetene (Addis Ababa University, Addis Ababa, Ethiopia). Their willingness to read and suggest improvements made the book better.

Krys Stave was supported by the Core Fulbright U.S. Scholar Program for her time in Ethiopia. She is immensely grateful to the program for the opportunity to learn from and collaborate with her many wonderful colleagues at Bahir Dar. The U.S. Embassy generously provided the grant that made the system mapping workshop described in Chap. 34 possible. She thanks her contacts at the Public Affairs Section of the U.S. Embassy in Addis Ababa, including Yohannes Birhanu, Tahra Vose and Learned Dees for facilitating her work with Bahir Dar University and Gedion Mamo for helping with the funding for the workshop. Thank you also to Tizezew Shimekach for her research assistance and friendship throughout the project.

Finally we thank our families. This has been an absorbing project that has taken time we otherwise would have been able to spend with them. Thank you Simon Wade, Dejen Gidey Stave, Tedros Gidey Stave, Banchi Amlak Worku, HaileAmlak Goraw, Hana Mariam Goraw, Dawit Goraw, Markos Goraw, Mazashwork Abebe, Marshet Shimelis, Samuel Shimelis, Haile-Gabriel Shimelis, and Hana Shimelis. Your love and support make our lives better every day.

Las Vegas, USA — Krystyna Stave
Bahir Dar, Ethiopia — Goraw Goshu
Bahir Dar, Ethiopia — Shimelis Aynalem

Contents

Editors and Contributors

About the Editors

Krystyna Stave is Professor of Environmental Studies at the University of Nevada Las Vegas. Her primary professional focus is to promote sustainable relationships between people and the environment. She uses a system dynamics perspective in her research and teaching to examine social-ecological systems, increase community understanding of environmental issues, and engage stakeholders in policy analysis and environmental decision-making. She teaches systems modeling, conducts research on the effectiveness of systems models for environmental decision-making, and works with stakeholder groups to develop models for specific issues. Recent applications of her participatory systems modeling research include projects in water policy, transportation-related air quality management, zero-waste planning and food system resilience. Her current research focuses on developing simulation models for discovery learning about sustainability and using participatory modeling to foster transdisciplinary collaboration on environmental and public problem-solving. She holds an undergraduate engineering degree in Operations Research, an M.S. degree in Systems Analysis and Policy Design, and Ph.D. from the Yale School of Forestry and Environmental Studies. She spent the 2013–2014 year as a U.S. Fulbright Scholar at Bahir Dar University, Ethiopia, working with the Blue Nile Water Institute.

Goraw Goshu is an Assistant Professor in the College of Agriculture and Environmental Sciences and Blue Nile Water Institute at Bahir Dar University. He was co-founder and director of the Blue Nile Water Institute (http://www.bdu.edu.et/bnwi/). His area of specialization is in environmental and water quality. Currently, he is a Ph.D. candidate at Wageningen University, Department of Environmental Sciences, since October 2015. He has a wide experience in environmental sciences—aquatic ecology and water quality management in particular and has worked and published as a development expert, researcher, project coordinator, environmental/water quality specialist. He earned his B.Sc. degree in Biology from Addis Ababa University (1997) and M.Sc. degree in Environmental Science (2007) from UNESCO-IHE, The Netherlands. His research areas include aquatic ecology and water quality management, water systems and global change, and pollution microbiology and microbial source tracking.

Shimelis Aynalem is the author of the field guide book the Birds of Lake Tana Area, Ethiopia. He is a lecturer in the College of Agriculture and Environmental Sciences, Bahir Dar Univeristy as well as a rearcher and co-founder of the Blue Nile Research Institute, where he has conducted and taken part in various ecological research projects. His professional focus is ecological and systematic Zoology and currently he is pursuing his Ph.D. degree at Addis Ababa University, College of Natural Science since 2013. He has a significant experience in field ecology, particularly with birds. He has published a number of articles across his areas of expertise (http://www.bdu.edu.et/bnwi). He has worked as a conservationist for EWNHS-NABU Project, Lake Tana and attended short term training in South Africa, 2008. He is also research associate of International Crane Foundation, Baraboo, Wisconsin 53913 USA. He holds two B.Sc and M.Sc from Haramaya and Addis Ababa University. He is passionate about cranes and wetlands.

Contributors

Biruk Abate School of Chemical and Food Engineering, Biotechnology Institute, Bahir Dar Institute of Technology, Bahir Dar University, Bahir Dar, Ethiopia

Azanaw Abebe College of Agriculture and Environmental Sciences, Bahir Dar University, Bahir Dar, Ethiopia

Wubneh Belete Abebe Amhara Design and Supervision Works Enterprise, Bahir Dar, Ethiopia

Merkuz Abera College of Agriculture and Environmental Sciences, Bahir Dar University, Bahir Dar, Ethiopia

Kefyalew Alemayehu Department of Animal Production and Technology, College of Agriculture and Environmental Sciences, Bahir Dar, Ethiopia

Mesfin Anteneh Department of Geography, Faculty of Social Sciences, Bahir Dar University, Bahir Dar, Ethiopia

Chalachew Aragaw Amhara Region Livestock Resources Development Promotion Agency, Bahir Dar, Ethiopia

Melak Mesfin Ayenew Division of Mathematics, Addis Ababa Science and Technology University, Addis Ababa, Ethiopia

Tenalem Ayenew Department of Earth Sciences, Addis Ababa University, Addis Ababa, Ethiopia

Shimelis Aynalem Blue Nile Water Institute and College of Agriculture and Environmental Sciences, Bahir Dar University, Bahir Dar, Ethiopia; College of Natural Science, Addis Ababa University, Addis Ababa, Ethiopia; Research Associate of International Crane Foundation, Baraboo, USA

Wubneh Belete Amhara Design and Supervision Works Enterprise (ADSWE), Bahir Dar, Ethiopia

Biazenlegn S. Beyene Blue Nile Water Institute and College of Natural Sciences, Bahir Dar University, Bahir Dar, Ethiopia

Emily Boersma Department of English Language and Literature, Bahir Dar University, Bahir Dar, Ethiopia

Nicholas Chisholm Department of Food Business and Development, University College Cork, Cork, Ireland

Baylie Damtie Washera Geospace and Radar Science Laboratory, Department of Physics, Bahir Dar University, Bahir Dar, Ethiopia

Daregot Berihun Department of Economics, Bahir Dar University, Bahir Dar, Ethiopia

J.J.M. de Klein Aquatic Ecology and Water Quality Management Group, Wageningen University, Wageningen, The Netherlands

Sewmehon Demissie Amhara Regional Agricultural Research Institute, Bahir Dar, Ethiopia

Patrick Enright Department of Food Business and Development, University College Cork, Cork, Ireland

Tesfahun G/Michael Amhara Design and Supervision Works Enterprise, Bahir Dar, Ethiopia

Genet Gebreegziabher Institute of Urban Planning, Bahir Dar, Ethiopia

Mekonnen Getahun Amhara Design and Supervision Works Enterprise, Bahir Dar, Ethiopia

Banchiamlak Getenet College of Agriculture and Environmental Sciences, Bahir Dar University, Bahir Dar, Ethiopia

Belachew Getenet Environment and Development Studies College, Addis Ababa University, Addis Ababa, Ethiopia; Faculty of Social Sciences and Blue Nile Water Institute, Bahir Dar University, Bahir Dar, Ethiopia

Goraw Goshu Blue Nile Water Institute and College of Agriculture and Environmental Sciences, Bahir Dar University, Bahir Dar, Ethiopia; Aquatic Ecology and Water Quality Management, Water Systems and Global Change, Wageningen UR, Wageningen, The Netherlands

Ellen Kalmbach The Nature and Biodiversity Conservation Union (NABU), Project Office Bahir Dar, Bahir Dar, Ethiopia; Michael Succow Foundation, Greifswald, Germany

Habtemariam Kassa Center for International Forestry Research, Ethiopia Office, Addis Ababa, Ethiopia

Dessalegn M. Ketema Department of Rural Development and Agricultural Extension, College of Agriculture and Environmental Sciences, Bahir Dar University, Bahir Dar, Ethiopia

A.A. Koelmans Aquatic Ecology and Water Quality Management, Water Systems and Global Change, Wageningen, The Netherlands

Birgit Kopainsky System Dynamics Group, University of Bergen, Bergen, Norway

Solomon Addisu Legesse Department of Natural Resources Management, College of Agriculture and Environmental Sciences, Bahir Dar University, Bahir Dar, Ethiopia

Elias Sime Leggesse Blue Nile Water Institute and Civil and Water Resource Engineering, Bahir Dar University, Bahir Dar, Ethiopia

James C. McCann Ethiopian Academy of Science, Boston University, Bahir Dar, Ethiopia

Minwyelet Mengist College of Agriculture and Environmental Sciences, Bahir Dar University, Bahir Dar, Ethiopia

Abebe Ameha Mengistu Program of Applied Biology, Adama Science and Technology University, Nazareth, Ethiopia; Department of Biology, Adama University, Nazareth, Ethiopia

Seyoum Mengistu College of Natural Sciences, Addis Ababa University, Addis Ababa, Ethiopia

Amare Sewnet Minale Department of Geography and Environmental Studies, Bahir Dar University, Bahir Dar, Ethiopia

Fenta Nigate Department of Earth Science and Blue Nile Water Institute, Bahir Dar University, Bahir Dar, Ethiopia; Department of Geology and Soil Science, Laboratory for Applied Geology and Hydrogeology, Ghent University, Ghent, Belgium

Mesenbet Yibeltal Sebhat Department of Water Resources and Irrigation Engineering, Bahir Dar University, Bahir Dar, Ethiopia

Alie Seid College of Natural Sciences, Bahir Dar University, Bahir Dar, Ethiopia

Yihenew G. Selassie College of Agriculture and Environmental Sciences, Blue Nile Water Institute, Bahir Dar University, Bahir Dar, Ethiopia

Krystyna Stave School of Public Policy and Leadership, University of Nevada Las Vegas, Las Vegas, NV, USA

Aklilu Tilahun Tadesse System Dynamics Group, University of Bergen, Bergen, Norway

Asaminew Tassew Department of Animal Production and Technology, College of Agriculture and Environmental Sciences, Bahir Dar University, Bahir Dar, Ethiopia

Berihun Tefera Agricultural Economics Department, College of Agriculture and Environmental Sciences and Blue Nile Water Institute, Bahir Dar University, Bahir Dar, Ethiopia; Department of Agricultural & Resource Economics, University of Connecticut, Storrs, USA

Kristine Walraevens Department of Geology and Soil Science, Laboratory for Applied Geology and Hydrogeology, Ghent University, Ghent, Belgium

Alemayehu Wassie College of Agriculture and Environmental Sciences, Bahir Dar University, Bahir Dar, Ethiopia

Ayalew Wondie College of Natural Sciences, Bahir Dar University, Bahir Dar, Ethiopia

Tadesse Terefe Zeleke College of Science and Blue Nile Water Institute, Bahir Dar University, Bahir Dar, Ethiopia

Part I
Overview

Chapter 1
Introduction: Regional Challenges and Policy Questions

Baylie Damtie, Emily Boersma and Krystyna Stave

Abstract The Lake Tana Basin is the headwater catchment of the Upper Blue Nile River. This highly dynamic region is experiencing significant population, economic, ecosystem, environment and social changes, raising concerns about sustainable development at regional, national and international levels. About 85% of the approximately three million people that live in the basin currently work in the agricultural sector, but factors such as population growth, environmental degradation, and rising education levels are driving migration from rural to urban areas and putting pressure on urban infrastructure and economy. The strong links between the region's human and environmental conditions mean that developing the region sustainably requires an understanding of its social and ecological characteristics as well as a framework for examining how the social system and the ecological system interact. This chapter describes key policy challenges in the region that motivated this book and lays out the structure of the sections and chapters that follow.

Keywords Upper Blue Nile River · Lake Tana Basin · Urbanization · Environmental conservation · Social-ecological system · Soil erosion

B. Damtie (✉)
Washera Geospace and Radar Science Laboratory, Department of Physics, Bahir Dar University, Bahir Dar, Ethiopia
e-mail: baylied@bdu.edu.et

E. Boersma
Department of English Language and Literature, Bahir Dar University, Bahir Dar, Ethiopia
e-mail: emily_boersma@yahoo.com

K. Stave
School of Public Policy and Leadership, University of Nevada Las Vegas, Las Vegas, NV, USA
e-mail: krystyna.stave@unlv.edu

K. Stave et al. (eds.), *Social and Ecological System Dynamics*,
AESS Interdisciplinary Environmental Studies and Sciences Series,
DOI 10.1007/978-3-319-45755-0_1

1.1 Overview

The Lake Tana Basin, the headwater catchment of the Upper Blue Nile River, is the most economically, historically, politically and environmentally important sub-basin of the Upper Blue Nile River System. It is a highly dynamic region, with drivers of dynamism rooted in both natural and human realms. About three million people live in this 15,114 km^2 catchment area, most in rural areas but increasingly migrating to urban settlements. The populations of major cities in the region like Gondar and Bahir Dar have been predicted to reach several million people in the near future. Depending on the trajectory of economic development in the near future, this urbanization may be driven by industrialization or poverty.

The region is experiencing significant population, economic, ecosystem, environment and social changes, raising concerns about sustainable development at regional, national and international levels. Locally, urbanization, intensification of rural land use, and climate change are resulting in environmental degradation and economic stress. On a positive note, nearly all school-aged children attend school and the majority of them will end up in colleges and universities. This means that the formerly high illiteracy rate is falling dramatically. It also means the new literate generation will demand different kinds of jobs than those we have at present in the region. With current agricultural practices, for example, the region will not be able to create enough jobs. Farmers are increasingly concerned about the lack of jobs in industries and other sectors of the economy for their sons and daughters who are graduating every year from technical and vocational schools, colleges and universities. The cultural practice of giving a fraction of their plot of land to their children is becoming an increasingly impossible option as plot sizes shrink. Illegal expansion of farmlands into community free grazing land is becoming more common, which in turn aggravates soil erosion and puts strong pressure on animal production. Migration of people from rural to urban areas in the Lake Tana basin makes it increasingly difficult to provide adequate basic services like housing, education, water supply and sanitation and health care. Creating enough jobs for this rapidly increasing population in urban areas is also a major challenge. The local dynamics are very complex and we need to gather detailed information to analyze and synthesize these dynamics.

At the national level, a major part of the basin has been designated as a key economic growth corridor in the Ethiopian Growth and Transformation Plan. An enormous amount of the nation's resources have been invested in this region, with expectations that it will sustain its present natural resources like the rivers and Lake Tana. A number of dams have been built for power generation as well as irrigation. The Koga dam, for instance, can irrigate about 7000 ha. The Tana-Beles hydropower facility, which was built at the shore of Lake Tana, can generate about 450 MW. Lake Tana Basin water, sediments, nutrients, and contaminants flow downstream across international borders into the Nile River and ultimately to the Mediterranean Sea. These flows affect downstream ecosystems and intercountry relations. The Grand Ethiopian Renaissance Dam (GERD) being constructed on the

Blue Nile River has implications regionally and for neighboring countries. It is expected to provide electrification to regions that have no power now.

The largest portion of the water that will fill the GERD, and also the water that one finds in the Nile when it reaches Sudan and Egypt, originates from the Tana Basin. This means that what happens in the Tana Basin has implications in a very wide region without borders. Due to the lack of information and sound scientific reasoning in the past, the Tana Basin has caused mistrust and unnecessary military rhetoric among countries in the region. Due to its unique ecosystem, the Tana Basin has recently been recognized as a UNESCO Biosphere Reserve.

In spite of all these indicators of the Lake Tana Basin's importance, there is no comprehensive, integrated, system-wide description of its characteristics and dynamics that are accessible to the region which can serve as a basis for its sustainable development. Although many institutions and researchers have given close attention to the region, their efforts are fragmented and are often not visible or useful to decision-makers and the people in the region.

1.2 Regional Challenges and Policy Questions

The dynamics of the region make sustainable management challenging. Regional population is increasing steadily. This leads to greater demand for resources, and conflict when competition for resources is high. There is inadequate understanding of resources and resource management, leading to decreasing resource conditions. Natural and human drivers are not well known. There is only patchwork knowledge of the system and this ignorance about resources fosters mistrust among resource users.

The challenge is how to manage the region proactively rather than reactively. This involves setting appropriate goals and steering development to help the region achieve its potential and optimize its resources. This kind of sustainable and proactive development must be knowledge-based.

For example, horticulture is increasingly becoming an important sector in the region in terms of job creation and export. Policy-makers want to formulate rules and regulations that govern horticulture like the kinds of pesticides, fertilizers and other chemicals that can be used in the region. Also, they want to make rules about how far the horticulture farms should be located from rivers and the lake and identify which places should be protected from such activities. In order to make all these rules and regulations, they need sound and scientific information organized in a holistic manner. The policy-makers will pay much more attention to research findings if government institutions, such as Bahir Dar University, present such information to them rather than having findings presented by individual researchers.

Agriculture dominates the Lake Tana Basin, with about 85% of the population working in the agricultural sector. Erosion from agricultural land is a critical and growing problem. Since all the rain ends up in Lake Tana, so too does much of the eroded soil. The population has been mobilized to carry out extensive environmental conservation interventions in order to minimize soil erosion. Now the government needs to know their effectiveness. This requires making an erosion susceptibility map of the region before and after the environmental conservation intervention. Such a map might contain specific information such as the number of grams of soil per litre of water present in specific rivers, in a specific period of time, in a specific season and how much of it was reduced due to the innovation. This demands a reliable and integrated system, which can monitor sediment loads and the direction and transport rates of sediment, and identify the farming and the environmental conservations practices that might exacerbate or mitigate the problem. Such a system is missing. Instead, many institutions have their own subsets of information that are not accessible or shared and, therefore, are unusable for any practical purpose.

This book is a key step toward developing a comprehensive database and integrated scientific information on the Tana Basin. The rationale for this initiative is to ensure the sustainability of the region for many generations to come. Sustainability has ecological, economic and cultural dimensions and it calls for a delicate balance among these dimensions. In order to find such a balance, Bahir Dar University has proposed three major tasks to be carried out systematically. The first major task is to establish a system that allows us to gather comprehensive data in a timely manner with modern data storage and management tools. The second task involves scientific analysis of the data gathered and the extraction of information usable in practice by researchers, government, policy-makers and general public. The third task is to take action in practice in accordance with the scientific recommendations. In sum, Bahir Dar University's contributions to the sustainability of the Tana Basin focus on finding the delicate balance among ecological, economic and cultural dimensions by establishing a comprehensive database infrastructure, carrying out intensive research and disseminating information for appropriate actions.

This book addresses the need for an integrated approach to understand and manage this complex social—ecological system. It presents both the social and ecological characteristics of the region and an integrated, system-wide perspective of the feedback links that shape social and ecological change in the basin. The book grew out of a recognition that management and policy actions to protect resources and promote healthy regional development must be research-based, and that Bahir Dar University, located on the shores of Lake Tana should take a more direct role in providing research supported and updated information about the Tana basin to decision-makers. The book lays a foundation for developing closer coordination and collaboration between decision-makers, the general public and researchers.

1.3 Structure of the Book

The book is divided into five parts. Parts I and V consider the region as a whole. Parts II and III focus on individual biophysical and socioeconomic components and Part IV presents details about policies and management practices that govern the links between people and the environment. Each of the chapters in Parts II–IV gives an overview of the chapter topic, a baseline characterization of what is known, a summary of research that has been done on the topic to date, and discussion of research questions and issues for further study.

Part I sets the stage for examining the Lake Tana Basin as a dynamic and complex social—ecological system. It introduces the problematic trends in the region and presents the theoretical framework that organizes the book. Chapter 2, following this introduction, gives an overview of the lake and the basin, and elaborates the challenges to this vibrant region's sustainability. It confirms that the problems are not simply environmental or purely social, but arise from the interconnections and interdependencies between human activity and environmental resources. Deforestation, soil erosion, water quality and fisheries degradation are the consequences of expansion and intensification of agriculture, improper waste management and overfishing, are in turn caused by changes in population and communities, and have further consequences for other parts of the system. These kinds of connections are not unique to the Lake Tana region, a point developed in Chap. 3, which sets this analysis in the broader context of social—ecological system (SES) analysis. Chapter 3 discusses existing SES frameworks and how they have been used to examine other human—environment systems. It proposes system dynamics as an analytical framework for the Lake Tana basin. System dynamics facilitates the representation of feedback across system sectors and collaboration among stakeholders. Chapter 4 further broadens our perspective, describing a long history in the region of intertwined social and ecological development. This story of how people and the landscape have co-adapted in the Upper Blue Nile basin over five distinct periods of economic, demographic, and spatial change in the early twenty first century from pre-1960 to the present since 1991 gives a long-term context for the changes we see in the region today.

Parts II and III provide a baseline characterization of the region's ecosystem and socioeconomic system, describing what is known and raising questions and issues for further study. Part II, Ecosystem Characteristics, contains 13 chapters describing characteristics of the catchment area as well as the lake itself. The section begins with an overview of the climatic conditions in the basin and trends in its temporal and spatial variability across the Lake Tana Basin. Further chapters describe the region's hydrogeology and groundwater occurrence, agricultural soil characteristics, hydrology and water quality, plankton, fish and fisheries in the lake, birds, herpetofauna and mammals in the basin, forest and wetland resources, and exotic and invasive plants. Part III's eight chapters describe the region's socioeconomic and land use characteristics, including demographics, gender and rural livelihood, stakeholder characteristics, the economy and role of natural resources in the basin, land use distribution and change, agriculture, farm animal productivity, and the extent and characteristics of urban areas in the basin.

Part IV discusses the regulatory and institutional context that shapes the management of environmental resource use in the region. The first three chapters describe existing environmental policies and laws as well as sustainable land use and watershed management practices. The next four chapters describe specific innovations in resource management structures and practices. Chapter 29 describes the establishment of the Lake Tana Biosphere Reserve as part of the UNESCO World Network of Biosphere Reserves. Chapter 30 describes an innovation in farming—the introduction of rice as a crop—that took advantage of an environmental challenge—frequent flooding—that was a long-standing problem for people in the Fogera district. With a change in resource use, this environmental characteristic has become an opportunity instead of a challenge. Chapter 31 examines the driving forces underlying the establishment of large areas of eucalyptus plantations in the Lake Tana Basin, and Chap. 32 discusses the growing problem of waste management.

Part V synthesizes the body of the book and looks forward. Chapter 33 reviews the region's problems, concluding that, although there is a tremendous amount of natural resource degradation in the region, efforts being made to mitigate the problems are encouraging and have the potential to be scaled-up toward sustainable natural resource management. Chapter 34 reports on an initial workshop that was held to apply the system dynamics concepts described in Chap. 3 to the Lake Tana basin. The purpose of the workshop was to create a systems map from the local stakeholder viewpoint of the key causal mechanisms leading to social and ecological change in the region as a framework for integrating researchers, decision-makers, and practitioners across disciplines. This chapter describes the process followed by the participants to develop a systems framework as well as the causal map produced from the workshop and insights and further research questions that emerged from the consolidated map analysis. Finally, Chap. 35 summarizes the research problems posed in the book's chapters and suggests a way forward.

1.4 Broader relevance of this Analysis to Other Regions

This book focuses on a specific region, establishing a foundation of baseline information and a framework for integrated analysis in the Lake Tana Basin. As such, it stands alone. However, it is more than just a case study. The questions of sustainable development faced in the Lake Tana Basin mirror issues in similar regions with tightly coupled socioeconomic and ecological systems. The approach shown here provides an example and framework that can be applied in other places.

Chapter 2
Problem Overview of the Lake Tana Basin

Goraw Goshu and Shimelis Aynalem

Abstract Lake Tana Basin is the second largest sub-basin of the Blue Nile which covers an area of 15,114 km^2. Lake Tana is a tropical Lake with surface area of 3111 Km^2. It is the largest fresh water resource of Ethiopia (50%). It is the source of the Blue Nile(Abay) River. Lake Tana basin and Blue Nile river provide economic, social, political, environmental, ecological and religious benefits also for downstream eastern Nile countries. The basin problems have also influence in downstream eastern Nile countries. Food security and environmental sustainability are grand challenges in the basin. Ensuring adequate supply and quality of water for water user sectors in the basin remains a challenge. The sanitation and hygiene coverage remains not significantly improved compared to the unprecedented population growth. The basin suffers from easily perceivable land, soil and water degradation which are manifested in different forms: Sedimentation, clearing of wetland, canalization of the tributaries, increased trend of eutrophication, toxigenic cyano bacteria, occurrence of invasive species like water hyacinth (*Eichornia crassipes*), stakeholders conflict, improper damming, construction of buildings in the Lake shore areas that are natural breeding and feeding grounds for some fish and bird species, poor waste management, increased prevalence of waterborne diseases especially in the riparian community which largely depend on raw water for drinking and recreation are major problems of the Basin. Climate change is also having its impact. Though the problems and challenges are known in the area, effective measures proportion to the magnitude of the problem are not yet taken sufficiently.

G. Goshu (✉)
Blue Nile Water Institute and College of Agriculture and Environmental Sciences, Bahir Dar University, P.O. Box 1701 (private) or 79, Bahir Dar, Ethiopia
e-mail: gorawha@gmail.com; gorawg@bdu.edu.et

S. Aynalem
Blue Nile Water Institute and College of Agriculture and Environmental Sciences, Bahir Dar University, P.O. Box 2205 (private), Bahir Dar, Ethiopia
e-mail: shimelisay@yahoo.co.in; shimelis.aynalem@gmail.com

S. Aynalem
College of Natural Science, Addis Ababa University, Addis Ababa, Ethiopia

S. Aynalem
Research Associate of International Crane Foundation, Baraboo, USA

K. Stave et al. (eds.), *Social and Ecological System Dynamics*,
AESS Interdisciplinary Environmental Studies and Sciences Series,
DOI 10.1007/978-3-319-45755-0_2

Keywords Canalization · Eutrophication · Hydrological alterations · Land degradation · Sedimentation · Tana basin · Water degradation

2.1 The Nile Basin

Nile River is a Trans Boundary River shared by ten riparian countries with different biophysical, socioeconomic and political settings. The downstream countries are dependent on Nile water and the demand is still increasing. Blue Nile (Abay) River, the only out flowing river from Lake Tana, and Atbara River together contributes about 85%. The Upper Blue Nile Basin is the largest one in terms of volume of water discharge - mean annual discharge of 48.5 km^3(1912–1997; 1536 m^3s^{-1}). It is the 2nd largest in terms of area coverage in Ethiopia, which is 17% of the landmass area of Ethiopia. The climate in the basin varies greatly. As the Blue Nile River drops down into the lowlands of Sudan area, the rainfall amount decreases but evaporation increases. The daily mean temperatures fluctuate between upper 15–18 Degree Celsius (°C) and lower 30 °C. In Sudan the irrigated land is more than 1.3 million ha, where as in Egypt, it is more than one million ha (BCEOM 1999). Ethiopia has a potential of irrigable land estimated to be 815,581 ha. The hydropower development potential is also estimated to be 3634–7629 Mega Watt (MW) that include about 120 identified potential sites and 26 investigated ones (BCEOM 1999). The Great Ethiopian Renaissance Dam (GERD), the largest hydropower dam in Africa with production capacity of 6000 MW, is under construction across the Blue Nile (Abay) river. This project not only increases the energy provision potential of Ethiopia but also brings economic integration among the riparian countries. It also brings a number of advantages to Sudan and Egypt. Flood risk avoidance and increasing the routing capacity of High Aswan Dam, controlled and uniform flow during dry period, drought mitigation through creating additional system storage, water saving and reduced transmission losses and Sediment control could be accounted as great advantages.

2.2 The Lake Tana Basin

Tana Basin is the second largest sub-basin of the Blue Nile, and covers an area of 15,114 km^2. The highest elevation is 4100 m a.s.l (meter above sea level) but 2025 m a.s.l is the average one; however, at Blue Nile out flow, at Lake Tana it is 1785 m a.s.l.

The average annual rainfall in Lake Tana is 1248 mm per year (mm yr^{-1}), which is 7% lower than the surrounding watershed. Atmospheric temperature decreases by 0.7 °C per 100 m and ranged from 13 to 22 °C. The main landforms in and around Lake Tana comprises plains, hills, mountains, mountain cliffs and depressions in different proportions. The flat plains and depressions mostly are wetlands. The land use in the Lake Tana basin is predominantly cultivable Land (71%), grazing (9%),

Infrastructure (6%), forest (3%) and others. The major type of land cover includes farm land, water bodies, wetlands, forest, wood land, shrubs, rangeland, grassland and settlements.

The soil type in the Lake Tana islands, peninsulas and wetlands and in the upland areas of the lake are dominated by Nitosols, Luvisols and Vertisols. Whereas flood plains of Fogera area, Libo Kemkem and Dembia places as well and the river mouth delta of Stumit, Kristos Semera, Nabega, Angara and Dirma area are dominated by alluvial deposits. This soil deposit has made shallow water depth and cultivated during the dry season following the receding or retreating lake water. Most of the transported high sediment loads are deposited and silted down in the water body of the lake. Thus, it can be said that Lake Tana is both a natural water reservoir and silt refinery for the Blue Nile River, but becomes detrimental to the long term existence of the lake to function as habitats for aquatic organisms.

Because of the significant importance of Lake Tana in supporting intensive irrigation based agriculture now and in the future, as a source of hydroelectric power(the source of water for Tana Beles and GERD Hydroelectric Power Plants supposed to produce 460 and 6000 MW respectively), fishery and tourist industry, bird habitats and biodiversity resources, it is worthwhile safeguarding its environment. However, the lake is receiving an ever increasing wastes from the point and diffuse sources.

There are 37 islands in lake and most of them have monasteries with historical, cultural, religious and touristic values. The riparian communities largely depend on the raw water for drinking, cattle watering, irrigation and recreation.

The wetlands of Lake Tana area are an integral part of the lake and play a significant role in sediment retention, flood protection, purification of water 'Kidney' of the landscape, important breeding grounds for birds and some fish species like Oreochromis niloticus. Lake Tana and its associated wetlands are also identified by IUCN as an important bird area (BirdLife International 2004). The Lake also provides valuable transport services to connect islands and Lake shore towns.

Lake Tana

Lake Tana is the largest Lake in Ethiopia which accounts for 50% of the fresh water resource of the country. The Lake has a surface area of ca. 3111 km^2, 284 km^3 volume, and has maximum length of 90 km and width of 65 km. Lake Tana is a shallow lake with a maximum depth of 14 m (m), but the average depth is eight m. The only out-flowing river is the Blue Nile River. This lake is the source of the Abay (Blue Nile) River and covers 20% of the surface area of the Lake Tana sub-basin. The Lake has an elevation of 1800 m.a.s.l and is fed by many streams and rivers with catchments in excess of 1000 km^2. The Lake basement comprises Pre-Cambrian, Metamorphic and Granitic rocks. These basement rocks are overlain by extensive deposits of Mesozoic sedimentary rocks that do not out crop in the Tana basin but are observed in Abay Valley.

Lake Tana use to be an oligotrophic lake (Wondie et al. 2007; Teshale et al. 2002; Wudneh 1998; Nagelkerke 1997) but its trophic status has changed gradually. Especially river mouths have experienced seasonal eutrophication (Goraw

2012). Lake Tana's bottom substrate is volcanic basalt mostly covered with a muddy substratum with little organic matter content 1% in 1994 (Howell and Allan 1994) and 14% in 2011 (Goraw 2011).

Basin potential and benefits

The basin in its natural state has high potential for agriculture, livestock, water resource, forest and wildlife, tourism, and fishery development besides too high biological diversity. There are animals, plants, fish, wetland and forest resources. The basin has also fertile soil and cultivable land for intensive agriculture. The agro-ecologies are also suitable to produce more than once per year. The population in the basin is increasing rapidly and there is no shortage of productive labor force in the region (Sewnet and Kameswara 2011). The basin provides multiple benefits like economic, social, political, religious ecological benefits. However, there seems to have imbalance between production and consumption pattern and conservation measures which has led to unsustainability (Teshale et al. 2002).

There are many polices and strategies in Ethiopia that dircctly and indirectly address the basin's challenges and threats. The polices were developed and have been implemented since last two decades The polices include Environmental Management proclamations, Agricultural and Rural Development Strategy, The Sustainable Development and Poverty Reduction Programs, Food Security Strategy (FSS), The National Biodiversity Conservation Policy, Science and Technology Policy, Water Resource Management Policy and Proclamations, Wildlife Policy, utilization of wildlife, Land use and land administration Proclamation, Forest Policy, Climate Resilient Green Economy (CRGE) Strategy and Fisheries Development and Utilization Proclamation. However, their implementation is weak and do not curb the problems caused by unsustainable management of resources in the basin.

Socioeconomic situation

The basin is a densely populated area due to high population growth rate and immigration (Sewnet and Kameswara 2011). This has resulted high dependency of the population on the basin resources. This high dependency in turn has already put high pressure on the basin resource.

To decrease threats and improve the management, proper identification of the stakeholders and ensuring active participation at all levels is indispensable. The stake holders can be broadly classified into governmental organizations, local communities, nongovernmental organizations, private (investors and enterprises) and international communities (Ketema 2013).

Lake Tana basin supports different economic activities, and agriculture is the major one. The basin has huge potential for socio economic development and because of this it has been identified as a major 'economic corridor'.

Biophysical situation

Lake Tana basin is the second largest sub basin of Blue Nile River basin and lake tana is the largest fresh water resource in the country. Regarding the hydrological characteristics of the basin, the lake water serves many functions apart from the ecosystem services. The annual out flow of water for the period 1976–2006 were estimated (Kebede et al. 2006). A total of more than 40 rivers drain water from Tana Basin into the lake. However, most of the hydrological analysis estimate was derived from gauged Lake Tana basin, which is only 42% of the basin (SMEC 2007). Due to this fact the estimated flow from the ungauged catchment highly varies across studies. There is limitation of data and diversification of the basin in terms of climate, geomorphology, and geology. This suggests that further research is needed to understand the hydrology of the ungauged catchment. Current research suggests how critical the soil erosion and sedimentation processes are in the basin. In addition, due to a number of projects planned in the basin the knowledge of soil erosion and sedimentation has become more important than ever. Hence, to understand the sediment transportation and deposition, the main factor driving the soil erosion should be investigated.

Most of the soils morphological, physical and chemical characteristics are identified and characterized but for a sustainable crop production, there is a need for guided inorganic fertilizer use and improved management practices in the area which will effectively minimize erosion and enhance and maintain soil quality and productivity (Mekonnen 2015). Further soil nutrient analysis and soil type identifications of the basin should be carried out at micro watershed level in order to recommend appropriate fertilizer application with a little impact on the environment or water quality of Lake Tana. The Vertisols at Fogera, Dembia and Alefa-Takusa plains have drainage problems due to inundation during the rainy season. Hence monitoring the water levels and developing appropriate irrigation schedules must be addressed in order to improve crop production.

The Lake Tana Basin biodiversity resource is so diverse and rich. The faunal diversity spans from the minute and cryptic puddle- and tree-frogs, to the medium sized Nile monitor lizard and African rock python, to the large-sized Nile crocodile, leopard and hippopotamus. So far 19 amphibians, 35 reptiles and 28 mammalian species have been identified. However, the total number of species listed could be underestimated. In addition, the bird fauna of the Lake Tana Basin is prominent. At varies studies a total of 437 birds are identified As well, several species of globally threatened, highland biome, winter migrant and endemic birds were recorded (Ash and Atkins 2009; Shimelis et al. 2011; Shimelis 2013). Areas that are important for breeding, roosting and feeding for globally threatened and migratory birds are identified. With regard to the flora, the major classification of forest resource types in the Amhara Region has been identified as Afroalpine and Sub-Afroalpine Forests, Dry Evergreen Montane Forest and Evergreen Scrub Combretum-Terminalia Woodland, Acacia-Commiphora Woodland, Bamboo Forests and Plantations. The size and quality of Afroalpine and Subafroalpine vegetation, the high Dry Afromontane forests, Bamboo forests and the Woodlands have decreased. These ecosystems have suffered considerable degradation in structural and species composition. This has resulted into a serious loss of indigenous biodiversity and created opportunities for invasive species.

The Lake Tana watershed wetlands are mainly occurring around the Lake Tana shore, major rivers that contribute to the lake and also in seasonal flood plains. The distribution and status of wetlands is not well studied. Some invasive and alien species are becoming a problem in the watershed. Water hyacinth infestation in Lake Tana can be a symptom of broader watershed management and pollution problems (Goraw 2011). Even though different alien species exists in the basin, water hyacinth becomes the most noxious and widespread weed in the basin. It is speculated that the biomass can be used for crafting, waste water treatment, heavy metal and dye remediation, of any invasive plant (EEA 2012). While researchers continue to investigate the perceived potential uses of water hyacinth, the current negative impacts of the weed far outweigh its benefits. The use of water hyacinth as raw material in cottage industry should not encourage propagation of the weed, but rather help control its growth. Changing land use practices in the catchment communities through watershed management will help reduce agricultural runoff as a mechanism for controlling the proliferation of water hyacinth. This is considered by many as one of the most sustainable long-term management actions.

Agriculture is the main stay of the Lake Tana Sub-Basin economy. The crop diversity and cropping pattern is widely known in the area. More than 80% of the cultivated land during the base-year is under rain-fed system, and the remaining are cultivated using irrigation and residual moisture, respectively. The farming system is characterized by crop-livestock mixed production system. Insect pests and diseases are one of the major production constraints.

Lake Tana, the biggest lake in Ethiopia, is very important water resource for community living and depending on the lake's resources. It plays a role in balancing of the microclimate of local areas in the catchment of the lake However, the recent development activities at the catchment areas have negatively affected the natural systems of the catchment. Natural resource degradation in the form of soil erosion, deforestation, and wetland resource depletion, is a major problem in Lake Tana Sub-Basin, Ethiopia. Lack of active participation of the local people, weak implementation of polices, lack of enforcement of some policies/laws, overlapping and conflicting responsibilities of different government institutions and conflict of interests accounted to failure to achieve the desired outcomes.

In Lake Tana basin, there are fragmented management interventions to conserve the natural resource. The major conservation effort so far has been the soil and water conservation campaign of Amhara region bureau of agriculture which mobilizes about 4–5 million farmers every day for two months every year since the last few years. These and other efforts do not stop the basin from degradation. There are also management interventions to control the spread of water hyacinth. Very recently, Lake Tana has been nominated as a new UNESO-Biosphere reserve under the UNESCO Man and the Biosphere program. It is Ethiopia's fourth natural heritage inscribed by UNESCO, after Simien National Park, Lower Valley of the Omo and Lower Valley of the Awash.

In the basin there is significant land use land cover change in the year between 1986 and 2013; built-up areas and cultivated land cover increased but the natural forest land and grassland covers decreased (Wubneh 2013 Unpublished).

The farming system in the basin is predominately crop production, mixed with animal production. Due to awareness and economic interest of the local community, new farming systems have been introduced, e.g. rice and eucalyptus production.

Bahir Dar and Gondar are rapidly urbanizing cites in the basin. Following the urbanization and immigration in search of better jobs, waste generation rate has increased. There is no integrated waste management system implemented in the basin. There is no liquid waste management at all. The only effort to manage the solid waste in Bahir Dar city is the effort of 'dream light' to collect solid waste door to door.

Problems in Lake Tana basin

There are a number of problems in the Lake Tana Basin watersheds that are easily perceivable. These problems mainly arise from imbalances of development interventions and environmental protection activities (Teshale et al. 2002). Unprecedented population growth, migration to Bahir Dar for better jobs, and urbanization exacerbate the problems. The problems include soil erosion, deforestation, hydrological interventions, wetland farming, habitat destruction, improper solid and liquid waste management, over grazing, lack of awareness, stakeholders' conflict and lack of decision support tool and inadequate organized data base system.

We began looking at the problems in the Lake Tana basin in a holistic way when the Blue Nile Water Institute (BNWI) was established in 2012 under the office of the Vice President for Research and Community Service, Bahir Dar University. The main justifications for establishing the institute include geographical proximity of the University to the largest water bodies of the nation, i.e., Blue Nile River and Lake Tana, a serious lack of basic information in the water bodies and a constantly increasing demand for information from internal and external stakeholders.

The vision of the Blue Nile Water Institute is to become one of the ten premier water research centers in Africa by 2025 recognized for its quality research, training, and service in support of sustainable management of water and water related resources in the upper Blue Nile basin. The mission is to conduct applied and basic research, provide tailor-made trainings, consultancy and advisory services, document and disseminate water and water related information and proven technologies. The institute shares the vision of Bahir Dar University and aligns its work with the different sustainable water resources development programs of the Ethiopian government. It functions under a multidisciplinary and project-based approach focused on applied research.

The BNWI consists of 13 research units across BDU from a variety of disciplines, including. Hydrology,watershed management, irrigation, hydraulics drainage, hydrogeology, socioeconomics, ecology, environment water supply, sanitation and hygiene, climate sciences. Since the establishment of BNWI, a grand National workshop has been planned as a platform for coordinating different institutions and individuals to discuss and produce useful information towards the interest of the institute. To this end, the second national Workshop on "Challenges and Opportunities of Water Resources Management on Lake Tana and its Environs, Upper Blue Nile Basin, Ethiopia" was conducted on 26–27 March 2012 at Bahir Dar, Ethiopia. The objectives of the workshop were: (1) To carry out awareness creation conference to disseminate important information on water resources of upper blue Nile Basin to relevant stakeholders (2) To identify research gaps towards sustainable utilization of water resources in the basin (3) To learn on the modalities of future communications between stakeholders (4) To establish a network of stakeholders and formulate inter-universities research taskforce that writes grant proposals (projects) and conduct research based on priority research gaps (5) To document fragmented research works on Lake Tana and its environs into compendium and organize into useful knowledge package. A team of experts in the water research and development arena were given task to review the state of the art knowledge in research, model development, policy recommendation gaps and capacity building of the institute following a holistic approach than a piece meal approach which had been the case before. The following issues were highlighted at the 2012 workshop.

2.2.1 Deforestation

Deforestation in the Lake Tana basin and other highlands of Ethiopia is a prominent watershed problem. It contributes to soil erosion and consequently to decline of soil fertility and agricultural productivity. The forest cover of the country was 46% in 1957 and decreased to 2.7% in 1987 (USAID 2004). However, due to watershed management and afforestation campaign this time, the forest cover reached 12% (Yihenew 2014 Wetland policy brief unpublished). In Lake Tana Basin, a total of 116358.5 ha of forest was lost or converted to another land cover during the period 1986–2013 (Wubneh and Goraw 2013 the land use land cover of Lake Tana area report, Unpublished).

2.2.2 Soil Erosion and Sedimentation

According to the Ethiopian highland reclamation study (FAO 1984), in the mid 1980's 27 million hectare or almost 50% of the highland area was significantly eroded, 14 million hectare seriously eroded and over 2 million hectare beyond reclamation. Soil erosion is a major watershed problem in Lake Tana basin causing significant loss of soil fertility, loss of productivity and environmental degradation. The Lake Tana basin is heavily affected by watershed management problems, caused by overpopulation, poor cultivation and improper land use practices, deforestation and overgrazing as a result sediment depositions in the lakes and reservoirs are becoming major issues (Setegn et al. 2008). According to the study by Setegn et al. (2009), around 12–30.5% of the watershed in Lake Tana basin is high erosion susceptible areas. There is a high silt concentration with loading rate of 8.96 to 14.84 Million tons of soil per year (Yitaferu 2007). Williams (2000a, b) reported deforestation in the Ethiopian highlands which has led to accelerated soil loss from the Ethiopian highlands to downstream reservoirs resulting siltation. The rate of erosion in the Ethiopian highlands today is one to two times higher than it used to be before

30 years (Hurni 1999; Williams 2000a, b). The long term rate of erosion averaged over 30 years is 10 to 15 million tons (McDougall et al. 1975 in Williams 2009).

2.2.3 Wetland Deforestation

In Lake Tana basin there are different forms of wetlands: lacustrine, riverine, floodplains, marshes and swamps. The wetlands in the Lake Tana basin comprise one of the largest wetlands in Ethiopia. They are mainly located surrounding the Lake and are flooded during the rainy season. The papyrus beds are one of the characteristic features of Lake Tana, but have declined in their extent dramatically. It is due to catchment degradation, over exploitation, expansion of invasive weed species and habitat fragmentation and loss of wetlands. The lake level drops during the dry season, the shore area become available for agriculture. The wetland areas are severely degraded due to siltation, conversion into cultivation and overgrazing.

The land use land cover change between 1986 and 2013 has indicated a total of over 52% of wetlands is converted to other land use forms; about 1259.1 ha (19.6%) is converted to farm villages with trees and cultivation, about 1388.9 ha (23.4%) is converted to intensively/moderately cultivated land, about 233.2 ha (3.9%) is converted to natural/plantation forest (this could be trees in swamps), about 193.52 ha (3.27%) is converted to water body (this could be the Lake is regulated by Chara-Chara weir) and about 501.23 ha (8.46%) is converted to woodland. The largest conversion is for cultivation and settlement (BNWI 2014 Wetland Management Plans Lake Tana Survey Report, unpublished).

2.2.4 Eutrophication and Problem of Toxigenic Bacteria

Water resources all around the world are under pressure and especially eutrophication is a major environmental problem. It is caused by excessive loading of dissolved and particulate organic matter and inorganic nutrients (Carbon, Nitrogen and Phosphorus). These nutrients are loaded by municipal and industrial sewage discharges and/or from diffuse sources in catchments areas. In addition to external nutrient loadings and even after their reduction, eutrophication can be maintained by internal loading, which corresponds to the release of nutrients from the lake sediments into the water column.

The concentration of nitrate and phosphate in the Lake Tana basin is increasing from time to time. Lake Tana use to be oligo-trophic (Wondie et al. 2007; Dejen et al. 2004; Wudneh 1998) but there is an increasing trend of increase in concentrations of nitrate and phosphorous especially in shore areas and river mouths that would eventually lead to eutrophication (Ilona et al. 2011; Wondie et al. 2007; Goraw 2012) (Fig. 2.1). These could happen as a result of improper disposal of waste water, and poor soil and water conservation practices in the catchment areas (Teshale et al. 2002; Yitaferu 2007).

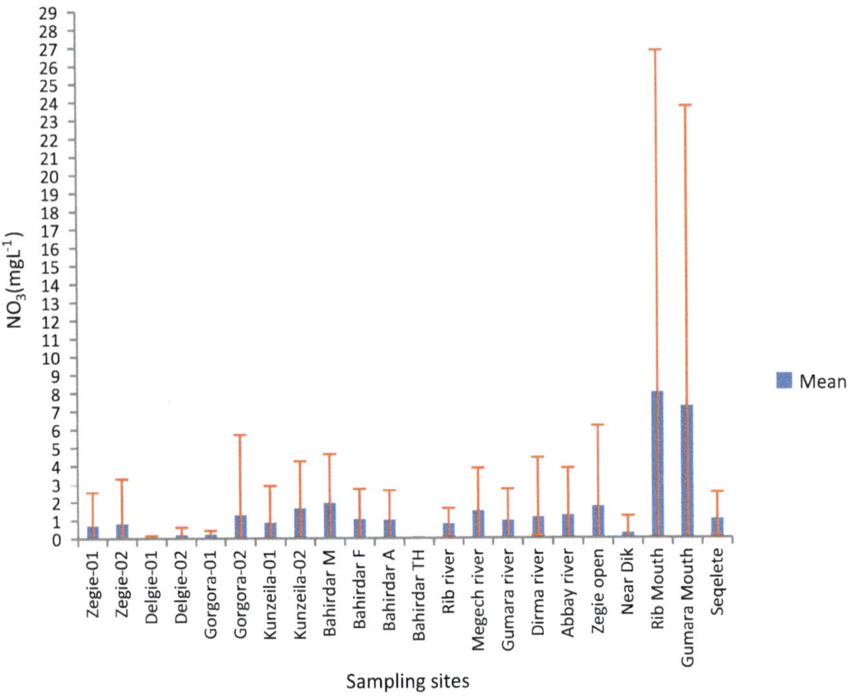

Fig. 2.1 Spatial patterns in nitrate concentration (mg L^{-1}) in Lake Tana samples collected from different littoral (Zegie 01, 02, Delgi 01, 02, Gorgora 01,02, Kunzila 01 and 02), city adjacent (Bahir Dar M, F, A and TH) Rivers (Megech, Gumara, Dirma, Abay, Ribb and Gumara) and pelagic (Near Dik and Sqelete) sites in March 2010 (*Source* Goraw 2012)

Toxigenic (potentially toxic) cyanobacteria were detected in Tana Lake on the basis of the new primers designed for mcyE gene (Ilona et al. 2011). The highest biomass of cyanobacteria (188.18 mg L^{-1}), with domination of Microcystis aeruginosa, and the highest concentration of microcystins (2.65 µg L^{-1}) were detected in the dry season, in November 2009 from Lake Tana and tributary rivers. The concentration of microcystins in the dry season exceeded the limit value for drinking water (1 µg L^{-1}) (WHO 2006) and indicated the first alarm level of risk for recreational activities (WHO 2003). Amount of total phosphorus appears to be an important parameter for the creation of the algal blooms in Tana Lake.

2.2.5 *Climate Change*

Ethiopia experiences persistent land, water and environmental degradation due to localized and global climatic anomalies. Climate change can cause significant impacts on water resources by resulting changes in the hydrological cycle. Climate change studies in the basin reported that the water balance can be significantly affected, which clearly amplifies its impact on sectors like agriculture, industry and urban development (Hailemariam 1999 cited in Sewnet and Kameswara 2011).

2.2.6 *Invasive Weeds*

There are many weeds that are already found in the Lake Tana Basin. The presence of water hyacinth (*Eichornia crassipes*) was reported and confirmed in Lake Tana at various times (Wondie et al. 2012). The presence of the floating weed, *E. crassipes*, was also reported in the area between Rib river mouth, Mitreha Abaworka kebele, Gondar Zuria Woreda and Dirma River mouth, Dembia Woreda (Shimelis and Getachew 2013 ecological effect of water hyacinth on waterfowls of Megech, Dirma and Rib River mouths at Lake Tana, Ethiopia unpublished paper). The areas where the weed was observed are mostly river mouths where the nutrient condition was relatively good and the water quality condition has started to deteriorate. The weed is very notorious and can cover the whole lake in a few years' time if immediate control strategies are not in place. It can destroy the fishery industry, create obstacles to navigation, clog canals of hydroelectric power plants and will generally create serious environmental imbalance.

2.2.7 *Improper Solid and Liquid Waste Management*

Bahir Dar and Gondar are the fastest growing urban cities in the basin with current population of 299,272 and 278,135 respectively (BOFED 2015, Amhara Region Bureau of Finance and Economic Development annual report, unpublished). It is expected that the expansion of Bahir Dar and Gondar may form a mega city. The Lake Tan Basin is identified as main economic corridor of the country. Large scale irrigation projects with a command area of more than 20,000 ha have been planned; most tributaries are dammed mainly for irrigation. In a nut shell, urbanization, population increase, agricultural intensification and industrial development will lead to generation of more solid and liquid wastes which has already created environmental problems. The solid and liquid waste management in the basin is very poor. None of the cities has waste water treatment plant and neither sanitary land fill.

2.2.8 *Stakeholders' Conflict*

Lake Tana is everybody's property which practically means nobody's property, "The tragedy of the common" applies to the resources utilization. There is no institution clearly responsible to manage Lake Tana. Different users would like to maximize their profit out of the Lake and its resources which lead to un attainability and conflict.

2.2.9 No Data Base System

There is no systematically organized data base system in the Lake Tana basin. The biophysical, socioeconomic and demographic characteristics of the basin are not monitored properly; no long-term and spatially representative data. The data is mainly of expeditious type and available here and there. Development intervention plans in the basin are not based on well-organized data. It is simply because of absence of adequate and well organized data. So documenting and conducting various researches to fill the gap and to build up the data system is mandatory.

2.2.10 Unsustainable Exploitation of Fisheries

The fisheries in the Lake Tana Basin is a livelihood for half million people (Gordon et al. 2007). This resource is not exploited sustainably; there is labeobarbs fish species stock decline (Degraaf 2003), off flavor of Nile tilapia (Workiyie and Goraw 2014) and decrease in distribution and abundance of other fish species over time. The immediate causes are introduction of monofilament gillnet which has a mesh size less than allowable(less than 10 Cm mesh size), habitat destruction, targeting spawning aggregation of migratory fishes at river mouths, pollution from the basin, sedimentation, hydrological regime change, invasive weeds, urbanization, and use of fish poisonous plants like *Millettia ferruginea*, which indiscriminately kills fishes of different type and age groups.

2.2.11 Lack of Enforcement of Environmental Related Polices and Strategies

Though there are polices like water and environment which in principle aims sustainable development, their enforcement is lacking resulting with many adverse impacts on the environment (Teshale et al. 2002).

2.2.12 Hydrological Alterations

There is intensification of large and small scale irrigation agriculture in Lake Tana basin with increasing trend of utilization of streams and rivers. Many farmers pump water from tributary rivers and streams to the extent of affecting the natural flow regime. Floriculture is also booming around all corners of the Lake, abstracting water from Lake Tana and ground water resources. Changing of the natural hydrological regime influences the biota and ecological integrity which intern make the water system unsustainable (Zalewski 2002).

2.2.13 Fecal Pollution

Water supply and sanitation coverage of Ethiopia in general and Amhara Region in particular is improving from time to time though still remains a lot to be done. Because large proportion of the rural population mainly depend on unsafe drinking and recreational water sources. The sanitation facilities are neither to the standard nor enough. The Lake Tana area is an area where frequent water borne disease outbreaks have occurred due to fecal pollution from point and diffuse sources. In 2006, the joint Government, WHO and UNICEF Rapid Assessment of Drinking Water Quality established that about 28% of all protected rural water supplies were contaminated with fecal micro-organisms (Rapid Assessment of Drinking Water Quality, Ethiopia Country Report, March 2007 unpublished). Goraw et al. (2010) also reported high level of fecal pollution of Bahir Dar gulf of Lake Tana which is higher than WHO standard for recreation. Total coliforms (TC), fecal coliforms (FC), *Escherichia coli* (EC) and *Clostridium perfringens* (CP) were detected in 100, 86, 82 and 90% of all sampling sites analyzed throughout the sampling period, respectively.

2.3 Summary

The Lake Tana basin faces multiple challenges. These challenges and problems should be solved in order to sustain the resource for now and the next generation. This chapter has given an overview of the major and foremost problems in the watershed today. Further chapters in this book expand on this overview, describing the ecological and socioeconomic context of the problems, what is known about the linkages in the system, and posing questions for further research.

Management of Tana basin resources traditionally concentrates on single issues and in a fragmented way such as one area specific activity, single season/period and on a targeted resource. The complexity of the basin and the negative influence and consequences for humans have brought this day a more integrated thinking that management of Lake Tana basin should follow. Since the basin has great concern especially of the water system due to the mere location in Ethiopia, the multipurpose role it has, and because it is the largest fresh water body in the country - and yet the basin is identified as growth corridor, its sustainable management in general calls an urgent integration of all stakeholders. The Government of Ethiopia needs to prepare and implement integrated water resources management plan in Lake Tana basin with full participation of all relevant stakeholders including the riparian community to sustain the water system in the basin. The background and baseline information provided in the following chapters is a resource for such integration.

References

Ash J, Atkins J (2009) Birds of Ethiopia and Eritrea: an Atlas of distribution. Christopher Helm, London

BCEOM and Associates (1999) A Bay River integrated development master plan project; Phase 2, Section II, Vol III: Water Resources: Part 1—Climatology and Part 2-Hydrology

BirdLife International (2004) Most endemic bird areas are in the tropics and important for other biodiversity too. http://www.birdlife.org/datazone/sowb/casestudy/61. Accessed 19 Aug 2014

Degraaf M (2003) Lake Tana's piscivorous barbs (Cyprinid, Ethiopia). PhD thesis, Wageningen University

Dejen E, Vijverberg K, Nagelkerke L et al (2004) Temporal and spatial distribution of micro crustacean zooplankton in relation to turbidity and other environmental factors in a large tropical lake (L. Tana, Ethiopia). Hydrobiologia 513:39–49

EEA (2012) The impacts of invasive alien species in Europe. EEA Technical reports No 16/2012. Publications Office of the European Union, Brussels, Luxembourg. http://www.eea.europa.eu/publications/impacts-of-invasive-alien-species. Accessed 12 Sep 2014

FAO (1984) Ethiopian Highland reclamation study (EHRS). Final Report, vol 1–2. Rome

Goraw G (2011) Annual Report on Lake Tana water quality base line monitoring. Tana Beles Integrated Water Resources Development Project, Bahir Dar

Goraw G (2012) Wetlands for sustainable development and climate change mitigation. In: Seyoum M, Tadesse F (eds) 2nd National Workshop, Bahir Dar, January 2012. Water quality deterioration as potential enabling environment for Proliferation of floating water hyacinth (*E. crassipes*) in NW-river mouths of Lake Tana, pp 162–190

Goraw G, Byamukama D, Manafi M et al (2010) A pilot study on anthropogenic fecal pollution impact in Bahir Dar Gulf of Lake Tana, Northern Ethiopia. Eco Hydrol Hydrobiol 10(2)4:271–280

Gordon A, Demissie S, Tadesse M (2007) Marketing systems for fish from Lake Tana, Ethiopia: opportunities for improved marketing and livelihoods. IPMS (Improving productivity and Market Success) of Ethiopian Farmers Project Working paper 2. ILRII (International Livestock Research Institute), Nairobi, Kenya, 49 pp

Howell PP, Allan P (1994) The Nile: sharing a scarce resource. Cambridge University Press, Cambridge

Hurni H (1999) Sustainable management of natural resources in African and Asian mountains. Ambio 28:382–389

Ilona G, Goraw G, Katarzyna K et al (2011) Detection of toxigenic cyanobacteria in Bahir Dar Gulf of Lake Tana–pilot study. In: Brook L, Abebe G (eds) Impacts of climate change and population on tropical aquatic resources. Proceedings of the Third International Conference of the Ethiopian Fisheries and Aquatic Sciences Association (EFASA), Haramya University, Haramya, 5–6 February 2011, pp 271–284. http://aau.edu.et/efasa/images/haramaya%20proceeding.pdf. Accessed 16 Feb 2017

Kebede S, Travi Y, Alemayehu T, Marc V (2006) Water balance of Lake Tana and its sensitivity to fluctuations in rainfall, Blue Nile basin, Ethiopia. J Hydrol 316(1–4):233–247

Ketema DM (2013) Analysis of institutional arrangements and common pool resources governance: the case of Lake Tana sub-basin, Ethiopia. PhD Thesis, University College Cork

McDougall I, Morton WH, Williams MA (1975) Age and rates of denudation of trap series basalts at Blue Nile gorge, Ethiopia. Nature 254:207–209

Mekonnen G (2015) Characterization of agricultural soils in cascape intervention Woredas of Amhara region. BDU-CASCAPE working paper 12.90 PP

Nagelkerke L (1997) The barbus of Lake Tana, Ethiopia. Morphological diversity and its implications for taxonomy, trophic resources portioning, and fisheries. PhD thesis, University of Wageningen

Setegn SG, Ragahavan S, Bijan D (2008) Hydrological modeling in the Lake Tana Basin, Ethiopia using SWAT model. Open Hydrol J 2:49–62

Setegn SG, Srinivasan R, Dargahi B et al (2009) Spatial delineation of soil erosion prone areas: application of SWAT and MCE approaches in the Lake Tana Basin, Ethiopia. Hydrol. Process Spec Issue Nile Hydrol 23(26):3738–3750

Sewnet A, Kameswara KR (2011) Hydrological dynamics and human impact on ecosystems of Lake Tana, Northwestern, Ethiopia doi:10.4314/ejesm.v4i1.7

Shimelis AZ (2013) Birds of Lake Tana Area Ethiopia. A photographic field guide. View Graphics and Printers, Addis Ababa

Shimelis A, Nowald G, Schroder W (2011) Observation on the biology and ecology of cranes: wattled cranes (grus carunculatus), black-crowned cranes (balearica pavonina), and eurasian cranes (grus grus) at Lake Tana, Ethiopia. INDWA. J Afr Crane Res Conservationist 7:1–12

SMEC (2007) Hydrological study of the Tana-Beles sub-basins, main report. Snowy Mountains Engineering Corporation, Australia

Teshale B, Lee C, Girma Z (2002) Development initiatives and challenges for sustainable resource management and livelihood in the Lake Tana Region of Northern Ethiopia international. J Technol Manage Sustain Develop 1(2):111–124

USAID (United States Agency for International Development) (2004) Ethiopia Land Policy and Administration Assessment. Final Report with Appendices. USAID Contract No. LAG-00-98-00031-00, Task Order No. 4

WHO (2003) Algae and cyanobacteria in fresh water. In: Bartram J (Coordinator) Guidelines for safe recreational water environments, Geneva, Switzerland, pp 149–151

WHO (2006) WHO guidelines for drinking water quality, incorporating first addendum third edition. Recommendations vol 1, World Health 697 Organization, Geneva, Switzerland, pp 407–408

Williams MA (2000a) Desertification: general debates explored through local studies'. Prog. Environ. Sci. 2:229–251

Williams MAJ (2000b) Desertification: general debates explored through local studies'. Prog. Environ. Sci. 2:229–251

Williams MAJ (2009) Lake Tana: source of the Blue Nile. In: Dumont HJ (ed) Human impact on the Nile Basin: past, present, future. Springer, pp 771–779

Wondie A, Mengistu S, Vijverberg K et al (2007) Seasonal variation in primary production of a large high altitude tropical lake (Lake Tana, Ethiopia): effects of nutrient availability and water transparency. Aquatic Ecol 41:195–207

Wondie A et al (2012) Wetlands for sustainable development and climate change mitigation. In: Seyoum M, Tadesse F (eds) 2nd National workshop, Bahir Dar, January 2012. Preliminary Assessment of Water hyacinth in Lake Tana, pp 117–135

Workiyie W, Goraw G (2014) Incidence of fish off-flavor for the first time from Lake Tana, Ethiopia: causes and possible management strategies. IJSRR 3(4):30–47

Wudneh T (1998) Biology and management of fish stocks in Bahir Dar gulf, Lake Tana, Ethiopia. PhD thesis, Wageningen University

Yihenew GS (2014) Wetland policy brief unpublished. Blue Nile Water Institute, 15 pp

Yitaferu B (2007) Land degradation and options for sustainable land management in the Lake Tana Basin (LTB), Amhara Region, Ethiopia. PhD thesis, University of Bern

Zalewski M (2002) Eco hydrology—the use of ecological and hydrological processes for sustainable management of water resources. Hydrol Sci J 47(5):825–834

Chapter 3
System Dynamics as a Framework for Understanding Human—Environment Dynamics

Krystyna Stave and Birgit Kopainsky

Abstract Understanding the dynamics of human—environment systems, and developing policies that promote their sustainability, requires a holistic, integrated approach. Although many frameworks have been developed that include social and environmental components, managing social and ecological systems as integrated systems has been difficult in practice. The analytical and practical challenge is to identify the interactions that underlie resource management problems, find leverage points where management or policy changes can effectively move the system in a more sustainable direction, and build cooperation among system stakeholders to implement change. This chapter gives an overview of existing frameworks for examining social—ecological interactions, then presents system dynamics as both a theoretical perspective and a practical method for integrating across disciplines. The system dynamics approach makes feedback relationships in the system explicit, and provides a platform to foster collaboration and coordination among stakeholders in the system. This chapter offers a systems framework for considering the connections among the individual chapters to follow. This approach was used for a collaborative mapping workshop on sustainability issues in the Lake Tana basin held in November 2014 as a first step toward integrating disparate research disciplines and stakeholders. Chapter 34 describes the workshop.

Keywords Social–ecological systems · SES · Analytic framework · Participatory modelling · System dynamics · System integration · Integrated system analysis

K. Stave (✉)
School of Public Policy and Leadership, University of Nevada Las Vegas, Las Vegas, NV, USA
e-mail: krystyna.stave@unlv.edu

B. Kopainsky
System Dynamics Group, University of Bergen, Bergen, Norway
e-mail: birgit.kopainsky@geog.uib.no

K. Stave et al. (eds.), *Social and Ecological System Dynamics*,
AESS Interdisciplinary Environmental Studies and Sciences Series,
DOI 10.1007/978-3-319-45755-0_3

3.1 Introduction

The Lake Tana region faces a number of challenges that arise from interactions between people and the environment. As Chaps. 1 and 2 described, environmental conditions such as soil and water quality and species diversity are deteriorating. Environmental degradation, in turn, threatens the quality of human life and shapes the types of natural resource use possible. This region is one of many in the world in which human activity and environmental conditions are closely interwoven. Though the details of each place are unique, the general relationships are similar. The influence diagram in Fig. 3.1 shows key relationships and feedback mechanisms linking human activity and environmental characteristics. Human activities are mediated by the environmental and social context in which they take place. Local environmental characteristics—the quantities and quality of environmental resources—constrain and provide opportunities for individual and community use of natural resources and ecosystem services. For example, the types of crops that

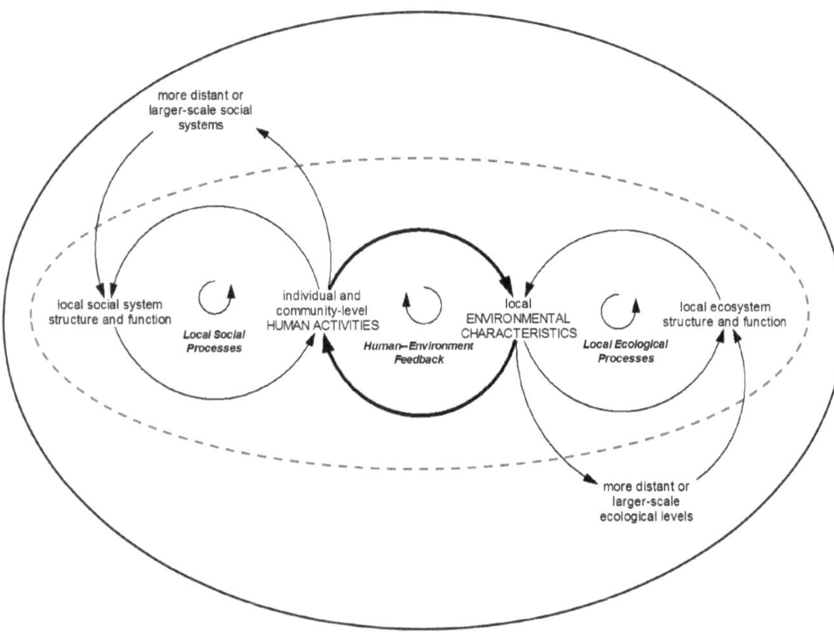

Fig. 3.1 A high-level view of feedback relationships in social–ecological systems. *Arrows* show the direction of cause-and-effect. Human activities lead to changes in environmental characteristics (quantity and quality of environmental components), which, in turn, constrain or provide opportunities for further human activities. Human activities also affect and are affected by social system characteristics. Changes in environmental characteristics in response to disturbances from human activities are shaped by and further shape ecosystem structure and ecological processes. These causal influences underlie the dynamic behavior of all parts of the system, including the central human—environment connection

can be grown or minerals that can be mined are a function of the resources that exist. Construction practices are influenced by climate and other factors such as bedrock geology and seismic activity. Human uses of environmental resources are further mediated by local social system structure and function. Social system factors including cultural norms, markets, institutional regulations, and governance structures shape decisions such as the types of crops farmers choose to grow, labor availability, farming practices, and land use patterns.

Human activities lead to further changes in environmental and social conditions. Farming practices such as land clearing or open grazing transform the landscape, changing forest habitats to open land, causing soil erosion in places and building up soil nutrients in others. Increasing rural population and decreasing land availability lead to agricultural intensification and pressure to convert marginal lands such as wetlands to farmland. Decreased wetland area leads to changes in vegetation and habitat near the lake shore, affecting the livelihoods of people who harvest papyrus and fish. Land use changes and farming outcomes can alter social relations among farmers. Changes in environmental conditions yield further changes in ecosystem structure and function such as microclimates, local hydrology, species composition and habitat. The direction and magnitude of such changes are further governed by ecological processes and environmental factors such as climate, topography, and geology.

Understanding such human—environment systems, and developing management strategies and policy decisions that promote their sustainability requires a holistic approach. Many conceptual frameworks have been developed that include social and environmental components. Integrating social and ecological systems in practice has been difficult to do, however. The analytical and practical challenge is to identify the interactions among components in such systems that underlie resource management problems, find leverage points where management or policy changes can effectively move the system in a more sustainable direction, and build cooperation among system stakeholders to implement change.

In this chapter we present system dynamics as a theoretical and analytical framework for addressing and formalizing this challenge in the Lake Tana region. System dynamics allows for integration of critical connections across disciplinary lines, highlights feedback relationships that underlie patterns of change in the system, and provides a platform that fosters collaboration and coordination among stakeholders in the system including researchers, policy-makers, and resource users. We first present an overview of existing frameworks for analyzing social—ecological systems and discuss some challenges in applying them. Then we describe what a system dynamics perspective offers for social and ecological system analysis and explain the steps in the approach. This approach was used as the framework for a collaborative mapping workshop on sustainability issues in the Lake Tana basin held in November 2014. Chapter 34 describes the workshop and the initial systems map that was produced.

3.2 Overview of Existing Frameworks, Their Application and Remaining Challenges

3.2.1 *Overview of Existing Frameworks for Social and Ecological System Analysis*

A number of conceptual and analytic frameworks exist for understanding the interrelationships between humans and nature. They include the three-legged stool or three-pillars model in which environment, economy, and society, often further described as ecological integrity, economic security, and social equity, must be balanced to achieve sustainability. This view is sometimes represented with three equal-sized circles for environment, economy, and society arranged so their edges overlap, with sustainability at the intersection of the three circles. Other versions show different arrangements, depicting unequal relationships among the sectors. Some argue economy and society cannot exist without the environment, so the environment must be the foundation for the stool, represented by a circle fully encompassing both the economy and society (e.g., Dawe and Ryan 2003). Such conceptual models help structure thinking about how people interact with the environment and promote an integrated view.

Other frameworks quantify specific aspects of human—environment interaction. Ecological footprint analysis (Wackernagel and Rees 1996) calculates the amount of land required to provide resources and process waste from human activities. Water footprint analysis (Hoekstra and Mekonnen 2012) determines how much water is used in agricultural or industrial production, how such embodied water flows around the world through international trade, and how water is consumed from sources internal or external to nations. Planetary boundary analysis (Rockström et al. 2009) describes nine dimensions of the environment and identifies a "safe operating space" within which human activity can take place. Comparing the effect of activity to thresholds gives a measure of how much development can take place before serious consequences arise.

These frameworks connect human activity and environmental conditions broadly, account for the resource demands of food and industrial production and the ecosystem services required to process waste, and begin to clarify how close we are to ecological limits. Other frameworks such as coupled human and natural systems (CHANS) (e.g., Liu et al. 2007) examine links among social and ecological system components. The water-energy-food nexus framework (World Economic Forum 2011) is a conceptual model of the links between water, energy, and food security. Biggs et al. (2015) extend the framework to include consideration of livelihoods. Liu et al. (2013) propose the idea of telecoupling, in which human—environment systems that may be spatially distant can be connected by flows of resources and products that have different effects on the sending and receiving systems.

3.2.2 The Social—Ecological System (SES) Approach

One approach that is particularly relevant to the resource management issues in the Lake Tana region is social-ecological system analysis (Ostrom 2009; Holling 1973). In this approach, natural resources used by humans are viewed as embedded in social-ecological systems, that is, in systems where cultural, political, social, economic, ecological, technological, and other components interact. A socio-ecological system provides essential services to society such as supply of food, fiber, energy and drinking water (Berkes and Folke 1998). Accordingly, fisheries, forests, pastures, coastal zones, and water bodies are social-ecological systems that are described widely in the SES literature (e.g., Anderies et al. 2006; Berkes et al. 2003; Cifdaloz et al. 2010; Ostrom 2009; Walker et al. 2002, 2006).

Like other types of systems, social-ecological systems consist of many different parts that interact in complex ways (Resilience Alliance 2010). SES subsystems span social, economic, political and environmental dimensions (Ericksen et al. 2009; Thompson and Scoones 2009; Cash et al. 2006) and several hierarchical levels within each dimension (Cash et al. 2006). The ecosystem dimension of an SES, for example, might be considered at five nested levels from micro-habitat to patch, reach, river, and biogeographical region (Dore et al. 2010: 41). In the economic dimension, local economies operate within the provincial, national, regional, and international economy. System activities take place across dimensions and across levels. Individual- and local-level human actions that happen on a daily or weekly basis can have significant implications for sustainable resource management when aggregated at higher spatial or temporal levels.

Through interactions and feedback effects across subsystems and levels in response to internal or external pressures, social-ecological systems can self-organize (i.e., adjust themselves through interactions among their components), novel configurations can emerge, and adaptation is made possible (Berkes et al. 2003). Interactions and feedback effects, however, often lead to the emergence of trade-offs between one set of services, e.g. food production, at the cost of another (often environmental services), e.g. cleaner water (Carpenter et al. 2009, Ericksen 2008, MEA 2005).

3.2.3 Analyzing Social-Ecological Systems

Analyzing a social-ecological system requires a systems approach, that is, a holistic approach that does not focus on a detailed understanding of parts, but on how links between key components contribute to the dynamics of the whole system. Integrated social—ecological systems cannot be analyzed with disciplinary approaches alone. Instead, complexity needs to be addressed in an inter- and transdisciplinary way (Carpenter et al. 2009; Ostrom 2009). A variety of methodologies for studying SES have been proposed (Binder et al. 2013). Although they differ with respect to their theoretical foundation and their conceptualization of the ecological and social subsystems and their interrelations, all of these methodologies aim at providing an integrative perspective on social-ecological systems.

The literature describes a variety of steps for analyzing specific SES (e.g., Cumming et al. 2005; Engle et al. 2013; Walker et al. 2002). These can be summarized into four generic stages: (1) defining the system and desired outcomes, (2) identifying drivers of change in the system, (3) identifying interventions to move the system toward desired goals, and (4) evaluating and implementing interventions. Analyzing a social-ecological system involves constructing a model of the system of interest, i.e., a simplification or distillation of the complex system into a conceptual map of the critical features of the place, issues, and people involved. Mapping the system components and flows is a way to *define the system and outcomes* (step 1). The map is used to *identify drivers of change* (step 2). Although some activities and questions address individual system components, these insights are meant to contribute to an understanding of the dynamics of the whole system. Assessing the impact of drivers of change and interventions along with the conceptual model helps to reveal factors that contribute to a sustainable future trajectory (steps 3, *identifying interventions,* and 4, *evaluating and implementing interventions*).

Computer simulation models for SES play an important role in the analysis of SES and consequently, SES modeling is an emerging field (Schlüter et al. 2014). Most existing SES models are largely theoretical, however, (Schlüter et al. 2012) and do not provide solutions to empirically measurable issues (Janssen and Ostrom 2006). Current SES models also have limited representation of feedbacks between the social and ecological systems (Schlüter et al. 2012).

3.2.4 Remaining Challenges and Opportunities for Integrated Social—Ecological System Analysis

While much progress has been made in understanding patterns and drivers of change in coupled human and natural systems, there are still several key challenges. These include: integrating social and ecological components across dimensions and hierarchical levels, accounting for feedback in the system, and coordinating and collaborating with stakeholders across disciplines (e.g., Liu et al. 2015, Hammond and Dubé 2012, Alberti et al. 2011). Interactions and feedback effects between ecological and social subsystems, settings and related ecosystems result in complex and often non-linear dynamics (Liu et al. 2007, Ostrom 2009). Reinforcing feedback loops amplify changes in SES components while balancing feedback loops have a stabilizing effect. Identifying feedback mechanisms and being able to examine the way changes in one part of the system feed through to affect other parts is critical for finding places to intervene with management and policies to reverse undesirable trends and promote sustainable ones.

Engaging and fostering collaboration among stakeholders presents different kinds of challenges. The first involves ensuring that the full range of resource users and other system actors, researchers with expert knowledge, and decision-makers contribute to framing the problems and goals of the analysis, describing the system, and analyzing policy options. Due to the multitude of disciplines involved in the study of an SES, model development in SES research increasingly uses participatory or transdisciplinary modes of operation (Etienne 2011, van de Fliert et al. 2011). Fostering coordination and collaboration among "siloed" researchers is difficult (Alberti et al. 2011; Hammond and Dubé 2012). Transdisciplinary work requires developing a common language across disciplines and between researchers and decision-makers. When the analysis is done, a further challenge is to translate findings into implementable policy recommendations (Liu et al. 2015).

In sum, the challenges are not necessarily to develop better models or measures of the human or environmental components, but, rather, to find better ways to understand and represent the connections between them, create tools that allow policy analysis, and integrate stakeholders into the process.

3.3 System Dynamics and SES Integration for Sustainability

System dynamics is a computer-aided approach for policy analysis and design in complex, dynamic systems. It applies to problems that can be framed as undesirable trends over time arising in systems characterized by interdependence, mutual interaction, information feedback, and circular causality (Richardson 1991). It is well suited for formalizing social-ecological systems analyses, because it provides a method for operationalizing the SES framework in Fig. 3.1, and includes well-developed techniques for addressing the challenges of system integration and stakeholder engagement.

The central principle of system dynamics is that a system's structure generates its behavior, where the structure of the system consists of variables describing system characteristics and the material and information flows among them that form feedback loops and cause the system to change. Operating over time, the structure generates dynamic behavior such as exponential growth or decline, s-shaped growth or decline, collapse or oscillations (Saysel et al. 2002). The purpose of a system dynamics study can be to explain observed trends, anticipate the system's likely behavior in response to disturbance, or find a solution for a pattern of behavior considered problematic (Stave 2015). A detailed description of the methodology is given in Ford (2010), Forrester (1961), and Sterman (2000).

System dynamics addresses issues of system integration in several ways, including an explicit focus on change over time and an emphasis on modeling a specific problem or set of problems rather than modeling a system. The explicit focus on dynamic behavior necessarily integrates over time. A system's behavior is a function of the interactions of system components over time. System dynamics simulation models keep track of the accumulation and depletion of stocks as the model steps through time. The problem orientation in system dynamics focuses attention on the causal relationships that generate the problematic behavior. The analysis process does not predetermine the disciplinary theories or data needed to understand an issue. Rather, it facilitates the identification of the causal pathways relevant to the particular behavior of interest, regardless of discipline or geography, and thus draws in and integrates the disciplines specific to understanding what is causing the problem at hand (Stave 2015). System dynamics also includes practices for engaging stakeholders in problem solving through participatory or group modeling (e.g., Andersen et al. 1997; Rouwette et al. 2011; Stave 2010).

3.3.1 System Dynamics Modeling Process

The fundamental premise of system dynamics—that a system's dynamic behavior is generated by its structure—drives the modeling and analysis process. Explaining an observed behavior, anticipating behavior after disturbance, or changing a problematic pattern of behavior all begin with a description of the behavior of interest, followed by identification of the components, causal connections and feedback relationships that make up the structure (Stave and Kopainsky 2015). Creating a computer simulation model of the structure involves translating the structure into a set of mathematical equations that represent the relationships between variables. The model is first tested to ensure it can produce the behavior of concern, and revised until it does. Then it is used to probe the system's potential response to planned policy interventions or unplanned disturbances.

The steps in the modeling process are summarized below (see, e.g., Richardson and Pugh 1981; Sterman 2000 for more detail). These steps build on each other, although the modeling process generally involves considerable iteration between steps:

- *Define the behavior of interest.* Identify the trend or set of trends over time that need to be explained, that might be expected in response to a disturbance, or that constitute the problem to be solved. Specify which quantities vary, over what time period, and with what pattern.
- *Develop a conceptual model of the structure underlying the behavior of interest.* System conceptualization is the development of a hypothesis about the structure that generates the dynamic behavior of interest. It is called a *dynamic hypothesis* because it is the structure that is proposed to be causing the behavior. System conceptualization can be qualitative, using diagrams to visually represent different types of variables and the relationships among them, or operational, in which mathematical equations describe relationships among variables. Different types of structural representations are used for different purposes. The process of system conceptualization is to work backward from the identified problematic

variable or variables to determine immediately previous causal connections. A system dynamics analysis traces from the problem behavior outward along chains of cause and effect, rather than from the system boundary inward. This is described as modeling the causes of the problem, rather than modeling the system.

- *Validate the proposed structure*. Model validation takes two main forms, first, testing the logic of the proposed model relationships against what is known in theory, data, or common sense. Second, if the model structure has been converted to an operational computer model, the model can be simulated and the output compared with observed or anticipated system behavior. If the model cannot produce the expected behavior, go back to the previous step and revise the structure.
- *Use the model for analysis*. Use the model to identify points of intervention, either where unplanned shocks or stressors might affect system variables, or where deliberate policy interventions could be targeted. The simulation model allows you to conduct "what-if" analyses to compare the potential effects of alternative policies.

3.3.2 Participatory System Dynamics Modeling

Participatory techniques include diagramming tools such as causal loop diagrams (CLD) or stock and flow diagrams (SFD) that make mental models public so they can be shared. Such visualization tools allow the meaning and importance of system elements to be negotiated. They provide a coherent grammar and syntax, a common language with which participants can communicate with each other about the system (e.g., Black 2013; Hovmand 2014). Engaging stakeholders in problem and system description promotes understanding of the feedbacks and system behavior. Diagramming tools can expand participant ideas about the range of potential solutions by showing system-wide points of intervention (Antunes et al. 2015). Participatory modeling and evaluation of proposed policies on problematic behavior promotes buy-in for implementing policies. A typical participatory systems mapping process consists of the following stages and activities (Antunes et al. 2015):

- Preparation. This stage includes the identification of stakeholders and establishment of a first contact with participants. At this point, guiding questions for initiating the discussion may be developed by the modeling team, as well as preliminary CLDs elaborated from preliminary interviews with the selected stakeholders.
- Workshops. Most of the collaborative construction of CLDs takes place during workshops. One or more workshops may be planned, with one mapping session typically lasting between 1.5–4 hours. The workshop format may include several activities occurring in plenary or small groups, such as identification of variables, establishing causal links, drawing reinforcing and balancing feedback

loops, identifying leverage intervention points, and documenting knowledge gaps.
- Post-production and follow-up. This stage includes tasks such as refinement and digitalization of CLDs, writing of narratives describing the main feedback loops, and use of CLDs as an input for construction of simulation models.

The high-level perspective on feedback relationships shown in Fig. 3.1 offers an analytical framework for considering the connections among the individual chapters to follow. It also lays the ground for the system dynamics-based integration effort that wraps up this book. In November 2014, a workshop was held to begin a process of integrated analysis and collaborative modeling for Lake Tana basin sustainability issues. The workshop used the system dynamics approach to complete a first pass of the first two steps of the system dynamics process and produce an initial causal map. This application of participatory systems dynamics to the Lake Tana SES is described in Chap. 34.

References

Alberti M, Asbjornsen H, Baker LA, Brozovic N, Drinkwater LE, Drzyzga SA, Jantz CA, Fragoso J, Holland DS, Kohler TA, Liu J, McConnell WJ, Maschner HDG, Millington JDA, Monticino M, Podestà G, Pontius RG, Redman CL, Reo NJ, Sailor D, Urquhart G (2011) Research on coupled human and natural systems (CHANS): approach, challenges, and ctrategies. Bull Ecol Soc Am 92(2):218–228

Anderies JM, Walker BH, Kinzig AP (2006) Fifteen weddings and a funeral: case studies and resilience-based management. Ecol Soc 11(1):21

Andersen DF, Richardson GP, Vennix JAM (1997) Group model building: adding more science to the craft. Syst Dyn Rev 13(2):187–201

Antunes MP, Stave KA, Videira N, Santos R (2015) Using participatory system dynamics in environmental and sustainability dialogues. In: Ruth M (ed) Handbook of research methods and applications in environmental studies. Edward Elgar Publishing, Cheltenham, UK; Northampton, MA, USA

Berkes F, Folke C (eds) (1998) Linking social and ecological systems: management practices and social mechanisms for building resilience. Cambridge University Press, New York

Berkes F, Colding J, Folke C (2003) Navigating social-ecological systems. Building resilience for complexity and change. Cambridge University Press, New York, NY

Biggs EM, Bruce E, Boruff B, Duncan JMA, Horsley J, Pauli N, McNeill N, Neef A, Van Ogtrop F, Curnow J, Haworth B, Duce S, Imanari Y (2015) Sustainable development and the water–energy–food nexus: a perspective on livelihoods. Environ Sci Policy 54389–397

Binder CR, Hinkel J, Bots PWG, Pahl-Wostl C (2013) Comparison of frameworks for analyzing social-ecological systems. Ecol Soc 18(4):26

Black LJ (2013) When visuals are boundary objects in system dynamics work. Syst Dyn Rev 29 (2):70–86

Carpenter SR, Mooney HA, Agard J, Capistrano D, DeFries RS, Diaz S, Dietz T, Duraiappah AK, Oteng-Yeboah A, Pereira HM, Perrings C, Reid WV, Sarukhan J, Scholes RJ, Whyte A (2009) Science for managing ecosystem services: beyond the millennium ecosystem assessment. Proc Natl Acad Sci USA 106(5):1305–1312

Cash DW, Adger WN, Berkes F, Garden F, Lebel L, Olsson P, Pritchard L, Young O (2006) Scale and cross-scale dynamics: governance and information in a multilevel world. Ecol Soc 11(2 C7–8)
Cifdaloz O, Regmi A, Anderies JM, Rodriguez AA (2010) Robustness, vulnerability, and adaptive capacity in small-scale social-ecological systems: The Pumpa irrigation system in Nepal. Ecol Soc 15(3C7–39)
Cumming GS, Barnes G, Perz S, Schmink M, Sieving KE, Southworth J, Binford M, Holt RD, Stickler C, Holt T (2005) An exploratory framework for the empirical measurement of resilience. Ecosystems 8(8):975–987
Dawe NK, Ryan KL (2003) The faulty three-legged stool model of sustainable development. Conserv Biol 17(5):1458–1460
Dore J, Robinson J, Smith M (eds) (2010) Negotiate—reaching agreements over water, gland. International Union for Conservation of Nature (IUCN), Switzerland
Engle NL, Bremond A, Malone EL, Moss RH (2013) Towards a resilience indicator framework for making climate-change adaptation decisions. Mitig Adapt Strat Glob Change 1–18
Ericksen PJ (2008) Conceptualizing food systems for global environmental change research. Glob Environ Change 18(1):234–245
Ericksen PJ, Ingram JSI, Liverman DM (2009) Food security and global environmental change: emerging challenges. Environ Sci Policy 12(4):373–377
Etienne M (2011) Companion modelling. A participatory approach to support sustainable development, Versailles, éditions Quae
Ford A (2010) Modeling the environment. Island Press, Washington, DC
Forrester JW (1961) Industrial dynamics. MIT Press, Cambridge, MA
Hammond RA, Dubé L (2012) A systems science perspective and transdisciplinary models for food and nutrition security. Proc Natl Acad Sci 109(31):12356–12363
Hoekstra AY, Mekonnen MM (2012) The water footprint of humanity. PNAS 109(9):3232–3237
Holling CS (1973) Resilience and stability of ecological systems. Annu Rev Ecol Syst 4(1):1–23
Hovmand P (2014) Community based system dynamics. Springer, New York
Janssen MA, Ostrom E (2006) Empirically based, agent-based models. Ecol Soc 11(2):37
Liu J, Dietz T, Carpenter SR, Alberti M, Folke C, Moran E, Pell AN, Deadman P, Kratz T, Lubchenco J, Ostrom E, Ouyang Z, Provencher W, Redman CL, Schneider SH, Taylor WW (2007) Complexity of coupled human and natural systems. Science 317(5844):1513–1516
Liu J, Hull V, Batistella M, DeFries R, Dietz T, Fu F, Hertel TW, Izaurralde RC, Lambin EF, Li S, Martinelli LA, McConnell WJ, Moran EF, Naylor R, Ouyang Z, Polenske KR, Reenberg A, de Miranda Rocha G, Simmons CS, Verburg PH, Vitousek PM, Zhang F, Zhu C (2013) Framing sustainability in a telecoupled world. Ecol Soc 18(2):26
Liu J, Mooney H, Hull V, Davis SJ, Gaskell J, Hertel T, Lubchenco J, Seto KC, Gleick P, Kremen C, Li S (2015) Systems integration for global sustainability. Science 347(6225)
MEA (2005) Millennium ecosystem assessment, ecosystems and human well-being: synthesis. Island Press, Washington, DC
Ostrom E (2009) A general framework for analyzing sustainability of social-ecological systems. Science 325(5939):419–422
Resilience Alliance (2010) assessing resilience in social-ecological systems: workbook for practitioners. Version 2.0. Resilience Alliance
Richardson GP (1991) System dynamics: simulation for policy analysis from a feedback perspective. In: Fishwick PA, Luker PA (eds) Qualitative simulation modeling and analysis. Springer, New York
Richardson GP, Pugh A (1981) Introduction to system dynamics modeling. Pegasus Communications, Williston, VT; Waltham, MA
Rockström J, Steffen W, Noone K, Persson A, Chapin FS, Lambin EF, Lenton TM, Scheffer M, Folke C, Schellnhuber HJ, Nykvist B, de Wit CA, Hughes T, van der Leeuw S, Rodhe H, Sorlin S, Snyder PK, Costanza R, Svedin U, Falkenmark M, Karlberg L, Corell RW, Fabry VJ, Hansen J, Walker BH, Liverman D, Richardson K, Crutzen P, Foley JA (2009) A safe operating space for humanity. Nature 461(7263):472–475

Rouwette EAJA, Korzilius H, Vennix JAM, Jacobs E (2011) Modeling as persuasion: the impact of group model building on attitudes and behavior. Syst Dyn Rev 27(1):1–21
Saysel AK, Barlas Y, Yenigün O (2002) Environmental sustainability in an agricultural development project: a system dynamics approach. J Environ Manage 64(3):247–260
Schlüter M, McAllister RRJ, Arlinghaus R, Bunnefeld N, Eisenack K, Hölker F, Milner-Gulland EJ, Müller B, Nicholson E, Quaas M, Stöven M (2012) New horizons for managing the environment: a review of coupled social-ecological systems modeling. Nat Resour Model 25(1):219–272
Schlüter M, Hinkel J, Bots PWG, Arlinghaus R (2014) Application of the SES Framework for model-based analysis of the dynamics of social-ecological systems. Ecol Soc 19(1):36
Stave KA (2010) Participatory system dynamics modeling for sustainable environmental management: observations from four cases. Sustainability 2(9):2762–2784
Stave KA (2015) System dynamics for environmental applications. In: Ruth M (ed) Handbook of research methods and applications in environmental studies. Edward Elgar Publishing, Cheltenham, UK; Northampton, MA, USA
Stave KA, Kopainsky B (2015) A system dynamics approach for examining mechanisms and pathways of food supply vulnerability. J Environ Stud Sci 5(3):321–336
Sterman JD (2000) Business dynamics. Systems thinking and modeling for a complex world, Boston et. al. Irwin McGraw-Hill
Thompson J, Scoones I (2009) Addressing the dynamics of agri-food systems: an emerging agenda for social science research. Environ Sci Policy 12(4):386–397
van de Fliert E, Hermann S, Olsson JA (2011) Integrated assessment of agricultural sustainability. Exploring the use of models in stakeholder processes. Earthscan, Oxford
Wackernagel M, Rees WE (1996) Our ecological footprint. New Society Publishers, Gabriola Island, Canada
Walker BH, Carpenter SR, Anderies JM, Abel N, Cumming GS, Janssen MA, Lebel L, Norberg J, Peterson GD, Pritchard R (2002) Resilience management in social-ecological systems: a working hypothesis for a participatory approach. Conserv Ecol 6(1):14
Walker BH, Anderies JM, Kinzig AP, Ryan P (2006) Exploring resilience in social-ecological systems through comparative studies and theory development: Introduction to the special issue. Ecol Soc 11(1):12
World Economic Forum (2011) Water security: the water–food–energy–climate nexus. World Economic Forum (WEF), Washington DC

Chapter 4
Bahir Dar and the Lake Tana Basin: Historical Phases of Growth and Ecology

James C. McCann

Abstract This essay sets out five distinct periods of economic, demographic, and spatial change that transformed the physical landscape and its effect on the region's hinterland on a local, national and international level – that would form the "urbanscape" of the early twenty first century. Lake Tana is at the Blue Nile's first major outflow and is the heartbeat of the watershed, registering its seasonal pulse. Lake Tana and its surrounding ecologies is the font of the Blue Nile's waters, sitting at 1800 meters above sea level from where it frames much of the watershed. The Blue Nile watershed's agriculture and its relationship with human settlement had evolved over many generations of smallholder agriculture and its regional role in trade. The Blue Nile basin, its geology and its geographies simultaneously shaped the cultures of several distinctive peoples. The ethnographic landscape included Christian highland farmers and aristocrats, Cushitic-speaking Agaw farmers, Muslim traders (who spoke Amharic), and Omotic-speaking Shinasha. These cultures traded places and bodies of knowledge on the local ecologies over time, resulting in a cereal-based agro-economy that supported livestock and small farms that managed them.

Keywords Lake Tana · Blue Nile · Urbanscape · Malaria · Dam construction · Nile diplomacy

4.1 Introduction

In 1905 Samuel Hayes, a British traveler passing by the southwestern edge of Lake Tana described there: "a village surrounded by a marsh of papyrus and a nearby village of two or three huts occupied by Wayto [hippo-hunting fisherfolk]." Three decades later, in 1938, the Italian Tourist Agency published a cursory description of that lakefront in their otherwise detailed guidebook to their newly conquered East

J.C. McCann (✉)
Ethiopian Academy of Science, Boston University, Bahir Dar, Ethiopia
e-mail: mccann@bu.edu

K. Stave et al. (eds.), *Social and Ecological System Dynamics*,
AESS Interdisciplinary Environmental Studies and Sciences Series,
DOI 10.1007/978-3-319-45755-0_4

African colony. That guide remains one of the most revealing snapshots of Ethiopia's state of infrastructure development in the year 1938. On page 383 of that guide is a description of the site of a seventeenth-century lakeside church Bahrdar (sic) Giyorgis and what the Italians had built as the site of their nearby headquarters—the Residenza del Tana Meridionale—, a dispersed rural district (not a town) that encompassed a c. 25,000 population across an agricultural landscape. The Italian official *Guida* listed official colonial buildings as a post office, a telegraph office, an infirmary, a grass air strip, and a landing for their single lake transport boat. What the *Guida* actually recorded was a minimal settlement: a few buildings superimposed on a lakeshore site, not a city or even a town.

The *Guida* literally tells us that Bahir Dar was, in fact, barely a "village," comprised of Abyssinian and Wayto thatched houses near the shoreline's sixteenth-century Orthodox church (Hayes 1905; Consociazione 1938).[1] And we have even earlier evidence of Lake Tana's absent human settlement on the southern shore. In the 1840s two French visitors to the lakefront described it as a "fatal zone" (Ferret and Galinier 1847–48). This comment referred to the area's disease ecology, rather than political security. The unhealthiness of the lakeside was a sign of malaria fevers resulting from seasonal floods that created ideal mosquito habitat and malaria transmission. Not much changed from the 1840s and between traveler Hayes' 1905 description and the 1938 *Guida* account. There were more rapid landscape changes to come, and we must not ignore the meaning and sequence of that transformation.

This chapter clarifies the historical human and landscape co-adaptation in the Upper Blue Nile basin in five distinct periods of economic, demographic, and spatial change in the early twenty first century. These five phases are:

A. Pre–1960: "A Fatal Zone"
B. 1930–1960: A Cat's Paw of Colonial Threats Haile Sellassie Era
C. 1960–1974: Era of the Dam and Cotton Factory
D. 1974–1991: Bahir Dar in the Age of Socialism
E. 1991–present: High Modern Ambitions: *Hedasse* Dam and Blue Nile Waters

A. *Pre–1960 "A Fatal Zone:" The Lake Tana/Upper Blue Nile Basin: Historical Antecedents*

In the years until 1960, Bahir Dar's modern urban site was sparsely settled and the threat to human health was a recurrent theme for the lake's edge. In the mid-twentieth century, sudden, and quite "deadly," malaria epidemics had broken out in the region in 1953, 1958, 1964, and 1998. Tens of thousands rural folk, possibly as many as 150,000 in 1958 alone (McCann 2015), in the lake's hinterlands and seasonal flood plains died in those epidemics, leaving little doubt that the lakeshore had never been a concentrated population center, urban economy,

[1]For visitors, Bahir Dar in those years was less a site of interest than an afterthought. The *Guida* map of the region misplaces Bahir Dar Giyorgis 100 km south of the lake.

or administrative core like that at seventeenth-century Gondar or, later at Addis Ababa in the twentieth century.

Lake Tana is the site of the Blue Nile's first major outflow and is the heartbeat of the watershed, registering its seasonal pulse, summer wet and winter dry.[2] Lake Tana and its surrounding tributary ecosystems are the font of the Blue Nile's waters, sitting at 1800 m above sea level from where they frame much of the watershed. The lake's geological origins were a volcanic blockage of the river's stream about two million years ago by a 50 km-long flow of basaltic molten rock that trapped water in the shallow basin that was to become Tana's remarkably shallow lakebed. Seasonal overflows of lake waters onto black vertic soils (*Qoticha* or *Merara*) on the eastern edge that allowed late season planting of legumes (vetch and lentils) once ox-plow agriculture spread there from the northern highlands, perhaps 3000 years ago.

The Blue Nile's historical headwater hydrology was a dynamic, changing flow and there is also evidence that the lake dried up 15,100 and 16,700 years ago and became a papyrus swamp. There was, in geological time, a fundamental change in lake ecology and biological habitat that took place then. In this dry phase historical geologists estimate that rainfall was less than 40% of what we expect has been the pattern in the last two centuries. Then, about 14,750 years ago, water in the entire Nile system seems to have increased and the lake overflowed the volcanic blockage into the Blue Nile riverbed again. This process seems also to have happened in the junior partner East Africa's White Nile when Lake Victoria, also in a dry phase, filled to overflowing and forming small lakes and ponds-allowing new fish sub-species diversity to evolve in those new ecologies. In its more recent geological history the Nile system has regained its permanent flow, though still with the strong seasonal, pulsing rhythm that drives the Blue Nile watershed's ecology and gives us the modern landscape of human health and disease that still frames the Bahir Dar urban hinterland (Lamb et al. 2007).[3]

In its recent history, including the 20th century, the lake is still a shallow vessel, with a maximum depth of only 14–16 m. Four small rivers drain into the lake, and the Blue Nile is the lake's only outflow. The lake's oxygen level is unusually high. Though the flow of water pumps the river's water and silt through the lake towards the Sudan border and Khartoum and into the great Nile flow; the lake's discharge at the outflow makes up only a total of 8% of the annual total of water contributed to the overall Blue Nile watershed. The Blue Nile's symbolic source at Gish Abbay south east of the lake is a cultural marker—its symbolic spring leading to the Gilgil Abbay (or Piccolo Abbay) river and thence entry to the lake. Yet, as it gathers momentum and other tributary flows along its journey to the south, southeast, the Blue Nile as a whole has made up as much as 84% of the total Nile flow.

---

[2]Tana, however, is not technically a pulse lake, like Cambodia's Tonle Sap on the Mekong river. The Blue Nile's waters rise and fall with the season, whereas the Tonle Sap has an annual backflow that sustains it distinctive aquatic ecology.

[3]Henry Lamb, et al. "Late Pleistocene desiccation of Lake Tana, source of the Blue Nile", *Quaternary Science Reviews* 26 (Lamb et al. 2007), 287, 296.

The Tana basin is itself a crèche of biodiversity. The river and lake's fish biodiversity results from the dramatic fall of 40 m at the Tissisat ("Smoke of fire") 30 km downstream from the Blue Nile's outflow, preserving the lake's remarkable fish ecology and floodplain breeding sites. Tana's biological endemism may soon, however, be a thing of the past. Twenty-eight fish species inhabit the lake, twenty-one of which are endemic. Seventeen of those species are large barbs, types unique to this place and its ecological cocktail of seasonal breeding grounds unique to the lake. Only 8 fish species inhabit the river basin both above and below the falls. Fish and other aquatic organisms did not ascend the river to inhabit the lake, keeping its endemism intact (Abebe Getahun and Eshete Dejen (2012)). Aquatic life—its fish species, plant life, and malarial habitat around the edges of the lake itself developed its own historical character (its endemism) isolated from the wider Nile system—what was a closed system maybe ended in 2006 by the opening of the 250 MW Tissisat Abay II dam at the falls' site.[4]

Other sub-ecosystems are also in a state of transition. The eastern and southern shores of the lake display a deeply seated malarial ecology in swamps dominated by papyrus, water lilies, and wet "black cotton" soils. Large and sprawling fig trees (*Warqa*) overhang the water's edge, providing nesting spots for African fish eagles whose shrieks are among the first sounds of morning. Yet, it is not the lake itself, but its seasonally flooded edges that form its malaria setting. The lake itself offers few habitats for mosquitos, but the surrounding wetlands and puddles formed by receding lake water in the early dry season (late September/early October) historically left habitat for *An. pharoensis* and *An. funestus* mosquitos that may have been primary vectors that made the lake area historically "a fatal zone" along its edges.[5] Lakeshore reeds and vegetation also offer a rich habitat for snails that are hosts for bilharzia's life cycle.

More than canary in a coal mine, malaria was not a sign of a problem but was the problem itself. Malaria's fevers and human sufferers gave us the human landscape. In the longer-range past these were the drivers of where people in the watershed lived, and when they could move about. The disease and appearance of its vectors (anopheline mosquitoes) marked seasonal change, almost as much as the rigors of the agricultural season, though in the higher areas malaria was an unstable, unpredictable visitor. Malaria outbreaks were unstable, i.e. periodic but unpredictable, with major eruptions in 1953, 1958, 1964, 1998, 2012 (Kebede 2002).

Diseases fundamentally molded human history in the watershed. But malaria in its epidemic forms was by far the most deadly local disease that shaped the movement of peoples and where they settled. Malaria was the aggressor and humans responded defensively; moving to safe areas but periodically enduring epidemic outbreaks that swept from lowland hotspots onto highland populations.

[4]The Blue Nile Falls' periodic closure to protect hydroelectric water flows has also encouraged farmers to claim land opened by reduced river bed flows.

[5]Another mosquito species found in the puddles and pools, *An. arabiensis*, has another story in recent years.

The southern edge of the lake was therefore not a prime area for human settlement. In those settings malaria was the lead partner in shaping where people settled and the risks they accepted to gain employment in the growing economy of lowland agriculture.

Malaria has long lived at the lake's edge, and framed human health and settlement for those who dared to make their living there. Through the millennia, the actions of Ethiopia's people on the highlands did little to alter these slow rhythms, compared to other, deeper effects of climate fluctuations like the Little Ice Age or the seasonal pulse of rainfall, surface water flow highland agriculture's annual scratching out human livelihoods on the non-malarial uplands (above 2300 m elevation) set the terms of rural livelihoods on what European visitors called Ethiopia's "salubrious" highlands (Harris 1844). The post-1960 lake ecology has evolved with urban growth of Bahir Dar town as a commercial center, industrial core (a cotton factory), and tourist destination—a change that altered mosquito habitat and human blood supply. Urban construction has, for the time being, laid out breeding sites for *An. arabiensis*, allowing malaria outbreaks in the urban setting, such as happened in 2012 in newly settled neighborhoods along the river on the northern edge of the city. As the city grows, it has drawn in public health facilities; urban growth brings, perhaps temporarily, mosquito habitat, but also the means to better manage malaria.

In terms of the human factor, the Blue Nile basin's geology and geographies simultaneously shaped the cultures of several distinctive peoples. The ethnographic landscape included Christian highland farmers and aristocrats, Cushitic-speaking Agaw farmers, agropastoralists, Muslim traders (who spoke Amharic), and Omotic-speaking Shinasha in the watershed's lowland areas to the southwest near the Sudan border. Over time, these cultures traded places and bodies of knowledge on the local ecologies, resulting in a cereal-based agro-economy that supported livestock and small farms that managed them. A few small trade towns sprouted—Burie, Dangla, Debre Tabor, Meshenti—but no genuinely urban population centers beyond Gondar in the high setting north of the lake (and above 2000 m elevation). Settled population centers have grown recently, and quickly. In the past two decades population growth and migration has turned crossroad villages into towns, towns into urban centers and rural markets into perennial settlements.

The Blue Nile watershed itself is not a political boundary, but it has a western lowland political border, since an obscure 1902 treaty between the British government and Emperor Menilk II included setting an international border with British-controlled Sudan. The Blue Nile/Tana waters and human ecology, however, have often tended to ignore such political ideas. Within the change of seasons and human movements, malaria and other diseases set their interactions in the ecologies of elevation, watery habitats, and human adaptations to them. Epidemic, (and "unstable") malaria is one of the deadly disease outcomes of this dynamic human/natural ecology, though so too were other enduring maladies of the watershed, exotic diseases of the sub-tropical world. We can count among these: rinderpest, kalazar, yellow fever, tick and louse-based relapsing fever, (bilharzia at the lake's edge), and lower intestinal tract infections. These were mainly lowland diseases that kept most human settlement away from lower elevations, but those infectious maladies periodically washed up and over highland zones, causing fatalities and sickness at an alarming rate. Growth of Bahir Dar as a regional city has changed this mix by attracting new population, setting a physical infrastructure, setting health care facilities, and a salary-based economy that created an urban market for agricultural products. Endemic human health conditions are, and have been, in the process of change.

Over the past millennia climate and its seasonal patterns were the primary determinates of human activities. The Blue Nile watershed was the first part of the northern highlands to receive rains in the late spring/early summer months from the band of heavy rainbearing clouds from the southwest, whereas rainfall shadows in the east side often curtail the arrival or duration of rains so essential for agriculture. Peoples of the Blue Nile/Tana watershed had, over time, married their annual cycles of food agriculture, local political life, and religious ritual to their moisture and ecological setting. Monasteries and major church sites to the east and south—Debre Tabor, Debre Marcos, Martola Mariam, Dembecha, Burie Mikael—dominated religious life, though Muslims had long plied the trade routes and livelihoods in small market centers where they built modest mosques in unobtrusive neighborhoods. Bahir Dar town, once established after 1960 had always had a strong Islamic presence, connected to a growing city and the older regional caravan trade. Trade between the region and the Nile and the Red Sea percolated across ecological zones that were malarial in relatively predictable seasonal rhythms.

Malaria in particular was an unwelcome, but only a sometimes visitor to the highland zones where people lived, and a more permanent resident of the lowland where only people on political or ethnic margins dared to live. Malaria on the edges of the region was, to some along the edges of ecological zones, a regular part of life. But it was a delicate balance that accepted the early death for infants. Those few lowland dwellers who survived to adulthood in the hot lowland zones on the west, likely had built acquired immunity—or died in childhood from *falciparum* malaria. But also for adults, since Ethiopia's "unstable" malaria meant that no highlanders—young or older—had developed that acquired immunity.

Climate at the lake's southeastern edge changed in the past 200 years, as it now entered another warming period post-Little Ice Age. But the ecology of the Blue Nile watershed that Scottish traveler James Bruce witnessed in the 1770s was a force of nature and human settlement in transition. Water and nighttime temperatures during those times were different than modern periods. Climate science, using soil profiles, lake sediment cores or ice cores, and glacier records on the landscape itself tells us that in 1770 parts of the planet were in a cooling period brought on by low solar radiation, changes in ocean temperature circulation, and perhaps volcanic activity—the Little Ice Age—when a "glacial expansion" that took place between about 1550 and lasted until the mid-nineteenth century-including the period when Bruce traveled in the Blue Nile region over several years. If Bruce tells us that in the 1770s he saw snow on the top of the areas highest peak (Mt. Ras Dashan), we should believe him; even if that snow that did not appear in the twentieth century as more observers arrived, but misunderstood past climate conditions. We might well read that as a sign of what the 2007 Intergovernmental Panel on Climate Change calls glacial expansion (Pachauri and Reisinger 2007). Snowy icecaps in Bruce's time might mean that nighttime temperatures would have been lower: fewer mosquitoes, fewer and less potent falciparum parasites maturing in mosquito guts, and perhaps later dates of harvest for grain crops. But what were the broader effects on health, politics, and the overall human condition in local settings in Ethiopia?

The end of the Little Ice Age, marked commonly at around 1850, coincided with climate change but also a rebirth of political centralization in Ethiopia. European arrivals into Ethiopia's malaria ecologies in the late-nineteenth century coincided with their new colonial ambitions. We know few details of climate conditions, but we can surmise that rising temperatures, especially at nighttime, and resurgence of regional trade across ecologies of elevation may have slightly changed the geography for disease transmission, even as political changes and conflicts like Tewodros's violent rule (1855–68) engulfed the central Ethiopian region. Politically, in those years the Tana basin was a rural hinterland to Gondar, not a seat of power to rival either Gondarine emperors of the eighteenth and mid-nineteenth centuries.[6] Addis Ababa, of course, emerged only in 1886.

While we know something about political change in the mid-1800s, we know next to nothing about the social or ecological life of disease or health and in the Upper Blue Nile basin or the later site of Bahir Dar town in that period. Travelers reported their fears of fever, but voices of the true status of economic change in the area were silent for decades. In this period the major source of economic and ecological stability at the local level was on the face of it a grain-based ox-plow agricultural ecology. Calamities like the arrival of cattle-killing rinderpest and great famine in 1889–91 affected the lake region, as did the Spanish Flu (*YaHedar Bisheta*) that came in 1919. But we have few descriptions of the local effects for the Tana region. As the Upper Blue Nile basin recovered from those health or economic disasters, all evidence is that it resumed a similar pattern of agro-ecology and politics as had existed before, though now under an expanding Addis Ababa-centered political structure. One wonders, for example, if those distinctive Fogera cattle who grazed on grassy floodplains had some resistance to rinderpest because of their local ecology of wetland black soil. The Tana basin was rural and gave few signs of the embryonic urban industrial core that would appear after 1960 (see below).

For 1953, we find a detailed, local view of the Lake Tana shore at Fogera, that flat, black-soil plain on the northeast side of the lake where the Gumara river feeds into the lake's edge. One local Ethiopian visitor—a loquacious traditional healer—there during the 1953 malaria outbreak offered poignant testimony. This observer Dabtara Asres—the healer/cleric—who knew the local landscape, its history of disease and it flood plain vegetative cover, tells us of the 1953 landscape he knew:

[6]The Blue Nile western lowlands did, in fact, offer shelter to bandits, such as the young Dej. Kassa who rose to the status of emperor, as Tewodros II (1855–68).

> The fevers caused misfortune in Foguera. This is extremely flat country and during this period nothing could grow there, not even eucalyptus like today. It was plain, absolutely bare. During the day the hyenas hid in Aferouannant, in Oudo, in Amora-Guedel, and in Dera, in the woods. They knew where to find carrion. At night, they did not stay in the hills; they descended on Foguera to feast on the dead cattle along the roads.... (Mercier 1988)

But this was a landscape that in two decades would emerge, tenuously at first, as fertile ground for economic development tied to nearby urban center at Bahir Dar.

B. *1930–1960 Lake Tana: A Cat's Paw for Nile Diplomacy*

The Lake Tana watershed took on an international focus in the early twentieth century when the Nile's greater basin came under the vision of both the British empire and Ethiopian centralization. And this vision of the hydraulic facts included knowledge that the Blue Nile at the turn of the twentieth century provided 84 percent of the water reaching Egypt—and virtually all of the silt that nourished agriculture in the Nile Delta (Garstin 1904; Grabham and Black 1925).[7] This understanding by British colonial strategists played a central role. The British idea of a Lake Tana dam emerged shortly after a 1902 treaty on the Nile and rising commodity prices that highlighted the value of Nile waters for development, or the cost of losing control of them. In era of Ethiopia's nation-building under a new emperor, British negotiations with Ethiopia over Blue Nile waters continued for a decade and a half, until the Italian invasion in 1935 (McCann 1981; Sutcliffe and Parks 1999).[8]

Yet, realities of infrastructure capacity also continued to isolate Lake Tana from regional and local political economies. Up to the 1935 Italian invasion, little had taken place to link Lake Tana to the nation or the world. During the June-September rainy season the river posed an impassable barrier, cutting off all to the lake region and its economy. The deep and difficult Blue Nile gorge would not have a bridge until 1933 (at the Shafartack crossing to Goha Tsion and thence Addis Ababa). The rains even interrupted the single-line telephone connection, that, in 1920, reached only to Burie, a town 180 km south of the lake. Beyond infrastructure, Ras Tafari (aka Haile Sellassie after 1930), faced political opposition from regional leaders who sought both a part of a financial agreement and to keep their local autonomy from a distant leader in Addis Ababa. By 1930, however, the newly crowned Haile Sellassie had used the Tana waters only as a political cat's paw, but had no real leverage on development of the lake's waters or its economic potential. The 1935–36 Italian invasion by Mussolini's troops and the Emperor's exile to Britain ended that gambit In 1941 Anglo-Ethiopian forces returned to Ethiopia, passing through the Blue Nile headwaters on their way to the capital. Neither the Emperor not the British declared any interest in Blue Nile hydrology.

---

[7]For example, Garstin reports a 1902 measurement at Khartoum, in August, the Blue Nile discharged 9.544 m^3 of silt-bearing water per second, while the White Nile offered only 710 (with little silt).

[8]Sutcliffe and Parks argue that regional data show that the Blue Nile's contribution to Nile waters overall has slightly declined in post WWII years.

The Bahir Dar region was until the early 1960s a rural economy, poorly serviced by health infrastructure and the penetration of a national economy. In 1953 the area of the lake and hinterlands suffered the devastating effects of a malaria outbreak, and another in 1958 when upwards of 150,000 rural folk had died. The Blue Nile watershed's agriculture and its relationship with human settlement had evolved over many generations of smallholder agriculture and its regional role in trade. The region was not yet a major grain supplier to the capital region. Gazing out at the agrarian landscapes that surrounded newly burgeoning towns in the 1970s, one could easily witness persistent practices that gave its fields and landscapes their visual character. The future Bahir Dar urban industrial economy had yet to appear.

C. *1960–74: The Era of Tissisat Dam and an Industrial Foothold*

What underlay a quiescent, but stable economy of the Upper Blue Nile were the annual rhythms of rainfall seasons that set agricultural cycles. In the early years of the 1960s a transformation had nonetheless begun, albeit slowly. Ethiopia's development plan for a new, planned city at the southern edge of the lake, Bahir Dar, linked the building of a textile factory, a smallish (11.5 MW) hydroelectric dam at the falls, and a longer-range plan to move the new city forward as an administrative capital. By 1974, the Ethiopian revolution had deposed Emperor Haile Sellassie and brought into play a military socialist plan for the regional economy and the town itself. At that point, Bahir Dar had a modest footprint. The town boasted 24 h-electric power (for the cotton factory and many town residents) a grass landing field for a once a day DC3 flight, a Russian-staffed polytechnic institute, a movie theater, a German-built hospital, and a gridwork of roads that marked the urban grid that planners imagined. Some nearby market towns, however, had 12-hour diesel-generated power to allow 25 W bulbs, a few had piped water, and the transport grid included small buses (Leoncino) on the main roads. Still, Debre Markos the then capital of the Gojjam region was an eight hour Leoncino ride, to Bahir Dar, stopping for local folks who flagged them down on the roadside.

Provincial capitals in Ethiopia's Tana/Blue Nile basin were sleepy burgs—Gondar north of the lake and Debre Marcos, 300 km on a gravel road toward Addis Ababa. The modern bridge over the Blue Nile river leading to Addis Ababa came only in 1960. As mentioned above, Bahir Dar's foundations as a planned urban settlement emerged in tandem with the first small (11.5 MW) hydroelectric dam at the falls and a new textile factory. Bahir Dar's street grid was planned, but not yet populated with population or much commerce. The urban footprint was small, reaching only a few kilometers beyond the official city limits, or across the Blue Nile's emergence from the lake.

Agriculture was the region's economic base, as it had been for generations in the past. The end result of the plow, farmers' local experience, and crop mixing was the patchwork quilt landscape of small plots growing diverse crops—teff, eleusine (*dagusa*), maize, wheat and pulses like lentils, field peas, and fava beans. From my own observations living in the area for two years in the mid-1970s; the countryside was a human landscape of dispersed homesteads surrounded by fenced gardens and crop fields scattered willy nilly across the rural scene—green, cropped fields in the summer and golden at harvest in the late fall; brown, stubbled fields by December and January.

Highland watershed farms sat on a mid-altitude plateau around 2000–2300 m high above a marginal lowland descending southward into the Abbay (Blue Nile) gorge where the population is sparsely settled, mainly along the new graveled road that crossed the Blue Nile river on the way to Addis Ababa. That road retraced the old caravan route that linked the coffee-rich areas on the Blue Nile's southern bank to market towns and thence to the old imperial capital at Gondar and destinations north. There was no bridge on that old coffee-trading route until the early 1980s.

The local agricultural system, as it turns out, had become a part of the region's economic stability. Tana watershed farmers on the upland plateau, like their compatriots in much of Ethiopia's basin middle-range elevations, used little chemical fertilizer prior to 1980, though they had long applied manure and ash on household garden crops such as maize, kale, capsicum (red pepper), and herbs cultivated inside their fenced homestead compounds, called *gwaro maret* (household garden plot). In this old farm landscape, maize plants were a minor partner: local, early maturing maize types germinated with the first rains that shed their pollen in the cool months of June and July when heavy rainfall washed pollen out of the fields, and the plants produced green ears in August and September. Roasted or boiled, these young ears (Amharic: *eshet*) offered welcome snacks when food shortages preceded the main harvests in November and early December. Those changes in crop patterns from the 1960–74 period were ephemeral at best for the Blue Nile watershed (and its urban settlement at the lake's edge). That was to change with the arrival of modern maize seed and national agricultural markets after 1980, beginning a new phase of Ethiopia's agricultural modernization

D. *1974–1991: The Watershed in the Era of Socialism*

Until the 1980s, maize was minor field crop in the northern highlands and appeared on farms mostly as a garden vegetable consumed green in the 'hungry season' (August/September). Farmers chose from among an array of early maturing local maize types, the best known ones they called *Mareysa*, *Harer*, and *Kafa*. Some of these probably arrived in Ethiopia via the Nile Valley in the nineteenth century, some from U.S. agriculture aid programs of the early 1950s, while others may have been descended from the original imports brought from the New World (via India) by Arab and south Asian merchants plying the Red Sea trade in the sixteenth century. Maize's most common name in northern Ethiopia was *YaBahar mashela* ("sorghum from [across] the sea").[9] At first, maize had a value mainly as a welcome snack for farms awaiting November–December cereal harvests.

[9]In the Amharic of Gojjam the word for maize became, simply, *mashela*, though in other parts of northern Ethiopia that term meant sorghum. *Boqolo*, was the other common term in middle and southern regions.

In the early 1980s political changes brought a transformation in Ethiopia's agrarian economy and agro-ecological balance. Under the socialist military government known as the Derg that ruled from 1974 to 1991, grain marketing policy, forced labor, insecurity over land holdings, and government food security programs brought increasing political chaos but also changes in Ethiopia's national crop patterns. Ethiopia's socialist government saw maize as a high yielding field crop to replace labor-intensive teff, a disease-prone wheat, and poor-yielding sorghum.

As in the Soviet Union, Ethiopia's socialist planners used maize as the ultimate product of an industrialized, scientific agriculture. For their part, farmers saw maize as a low labor, quicker maturing crop that provided food in insecure times when the socialist state forced their labor onto public works projects like tree planting or fixed the prices for their other farm produce. Farmers were confused about imposed national policies and maize was a rational choice in troubled times for them. By the mid-1980 s maize had unceremoniously surpassed teff and barley as the major grain crop produced in the region. This agricultural ecology based itself in small farms and still formed the basis for politics, local economies, and formed the canvas on which national issues played. Development on the Upper Blue Nile was part of that plan

Socialism and an interventionist military government had attempted to bring its policies to a rural landscape in the Blue Nile watershed. It largely failed. In 1992 the road from Bahir Dar south to Addis Ababa had deteriorated to a rutted roadbed barely worthy of being called a road. Retreating government soldiers and invading rebel troops in 1991 had turned Bahir Dar's German mission hospital into a morass of files strewn on the hospital's open courtyard with patients wandering around empty wards. Diesel fuel in town was in very short supply, and non-existent elsewhere in the area.[10]

Foreign investment had dried up. In the late 1980s socialist Ethiopia had the lowest per capital international aid of any country in the world. Yet, the government had nonetheless pursued its own style of development in the countryside of the watershed. At Pawe, an Italian firm contracted to build the Tana Beles dam to channel water for agricultural development out of the Nile system and the lake. The Green Revolution (*Aranguade Zamacha*) had marched students out of towns to new lowland settlement areas on the lowlands to build villages. This program was an abject failure that resulted in massive outbreaks of malaria, dislocation, and forced settlement of highland people onto empty landscapes. To its credit, the military/socialist government had expanded rural roads and at least had begun the development of a rural administration that set the stage for Bahir Dar's emergence, under the present government as an administrative center in the next decade. Bahir Dar took on a new status as capital of Amhara region and absorbed the benefits of a new class of administrators, a university, and agro-industrial businesses. By the early 1980s the Blue Nile basin began a period of sustained growth in its agricultural base and in Bahir Dar's growing urban footprint.

[10]Passing through town in 1992 my Toyota Hilux diesel pick-up obtained its supply from under the bed of a woman of ill-repute who sold it at a premium price to those who could pay.

E. *1991–Present: High Modern Ambitions: Hedasse Dam and Blue Nile Waters*

In 2012 when I traveled the tarmac road from Bahir Dar, the new city, to crossroads town of Warata (and then on to Gondar) across the egress of the Gumara river a very different landscape presented itself from what Asres had described in 1953. On the right side of the road were smallholder plots which had been vetch (*gwaya*), a mildly toxic pulse crop that tolerated black soils and flooding, were now small plots of irrigated medium-grain rice, cultivated by ox-plows and using diesel pumps extend their planting/harvesting seasons. That change had begun with an abortive North Korean development project under the socialist government that had returned to smallholder ox-plow agriculture that included new varieties of rice within its crop repertoire—a remarkable local innovation. The area sold its tomatoes, red onions, and lettuce to tourist hotels and market shops that served tourism but also new urban residents on government salaries. The city had emerged as the region's economic engine.

References

Consociazione Turistica Italiana (1938) Guida dell'Africa Orientale Italiana. Milan

Ferret PV, Galinier JG (1847–48) Voyage en Abyssinie. Paris

Garstin W (1904) Report upon the basin of the Upper Nile. Cairo

Getahun A, Dejen E (2012) Fishes of Lake Tana: a guide book. Addis Ababa

Grabham GW, Black RP (1925) Report of the mission to Lake Tana 1920–1921. Cairo

Harris WC (1844) The highlands of Ethiopia, 2 vols. London

Hayes S (1905) Source of the Blue Nile. London

Kebede A (2002) Overview of the history of malaria epidemics in Ethiopia. Paper presented at the workshop on capacity building on malaria control in Ethiopia, Addis Ababa

Lamb H et al (2007) Late Pleistocene desiccation of Lake Tana, source of the Blue Nile. Quat Sci Rev 26:287–296

McCann J (1981) Ethiopia, Britain, and negotiations for the Lake Tana Dam, 1922–1935. Int J Afr Hist Stud 14(4):670–671

McCann J (2015) The historical ecology of malaria in Ethiopia: deposing the spirits. Ohio University Press, Athens, OH

Mercier Jacques (1988) Asrès, le magician éthiopien: Souvenirs 1895–1985. Lattés, Paris

Pachauri RK, Reisinger A (eds) (2007) Contribution of working groups I, II and III to the fourth assessment report of the intergovernmental panel on climate change, Geneva, Switzerland

Sutcliffe JV, Parks YP (1999) The hydrology of the Nile. Colombo

Part II
Ecosystem Characteristics

Chapter 5
Climate of Lake Tana Basin

Wubneh Belete Abebe, Tesfahun G/Michael, Elias Sime Leggesse, Biaznelign S. Beyene and Fenta Nigate

Abstract Lake Tana basin, the subbasin of Abbay (Blue Nile River), covers an estimated area of 15,114 km^2 of which the Lake Tana accounts 20.47%. It is located in Amhara Region in the north western Ethiopian highlands. The mean elevation of the Basin is 2025 masl (meter above sea level), with the highest elevations at some 4100 masl in the Simien Mountains, in the north-eastern part of the basin and the lowest at the outlet of the lake into the Blue Nile at 1786 masl. The topography in the eastern side of the basin is dominated by the presence of two large shield volcanoes, Mt. Choke and Mt. Guna, while in the west it drops sharply to the adjacent Beles and Dinder basins. Bahir Dar city is located in the southern tip of the lake where the outlet of the lake into the Blue Nile River. The climate of the region is tropical highland monsoon with one rainy season between June and September. The air temperature shows a large diurnal but small seasonal change with an annual average of 20 °C. The distribution of rainfall of the region is controlled by the movement of the inter-tropical convergence zone (ITCZ) to the northward and southward. Based on rainfall variability analysis, 57.5% of the basin is found under less rainfall variability. Little has been done so far on understanding the effect of climate change on rainfall amount and distribution and hence of the Basin Rivers flows. The largest area of the basin (88.5%) is not affected by frost hazard.

Keywords Climate · Temperature · Rainfall · Lake Tana · Blue Nile River

W.B. Abebe (✉) · T. G/Michael
Amhara Design and Supervision Works Enterprise, Bahir Dar, Ethiopia
e-mail: wubnehb@yahoo.com

E.S. Leggesse · B.S. Beyene · F. Nigate
Blue Nile Water Institute, Bahir Dar University, Bahir Dar, Ethiopia
e-mail: 2003sime@gmail.com

B.S. Beyene
e-mail: biazenlegns@bdu.edu.et; biazenlegn@yahoo.com

F. Nigate
e-mail: fentanig_2003@yahoo.com

K. Stave et al. (eds.), *Social and Ecological System Dynamics*,
AESS Interdisciplinary Environmental Studies and Sciences Series,
DOI 10.1007/978-3-319-45755-0_5

5.1 Introduction

Climate is an active factor in the physical environment of all living things. The principal direct effects of climate on plants are exerted by elements of water and heat budgets, precipitation and soil moisture, humidity, temperature, sun light and winds. Rainfall, temperature and solar radiation have a direct impact on plant growth either through the uptake of soluble nutrients from the soil and the root zone, the promotion of chemical reactions and photosynthesis in the plant, and the build-up of biomass. The distribution of rain, low and high temperatures and sunshine over the year determine day length, the importance and nature of the growth cycle and the time and the conditions under which vegetative development can take place. Rain water can infiltrates into the soil emerges as ground water or as springs. Shortage of water or too low temperature determines what parts of the vegetation can survive what yield reductions can be expected. Climatic hazards or extreme conditions lead to damage of crops.

Lake Tana Basin is part of the Blue Nile basin, which lies in a natural drainage basin of about 15,114 Km^2. Among which about 20.47% is covered by the Lake Tana. Lake Tana basin is found in North West part of Ethiopia and it extends between 10.95°N to 12.78°N latitude and from 36.89°E to 38.25°E longitude as shown in Fig. 5.1.

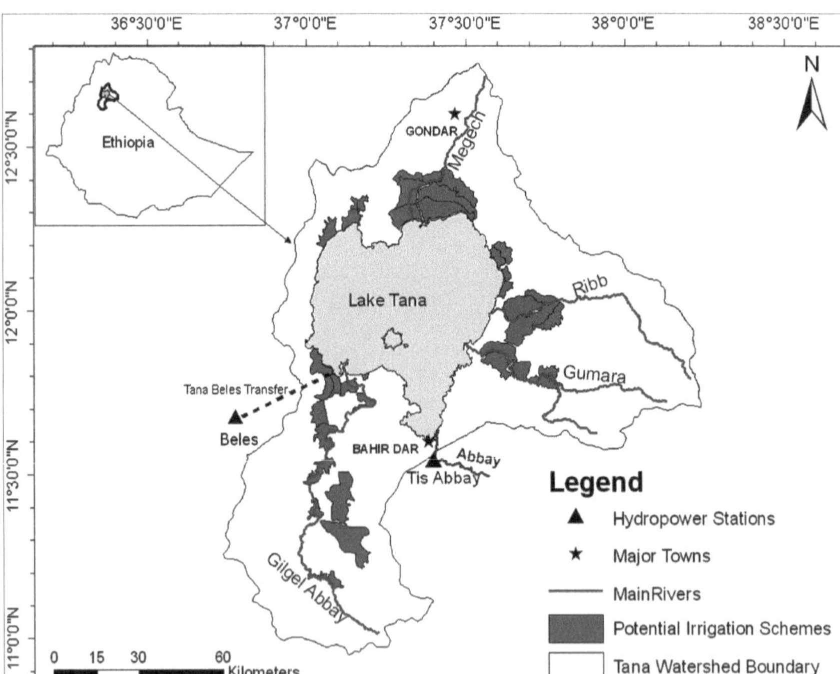

Fig. 5.1 Location map of Lake Tana catchment showing catchment area, major inflowing rivers and planned irrigation and hydropower sites as well as Bahir Dar and Gondar towns (*Source* Tadesse et al. 2010)

5.2 Climate

5.2.1 Rainfall

The rainfall pattern is associated with the Intercontinental Convergence Zone (ITCZ). Dynamics of the moist air coming from Atlantic and Indian oceans following the north-south movement of the ITCZ produces inter-annually varying rainfall over Lake Tana basin (Mohamed et al. 2005). The northwards movement results the major rainy season, whilst the southwards movement of the ITCZ between October and May results in dry climatic condition.

The annual climate in the region can be divided into distinct rainy and dry seasons. The rainy season can be subdivided into one minor and one major rainy season. The major rainy season (from June to September) bears about 70–90% of the annual rainfall amount (Uhlenbrook et al. 2010). The amount of rainfall is maximum during July/August, when it reaches 250–330 mm per month. Mean annual rainfall nearly 1280 mm (Setegn et al. 2008) (Fig. 5.2).

The mean annual evapotranspiration is ca. 773 mm (Setegn et al. 2008). Mean annual relative humidity at Bahir Dar is 58%.

Considerable amount of research has been done on the rainfall variability across the basin. Rainfall is the major contributor of the water resource not only in the basin's premises but also across the Blue Nile River. However, due to limited and uneven distributed gauging stations, spatiotemporal distributions of hydrometeorological variables require extrapolations (Kebede et al. 2006).

The spatial variability of rainfall in Lake Tana Basin is highly influenced by terrain orientation in the region (Buytaert et al. 2006; Gebremichael et al. 2007). The impact can be described in terms of features such as altitude, slope and the associated diurnal wind direction. However, spatial pattern on larger scale indicates that the annual amount of rainfall decreases from the south (ca. 1600 mm) to the north (ca. 1200 mm). Thus the southern part of the basin is wetter than the western and the northern parts (Kebede et al. 2006).

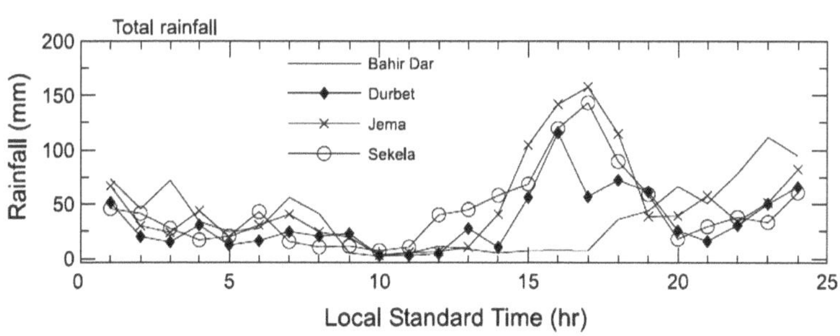

Fig. 5.2 Diurnal rainfall cycle in local standard time, averaged over years for four stations in Lake Tana basin (*Source* Setegn et al. 2008)

5.2.2 Temperature

Ethiopia is located in the tropics, but in the Lake Tana catchment its tropical temperature is modified by its higher altitudes. The mean annual average air temperatures of Lake Tana basin is 22 °C. Four different thermal zones could be distinguished in the Lake Tana Basin, namely: warm, tepid, cool, and cold (Fig. 5.3). Calculations were based on mean annual air temperature during the growing period and their corresponding altitudes. Tepid zone represented the largest part of the basin area (91%), followed by cool (7.8%) and warm (0.64%).

Length of growing period (LGP) is the continuous period of the year when precipitation exceeds half of Penman evapotranspiration plus a period required to evapotranspire an assumed soil moisture reserve and when means daily temperature exceeds 6.5 °C (FAO 1996). The length of growing period (LGP) in the basin belongs to single growing zone which varies from 131 to 221 days. As indicated in the Fig. 5.4, the LGP ranges from 131–180 days which is defined as moist zone (85%) whereas the remaining part of the basin is defined by sub humid zone which ranges from 181 to 221 days (14.8%).

Agro-climatic zone (ACZ) is a land resource mapping unit defined in terms of climate, and having a specific range of potential and constraint for land use. The significant elements in defining agro climate zones are length of growing period and thermal zones. Based on constraints of agro-climate resources of each land unit, some recommendations could be given for agriculture development.

The agro-climatic zones are produced using the map of the growing period zones is superimposed over the thermal zones. The project area has been divided into eight different agro-climatic zones namely, moist tepid, sub-humid tepid, moist cool, moist warm, moist cold, moist very cold, sub-humid cool and sub-humid cold. As indicated in the Fig. 5.5, most of the project area (79.4%) is found in moist tepid agro climatic zone followed by sub-humid tepid, moist cool and sub-humid cool which account for 12, 5 and 3% respectively.

The largest area coverage, moist tepid ACZ, has a wide variety of crop types which is suitable for maize, potato, garlic, onion, pepper, head cabbage, barley, chickpea, wheat, finger millet, teff, niger seed, hope, coffee, faba bean, field pea, grass pea, black cumin, 'bulea' and 'cuche' sorghum. As to Pest and Disease the area has identified Cut worms (on wheat, chickpea, teff, maize, niger seed), root rot (pepper), custa, smut, African ball worms, chekolet spot (on faba bean), striga on sorghum, teff worms, locust and stalk borer as agriculturally important pests. Besides, Livestock Types including goats, cattle, sheep and donkeys are being reared in the basin.

Understanding of the year-to-year rainfall variation is very important not only for crop production but also for planning in terms of agricultural development, hydrological management aspects etc. According to Hare (1983) in areas with coefficient of variation greater than 30% the rainfall is highly variable, and the areas are vulnerable to drought. The availability of natural vegetation for grazing and browsing for their animals is uncertain and water sources may be scarce or

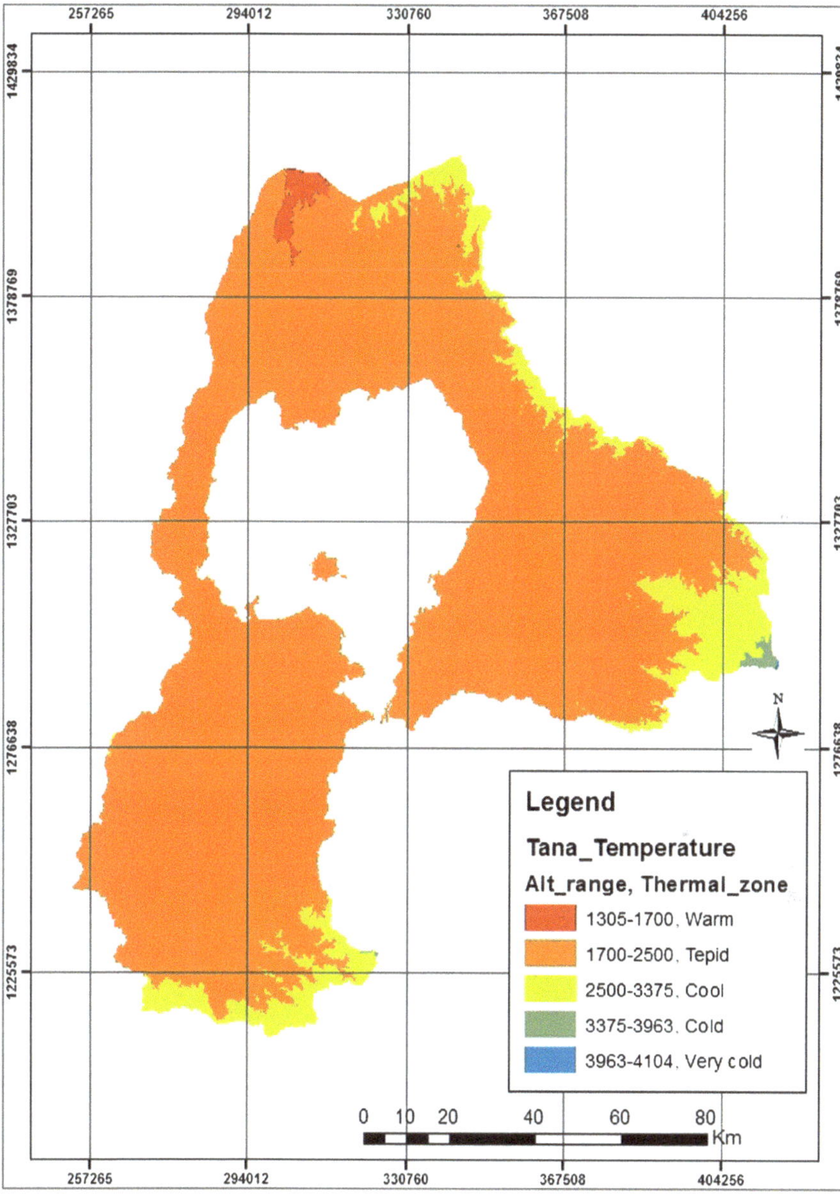

Fig. 5.3 Thermal zones of Lake Tana Basin (*Source* ADSWE 2015)

abundant. Whereas rainfed cropping is often in dangers with a high risk factors. Based on rainfall variability analysis, 57.5% of the basin is found under less rainfall variability. Little has been done so far on understanding the effect of climate change on rainfall amount and distribution and hence of the Basin Rivers flows.

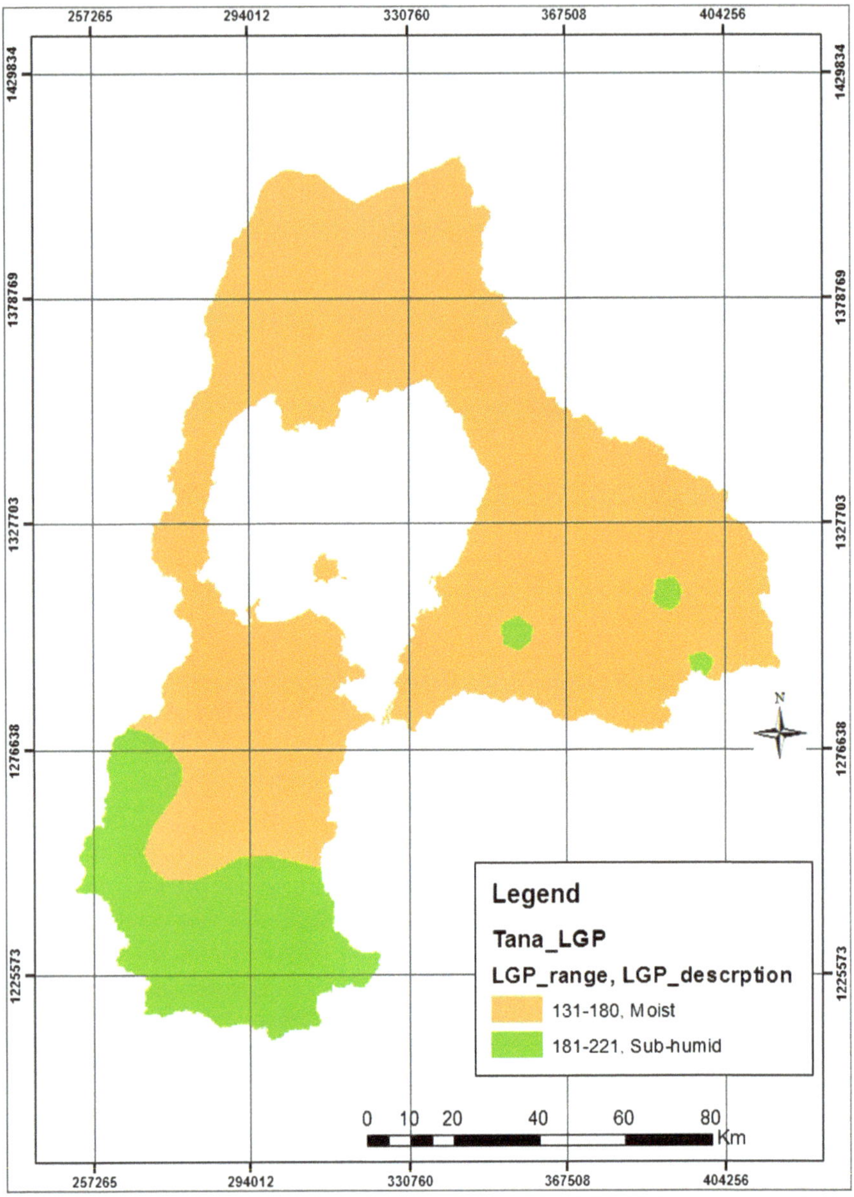

Fig. 5.4 Length of growing period in the Lake Tana Basin (*Source* ADSWE 2015)

Frost is major problem for many crops. At low temperatures most physiological functions are stopped, and in some cases this temperature level is even lethal. Therefore, the frost free period is a common parameter in agro-climatic classifications, especially in high-latitude countries. Frost is a critical factor especially in

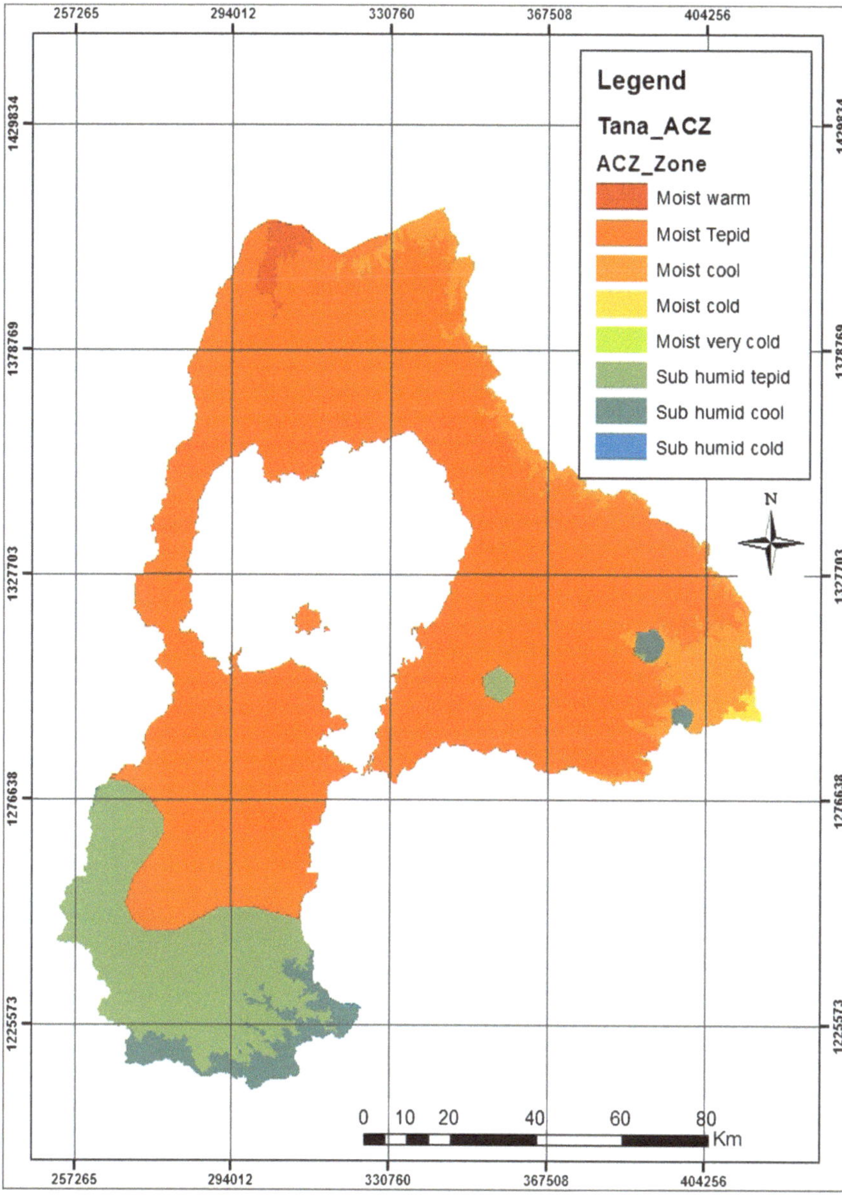

Fig. 5.5 Agro-climatic zones of Lake Tana Basin (*Source* ADSWE 2015)

fruit vegetable production. Frost corresponds generally to a temperature of 0 °C (32 °F). Frost is divided into four classes according to its severity: light frost between 0 and −2.5 °C, moderate frost between −2.5 and −5 °C, severe frost between −5 and −10 °C, and very severe frost below −100 °C (UNDP/FAO 1984). The largest area of the basin (88.5%) is not affected by frost hazard.

References

ADSWE (2015) Land use planning of Lake Tana basin development corridor, agro-climatic resources report. Bahir Dar, Ethiopia (unpublished)

Buytaert W et al (2006) Spatial and temporal rainfall variability in mountainous areas: a case study from the south Ecuadorian Andes. J Hydrol 329:413–421

FAO (1996) Agro-ecological zoning guideline. Food and Agriculture Organisation, Rome, Italy

Gebremichael M, Vivoni ER, Watts CJ et al (2007) Sub mesoscale spatiotemporal variability of North American monsoon rainfall over complex terrain and climate 20:1751–1773

Hare FK (1983) Climate and desertification, pp 5–20, Revised analysis (WMO-UNDP) WCP-44

Kebede S, Travi Y, Alemayehu T et al (2006) Water balance of Lake Tana and its sensitivity to fluctuations in rainfall, Blue Nile basin, Ethiopia. J Hydrol 316(1–4):233–247

Mohamed YA, Van den Hurk BJ, Savenije HG et al (2005) Hydroclimatology of the Nile: results from a regional climate model. Hydrol. Earth Syst. Sci. 9:263–278. doi:10.5194/hess-9-263-2005

Setegn SG, Srinivasan R, Dargahi B (2008) Hydrological modelling in the Lake Tana Basin, Ethiopia using SWAT model. Open Hydrol J 2:49–62

Tadesse A, Matthew M, Seifu K (2010) The water resource implications of planned development in the Lake Tana catchment Ethiopia. Eco-hydrol Hydrobiol 10(2–4):211–222

Uhlenbrook S, Mohamed Y, Gragne AS (2010) Analyzing catchment behavior through catchment modeling in the Gilgel Abay, Upper Blue Nile River Basin, Ethiopia. Hydrol Earth Syst Sci 14:2153–2165

UNDP/FAO (1984) Agro-climatic resources inventory for land use planning (Vol. 1: Main text and Appendix 1). Rome

Chapter 6
Temporal and Spatial Climate Variability and Trends Over Abay (Blue Nile) River Basin

Tadesse Terefe Zeleke and Baylie Damtie

Abstract The spatial and temporal variabilities of climate in Abay (Blue Nile) river basin from 1979 to 2014 have been studied using both station and satellite based observations. Rotated empirical orthogonal function (REOF), regression and wavelet analyses were used to investigate the trend, frequency and intra-annual variability of climate over the Abay (Blue Nile) river basin. High variability of rainfall has occurred over the western regions during spring and southern regions during summer seasons from 1979–2014. The results have shown that, the variability over the regions is dominated by inter-decadal signals, which is similar periodic variability with large-scale circulation. Different conditions of atmospheric and oceanic anomalies are responsible for the observed negative/positive anomaly years in the two seasons. For instance, negative anomaly years during spring season are significantly associated with the negative phase of El Niño Southern Oscillation (ENSO), the weakening of the Arabian High and the subtropical westerly jet streams whereas during summer they are linked to the positive ENSO phase, a weakening of the upper level jet streams, and a lower level moisture influx from the Atlantic and Indian oceans. Un-even heating of equatorial Pacific, Atlantic and Indian Ocean significantly affects extreme anomaly years during both seasons via altering the rain producing large-scale circulations over Abay river basin.

Keywords Abay (Blue Nile) · Climate variability · Large-scale circulations · REOF · Wavelet

---

T.T. Zeleke (✉)
College of Science and Blue Nile Water Institute, Bahir Dar University, Bahir Dar, Ethiopia
e-mail: tadessekid@yahoo.com

B. Damtie
Washera Geospace and Radar Science Laboratory, Department of physics, Bahir Dar University, Bahir Dar, Ethiopia
e-mail: baylied@bdu.edu.et

K. Stave et al. (eds.), *Social and Ecological System Dynamics*,
AESS Interdisciplinary Environmental Studies and Sciences Series,
DOI 10.1007/978-3-319-45755-0_6

6.1 Introduction

The desire to study the variations and trends in extreme climate events increased during the last few years, due to the often large loss of human life, material properties and environmental functions (Easterling et al. 2000; Douben 2006; WMO 2009) and the presence of very high confidence that extreme climate events (such as heat waves, droughts and floods) and the associated impacts are expected to increase in the future (IPCC 2007). In addition to changes in mean climate conditions, climate change is expected to cause changes in climate variability, in particular to the frequency, severity and duration of extreme climate events, such as floods, droughts and heat waves. It was reported that damage caused by extreme climate events has already been growing over the last a few decades (Douben 2006). Short-term duration extreme heat or cold are often responsible for the major impacts on human health and life. For instance, heat wave incidence has killed thousands of people in 1995 in USA (Easterling et al. 2000) and in 2003 in Europe (Smith and Lawson 2011, 2012). The changes in annual and seasonal temperatures have also put significant effects on environmental functions, agricultural production and spread of some vector-borne disease (Stern et al. 2011). Temperature change has also significant effects on surface and soil water availability by affecting the rate of evaporation and evapotranspiration, particularly in the water stressed area of arid and semi-arid regions.

As noted in IPCC (2007), in many regions of the world it is not yet possible to make firm assessments of how global warming has affected exterme events. This means that we know less about past changes in extreme events than past changes in mean climate. This is because observational time series are generally not available in the required high time resolution digital form. Consequently, the limited spatial coverage of the available datasets with high enough resolution (at least daily values) often hampers regional assessments. Even where the necessary data are available, systematic changes in exterme events may be difficult to detect locally if there is a large amount of natural inter-annual variability in extremes (WMO 2009).

Studies conducted in Ethiopia reported a consistent warming trend in the maximum and minimum temperatures over the past few decades in Ethiopia. NMA (1996, 2001, 2007) reported increases of 0.1 °C per decade for maximum temperature and 0.25–0.37 °C per decade for the minimum temperature. McSweeney et al. (2008) found an increasing trend of about 0.28 °C per decade for the mean annual temperature across Ethiopia. Studies mentioned above found that the minimum temperature has been increasing faster than the maximum temperature. Studies on rainfall, on the other hand, do not show any consistent pattern or trends. However, NMA (2001) reported a declining trend in rainfall for the northern and southern parts and an increasing trend for the central part of the country. In contrast, Osman and Sauerborn (2002) observed a declining trend in rainfall for the central highlands. Seleshi and Zanke (2004) and Cheung et al. (2008) did not find a trend in rainfall for the central highlands. Instead, Seleshi and Zanke (2004) reported a declining trend in the east, south and southwest parts of Ethiopia. Many other recent studies Bewket and Conway (2007); Rosell and Holmer (2007); Chueng et al. (2008) reported the absence of upward and downward annual and seasonal rainfall trends over central and northern parts of Ethiopia. The possible reasons for such discrepancies according to Bewket and Conway (2007) are variation of temporal data that we considered for analysis, region and density of recording stations used.

However, all the above studies reported the presence of high inter-annual and intra-seasonal rainfall variability accompanied with severe drought and flood hazards. Major drought hazards with high socio-economic and environmental damage were observed in 1952, 1959, 1965, 1972, 1973, 1978, 1984, 1991, 1994, 1999 and 2002, while major floods were observed in 1988, 1993, 1994, 1995, 1996 and 2006 (NMA 2007). According to studies Gissila, et al. (2004), Korecha and Barnston (2007), Segel et al. (2009b), Diro et al. (2011a, b), Zeleke (2013), Zeleke and Damtie (2015a, b) ENSO (El Nino and La Nina phenomena) is responsible for the inter-annual Kiremt rainfall variability over Ethiopia. Ethiopia has suffered from periodical extreme climate events, manifested in the form of frequent drought (1965, 1974, 1983, 1984, 1987, 1990, 1991, 1999, 2000, 2002, 2011) and occasional flooding (1997 and 2006) (Zeleke and Damtie 2015a). Rainfall variability and associated droughts have been the major causes of food shortage and famine in Ethiopia. At the national scale, the link between drought and crop production is widely known.

There are also some studies conducted on the variability and trend of extreme rainfall events in different parts of Ethiopia (Seleshi and Zanke 2004; Seleshi and Camberlin 2006; Bewket and Conway 2007; Rosell and Holmer 2007; Kebede and Bewket 2009; Zeleke 2013, Zeleke and Damtie 2015a, b) and reported complex and local scale trends in extreme rainfall trends for the last four decades. However, most of them have used very few meteorological stations, only focus on rainfall, and few extreme indices.

There is a general understanding that climate variability mainly precipitation depends on multi-scale temporal and spatial variability of water sources. For instance, the duration and intensity of rainfall in Ethiopia is influenced by fluctuations of the macro-scale pressure systems and large scale atmospheric circulations (Shanko and Camberlin 1998; Segele and Lamb 2005; Yeshanew and Jury 2007; Segele et al. 2009a, b; Diro et al. 2011a, b; Zeleke et al. 2012; Viste et al. 2012). The Tropical Easterly Jet (TEJ), Inter Tropical Convergence Zone (ITCZ), Quasi Biennial Oscillation (QBO), African Easterly Jet (AEJ), winds from the Atlantic and Indian Ocean, the East African Low Level Jet (EALLJ) and the high pressure systems over Mascarene; St Helena and North Atlantic (Azores) are the most likely controls of Ethiopian rainfall during the summer season (Segele and Lamb 2005; Segele et al. 2009a, b; Diro et al. 2011a, b; Zeleke et al. 2012). Conversely, spring rain over southern Ethiopia is affected by the ITCZ, the Subtropical Westerly Jet Streams (SWJS), easterly anomalies and tropical cyclones from the Indian Ocean, the frequency of tropical cyclones over the southwest Indian Ocean (Kassahun 1987; Camberlin and Philippon 2002; Camberlin 1995, 1997; Segele and Lamb 2005; Diro 2008; Segele et al. 2009a, b; Diro et al. 2011a, b).

These atmospheric circulations could be triggered by the Sea Surface Temperature (SST) anomalies over different oceanic basins. For instance, summer rainfall over the Ethiopian highlands is positively correlated with the equatorial East Pacific sea level pressure and the southern oscillation index and negatively correlated with the SST over the tropical eastern Pacific Ocean (Diro et al. 2011a; Zeleke 2013; Zeleke and Damtie 2015a, b). High positive SST (El Niño) anomalies during the summer are associated with high drought probability over most of agricultural productive land and major water reservoir areas of the country (Degefu 1987; Tadesse 1994; Seleshi and Demaree 1995). This phenomenon (variability of ocean basins) has significant impact on the displacement and weakening of the rain-producing mechanisms in Ethiopia (Korecha and Barnston 2007; Segele et al. 2009a; Diro et al. 2011a). Overall, SST anomalies in the equatorial Pacific Ocean are significantly correlated with East African rainfall variations, but the signs of the correlations and their phase vary from region to region (Camberlin 1995; Nicholson 1996, 1997; Segele and Lamb 2005; Diro 2008; Diro et al. 2011a, b).

This study investigated the variability and driving forces of extreme climate events in the Abay river basin. Most of the previous research works conducted in the basin are based on annual and seasonal time series (Conway 2005; NMA 2007; Chueng et al. 2008; Zeleke 2013) and the others that conducted on extreme climate events (Bewket and Conway 2007) were used either very few stations to cover the whole basin or very few rainfall parameters and did not account temperature extremes in their studies. These situations have created information gaps for the development of local or basin scale water management and climate change adaptation planning in the Abay (Blue Nile) river basin. In addition, the Blue Nile river basin has been selected by the government of Ethiopia as one of the economic "growth corridor" of the country (McCartney and Girma 2012). Ethiopia has five hydropower plants in this basin namely Tis-Abay I, Tis-Abay II, Fincha and Beles with 12 MW, 73 MW, 134 MW and 460 MW installed capacities, respectively. Currently, the government of Ethiopia is building large scale hydropower plant; 'The Grand Ethiopian Renaissance Dam' with 6000 MW installed capacity. At the same time Tana Beles, Koga, Megech, Gummera, Jemma and Gilgel-Abay irrigation projects are under construction. All these development projects are sensitive to climate variability and any long-term changes in extreme events, specifically drought and flood hazards (McCartney and Girma 2012). Thus understanding the current climate variability and driving forces for the Blue Nile Basin could help to develop local or basin scale water management and climate change adaptation planning works.

6.2 Methodology

The station rainfall dataset that used to calculate the variability was obtained from the Ethiopian National Meteorological Agency (NMA). The monthly mean dataset included 430 unevenly distributed meteorological stations (Fig. 6.1) throughout the

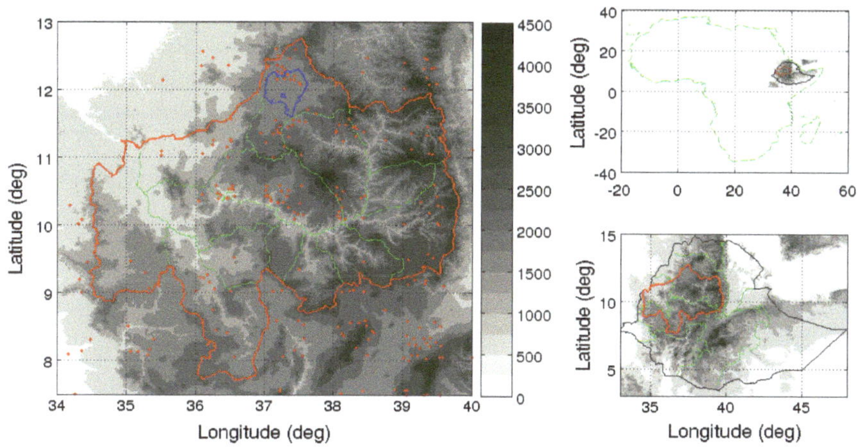

Fig. 6.1 Topography (in meters) of the study area. The *red dots* represent the rainfall stations used in this study

country for the period 1979–2014 (Zeleke et al. 2012). The distribution of the gauges and quality control methods for the observational rainfall data are discussed in detail by Zeleke et al. (2012).

In addition to the station data that described above, we use a blended gauge and satellite product and a station-based on 2.5° × 2.5° gridded dataset: the Global Precipitation Climatology Project (GPCP) (Adler et al. 2003).

6.3 Empirical Orthogonal Function (EOF)

Principal Component Analysis (PCA) was used to capture patterns of co-variability of rainfall and other atmospheric variables at different stations in the Abay (Blue Nile) river basin. The PCA method consists in computing the covariance matrix of the analyzed atmospheric variable dataset with the corresponding Eigen-values and Eigen-vectors (Navarra and Simoncini 2010). The projection of the analyzed atmospheric fields (e.g. rainfall, temperature, wind, etc.) into the orthonormal Eigen-functions provides the Principal Components (PC) score time series. The spatial patterns (Eigen-vectors), properly normalized (divided by their Euclidean norm and multiplied by the square root of the corresponding Eigen-values), are called Empirical Orthogonal Function (EOF) or simply "loadings". The loadings in this study are the correlation values between the original data time series at each grid point and the corresponding principal component time series.

In order to extract more localized spatial patterns of variability, the Varimax rotation to the loadings method was applied (Richman 1986; Jolliffe 1987; Rencher 1998; Von Storch and Zwiers 1999). To remove the influences of location and spread from a set of data, all atmospheric variables time series were standardized by subtracting the mean and dividing by the standard deviation. For each mode, a spatial pattern of loadings describes its area of influence and time scores that reveal the amplitude and wavelength of oscillation. Hence, we used standardized anomalies of time scores of Rotated Empirical Orthogonal Function (REOF), or Rotated Principal Components (RPCs) for wavelet and correlation analysis for the dominant modes of atmospheric variables.

6.4 Wavelet

Wavelet analysis is a common tool for analyzing the power of local variation within a time series by decomposing a time series into time versus frequency space. Hence, it describes the variability of climate in terms of pattern and representing the contribution of each period/scale for the overall variance (Torrence and Compo 1998). The wavelet transform is able to measure both the dominant modes of variability and how those modes vary in time versus frequency. The wavelet transform, therefore, expresses a time series into three-dimensional space:—time (x), scale/frequency (y), and power (z) (Torrence and Compo 1998). Hence, the periodic characteristics of atmospheric variables used in this study are explored using wavelet analysis.

We can characterize a wavelet by how localized it is in time (Δt) and frequency ($\Delta\omega$ or the bandwidth) by compromising Heisenberg uncertainty. An example is the Morlet wavelet, defined as

$$\psi_o(\eta) = \pi^{-1/4} e^{i\omega_o \eta} e^{-\frac{1}{2}\eta^2}$$

where ω_o is dimensionless frequency and η is dimensionless time. When using wavelets for feature extraction purposes the Morlet wavelet (with $\omega_o = 6$) is a good choice, since it provides a good balance between time and frequency localization. Assume that we have a time series, x_n, with equal time spacing δt and $n = 0, \ldots, N-1$. We assume also a wavelet function $\psi_o(\eta)$ that depends on a non-dimensional "time" parameter η. Torrence and Compo (1998) then derive the continuous wavelet transform of the discrete sequence x_n as the convolution of x_n with the scaled and translated version of $\psi_o(\eta)$:

$$W_n(s) = \sum_{n'=0}^{N-1} x_{n'} \psi * \left[\frac{(n' - n)\delta t}{s} \right],$$

where * is the complex conjugate of $\psi_o(\eta)$. In the Fourier space the above expression becomes:

$$W_n(s) = \sum_{k=0}^{N-1} \hat{x}_k \hat{\psi} * \left[(s\omega_k) e^{i\omega_k n\delta t} \right]$$

where the angular frequency is defined as

$$\omega_k = \begin{cases} \frac{2\pi k}{N\delta t}, k \leq \frac{N}{2} \\ \frac{-2\pi k}{N\delta t}, k > \frac{N}{2} \end{cases}$$

and the discrete Fourier transform.

$$\hat{x}_k = \frac{1}{N} \sum_{n=0}^{N-1} x_n e^{-\frac{2\pi ikn}{N}}$$

Finally the wavelet power spectra is defined as $[W_n(s)]^2$.

6.5 Rainfall Seasonal Variability and Annual Cycle

In this section, we examine the spatial patterns of spring and summer seasons mean precipitation and its intra-annual variability. The observed precipitation (Fig. 6.2a, b) over the Abay basin during spring season is mostly confined to Southwest regions (Didessa and Anger) of the basin, while the northern, northwestern and eastern regions are predominantly dry. The pattern is reasonable because of the meridional component of ITCZ during this season. The southwestern regions receive on average 5 mm/day of rainfall; however, small precipitation rates (1–2 mm/day) are located in western and central regions (see Fig. 6.2a). It exhibits also a southwest-northeast gradient where rainfall decreases from ~5 to less than ~1 mm/day.

As one moves from southwestern towards northern and northeastern the pattern of spring rainfall decreases and become dry over north and extreme northeast regions. The GPCP also shows this pattern of rainfall; however, the satellite based observation slightly overestimated in most regions of the basin, except underestimated in the northeastern tip regions (Fig. 6.2c).

Mean June, July, August (JJA) rainfall over Abay basin for the period 1979–2014 shows that central mountainous regions, S. Gojjam, Beles, Wonbera, Anger, southern of Dabus, Didessa and Tana basin receive on average more than 12 mm/day of rainfall (Fig. 6.2b, d). The Beshilo, Welaka, and Jemma regions, which are semiarid, receive comparably less precipitation during this season. Similar climatological pattern of rainfall is shown using GPCP, although with positive and negative biases over the western mountainous regions and some isolated lowland areas, respectively (Fig. 6.2d).

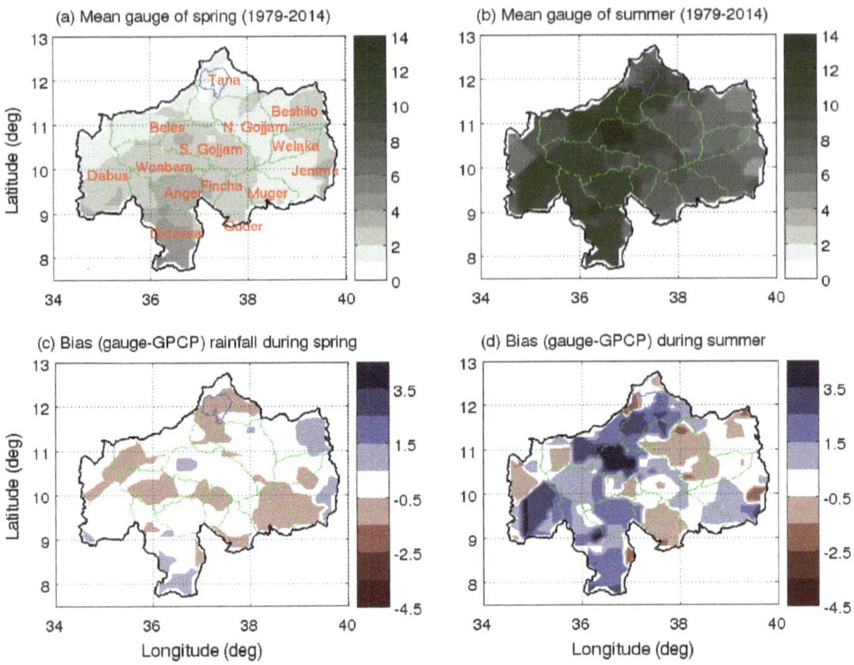

Fig. 6.2 Mean rainfall: **a** mean rainfall during spring (gauge), **b** mean rain fall during summer (gauge), **c** Bias (gauge-GPCP) rainfall during spring, **d** Bias (gauge-GPCP) rainfall during summer. Units are mm/day

Figure 6.3 show the mean annual cycle rainfall for the sub regions of Fig. 6.2a using different sets of observations (gauge, GPCP). The annual cycle values are averaged for each sub regions Fig. 6.2a over the whole observations. Over Tana basin, Beshilo and S. Gojjam (Fig. 6.3), which are on the north and central part of the basin, some slight differences in the intensities between the observed data sets are observed. Both observations showed the unimodal precipitation cycle and the July-August maximum. The central regions show a precipitation peak in July/August, and GPCP underestimate summer rain. For Fincha and Guder regions both observation dataset indicates uni-modal longer (May-October) rainy season in similar way. For these regions, the summer monsoon rainfall and the pre-(May) and post-monsoon rain, are observed.

The southwestern regions of the basin (i.e. Didessa) show unimodal, but longer rainy season almost throughout the year maximum between May-September. However, there is a wide spread in the magnitude and phase of the precipitation maxima between these datasets, with the gauge showing the largest magnitudes and GPCP the smallest.

For Jemma and Welaka regions eastern part of the basin, gauge and GPCP observations exhibit a June break between the spring (FMAM) and summer

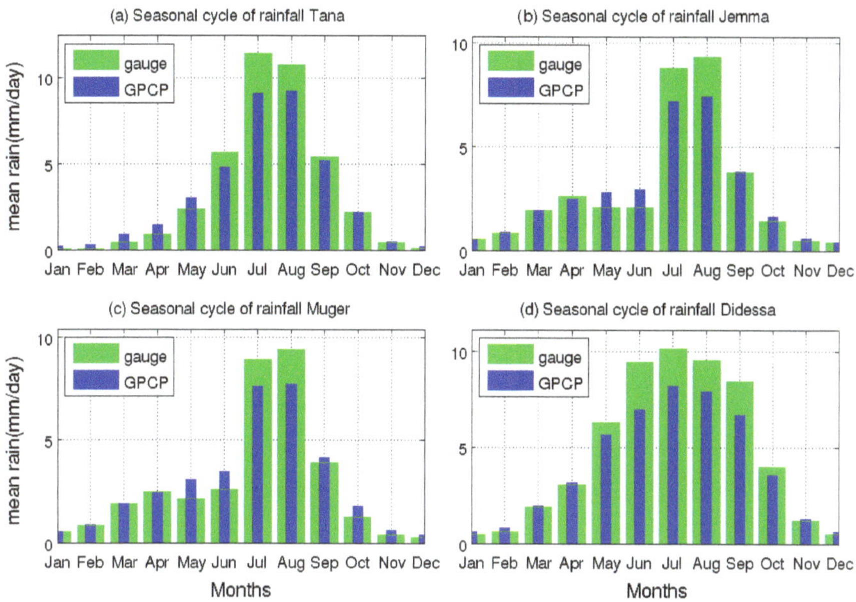

Fig. 6.3 Sub-regional of the Abay basin means seasonal cycle of precipitation (mm/day) for selected region

Table 6.1 Correlation of the areal averaged standardized anomalies and dominant RPCs of gauge versus GPCP of each sub-regions of Abay (Blue Nile) river basin

Sub-region	Anomalies	RPCs	Sub-region	Anomalies	RPCs
Tana	0.69	0.53	Welaka	0.75	0.78
Beshilo	0.79	0.69	Wonbera	0.88	0.98
Beles	0.58	0.29	Fincha	0.63	0.38
N. Gojjam	0.77	0.52	Anger	0.74	0.48
Dabus	0.89	0.97	Muger	0.63	0.43
S. Gojjam	0.63	0.48	Didessa	0.74	0.53
Jemma	0.82	0.65	Guder	0.32	0.24

(JAS) rain. The rainfall amount is generally underestimated by the GPCP (compared to gauge dataset). We note the rainfall in almost all months is systematically underestimated over the eastern regions (Table 6.1).

6.6 Intra-annual Variability

Figure 6.4a, b shows the spring season intra-annual dominant variability patterns; which explain ~26% of the total variance of rainfall. The pattern is confined over the western regions (Dabus, Wonbera, Anger, Didessa, Beles and western part of

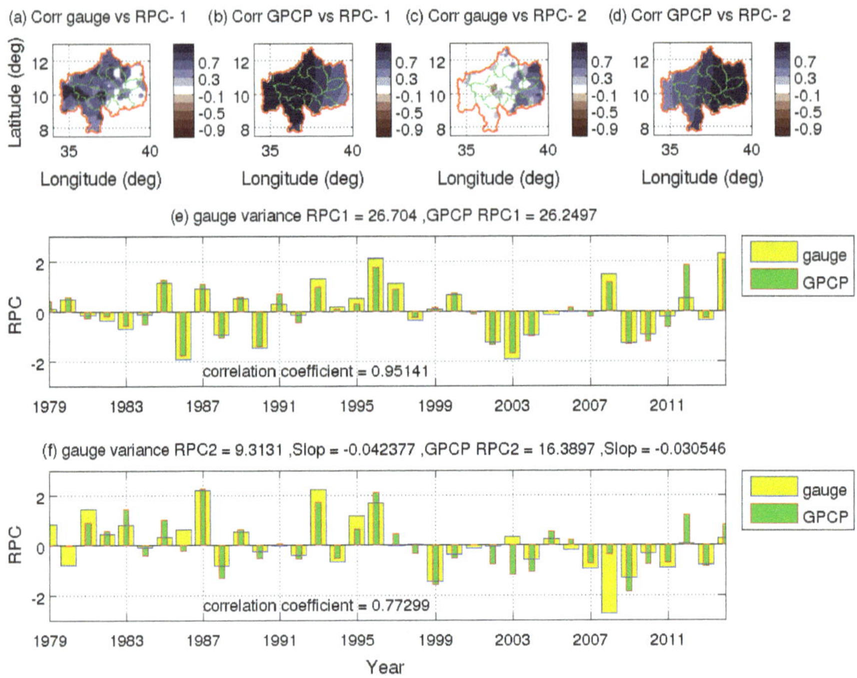

Fig. 6.4 Correlation of dominant RPCs with corresponding raw datasets during spring, **a** gauge dominant correlation pattern, **b** GPCP dominant correlation pattern, **c** gauge second dominant correlation pattern, **d** GPCP second dominant correlation pattern, **e** the time of evolution of the first dominant RPCs, **f** the time evolution of the second dominant RPCs

Tana Lake basin, S. Gojjam and Fincha). The corresponding rotated principal component of GPCP is correlated significantly with gauge dataset (i.e. correlation gauge-RPC and GPCP-RPC ~0.95). Observed extreme events of spring season for example 1985, 1987, 1993, 1996, 1997, 2008 and 2014 which are positive extreme anomaly years and 1986, 1988, 1990, 2002, 2003, 2009 and 2010 negative extreme years from first dominant Rotated Principal Component (RPC) are shown. Overall this dominant RPC shows variability without significant increasing/decreasing trend in spring season over western part of the basins (Fig. 6.4c).

Similarly, the second dominant variability patterns are similar in both gauge and GPCP datasets (Fig. 6.4b) which shows the eastern regions of the Abay river basin (Beshilo, Welaka, Jemma, Muger, N. Gojjam and eastern part of Tana Lake basin, S. Gojjam and Fincha) and explained ~9.3, 16% of gauge and GPCP variances, respectively. The second dominant RPC-2 (intra-annual variability) of GPCP on spring rainfall is significantly correlated (Fig. 6.4c, GPCP vs. gauge ~0.77) and coincides with extreme years of RPC-2 of gauge observation with significant decreasing trend (rate of ~−0.04/a gauge and ~−0.03 GPCP).

Observed areal averaged standardized anomalies and dominant RPCs of each sub-regions of Abay (Blue Nile) river basin are significantly (a significance level of 95%) correlated with the corresponding satellite based observation (GPCP) dataset and summarized in Table 6.1. The monthly anomalies were calculated with respect to the monthly precipitation mean derived from the 36 year period (1979−2014) and mean rainfall from the study period. The area averages of these anomalies were normalized by the standard deviation that derived from the 1979–2014 time series. This result corresponds with the work of Zeleke and Damtie (2015b), who concludes mainly the variability are forced by large scale circulations than local variables. On the contrary weak correlation over Beles and Guder regions are observed. This may be because of number of station differences used by the datasets to represent the region. Similar results are reported by Mengistu (2012), who compared GPCP with gauge and got weak correlation in different boarder regions of Ethiopia, even if, GPCP represented better than Climatic Research Unit (CRU), of the University of East Anglia (Mitchell et al. 2005).

Figure 6.5a shows the spatial patterns of correlation coefficient between the first dominant summer RPC1 of observational rainfall and the corresponding raw summer mean rainfall time series at each grid point over Abay (Blue Nile) river basin region. The results of this study showed that the western regions have high coefficient of variation in rainfall during summer season using gauge and second dominant variation region using GPCP. The variance explained over this region is ~14% and ~10% of the total variance using gauge and GPCP, respectively. The pattern of GPCP is wider when we compare with the gauge (Fig. 6.5a, b). The corresponding rotated principal component (RPC1) of gauge significantly correlated with RPC2 of GPCP (correlation between RPC1 of gauge and RPC2 of GPCP is ~0.93). Similarly, the extreme positive years (e.g. 1981, 1988, 1994, 1998, 1999 and 2001) and extreme negative years (e.g. from 1982, 1984, 1987, 1990, 1993, 2009 and 2014) were observed by both observational data sets over these regions.

The second patterns of correlation between second RPC2 of summer rainfall and its raw data by gauge and the first variability region using GPCP are shown in Fig. 6.5c, d. The patterns in both datasets indicate a strong correlation over southern regions of Abay river basin. Out of total variance over southern region, ~9 and ~43% variances were explained by the gauge and GPCP, respectively. The second RPC of gauge is significantly correlated with the first GPCP-RPC1 (correlation coefficient, ~0.76). Most of positive/negative extreme years are shown during the first/recent heptads, in both datasets in similar way.

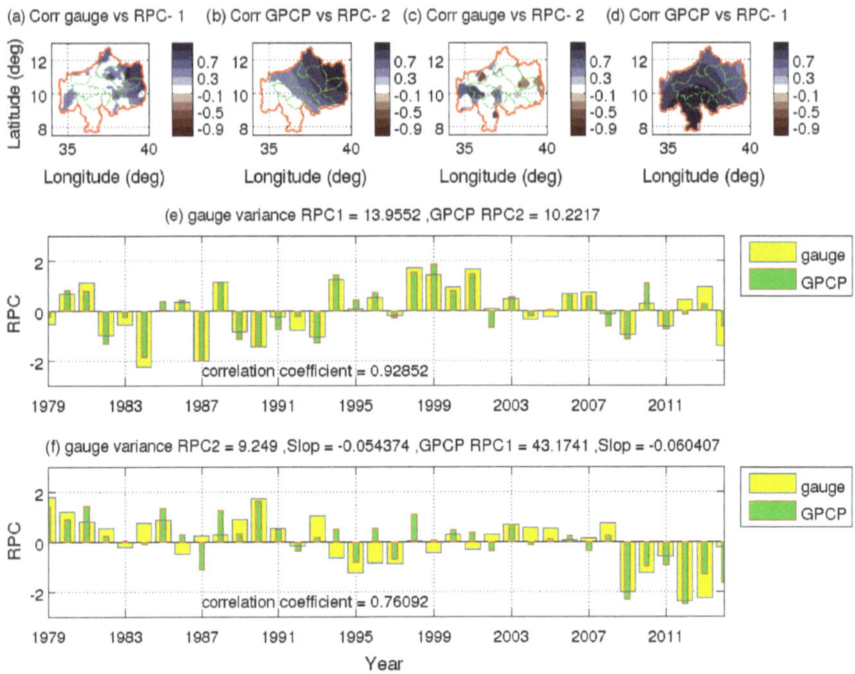

Fig. 6.5 Correlation of dominant RPCs with corresponding raw datasets during summer, **a** gauge dominant correlation pattern, **b** GPCP second dominant correlation pattern, **c** gauge second dominant correlation pattern, **d** GPCP dominant correlation pattern, **e** the time evolution of the second dominant RPC of gauge and GPCP-RPC1

6.7 Periodicity of Dominant RPCs

To analyze the non-stationary characteristics of climate in the observational time series, we performed a wavelet analysis by decomposing the time series into time–frequency signals. In this way, we compared both the dominant modes of variability and the evolution of these modes in time. The wavelet spectrum of the linear de-trended and normalized time series of rainfall anomalies of the gauge (Fig. 6.6a, b) and GPCP (Fig. 6.6a, c) shows a strong non-stationary behavior. The top panels of Fig. 6.6 show the time evolution of the first dominant RPCs, which explains ~13 and 21% variances, of gauge and GPCP, respectively, over the southwestern regions of the basin (Fig. 6.6c, e). While the lower panels (Fig. 6.6b, d) show the gauge and GPCP rainfall wavelet Power Spectrum (WPS). These dominant RPCs are significantly correlated each other (correlation coefficient ~0.88).

These results show the dominant RPCs that exhibit significant energy peaks with distinctive periodicity of 1–2, 4–5 and 7–8 years, in both datasets. GPCP agree on the peak power spectrum and periodicity of gauge dataset. The dominant RPCs of both datasets show significant power peak at the 1–2 year period ~1980–1985,

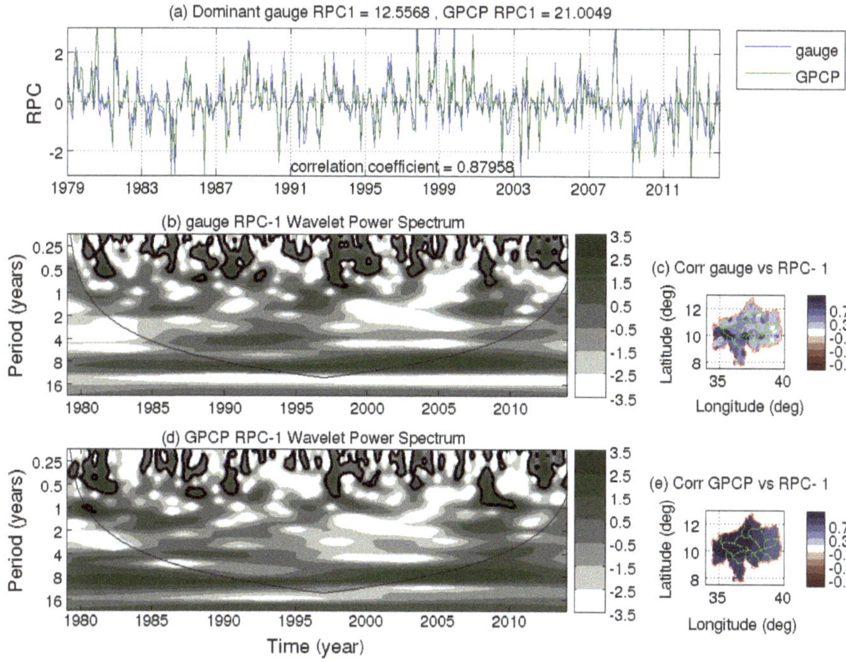

Fig. 6.6 Significant power spectra (high variance) of the dominant RPCs of monthly mean gauge and GPCP rainfall, for the period 1979–2014, with 95% significance contour and cone of influence, **a** the time evolution of the dominant rotated principal components of both dataset, **b** gauge RPC-1 power spectrum **c** The significant correlation pattern of gauge RPC-1 versus its raw data anomaly, **d** GPCP RPC-1 power spectrum, **e** The significant correlation pattern of GPCP RPC-1 versus its raw data anomaly

1995 and 2009 and at the 4–5-year period in ~1985–1990. 7–8 years periodicity seems dominant throughout the study period for these dominant RPCs of both dataset, which confirms the variability of the region is dominated by low frequency (7–8 years periodic) signals. Like western and southern region, the eastern region of Abay river basin are showed periodicity of 1–2 and 4–5 were significant energy peaks (Fig. 6.7). Similar result was reported by Zeleke and Damtie (2015b), who discussed the variability and corresponding large scale circulation forcing, and got high correlation between rainfall variability and east pacific equatorial region. For instance, negative anomaly rainfall years during spring are significantly associated with the negative phase of ENSO, the weakening of the Arabian High and the subtropical westerly jet streams. However, during summer the negative anomaly rainfall years are linked to the positive ENSO phase, a weakening of the upper level jet, Tropical Easterly Jet (TEJ) streams, and a lower level moisture influx from the Atlantic and Indian oceans. Uneven heating of equatorial Pacific, Atlantic and Indian Ocean significantly affects negative anomaly conditions during both seasons via altering the rain producing large-scale circulations over the basin (Zeleke and Damtie 2015b).

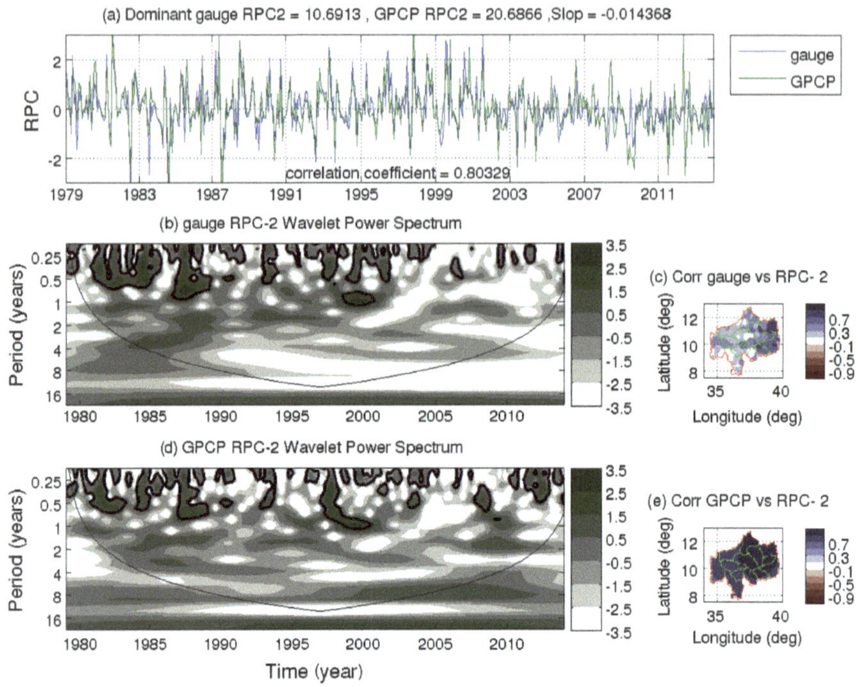

Fig. 6.7 Significant power spectra of the second dominant RPCs of monthly mean gauge and GPCP rainfall, for the period 1979–2014, with 95% significance contour and cone of influence, **a** the time evolution of second dominant rotated principal components of both dataset, **b** gauge RPC-2 power spectrum **c** the significant correlation pattern of gauge RPC-2 versus its raw data anomaly, **d** GPCP RPC-2 power spectrum, **e** the significant correlation pattern of GPCP RPC-2 versus its raw data anomaly

6.8 Summary and Conclusions

In this study, river basin climate variability over the Abay (Blue Nile) river basin was assessed from different observation datasets. The observed rainfall datasets indicated that the central mountainous regions of the basin received more than 12 mm/day rainfall during the summer season. However, the rainfall amount showed decreasing trends towards the northeastern, southeastern, and southern parts of the Abay river basin following the slopes of the western and eastern plateaus. While, during spring season the rainfall climatology exhibits a southwest-northeast gradient which rainfall decreases from ~5 to less than ~1 mm/day. The southwestern regions of the basin have received rainfall for more than 7 months per year. However, the duration of the rainy season decreases to 2 months in the northern parts of the Abay river basin. High variability of rainfall has occurred over the western regions during spring and over southern regions during summer from 1979–2014. This is mainly due to the variability of the large-scale circulations. The result also confirms that the variability of rainfall in the region is dominated by low frequency (7–8 years periodic) signals. Zeleke and Damtie (2015b) also reported the variability rainfall over Abay river basin regions have high correlation with east pacific equatorial region. For instance, negative anomaly rainfall years during spring are significantly associated with the negative phase of ENSO, the weakening of the Arabian High and the subtropical westerly jet streams whereas during summer they are linked to the positive ENSO phase, a weakening of the upper level jet, Tropical Easterly Jet (TEJ) streams, and a lower level moisture influx from the Atlantic and Indian oceans. Uneven heating of equatorial Pacific, Atlantic and Indian Ocean significantly affects negative anomaly conditions during both seasons via altering the rain producing large-scale circulations over basin (Zeleke and Damtie 2015b). This study confirms that, the variability over the region is dominated by inter- decadal signals, rather long term changes (insignificant trends), which shows similar periodic variability with large-scale circulation (Zeleke and Damtie 2015b).

Acknowledgements The authors would like to acknowledge greatly National Meteorological Services of Ethiopia. The authors would like to thank NOAA/OAR/ESRL for providing the GPCP data via Web site at http://www.esrl.noaa.gov/psd/, ECMWF for providing the ERA-Interim dataset. The SST data are available from the National Ocean and Atmosphere Administration website at http://www.cdc.noaa.gov. The authors would like also to extend their gratitude to Blue Nile Water Institute (BNWI) for financial support to complete this work.

References

Adler RF, Huffman GJ, Chang A et al (2003) The Version 2 Global Precipitation Climatology Project (GPCP) monthly precipitation analysis (1979–present). J Hydrometeorol 4:1147–1167

Bewket W,Conway D (2007) A note on the temporal and spatial variability of rainfall in the drought-prone Amhara Region of Ethiopia. Int J Climatol 27:1467–1477. doi:10.1002/joc.1481

Camberlin P (1995) June-September rainfall in northeastern Africa and atmospheric signals over the tropics: a zonal perspective. Int J Climatol 15:773–783

Camberlin P (1997) Rainfall anomalies in the source region of the Nile and their connection with the Indian summer monsoon. J Clim 10:1380–1392

Camberlin P, Philippon N (2002) The East African March-May rainy season: associated atmospheric dynamics and predictability over the 1968–97 period. J Clim 15:1002–1019

Cheung WH, Senay G, Singh A (2008) Trends and spatial distribution of annual and seasonal rainfall in Ethiopia. Int J Climatol 28:1723–1734. doi:10.1002/joc.1623

Conway D (2005) From headwater tributaries to international river: observing and adapting to climate variability and change in the Nile basin. Glob Environ Change 15:99–114

Degefu W (1987) Some aspects of meteorological drought in Ethiopia. In: Glantz MH (ed) Drought and Hunger in Africa: denying famine a future. Cambridge University Press, Cambridge, pp 23–36

Diro GT (2008) Seasonal forecasting of Ethiopian rainfall. Dissertation, Department of Meteorology, Reading University, Reading

Diro GT, Black E, Grimes DIF (2008) Seasonal forecasting of Ethiopian spring rains. Meteorol Appl 15:73–83

Diro GT, Grimes DIF, Black E (2011a) Teleconnections between Ethiopian summer rainfall and sea surface temperature: part I observation and modeling. Clim Dyn 37:103–119. doi:10.1007/s00382-010-0837-8

Diro GT, Grimes DIF, Black E (2011b) Large scale features affecting Ethiopian rainfall. In: Kniveton DR, Williams CJR (eds). Springer, Dordrecht, pp 13–50

Douben KJ (2006) Characteristics of river floods and flooding: a global overview. doi:10.1002/ird.239

Easterling DR, Evans JL, Groisman PY et al (2000) Observed variability and trends in extreme climate events: a brief review. Bull Amer Meteor Soc 81: 417–425

Gissila T, Black E, Grimes DIF et al (2004) Seasonal forecasting of the Ethiopian summer rains. Int J Climatol 24:1345–1358

IPCC (2007) Synthesis report. Contribution of working Groups I, II and III to the fourth assessment report of the intergovernmental panel on climate change (Core Writing Team, Pachauri RK, Reisinger A, eds). IPCC, Geneva, Switzerland, 104 pp

Jolliffe IT (1987) Rotation of principal components: some comments. J Climatol 7:507–510. doi:10.1002/joc.3370070506

Kassahun B (1987) Weather systems over Ethiopia. In Proceedings of first tech. conf. on meteorological research in Eastern and Southern Africa, Nairobi, Kenya, Kenya Meteorological Department, pp 53–57

Kebede G, Bewket W (2009) Variations in rainfall and extreme event indices in the wettest part of Ethiopia. SINET Ethiop J Sci 32(2):129–140

Korecha D, Barnston A (2007) Predictability of June-September rainfall in Ethiopia. Mon Weather Rev 135:628–650

McCartney MP, Girma MM (2012) Evaluating the downstream implications of planned water resource development in the Ethiopian portion of the Blue Nile River. Water Int 37(4):367–379

McSweeney C, New M, Lizcano G (2008) UNDP climate change country profiles: Tanzania. Available at http://country-profiles.geog.ox.ac.uk/index.html?country=Tanzania&d1=Reports. Accessed 30/06/10

Mengistu T (2012) High-resolution monthly rainfall database for Ethiopia: homogenization, reconstr gridding. J Clim 25:8422–8443

Mitchell TD, Carter TR, Jones PD, Hulme M, New M (2005) A comprehensive set of high-resolution grids of monthly climate for Europe and the globe: the observed record (1901–2000) and 16 scenarios (2001–2100). Tyndall Centre for Climate Change Research, Norwich, working paper 55

Navarra A, Simoncini V (2010) A guide to empirical orthogonal functions for climate data analysis. Springer, Dordrecht. doi:10.1007/978-90-481-3702-21

Nicholson SE (1996) A review of climate dynamics and climate variability in Eastern Africa. In: Johnson TC, Odada EO, (eds) The limnology, climatology and paleoclimatology of the East African Lakes. Gordon and Breach, pp 25–26

Nicholson SE (1997) An analysis of the ENSO signal in the tropical Atlantic and western Indian Oceans. Int J Climatol 17:345–375

NMA (National Meteorological Agency of Ethiopia) (1996) Climatic and agroclimatic resources of Ethiopia. National Meteorological Services Agency of Ethiopia, Meteorological Research Report Series 1(1):1–137

NMA (National Meteorological Agency of Ethiopia) (2001) Initial national communication of Ethiopia to the United Nations framework convention on climate change. National Meteorological Services Agency of Ethiopia, Addis Ababa. pp. 127

NMA (National Meteorological Agency of Ethiopia) (2007) Climate change national adaptation program of action (NAPA) of Ethiopia. Ministry of Water Resources, Addis Ababa

Osman M, Sauerborn P (2002) A preliminary assessment of characteristics and long-term variability of rainfall in Ethiopia basis for sustainable land use and resource management.

Paper presented at conference on international agricultural research for development, Witzenhausen, Germany, 9–11 Oct 2002
Rencher AC (1998) Multivariate statistical inference and applications. Wiley, New York
Richman MB (1986) Rotation of principal components. J Climatol 6:293–335
Rosell S, Holmer B (2007) Rainfall change and its implications for Belg harvest in South Wollo, Ethiopia. Geogr Ann 89(4):287–299
Segele ZT, Lamb PJ (2005) Characterization and variability of Kiremt rainy season over Ethiopia. Meteorol Atmos Phys 89:153–180
Segele ZT, Lamb PJ, Leslie L (2009a) Large-scale atmospheric circulation and global sea surface temperature associations with Horn of Africa June–September rainfall. Int J Climatol 29:1075–1100
Segele ZT, Lamb PJ, Leslie LM (2009b) Seasonal-to-inter annual variability of Ethiopia/Horn of Africa monsoon. Part I: associations of wavelet-filtered large-scale atmospheric circulation and global sea surface temperature. J Clim 22:3396–3421
Seleshi Y, Camberlin P (2006) Recent changes in dry spell and extreme rainfall events in Ethiopia. Theoret Appl Climatol 83:181–191
Seleshi Y, Demaree G (1995) Rainfall variability in the Ethiopian and Eritrean highlands and its links with the Southern Oscillation. J Biogeogr 22:945–952
Seleshi Y, Zanke U (2004) Recent changes in rainfall and rainy days in Ethiopia. Int J Climatol 24:973–983
Shanko D, Camberlin P (1998) The effect of the Southwest Indian Ocean tropical cyclones on Ethiopian drought. Int J Climatol 18:1373–1388
Smith CL, Lawson N (2011) Identifying extreme climate thresholds for greater Manchester. Examining the Past to Prepare for the Future, UK. Meteorol Appl. Published online in Wiley Online Library (wileyonlinelibrary.com). eScholarID:99790 doi:10.1002/met.252
Smith C, Lawson N (2012) Exceeding climate thresholds: extreme weather impacts on the environment and population of greater Manchester. North West Geogr 12(1):1–9. eScholarID:169851
Stern DI, Gething PW, Kabaria CW, Temperley WH, Noor AM, Okiro EA et al (2011) Temperature and malaria trends in highland East Africa. PLoS ONE 6(9):e24524. doi:10.1371/journal.pone.0024524
Tadesse T (1994) The influence of the Arabian Sea storms/depressions over the Ethiopian weather. In: Proceedings of the international conference on monsoon variability and prediction, WCRP-84 and WMO/TD No. 619, World Meteorological Organization, Geneva, pp 228–236
Torrence C, Compo GP (1998) A practical guide to wavelet analysis. Bull Am Meteorol Soc 79:61–78
Viste E, Korecha D, Sorteberg A (2012) Recent drought and precipitation tendencies in Ethiopia. Theor Appl Climatol 112:535–551
Von Storch H, Zwiers FW (1999) Statistical analysis in climate research. Cambridge University Press, p 484. ISBN 0521 450713
WMO (World Meteorological Organization) (2009) Guidelines on: analysis of extremes in a changing climate in support of informed decisions for adaptation. Climate Data and Monitoring WCDMP-No. 72
Yeshanew A, Jury M (2007) North African climate variability. Part 2: Tropical circulation systems. Theoret Appl Climatol 89:37–49
Zeleke T (2013) Assessment of drought variability and trend over Ethiopia using observational data analysis and regional climate model experiments. Addis Ababa University, Addis Ababa Ethiopia
Zeleke T, Damtie B (2015a) Periodicity of extreme events and analysis of rainfall trend over Ethiopia (unpublished)
Zeleke T, Damtie B (2015b) Multi-scale climate variability and associated driving forces over Ethiopia (unpublished)
Zeleke T, Giorgi F, Mengistu T et al (2012) Spatial and temporal variability of summer rainfall over Ethiopia from observations and a regional climate model experiments. Theor Appl Climatol. doi:10.1007/s00704-012-0700-4

Chapter 7
Overview of the Hydrogeology and Groundwater Occurrence in the Lake Tana Basin, Upper Blue Nile River Basin

Fenta Nigate, Tenalem Ayenew, Wubneh Belete and Kristine Walraevens

Abstract The Blue Nile (Abay) River is originating in the Lake Tana basin where many perennial and seasonal streams feed the largest lake of Ethiopia (Tana). The basin is characterized by different volcanic formations covered by thick alluvial and residual Quaternary sediments at the center of the basin around Lake Tana. In this study, an attempt was made to outline the hydrogeology of the basin based on limited information from well lithologic logs and previous relevant works. The Lake Tana basin is considered as a potential area for surface water and groundwater development corridors. The hydrogeology and groundwater occurrence is complex due to tectonism and existence of different volcanic rocks covered with thick alluvial sediments at the center of the basin. The shallow aquifer systems in the middle of the basin from the alluvial sediments are tapped by local communities using shallow wells fitted with hand pumps. The volcanic aquifers are multi-layer and tapped from different volcanic layers. The latter are highly productive depending up on the degree of fracturing and weathering. In few cases, as in the case of Kola Diba area there are artesian wells. The groundwater system converges towards the center of the basin. The Lake Tana basin leaks groundwater to the adjacent Beles basin and through the Blue Nile outlet.

Keywords Abay River · Groundwater occurrence · Lake Tana · Aquifer · Northern Ethiopia

F. Nigate (✉)
Department of Earth Science and Blue Nile Water Institute, Bahir Dar University, P.O. Box 79, Bahir Dar, Ethiopia
e-mail: fentanig_2003@yahoo.com

T. Ayenew
Department of Earth Sciences, Addis Ababa University, P.O. Box 1176 Addis Ababa, Ethiopia

W. Belete
Amhara Design and Supervision Works Enterprise (ADSWE), Bahir Dar, Ethiopia

F. Nigate · K. Walraevens
Department of Geology and Soil Science, Laboratory for Applied Geology and Hydrogeology, Ghent University, Krijgslaan 281/S8, 9000 Ghent, Belgium

K. Stave et al. (eds.), *Social and Ecological System Dynamics*,
AESS Interdisciplinary Environmental Studies and Sciences Series,
DOI 10.1007/978-3-319-45755-0_7

7.1 Introduction

The Lake Tana basin is located within the northern Ethiopian highlands, in the upper Blue Nile (Abay) river basin. The total area surface area is 15,000 km^2, out of which 3060 km^2 is covered by Lake Tana, which is the largest lake in Ethiopia (Wale 2008). The basin is formed by volcano-tectonic activity followed by erosion and sedimentation. The watershed boundary is dominantly shield volcanic mountains. The center of the basin is covered with thick sediments that extend below the lake bed.

Three major stratigraphic units can be identified in the Lake Tana basin. These are Tertiary Tarmaber flood basalts, Quaternary scoracious basalts, and Quaternary to recent alluvial lacustrine sediments (Kebede et al. 2011). Some of these formations are subdivided into different lithologic units (Fig. 7.1).

Groundwater in the basin is a source of fresh water for the fast expanding cities and towns in the basin. Water from springs, shallow and deep wells is the main source for community water supply and in few cases local irrigation. For example 71% of the public utility water supply for the city of Bahir Dar comes from groundwater (Tropics Consulting Engineer's 2009 Water Resources Verification Report, volume 1, unpublished). The city of Gondar and other small towns get water from drilled boreholes. In fact springs, which yield also shallow groundwater, provide sustainable water sources for the great majority of the rural community in highland areas.

One of the challenges of utilizing the potential groundwater system of the basin is the complexity of the hydrogeology. This is especially the case in the plain areas

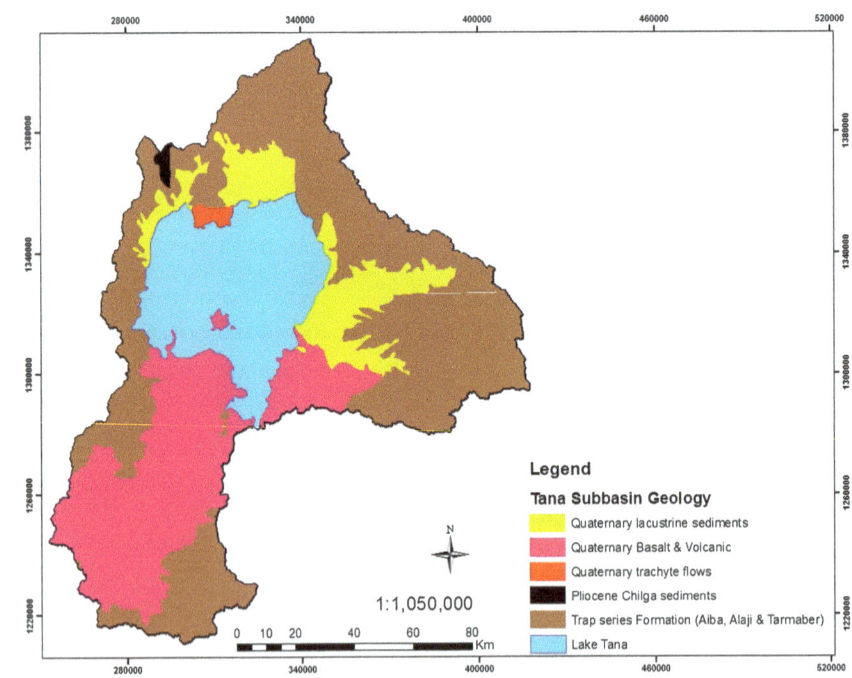

Fig. 7.1 Simplified geological map of the Lake Tana basin

surrounding Lake Tana where visible geological structures are not observed. The complexity is also well reflected in the hydrochemistry and chemical evolution (Kebede et al. 2005). In some parts of the basin there are unusually saline waters that cannot be readily used for drinking (Sileshi 2015). Hence it is vital to understand the hydrogeology of the basin for future utilization of groundwater for community water supply and irrigation.

Its high potential in groundwater and surface water draws the attention of the Ethiopian Government to consider the basin as one of major development corridors of the country. In collaboration with World Bank, the Government of Ethiopia launched the Tana-Beles integrated water resource development program (TBIWRDP) to promote sustainable water resource development and management in the Lake Tana and Beles sub-basin. For such programs understanding the complex hydrogeology of the basin contributes a lot. In this work, an attempt is made to highlight the hydrogeology of the Lake Tana basin based on limited field observations and secondary data. The complexity of the hydrochemistry is not accounted for in this work.

7.2 General Description of the Lake Tana Basin

The Lake Tana basin is located within UTM zone 37, between 252,788 and 396,487 mE longitude, and 1,218,577 and 1,410,176 mN latitude (Fig. 7.2). Rugged mountainous volcanic terrain, moderate to gentle slopes in volcanic rocks

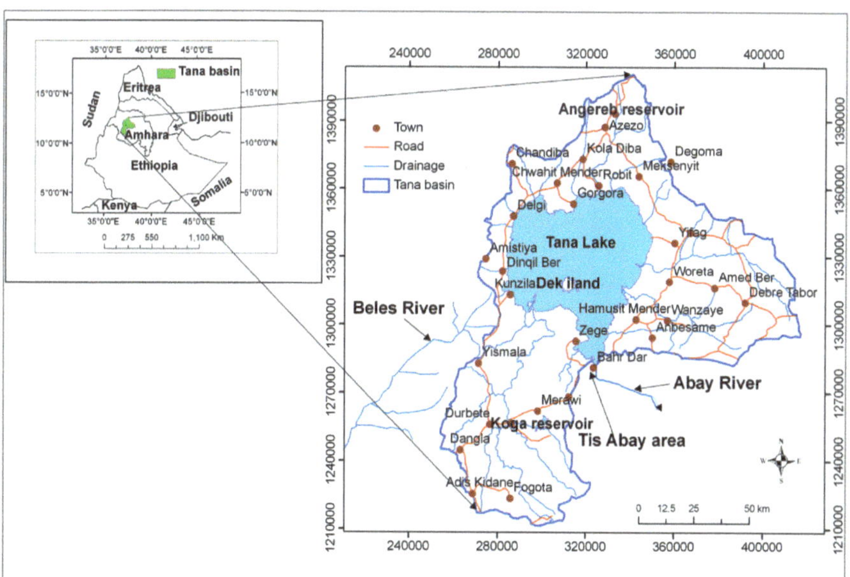

Fig. 7.2 Location of Tana basin

with some isolated hills, escarpment and plain constitute the physiographic units of the area.

Most part of the Lake Tana basin has temperate climate while some parts have sub-tropical, sub-Afro Alpine and Afro Alpine climatic conditions. The area is characterized by uni-modal type of rainfall pattern. The main rainy season lasts from June to September. The areal average precipitation and potential evapotranspiration of the Tana basin are 1400.4 and 1329 mm y^{-1}, respectively (Sileshi 2015). The mean annual air temperature varies between the lowest 15.8 °C at Debre Tabor meteorological station, and the highest 20.6 °C at Gonder station. Rivers that flow into ten sub-basins drain the area. Most of the streams in these sub-basins are perennial. These rivers flow into the Tana Lake. Lake Tana outlet forms the Blue Nile or Abay River.

7.3 General Geological Setup

The Lake Tana basin is characterized by different volcanic rocks of both Tertiary and Quaternary age (Chorowicz et al. 1998). The Tana basin can be related to the Afar mantle plume uplifted dome (Pik et al. 2003). According to Hautot et al. (2005), the basin was formed by faulting of mid-Tertiary Basalts. As confirmed from structural data, Lake Tana occupies a center of subsidence and graben convergence (Chorowicz et al. 1998). Chorowicz et al. (1998) reported that there are three grabens that form the Tana basin, which are the Gondar graben evidenced from satellite images located north of the basin; the Dengel Bergraben in the south; and the Debre Tabor graben located in the eastern side of the basin. Furthermore, the Tana basin is perched on a horizontal topographic high that is individualized within the Ethiopian plateau (Chorowicz et al. 1998). Uplift, subsidence, block faults, recent volcanism and organic rich alluvial sediment depositions at the center characterize the basin. Figure 7.2 shows the simplified geologic map of the basin.

The Lake Tana basin is characterized by complex lithologic and tectonic features unlike part of headwater system of the upper Blue Nile basin (Kebede 2013). Tertiary continental flood basalt covers extensively the northern western plateau of Ethiopian highlands and this masks the underlying formations (Kebede 2013). The underlying Mesozoic and Tertiary sediments are exposed in limited locations, especially north and east of Lake Tana.

Generally, the Tana basin consists of Tertiary Trap Series volcanic formations, Quaternary olivine volcanic rock exposed south and south-west of Lake Tana, Pliocene sedimentary formations (north of Lake Tana) in Chilga area and thick Quaternary lacustrine and alluvial deposits are common surrounding Lake Tana in the vast Fogera and Dembia plains and at the mouth of the Gilgel Abay river. The Quaternary volcanics are patchy and also form islands.

Normal faults and lineaments of northeast-southwest, northwest-southeast, north-south and east-west orientations, and multi sets of macroscopic joints that are interconnected cut the volcanic rocks.

7.4 Hydrogeology and Groundwater Occurrence

The detailed account of the hydro stratigraphic units of the Lake Tana basin has been presented by Sileshi (2015). Here a simplified overview of the aquifer systems is given.

Within the Lake Tana basin, three aquifer systems can be identified. These include:

(a) Aquifers in Tertiary volcanics (mostly those in the Ashangi, Aiba and Tarmaber basalts),
(b) Aquifers in Quaternary volcanics (laterally limited),
(c) Aquifers in Quaternary alluvial and lacustrine deposits.

The Tertiary aquifer system is mainly controlled by fracturing and weathering, whereas the Quaternary basaltic aquifers are fractured, weathered and scoracious systems in which groundwater flow is largely related to the intensity of fracturing and the vesicles. The yield of boreholes from the different aquifers is highly variable depending on the potential of recharge and nature of the rock. In general the yield of the Tertiary basaltic aquifers at depth is very high as in the case of the Koladiba well field. While the yield of the shallow wells from the Quaternary volcanics is low. The intergranular aquifers of the Quaternary sediments form aquifers with highly variable yields depending upon the thickness and grain size of sediments. Most of the wells in these aquifers are shallow. The detailed description is presented below.

7.4.1 Aquifers in Tertiary Volcanic Formations

In the Tana basin, Tertiary volcanic formations cover wide areas in the basin. The most important lithological units that could be distinguished within the Tertiary volcanic formations are: Ashangi, Aiba, Alaji and Tarmaber units/plateau formations and the trachyte/tuffs, flows and plugs (associated with shield volcanism). These distinctions are made based mainly on age.

In most cases the hydrostratigraphic features of these formations are very much related to degree of weathering and fracturing rather than stratigraphic positions. Those rocks close to major lineaments are highly porous and wells drilled in these areas have higher yields.

7.4.1.1 Groundwater in Ashangi Formation—Lower Basalts

Ashangi Formation comprises the oldest basalts that correspond to the earliest fissural flood basalts. It is deeply weathered and mostly associated with fine materials, and it has low permeability in the basin (SOGREAH 2013). This hydrostratigraphic unit is found in the rugged areas of the Beles adjacent sub-basin and the cliffs of

Abay canyon. In these areas the formation does not form a laterally extended aquifer. Primary and secondary pore spaces are sealed by secondary mineralization. Brecciate parts of the Ashangi Formation are characterized by lower permeability. This reduces the water holding and well yielding capacity of the formation. In contrast, in the Beles basin, there are highly yielding wells in the same formation.

7.4.1.2 Groundwater in Aiba and Alaji, Formation—Middle Basalts

These formations correspond to layered flood basalts (Aiba and Alaji types) and acidic flows with large amounts of associated tuffs, scoracious basaltic lava flow and paleosoils. The occurrence of groundwater is limited to the weathered and fractured parts of the formations, as well as to columnar joints. Aquifers can be relatively extensive. Springs can be found at the contact between the lower basalt (Ashangi) and these formations.

7.4.1.3 Groundwater in Shield Volcanic Rocks—Upper Basalts

This thick volcanic succession forms an alternation of pervious and impervious strata, which makes groundwater conditions highly variable from place to place. Thus favorable aquifers may be found only where the rocks are weathered and fractured. The groundwater potential of the area is limited to zones of high fracturing and weathering in the large shield volcanoes in high altitude areas. The transmissivity of the Tertiary basalt aquifers is relatively low. Pumping test data from these formations in Gonder area give transmissivity of 32.8 m^2day^{-1}, the remaining 9 wells have a range of <1–11 m^2day^{-1} (SMEC 2008). The same report displays that the yield of the wells from this formation ranges from 0.7 to 17 L per second (ls^{-1}). Generally, specific capacity is low: the average reported in different areas is 0.25 and 0.27 $ls^{-1}m^{-1}$ (SMEC 2008).

7.4.2 *Aquifers in Quaternary Volcanic Formations*

Quaternary volcanic formations are mainly encountered south of Bahir Dar, and in the Gilgel Abay catchment. They are characterized by massive and weathered basalts, possibly scoracious or vesicular. To a limited extent, these formations are overlain by more recent Quaternary lava flows forming vesicular and highly pervious basalts, possibly slightly weathered. Aquifers in Quaternary basalts can be relatively productive. For instance, boreholes in Dangila (yield about 20 ls^{-1}), boreholes around Bahir Dar yields greater than 20 ls^{-1} were reported. There are also high discharge springs emanating from this formation including the Lomi spring (>60 ls^{-1}), and Areke (>150 ls^{-1}), Tikurit (>40 ls^{-1}), Infranz springs (three springs with 30 ls^{-1}each) south-west of Bahir Dar, Andasa (Tikurit) spring (>50 ls^{-1}) and Merawi (Burka) spring (>60 ls^{-1}) springs were reported. In fact some of the groundwater for these springs may also be contributed from the underlying Tertiary volcanics.

The most permeable zones appear to be localized in the Bahir Dar region where transmissivity values commonly greater than 20 m^2day^{-1} and as high as 180 m^2day^{-1} were reported (SMEC 2008).

7.4.3 Aquifers in Alluvial Formations

Major alluvial aquifers are found along the eastern and northern Lake Tana edges: in the Rib-Gumara and Infranz catchments in the Megech area. The productivity and yield of wells and springs very much depends on the grain size. Groundwater potential also depends on the lateral extent and thickness of sediments and their relation with the volcanics in elevated areas. Groundwater recharge in these formations is high. Less boreholes were drilled in this formation and the transmissivity is not well documented. Transmissivity values ranging from 1 m^2day^{-1} to 700 m^2day^{-1} were reported (Kebede et al. 2005; Alemayehu and Kebede 2011).

7.4.4 Aquifer Hydraulic Characteristics and Spatial Distribution

Aquifers with very low productivity mainly correspond to those located in steep slopes and mountain foot of the Adama and Guna volcanoes. In these mountain zones, the topography favors high runoff and low groundwater retention. Hence specific storage in these areas is very low owing to topographic factors.

The Gish Abay spring at the head water of Gilgel Abay River, top of Adama Mountain, emerges from thick brown soil in a valley bottom, due to topographic factors. The yield is estimated to be lower than 1 ls^{-1}. In this highland area, people use low discharge springs and hand dug wells. A borehole drilled around Sekela town to depth of 90 m shows static water level of 1.5 m below the ground surface, and a yield of 1.5 ls^{-1}. Lithologic log shows hard rock at shallow depth which reduces the groundwater storage potential. In the same area, some boreholes went dry due to hard nature of the rocks. However, a borehole drilled to depth of 60 m north of Sekela is productive and the yield is estimated to be 15 ls^{-1}.

At the top of the Guna volcano, outcrops of trachyte flow and tuff are mapped (SOGREAH 2013). Due to the steep slopes and the less pervious nature of the rocks, groundwater storage is not high. In Gasay and Kimr Dingay, where the topography is less rugged, hand dug wells (4–14 m depth) and shallow wells (40–70 m depth) present yields lower than 1 ls^{-1} in these formations (SOGREAH 2013).

7.4.4.1 Tertiary Volcanic Aquifer Formations

Tertiary volcanic formations mostly consist of an alternation of massive and weathered/fractured basalts (slightly, moderately to highly weathered layers). This geologic unit has wide areal distribution in the highlands of the Lake Tana Basin. Locally, paleosoils, clay layers, and ashes may also be intercalated. Such basalts form stratified aquifers with widely variable hydrogeological characteristics. Hydrogeological characteristics depend on (1) the fracture penetration within each stratified layer, (2) the degree of weathering, (3) the presence or absence of conductive or impervious layers that either serves as conduits or barriers to groundwater flow.

The overburden soil type and thickness as well as the geomorphology of the surrounding land have strong impact on the amount and distribution of the groundwater in the Tertiary volcanics (when not covered by Quaternary volcanic formations).

Most of the deep borehole information is limited to 150–200 m depth in the Tana basin (with existing accessible groundwater resources mostly contained in the upper 80–100 m below ground level). Groundwater information at greater depth is scarce. Tertiary volcanic aquifers are found all over the basin. The deep boreholes of Koladiba well field have exceptionally high yields. Some of the wells drilled more than 300 m are artesian with a yield of about 45 ls^{-1}. The groundwater potential in this area is exceptionally high. In the deep aquifers high groundwater potential is expected to exist in the Fogera and Dembia plains in these formations.
The Northern Lake Tana sub-basin: In this region visible regional faults localized groundwater as in the case of the Koladiba and Azezo well fields. The deep volcanic formations along these fault zones form the most productive aquifers. The aquifers are multilayer and in some cases form artesian systems. The degree of weathering and fracturing of the volcanic rocks is high. The Kola Diba and Teda areas are affected by many faults as indicated in the geological map of Ethiopia (EIGS 1996) and geophysical investigations carried out in these two areas by SOGREAH (2013).
The Megech–Seraba areas: Most of well completion reports in these areas indicate the existence of thick weathered basalts with different degree of weathering and clay layers mixed or intercalated within basalts. The borehole of Chuahit, located 3 km SE of Chuahit town revealed the existence of 3 m thick clay layers within the basalt. These layers form confined and in some cases artesian wells in the Fogera plains. Whereas, the other two well reports indicate alluvial deposits (fine to coarse grained) in the top 35 m depth below ground level, overlying fractured and fresh basalts. The range of yields of boreholes varies from 0.5 to 3 ls^{-1}. Around Kola Diba town, many boreholes were drilled. Boreholes drilled north-west and west of the town, which were drilled in volcanic rocks, have lower yields and sometimes become dry, whereas boreholes drilled south of the town in the Megech plain have very good yield ranging from 4 to 45 ls^{-1}. This specific area is one of the most promising for large scale groundwater development.

Many boreholes were drilled in Gonder area. The depth of the wells ranges from 40 to 200 m. The boreholes yield ranges from 10.2 to 6.5 ls^{-1}. The yield is mainly good in the top 100 m. These wells were drilled in Tarmaber Formation. In contrast with the Koladiba well field, the yield decreases with depth. The boreholes drilled for Gonder University in the city found to be low yielding. Some of them were abandoned.

Drawdown and transmissivity data are relatively scarce in the area. However, drawdown values range between 15 and 50 m (SOGREAH 2013). Pumping test data analysis from data of limited area show that the transmissivity value is highly variable in Gonder area (0.02 m^2d^{-1}– 2.1 m^2d^{-1}) (SMEC 2008).

Ten boreholes were drilled south of Azezo for Dashen Brewery. The lithological logs show alternation of fresh, weathered and fractured basalts. Most of the weathered basalts are slightly to moderately weathered, though some highly weathered horizons were encountered. The boreholes yield ranges from 1.5 to 20 ls^{-1}. Similarly, about 5 boreholes were drilled around Ayimba area for Ayimba and Chilga town and the average yield was found to be 5 ls^{-1}.

Generally, the most important aquifer in the northern part of Tana basin is the basaltic Tarmaber formation. From practical experience in the field, it is possible to conclude that the aquifer is quite heterogeneous and the aquifer yield is not good except few patchy areas, like Kola Diba and Ayimba areas.

The Eastern part of Lake Tana sub-basin: major faults and fractures form the Rib-Gumera graben. This structure favours localization of groundwater. The mountain area between Addis Zemen and the Rib River shows intense weathering in its upper part. It is underlain by relatively massive basalts with slight fracturing. Many shallow wells exist along the road between Addis Zemen and Ibnat. Shallow wells were drilled to a depth of 35 to 50 m. But there is no pumping test data to characterize these aquifers in the area.

A deep borehole located in Addis Zemen town makes an exception: it was drilled up to 137 m and the estimated yield is about 6.5 ls^{-1}. No lithological log is available for this borehole. The presence of local faults and fracturing and weathering may be the cause of the higher yields. However, other boreholes drilled in the Tarmaber volcanic aquifer gives low yields. The aquifer characteristics in the area are highly variable. Boreholes drilled around Yifag have low yield and some are abandoned. The boreholes close to major faults have relatively better yield. From the geological map of Ethiopia it is possible to understand that the area is affected by NW-SE trending faults.

Several deep boreholes exist around Debre Tabor. Depth mostly ranges between 80 and 160 m. The available information indicates that well yields range from 1.5 to 15 ls^{-1}. Fewer boreholes drilled around Debre Tabor town were abandoned and whereas others are still productive. In the area the yield ranges from 1.5 to 4 ls^{-1}. Boreholes drilled in Gasay and Kimir Dingay area show similar characteristics as Debre Tabor area. However, boreholes drilled south of Gafat area in Qanat locality give very high yields ranging from 6 to 20 ls^{-1}, except one borehole with a yield of 3 ls^{-1}. These boreholes are located along the small stream and this high productivity may be related to local structures and weathering of rocks as observed from geological logs during drilling in the area. The high rainfall also favors high recharge in the area.

Boreholes were drilled at Alem Ber and Awura Amba village. The borehole log drilled near to Awura Amba village has shown massive basalt and become abandoned. Whereas, boreholes drilled at Alem Ber village and one borehole drilled at the junction of the road to Awura Amba village was found to be productive. In areas where massive basalts are present, the yield is very low or the wells get dry.

The boreholes drilled in Hamusit and Zenzelma to a depth of 120 m were found to be productive. While a well drilled to a depth of 150 m was found to be dry. This variation is related to local structures and/or differences in degree of fracturing or lithology. The first borehole at Hamusit is giving a discharge of 2.5 ls^{-1} whereas the borehole drilled around Zenzelma is abandoned and the lithologic log indicates massive and fresh basalt. The Zenzelma-Hamusit-Ambesame areas are covered by hard Quaternary basalt. In this case the rocks do not form productive aquifer.

Southern part of the Lake Tana sub-basin: In this area the most common rock is Quaternary volcanics. The Quaternary volcanics are mainly scoracious and hard lava flow. Local lithologic logs in this area indicate fresh massive (mostly fine grained) basalts, slightly to moderately fractured and weathered basalts, highly weathered basalts associated with paleosoils, and vesicular and scoracious basalts showing high spatial variability.

As mentioned earlier, better groundwater conditions are expected in more pervious layers such as in moderately to highly weathered basalts, fractured and in scoracious basaltic layers. Fracture pattern, fracture penetration depth, type of overlying soil, thickness of Quaternary basalts present control the amount and distribution of groundwater. As confirmed from different borehole lithologic logs, thick massive basalt layers, clay, baked basalts and paleosoils are present, which considerably limit groundwater flow as well as the groundwater recharge process. Scoria formations, which are pervious, have generally limited areal coverage (around Bahir Dar city, Meshenti, Merawi, and the way to Chimba) which also limits the groundwater potential. Their boundaries act as local barriers to groundwater flow.

Gilgel Abay catchment comprises at least two geologic units: Tarmaber basalt in the upper part of the sub-basin and Quaternary basalt in the lower part of the sub-basin. Geomorphologically, the upper part of Gilgel Abay is mountainous and steep, consisting of Adama and Sekela Mountains with main streams originating from them. High rainfall and surface runoff are common in the area. However, the geology of the lower Gilgel Abay catchment is characterized by extensive Quaternary volcanic formations mainly composed of massive and weathered basalts, possibly scoracious and/or vesicular formations. To a limited extent, these formations are overlain by more recent Quaternary lava flows which are highly pervious.

Field experience and analysis of lithological log reports available in the Gilgel Abay catchment (Meshenti, Merawi, Durbete and Dangila well fields) indicate coarse and fine alluvium is encountered at depth. The coarse paleosoils intercalated within volcanic formations are highly productive. The rough and blocky contact of the Quaternary basalts provides good opportunity for the infiltration of rainfall. Infiltration is estimated at 20% of precipitation in the area (BCEOM 1999).

As depicted from lithologic logs (massive basalt, weathered and fractured basalt, some layers of vesicular basalt and fine scoracious basalt), the boreholes drilled in Meshenti, Merawi, WetetAbay, Durbete and Dangila area show approximately similar hydrostratigraphy.

Several deep boreholes were drilled around Meshenti for flower and fruit farms. Depth of boreholes ranges from 75 to 245 m. Yields are moderate to high within the range of 3–8 ls^{-1} (Yinesa Sostu wells). However, the Yishef borehole, located on the other side of these boreholes (in Andasa catchment) presents a very high yield ranging from 15 to 20 ls^{-1}. During pumping tests, wells (except Yishef well) show high drawdown.

In Wetet Abay area, one borehole located close to the Gilgel Abay and Koga Rivers confluence supplies the town from a shallow aquifer (total depth of well is 60 m). This well is located close to two streams that favour recharge. The yield is about 4.5 ls^{-1}. In the vicinity of Durbete, water point analysis indicates low well yields (<3 ls^{-1}). A new borehole, drilled in 2011 in the area for the town water supply, has a yield of 20 ls^{-1}. The depth of the well is 94 m. However, boreholes drilled near the existing well (on the other side of the new well) to a depth of 15 0 m become abandoned. The lithological log indicates slightly weathered—fractured basalt (95 and 105 m).

Similarly, boreholes drilled in Dangila area show different yields. Boreholes drilled to a depth of 120 m have a yield of 3 ls^{-1}. However, in the same well field, new boreholes were drilled recently to a depth of 200 m and the yield was found to be 24 ls^{-1}.

In the western side of the Gilgel Abay catchment, the productivity of the aquifers is low. This is probably related to the existence of rhyolite formations which seem to be only slightly weathered and fractured. The Liben borehole makes an exception with a moderate yield of 5 ls^{-1}. Many boreholes drilled in this area become abandoned or give small yield, often less than 2 ls^{-1}.

7.4.4.2 Aquifers in Quaternary Lava Flows South of Lake Tana

The occurrence and movement of groundwater in the area as well as the structure of the aquifer system is controlled by the geological and geomorphological setting of the regional as well as local structures and recharge from Lake Tana. Among different types of volcanic units found in the area, the recent Quaternary basaltic flows, which are the dominant units around Bahir Dar area and its surroundings, are the most productive and extensive type of aquifers. These are the young volcanic rocks with porphyritic and vesicular types. Interconnection of the vesicles combined with fractures developed in the unit as well as its inherent scoracious nature, result in high porosity and permeability. The drainage density in these rocks is low due to the high permeability. Surface runoff is very low.

Five high discharge springs are located in the lava flow: the Lomi spring (75–100 ls^{-1}) and Areke spring (100–150 ls^{-1}, possibly 220 ls^{-1}), Tikurit (40–60 ls^{-1}), three Infranz springs with discharge estimated at 30 ls^{-1} each. Probably the high discharge of the springs is associated with geological structures and recharge from elevated areas.

The groundwater bearing formations are moderately to highly productive aquifers. Indeed, shallow wells drilled up to a depth of 60–80 m have discharge value of about 3–4 ls^{-1}. Other boreholes drilled up to 70–130 m depth, have higher yields with estimated discharge of 10–20 ls^{-1}.

In general the volcanic rocks have wide variability in aquifer hydraulic properties. For example analysis of pumping test data of 70 boreholes from Ethiopian plateau, including the Tana basin in Tarmaber basalt, has resulted in transmissivities ranging from 0.1 to 32 m^2d^{-1} (Kebede et al. 2011). The transmissivity value of the scoracious Quaternary basalt in the Tana basin has a wide range from 2 to 500 m^2d^{-1}. Borehole yields range from 3 to 6 ls^{-1}. However, in some localized areas, greater discharges can be encountered (10–20 ls^{-1}) due to favorable local geological structures.

7.4.4.3 Intergranular Aquifer in the Lake Tana Flood Plain

Extensive aquifers with intergranular permeability are mainly mapped in large alluvial plains east and north of Lake Tana: in the Rib-Gumara, Infranz and Megech areas, and around Delgi. Alluvial sediments can be found overlying the volcanic formations (e.g. Fogera and Dembia plains). The alluvial deposits are thick at the center of the basin. The lateral extent at depth is unknown.

Coarse alluvial layers and weathered/fractured basalts, scoria and volcanic ashes layers could be very good aquifers. The alluvial sediments and river gravels in the Tertiary volcanics are important contributors of groundwater for high yielding wells. On the contrary, the presence of fine lacustrine sediments and paleosoils would lower the storage capacity.

The Rib and Gumara plains (Fogera plain) correspond to large plains extending 30 km from north to south and 15 km from east to west (SOGREAH 2013). The aquifer extends from Lake Tana to the Guna volcano western flanks. The thickness of this geologic unit is estimated to be 200−300 m. The thickness may increase toward the lake (SOGREAH 2013). Two boreholes were drilled near Wereta town, the old borehole has 75 m depth, which with a yield of 7.5 ls^{-1} and the new borehole drilled to a depth of 88 m has a yield of 40 ls^{-1}. The lithologic log of the latter well shows that the top 84 m is alluvial deposit (sandy gravel, compacted clay and sand) and slightly weathered vesicular basalt from 84 m to the bottom of the well. Three boreholes were drilled at Bura Lideta (near Shini River) for Addis Zemen town to depths ranging from 80 to 105 m. The lithologic log shows that the top 40 m is alluvial deposit (sandy gravel and clay) with underlying fractured and weathered basalt. Both formations act as an aquifer. But the alluvial deposit is more productive. The yield was estimated between 6 and 20 ls^{-1}.

Groundwater bearing formations of the Megech plain (part of Dembia plain) have significant coarse alluvial layers, scoracious basalts and weathered/fractured basalts, with important variations in their spatial distribution. Lithologic log for boreholes of Kola Diba well field (Megech flood plain) show an alternation of alluvial deposit and fractured volcanic rocks. Wells with a depth of less than 100 m have a yield of 4–6.5 ls^{-1} whereas deep wells provide a yield that ranges from 18 to 37.5 ls^{-1}. However, deep boreholes drilled north and west of Kola Diba town, where there is shallow and fine alluvial deposit, provide a discharge less than 4 ls^{-1}. Recently wells having a yield of more than 35 ls^{-1} have been drilled in Koladiba well field.

7.4.5 Groundwater Circulation and Flow System

The Quaternary tectonic activity of the Lake Tana basin plays a significant role in governing the occurrence and circulation of groundwater (Alemayehu and Kebede 2011). The flow is directed toward the center and converges to Lake Tana (Kebede et al. 2005). Recent integrated hydrogeological investigation revealed that the Lake Tana basin leaks to the Beles River basin and across the basin along the drainage line of the Blue Nile (Abay) River outlet (Sileshi 2015). Therefore, the groundwater and surface water divide do not correspond in the basin.

The hydrogeochemical and isotope data collected in the basin by different researchers have shown that there are two types of groundwater. These are low salinity, Ca–Mg-HCO_3 type and isotopically relatively enriched shallow groundwater from basaltic aquifers in the plateau and lowland area and springs from the same geological setting (Nigate et al. 2016; Kebede et al. 2005); high TDS, Na-HCO3 type, isotopically relatively depleted thermal groundwater along the regional fault systems (Kebede et al. 2005; SMEC 2008; SOGREAH 2013). In the latter case, most of the water is hotter.

There are also unusually high saline ground waters in the Dembia and Fogera plains. These issues have been addressed in detail by recent study made by Seleshi (2015). Two groundwater flow systems can be identified. These are shallow waters that flow from the highlands to the lowland plains, characterized by flushed isotopically depleted waters, and deep circulations localized along regional faults. In general most of the water in the basin is a mixed system. However, the presence of artesian wells in the plain displays the presence of deep groundwater systems localized in Tertiary volcanics.

7.5 Conclusion and Recommendations

The Lake Tana basin is underlain by three hydrostratigraphic units: CenozoicTarmaber flood basalts, Quaternary scoracious basalts, and Quaternary to recent alluvial-lacustrine sediments.

Generally the Tana basin consists of a heterogeneous aquifer system. The Tarmaber Formation is less productive than the Quaternary scoracious basalts and Quaternary alluvial deposits. The Tarmaber basalt aquifer system has a transmissivity ranging from 0.1 to 32 m^2d^{-1} while the Quaternary scoracious basalt has transmissivity values ranging from 2 to 500 m^2d^{-1}. Due to limited data, the transmissivity of Quaternary alluvial deposits is not well understood. In few cases, the volcanic formations have transmissivities as high as 700 m^2d^{-1}. This displays the extreme heterogeneous nature of the volcanic aquifers in the Tana basin.

In general two aquifer systems can be identified. The shallow system is relatively flushed and low yielding, with the exception of Quaternary basalts. The deep aquifer system is for the most part multi-layer and in few cases artesian, with relatively isotopically depleted waters.

The most productive aquifers are associated with the Tertiary flood basalts covered with thick alluvial and lacustrine sediments localized in wide plains such as the Fogera and Dembia plains.

The volcanic rocks are interspersed by large regional faults. These faults strongly controlled the distribution and movement of groundwater. Most wells close to regional faults have better yields.

In terms of groundwater movement, the groundwater converges towards Lake Tana. However, the groundwater partly moves out of the basin to the adjacent Beles River basin to the west and along the course of the Abay River to the south. This indicates that the groundwater and surface water divides do not coincide.

References

Alemayehu T, Kebede S (2011) The role of geodiversity on the groundwater resources potential in the upper Blue Nile River basin. Ethiop Environ Earth Sci 64:1283–1291

BCEOM (1999) Abay River Basin integrated master plan, main report. Ministry of Water Resources, Addis Ababa

Chorowicz J, Collet B, Bonavia FF et al (1998) The Tana basin, Ethiopia: intra-plateau uplift, rifting and subsidence. Tectonophysics 295:351–367

Ethiopian Institute of Geological Surveys (EIGS) (1996) Explanation of Geological Map of Ethiopia

Hautot S, Whaler K, Gebru W et al (2005) The structure of a Mesozoic basin beneath of the Lake Tana area, Ethiopia, revealed by magnetotelluric imaging. J Afr Earth Sc 44:331–338

Kebede S (2013) Groundwater in Ethiopia, features, numbers and opportunities; ISBN 978-3-642-30390-6. Springer, Publisher, p 282p

Kebede S, Yves T, Alemayehu T et al (2005) Groundwater recharge, circulation and geochemical evolution in the source region of the Blue Nile River. Ethiopia, Applied Geochemistry 20:1658–1676

Kebede S, Admasu G, Travi Y (2011) Estimating ungauged catchment flows from Lake Tana flood plains. Ethiopia: An Isotopic hydrological approach, Isotopes in Environmental and Health studies 47(1)1:71–86

Nigate F, Camp VM., Kebede S et al (2016) Hydrologic Interconnection between the Volcanic Aquifer and Areke-Lomi and other springs, Lake Tana Basin on the Upper Blue Nile J. Afr Earth Sc 121: 154–167

Pik R, Marty B, Carignan J et al (2003) The stability of the Upper Nile drainage network (Ethiopia) deduced from (from (U-Th)/He thermochronometry: implications for uplift and erosion of the Afar plume dome. Earth and Planet Sci lett 215:73–88

Sileshi M (2015) Integrated Hydrological and Hydrogeological System Analysis of the Lake Tana Basin, Northwestern Ethiopia. PhD thesis, Addis Ababa University

SMEC (2008) Hydrological study of the Tana-Beles sub basins, Groundwater investigation, unpublished

SOGREAH (2013) Groundwater investigation and monitoring in the Tana and Beles sub basins stage 1, final report, Vol 11, part 6. hydrogeologic survey, unpublished

Wale A (2008) Hydrological balance of Lake Tana, the Upper Blue Nile. Thesis, ITC, the Netherlands, Ethiopia, MSc

Chapter 8
Characterization, Classification and Mapping of Soils of Agricultural Landscape in Tana Basin, Amhara National Regional State, Ethiopia

Mekonnen Getahun and Yihenew G. Selassie

Abstract A study was conducted at different locations of Tana sub basin to understand their physico-chemical and mineralogical properties of the soils and characterize, classify and map them. Six major soil units were identified in the area including Luvisols, Vertisols, Nitisols, Leptosols, Gelysolsl and Fluvisols on the basis of land forms and surface texture. Morphological properties of the soils reveal that Luvisols are deep, well drained and generally loamy sand in texture. Vertisols (VR) were poorly drained with deep rooting zone and texture varies from clay on the surface to heavy clay in the subsoil. Lithic Leptosols have an impermeable layer at less than 50 cm depth with generally gravelly sand texture. The texture of the soils in the study area are generally characterized by clay to heavy clay and have workability problems during rainy seasons. The dominant exchangeable bases were calcium and magnesium with the soils having high base saturation. The Vertisols have neutral to slightly alkaline reaction (pH). The soil reaction of Nitisols at Jema and Sekela sampling sites were 6.5 and 5.5, respectively. Soil reaction of Luvisols was 5.3. Available phosphorous in all soils are low to very low. Total nitrogen and organic carbon contents also are low to very low. The soils found within the Tana sub-basin were classified as Haplic/Chromic Luvisols, Haplic/Rhodic Nitisols, Lithic Leptosols, Pellic/Chromic Vertisols and Cambic Cambisols and other association soil types such as Fluvisols and Regosols.

Keywords Characterization · Classification · Landscape · Slickenside · Morphological properties

---

M. Getahun (✉)
Amhara Design and Supervision Works Enterprise, Bahir Dar, Ethiopia
e-mail: mekonnengetahun2001@gmail.com

Y.G. Selassie
College of Agriculture and Environmental Sciences, Blue Nile Water Institute, Bahir Dar University, Bahir Dar, Ethiopia
e-mail: yihenewgs@gmail.com

K. Stave et al. (eds.), *Social and Ecological System Dynamics*,
AESS Interdisciplinary Environmental Studies and Sciences Series,
DOI 10.1007/978-3-319-45755-0_8

8.1 Introduction

Soil characterization, soil classification and soil mapping provides a powerful resource for the benefit of mankind especially in an area where food security and environmental sustainability are grand challenges. Soil characterization provides information for our understanding of the physical, chemical, mineralogical and microbiological properties of the soils we depend on to grow crops, sustain forests and grasslands as well as support homes and society structures (Ogunkunle 2005). Soil classification, on the other hand, helps to organize our knowledge, facilitates the transfer of experience and technology from one place to another and helps to compare soil properties. According to Eswaran (1977), some different uses of soil characterization data include to aid in the correct classification of the soil and enable other scientists place the soils in their classification systems and to serve as a basis for more detailed evaluation of the soil as well as gather preliminary information on nutrient, physical or other limitations needed to produce a suitability class. A soil characterization study, therefore, is a major building block for understanding the soil, classifying it and getting the best understanding of the environment (Esu 2005).

Topography is one important major factor which control most surface processes taking place on the study area, i.e. soil formation and soil development. Topography has influence on soil chemical and physical properties and also on pattern of soil distribution over landscape (Kalivas et al. 2002; Esu et al. 2008). For instance, the impact of rainfall is great where landscape is sloppy with regards to erosion and deposition. Soils on hill slopes differ from those at summits or valleys in terms of moisture distribution, soil depth, cations distribution; organic matter contents (Asadu et al. 2012).

The Tana basin is one of the biggest watersheds in the Abbay Basin. The geographical location of the Tana basin extends from 253,277 to 416,782 m east and from 1,211,053 to 1,410,446 m north. The mean elevation of the sub-basin is 2024 masl. The sub basin embraces four zones (North and South Gonder, West Gojjam and Awi zones) and 31 Woreda and covers an area of 15,054 km^2, of which about 3078 km^2 is represented by the Lake Tana. The upstream of Tana sub-watershed, where most of tributaries are originating has very steep slope of mountainous to escarpment topography. In terms of bio-physical environment, the Tana sub-basin has diverse environmental characteristics namely, Dega (2300–2408 m) and Woyinadega (1500–2300 m), in which the elevation drastically falls to the end of the watershed.

Tana basin, one of the development corridors in Amhara National Regional State, is of paramount importance for economic growth at national level. The Tana basin in its natural state has high potential for agriculture, livestock, water resource, forest and wildlife, tourism, and fishery development but the land use system that has been practiced in the basin adversely affects the environment significantly. Problems observed in the basin includes soil erosion and soil fertility degradation like loss of plant nutrients, chemicals, physical and biological degradations, etc., water resource pollution, invasion of aquatic weeds, cultivation of marginal lands and wetlands, deforestation, overstocking, population pressure and unplanned settlements. It is obvious that natural resources are limited, scarce, and not spatially distributed in relation to the population needs and hence, proper management of these resources is essential to satisfy needs of the ever increasing population and maintain sustainability. To address all opportunities and problems and underpin agricultural productivity around basin providing of soil information, erosion control and nutrient cycling are very vital elements.

Soil is a vital natural resource on whose proper use depends the life supporting systems of a country and the socio-economic development of its people. Soils provide food, fodder and fuel for meeting the basic human and animal needs. With the increase in human and animal population, nutrient depletion of soils for more food production is on the increase. However, the capacity of a soil to produce is limited and the limits to production are set by intrinsic characteristics, agro-ecological settings, use and management. This demands systematic appraisal of our soil resources with respect to their extent, distribution, characteristics, behavior and use potential, which is very important for developing an effective land use system for augmenting agricultural production on a sustainable basis. Therefore detailed information from soil resources will be helpful for researchers, decision makers on the environmental consequences of proposed actions, private investors and farmers in general. Hence this review is to bring together all the available soil information in the basin.

8.2 Major Soil Types in Lake Tana Basin

The major soil types in Gumara, Rib and Megech exhibits a general relationship with altitude and slopes. Vertisols and Fluvisols are generally dominating the lowland flat plains, valley bottoms and river terraces. Texturally, these soils are heavy clay and sandy loam, respectively. Shallow Leptosols are the dominant soil types found in the mountain and hills of the whole Tana basin. Luvisols dominate the southern and central part of the Tana sub-basin. In the study area, there are five major soil classifications including some minor ones generalized as others in the soil classification for the sake of compatibility to the weighted overlay analysis.

The catena or topo-sequence approach has played an important role in the study of Tana sub-basin watershed areas especially on the south-east side (Megech, Ribb and Gumara) micro watersheds. The central concept is to establish the relationship of soil types occurring at various slope positions along a topo-sequence. Soil identification and characterization was done by considering the geology, parent materials, topographic and hydrological factors in the micro-watershed areas. The soil topo-sequence of south-eastern part of the basin have four land elements including the crest (t), the upper and middle slopes (u), the lower slope (l) and the valley bottom (v). The soils on the crest position are young, shallow to very shallow lithic Leptosols, soils in the upper and middle slopes are relatively deep, weathered soils (kaolinitic haplic Luvisols). The seasonally wet but nutrient enriched soils (wet Vertisols and Wet Fluvisols) in the valley bottom usually contain substantial amounts of smectite in clay formation. As indicated in the lab result, the soil texture of the surface horizon of all upland positions is either sandy or loamy sand in Leptosols; because the parent rocks are coarse-grained granite and gneisses. Generally, the high base status kaolinitic soils (Haplic/Chromic Luvisols) occurring in the upper and middle slope of are prone to erosion and compaction which make them to be less suitable for large scale annual food production. The smectitic and base enriched soils in the valley bottom of Fogera, Denbia and Kunzila-alefa-takusa plains are suitable for intensive vegetable production (such as onion, pepper and tomato) during the dry season, and for rice in the wet season.

8.3 Soil-Landscape Relationship

Soil-Landscape relationship: To understand the general soil landscape relationship, Tana sub-Basin were divided into different landscape units. Detailed descriptions are presented as follows.

The Plateau landscape (Pu): The plateau landscape is located at the southwest and southeast of the Tana basin. The Gumara, Rib and Megech micro watershed ranges are the continuation of the peaks coming from the south Gonder to north Gonder zones. The highest altitude is about 2100 masl (meter above sea level). The predominant relief correspond to Cuesta where the dominant geological material is sandstone. Hence, three landforms (cliff, very steep and steep talus) are the dominant of this unit. This landscape is geo-structurally controlled and forms the plateau in its highest point out of the Tana sub watershed bordering lower ridges starting from the hill land landscape. Slopes in this landscape range from 0 to 40%. Some strong coarse fragments can be found in the escarpments. The topo-sequences are result of the elevation and principally the slope as the most important pedogenic factor. Natural vegetation such primary forest can be found due to lack of accessibility but there is also some grassland. Soils were studied very generally in areas where no topo-sequences are located. In general, soils of this landscape are Haplic/Chromic Luvisols, Haplic/Rhodic Nitisols, Lithic Leptosols, and other association soil types such as Regosols.

The Hill land landscape (LM): The hill land landscape is dominant in the study area and has its highest peak in the hill both at Gilgel abbay, Gumara, Ribb, Megech and Infranz. It is composed of five relief types: escarpments, ridge, high hills, moderate high hills, and low hills. The slope varies from 12 to >65% in the very high hills and ridges. As a pattern of the landforms for this landscape, we could find the principal components of the slope facets model: summit, shoulder, back slope, foot slope and toe slope. Human influence through deforestation in this landscape lead to a land degradation process where agricultural activities are present. Soils in this area range from very shallow to very deep soils with very fine textured clay horizons.

The Piedmont landscape (Pi): The Piedmont landscape in the Tana sub watershed is composed of small remnants that become continuous in the surrounding areas. This landscape is composed by higher glacis terraces as principal relief type where the dominant material is the siltstone making the soil susceptible to erosion and this situation shows why we can find only very dissected remnants. The soils identified in this landscape are Lithic and Eutric Leptosols, Regosols and some Cambisols.

The Valley landscape (Va): The Gilgel abbay, Gumara, Ribb, Megech and Infranz Rivers are the tributary of the Tana basin that goes to the Abbay River. Small valleys passing through the area and forming the flood plain relief type characterize this landscape. The principal parent material is the alluvial material transported from the origins of the affluent and some colluvial local slides at the toe slope. The dominant soils in the area appear to be Fluvisols and Vertisols. Most of the flood plains are used for agriculture purposes and also some settlements are there. These soils have clay to heavy clay texture and are very deep.

The lower micro-toposequence in eastern part of Lake Tana is located in a traditionally agriculture area in the Gumara, Ribb, and Megech Rivers. Slope complex are the representative landform facets in the ridge relief type located in the hill land landscape. Five different land units were identified corresponding to slope facets. The erosion process in the toposequence goes from slight to severe sheet and rill soil erosion. Gully erosion is also sever at all above mentioned micro watershed areas. Soil erosion is related to the slope grade and the different land cover present in each unit.

The slope grade goes from 0 to greater than 30% and heavily determined the soil types that are in general shallow (summit and shoulder facets). However, deeper soils with a very fine texture dominated by clay materials are more prevalent in the rest of the land units. Land use through the toposequence is related to the land resources characteristics going from low level of utilization (bushes) to an intensive agriculture use (Vegetable crops combined with annual crops).

The micro-toposequence in the south part of Lake Tana is located in a traditionally agriculture area in the Gilgel Abbay catchment, south west of the Tana sub-watershed, west Gojam zone. (It consists of a short microtoposequence in a moderately high hill relief type into the hill and landscape and is very representative of it. Land units are represented by the slope complex facets (summit, shoulder, back slope, foot slope and toe slope). The erosion process in the toposequence goes from slight to severe sheet and rill soil erosions. Moreover, gully erosion is also sever in the upper and middle part of the toposequence. Tuff parent material is the dominant one followed by local alluvial colluvial depositions. This relation determined the different landforms and created soils characterized by shallow soils (summit and shoulder facets) to deep soil (lower foot slope and valley bottom) with a very fine texture that are dominated by clay materials where vertic characteristics are present.

8.4 Morphological, Physical and Chemical Characteristics of the Tana Sub Basin Soils

8.4.1 Vertisols

These soils occupy the down slope positions, under the warm and alternating wet-dry climate the bases have synthesized with silica to form montmorillonitic clay. These deep cracking clays (Vertisols) occupy stable lower (relative to the red soils) topographic positions. In the soil pedons opened on different fields of Vertisols, clear soil horizon differentiation and boundaries could not be obtained. It is worth-mentioning that Vertisols exhibit minimal horizon differentiation as a result of pedoturbation and reports from different scholars (Probert et al. 1987; Kamara and Haque 1988; Mitiku 1987, 1991) revealed similar findings.

Vertisols are formed in basins and lower landscape positions of Dembia, Fogera, mecha and Kunzila on convex slopes. This soil type consists of very deep, poorly and very poorly drained soils formed on convex depression plains. The parent material for these soils is primarily the alluvial/colluvial veneer overlying the solid geology although the reddish clays and Vertisols can develop in situ from basalt and basic metamorphic rocks. Slope ranges from 0 to 3 percent. The clay fraction of these soils is dominated by exchangeable 2:1 lattice clays, mainly montmorillonite. They have clay texture throughout the profiles (Table 8.1) and in most cases the proportion of clay fractions are very high (more than 60%). The soils cracks widely during the dry season, but the cracks close up again on re-wetting. The resulting expansion and contraction is responsible for the formation of grooved shiny faces (slickensides) and wedge shape structures at depth (25–100 cm). during the dry season animals, as well

Table 8.1 Morphological and physical property of sample profile of Pellic Vertisols at Fogera plain

Depth (cm)	Texture (%)				Colour	Structure	Consistence	Stickiness and plasticity
	Sand	Silt	Clay	Textural class				
0–20	3	16	80	HC	Black (2.5Y2.5/1)	st Co SAB	F	VST/VPL
20–95	2	12	86	HC	Black (2.5Y2.5/1)	st Co AB	VF	VST/VPL
95–140	2	10	88	HC	Black (2.5Y2.5/1)	st Co AB	VF	VST/VPL
140–200	0				Black (2.5Y2.5/1)	st Co AB	VF	VST/VPL

*Soil colour determined at wet condition and codes are according to FAO (2006): *dyb* dark yellowish brown, *dr* dark red, *vrb* very reddish brown, *vdgb* very dark greyish brown, *b* brown, *gb* greyish brown, *dg* dark grey. Structure**: *stcosab* strong coarse subangular blocky, *stcoab* strong coarse subangular blocky, *mostsab* moderate and strong subangular blocky, Consistence***: *fr* friable, *vfr* very friable;. *ha* hard, *f* firm, *sha* slightly hard. Sickness & plasticity****: *st/pl* sticky/plastic, *vst/vpl* very sticky/very plastic. Mottling: *dccs* distinct.continuous clay skins, *fdyr* FD yellowish red mottles, *cfdrm* common fine distinct reddish mottles as wind, cause the fine surface material to fall in the widely opened cracks (pedotutbation) causing soil heaving as the soil expands during the rainy season, developing characteristic hollows and mounds microrelief known as *gilgai*.

The soils are developed on alluvial parent material derived from volcanic rocks. The unit is mainly intensive agricultural crop production. The soils are very deep and have distinctly developed AB and C-horizons. They are poorly drained. Top soils have Black (2.5Y2.5Y/1), moist, very dark greyish brown (10YR3/2), moist and very dark gray (2.5YR3/1) when mist at Fogera, Dembia (Table 8.2) and Kunzila-Alefa takusa plains (Table 8.3), respectively. The sub-surface soils have very dark gray to gray (10YR 4.5/1) colour when moist. The texture is clay throughout the profile and show shrinking and cracking behavior during dry season. The structure is moderate, medium sub-angular blocky and prismatic, with wide cracks up to 50 cm depth. The consistency is hard when dry, very sticky and very plastic when wet. Tillage and other farm operations will not be easy as the soils become very sticky when wet and hard when dry.

The soil reaction of the top soil is moderately acid with PH 5.5 and 5.7 at Fogera, Dembia and Kunzila-Alefa Takusa plains, respectively and increasing throughout the profiles. As it is shown in Table 8.4, the soil pH increased as depth increased due to an increase in the concentrations of base forming cations such as Ca, Mg and Na and a decrease in the soil OM. However, exchangeable K decreased with an increase in depth, which might be attributed to an increase in the K fixing capacity of the expanding clay minerals of Vertisols and the tendency of K to move downwards in the soil profile along with the infiltrating water due to the high rate of mobility of K. The organic carbon and total nitrogen values are 2.3% and 0.29, 1.6 and 0.14%, and 1.4 and 0.2% for Fogera, Dembia and Kunzila-Alefa takusa plains, respectively in surface soils; and 1.2% and 0.12%, 1.0 and 0.09 and 0.8 and 0.1 in

Table 8.2 Soil physical characteristics of Chromic Vertisols at Dembia plain

Depth (cm)	Texture (%)			Textural class	Colour	Structure	Consistence	Stickiness and plasticity
	Sand	Silt	Clay					
0–25	2	18	80	Clay	VDGB (10YR3/2)	st Co SAB	VF	VST/VPL
25–60	2	10	88	Clay	VDGB (10YR3/3)	st CoS AB	VF	VST/VPL
60–100	2	8	90	Clay	DGB (2.5YR4/2)	st Co AB	VF	VST/VPL
100–200	0			Clay	VDG (2.5Y3/1)	st Co AB	F	VST/VPL

Table 8.3 Soil physical characteristic of Chromic Vertisols at Kunzila-Alefa plains

Depth (cm)	Texture (%)				Colour	Structure	Consistence	Stickiness and plasticity
	Sand	Silt	Clay	Textural class				
0–30	2	13	85	Clay	VDG (2.5YR3/1)	MC SAB	F	VST/VPL
30–80	2	14	85	Clay	DG (5YR4/3)	st Co AB	VF	VST/VPL
80–145	2	9	88	Clay	DG (5YR4/1)	st Co AB	VF	VST/VPL
145–200				Clay	DG (5YR4/1)	st Co AB	VF	VST/VPL

sub-surface soils, respectively. Both the organic carbon and total nitrogen values decrease gradually with depth.

The electrical conductivity (ECe) at the surface horizon of the (0–20 cm) is 0.2 ds/m and the exchangeable sodium percentage (ESP) is 1.0 at the surface and increase with depth. The available phosphorous content of surface soils is 1.2, 2.4 and 3.5 ppm at Fogera, Dembia and Kunzila-Alefa takusa plains respectively (Table 8.4). Most sites have less than 5 ppm (Olsen) phosphorus content, which is below the critical limit for the growth of most crops based on Sims (2000). Tekalign and Haque (1991) reported that 8 ppm P was the critical level for Ethiopian soils when assessed by the Cate and Nelson method. Yihenew (2004) also reported P critical levels of 11 and 14 mg kg^{-1} for Olsen and Bray-2 methods, respectively for maize crop for Alfisols of North-western Ethiopia.

The Vertisols of Fogera plain have the highest CEC levels (76 cmol(+)kg^{-1}) followed by the same soils of Kunzila-Alefa Takusa (58 cmol(+)kg^{-1}) as compared to soils in other site. Vertisols are characterized by their high content of expanding smectite clay minerals, which are known for their substantial cation exchanging permanent negative charge sites. The cation exchange capacity of the soils is generally very high according to Metson (1961) using the rating of <6 very low, 6–12 low, 12–25 moderate, 25–40 high and >40 very high. The base saturation percentage of these soils is 97, 94 and 95% at the surface of Fogera, Dembia and Kunzila-Alefa Takusa plains (Table 8.5), respectively. Based on the physical field information the soils are classified as Pellic (Eutric) Verisols at Fogera and Some part of Dera plains (Jigna kebele), Chromic Vertisols at Dembia and Kunzila-Alefa Takusa plains according to FAO-UNESCO-ISRIC revised legend of the soil Map of the World (FAO 1988, 1994) ISSS-ISRIC-FAO; FAO (1998) ISSS-ISRIC-FAO; (Mekonnen 2011; WRB 2014).

In Fogera, Dembia and Kunzila-Alefa-Takusa plains, the exchangeable cations are turn out to be in the order of Ca > Mg > Na > K (Table 8.6). This might be due to the parent material from which the soils have developed. Mesfin (1998) reported

Table 8.4 Soil chemical characteristic of Pellic Vrtisols Fogera plain

Depth (cm)	pH 1:2.5		ECe (dS m^{-1})	ESP (%)	Olsen P mg kg^{-1}	Exchangeable cations, (cmolc kg^{-1})				Sum	CEC (cmolc kg-^{1})	BS (%)	OC	TN
	Water	KCl				Ca	Mg	K	Na				(%)	
0–20	5.5	4	0.2	1	1.2	50.2	22.8	0.2	0.6	73.8	76	97	1.6	0.14
20–95	6.1	4.8	0.2	1	1.1	50.6	25.3	0.4	1.4	77.7	82	95	1	0.09
95–140	7.1	5.8	1.4	2	0.9	55.6	44.7	0.3	1.3	101	75	136	0.8	0.07
140–200	7.6	6.5	0.6	–	–	13.6	49.7	0.4	0.9	64.5	66	97		

Table 8.5 Soil chemical characteristic of Chromic Vertisols at Kunzila-Alefa plains

Depth (cm)	pH 1:2.5		ECe (dS m^{-1})	ESP (%)	Olsen P mg kg^{-1}	Exchangeable cations, (cmolc kg^{-1})					CEC (cmolc kg–^{1})	BS (%)	OC	TN
	Water	KCl				Ca	Mg	K	Na	Sum			(%)	
0–30	6.4	4.7	0.2	1	3.5	31.4	18.8	0.3	0.6	51	54	95	1.4	0.2
30–80	6.9	5.4	0.2	2	3	22.4	4.5	0.2	1.3	28.4	56	51	0.8	0.1
80–145	8.2	6.4	0.3	3	2.9	40.3	17	0.4	1.9	59.7	62	96	0.6	0.08
145–200	8.1	6.3	0.5	1	–	34.9	9.9	0.2	2	47	63	74		

a similar order (Ca > Mg > Na > K) for Alfisols of Debre Markos and Bahir Dar. Similar result was recorded to the findings of by Yihenew (2002) for north-western Ethiopia. Especially with high levels of exchangeable calcium, care is needed with the selection and amounts of phosphate fertilizer. In soils having high exchangeable calcium levels, P fertilizer tends to be converted to calcium phosphate which is not available to plants (due to a change to the less soluble apatite form).The medium values of exchangeable Na is obtained from.

8.4.2 Cambisols

Cambisols are soils that are in the incipient stages of formation towards one of the mature soils but have not fully developed the diagnostic properties. Their agronomic qualities are good. It is relatively elevated plain land with slope of 5–8%. The soils are developed on alluvial and colluvial parent materials.

The colour of the surface soil varies from very dark greyish brown (10YR3/2) to very ark brown (10YR2/2). The soil textures are clay. They have deep to very deep soil profile (>200 cm), with moderate drainage (Table 8.7). The have moderate and coarse surface structure; cracks are abundant with widths ranging from medium to wide. Their width reaches 2–5 cm while their width reaches 80 cm. They possess low to medium *gilgai* formation with 10–20 cm height. Most of the unit is under cultivated land use system. The subsoil has medium to coarse angular and sub angular blocky structure and the profiles have diffused and sometimes wavy boundary.

The reaction (pH) of the topsoil of Cambisols have moderately to strongly acidic with values ranging from 4.9 to 5.5, with an average of 5.1. The soils are none-saline and non-sodic with average ECs of <0.2 dS m^{-1} and exchangeable sodium percentage (ESP) of <2% (Table 8.8).

Cambisols of Mecha show the very low levels of Na according to the rating of Metson (1961) which is <0–0.1 very low, 0.1–0.3 low, 0.3–0.7 moderate, 0.7–2.0 high and >2 very high. The value of CEC at the surface is 50.2 cmol(+)/kg of soil on the top and the value decreasing down with soil depth. Based on the field investigation and laboratory result the soils are classified as Cambic Cambisols (CMcm).

8.4.3 Luvisols

The soils are found within the study area have Reddish brown (2.5YR4/3) colour at the surface followed by dark reddish brown (2.5YR3/3) at sub surface and very dusky red below 90 cm. These major soil groups are distributed within south

Table 8.6 Soil chemical characteristics of Chromic Vertisols at Dembia plain

Depth (cm)	pH 1:2.5		ECe (dS m^{-1})	ESP (%)	Olsen P mg kg^{-1}	Exchangeable cations (cmolc kg^{-1})					CEC (cmolc kg^{-1})	BS (%)	OC	TN
	Water	KCl				Ca	Mg	K	Na	Sum			(%)	
0–25	6.6	5.2	0.1	0	2.4	28.7	11.6	0.3	0.5	41.1	44	94	1	0.12
25–60	7.1	5.9	0.1	0	2.4	25.1	28.7	0.2	0.5	54.5	56	98	0.8	0.11
60–100	6.9	5.4	0.2	1	2.9	35.8	19.7	0.2	0.8	56.6	58	98	0.7	0.09
100–200	8.2	6.8	0.5	4		22.6	20.8	0.2	1.2	44.8	52	86		

Table 8.7 Soil physical characteristics of (Eutric) Vertic Cambisols at Mecha Woreda

Depth (cm)	Texture (%)				Colour	Structure	Consistence	Stickiness and plasticity
	Sand	Silt	Clay	Textural class				
0–15	5	14	81	Clay	VDGB (10YR3/2)	MC SAB	FR	ST/PL
15–50	7	11	82	Clay	B (10YR4/3)	st Co SAB	FR	ST/PL
50–120	8	12	80	Clay	GB (10YR5/2)	st Co AB	FR	ST/PL
120–00	6	12	82	Clay	DG (10YR4/1)	st Co AB	FR	ST/PL

Gonder and West Gojjam zones of the Lake Tana sub-watershed. They have argic horizons and have high activity clays and high base saturation throughout the profile (Table 8.9). These soils have potentially high soil fertility, provided that the land is not eroded. Exposure to raindrop impact and direct sunlight often leads to surface crusting, runoff and sever erosion.

The soils are developed in well-drained areas at higher altitude. They are mainly developed in variable parent materials such as acid to intermediate volcanic rocks, basalts, alkali trachytes and rhyolite. The soils are generally deep, predominantly heavy clays; their structure is moderately developed, medium sub angular blocky. Consistence is hard (dry), friable to firm (moist) and sticky and slightly plastic (wet). The soils have good permeability. Their typical characteristic is that they are found in areas where climatic conditions permit clay movement. They are found commonly on flat and gently sloping topography.

The pH of the topsoil is 5.3. Its value increases in depth to 6.6 in sub-surface. Organic carbon and the total nitrogen content percentage of the topsoil is 2.18 and 0.27 respectively. The available P content of the surface soil is 8.1 ppm which is low. The soils are non-saline and non-sodic with ECs value of <0.5 dS m^{-1} and ESPs of <2%. The cation exchange capacity of Luvisols at the topsoil at 0–14 cm depth is 54.6 meq/100 gm of soil. This value increases to 61.20 meq/100gm of soils in sub-surface, while the base saturation percentages of the soils are greater than 35.5.4% (Table 8.10). Based on the information available, the soils are classified as Rhodic Luvisols (LVro).

8.4.4 *Nitisols*

The Nitisols have very dark brown (7.5YR2.5/2) color at the surface and become dark yellowish brown (10YR3/4) at subsurface of Adama site and dark red (2.5YR3/6) at the Sekela sampling site. Nitisols are soils with nitic horizons generally developed in materials derived from mafic rocks (Table 8.11). They are very

Table 8.8 Soil chemical characteristics of Eutric Vertic Cambisols at Mecha Woreda

Depth (cm)	pH 1:2.5		ECe (dS m^{-1})	Olsen P mg kg^{-1}	Exchangeable cations, (cmolc kg^{-1})					CEC (cmolc kg–[1])	BS (%)	OC	TN
	Water	KCl			Ca	Mg	K	Na	Sum			(%)	
0–15	4.7	3.9	0	6.4	13.6	13.6	0.2	0.1	27.4	47	58	2	0.17
15–50	5	3.9	0	1.3	11.8	10.8	0.1	0.1	22.8	37	61	1	0.11
50–120	5	4	0	3.3	12.7	14.5	0.1	0.1	27.3	36	75	0.9	0.05
120–200	5.2	4.2	0	12.2	12.7	19	0.1	0.1	31.8	41	77	0.8	0.06

Table 8.9 Soil physical characteristics of Rhodic Luvisols at Dera Woreda

Horizon	Depth (cm)	Texture (%)				Colour	Consistence	Stickiness and plasticity
		Sand	Silt	Clay	Textural class			
AP	0–14	12	38	50	Clay	RB (2.5YR4/3)	FR	SST/SPL
Bt1	14–55	14	30	57	Clay	DRB (2.5YR3/3)	FR-F	ST/PL
Bt2	55–90	12	24	64	Clay	DR (2.5YR3/6)	F	ST/PL
Bt3	90–135	13	23	64	Clay	VDR (2.5YR2.5/2)	F	ST/PL
BC	135–200	10	28	62	Clay	VDR (2.5YR2.5/2)	F	ST/PL

deep, dark reddish brown, strongly weathered clay soils characterised by shiny ped faces, gradual boundaries (which makes identification of horizons difficult), uniform subsoil clay content and moderately developed structures. Clay skins are frequently observed in the subsoil, but not always. When dry, the soils are hard, compact and difficult to auger. When moistened they are friable and moderately permeable. They are relatively fertile and extensively cropped. These soils have been developed in well-drained areas with good permeability and a favourable structure. They are found dominantly on flat to rolling topographies. They are characterized by very thick and more or less uniform clay distribution. The soils are reddish in colour, derived from different parent materials. As observed during field investigation, Nitisols are mostly distributed in Mecha, Yilmana Densa, Dangila and Sekela woredas.

The available phosphorus of Nitisols is low with an average value of 9.4 ppm at the surface, which is low. Ethiopian soils particularly Nitisols, Luvisols and other acidic soils are reported to have low phosphorus content. This is might be not only the inherently low available P content, but also the high fixation capacity of the soils (Table 8.12).

The pH of the Nitisols at Jema and Sekela sampling sites at surface soils is 6.1 and 5.5, respectively and decreases down in the subsurface at two sites (Table 8.13). The soils are non-saline and non-sodic with ECs of <0.2 dS/m and ESPs of <2%. The organic carbon and total nitrogen contents of the surface soil are very low and medium on an average of 2.1 and 0.11 and 3.8 and 0.18 at two above mentioned sites respectively. Similarly, the C/N ratio value are 6 and 9.7 at above mentioned two sites, indicating, that the organic carbon percentage is relatively high than the total nitrogen percentage (Table 8.14). Nitosols have very low levels of Na according to the rating of Metson (1961) which is <0–0.1 very low. Based on the field study the soils are classified as Haplic Nitisols (NTha) and Luvic (Rhodic) Nitisols (NTro) at Adama and sekela, respectively.

Table 8.10 Soil chemical characteristics of Rhodic Luvisols at Dera Woreda

Depth (cm)	pH 1:2.5		ECe (dS m^{-1})	ESP (%)	Olsen P mg kg^{-1}	Exchangeable cations, (cmolc kg^{-1})					CEC (cmolc kg^{-1})	BS (%)	OC	TN
	Water	KCl				Ca	Mg	K	Na	Sum			(%)	
0–14	5.3	4.3	0.1	1	5.1	10.7	5.4	0.8	0.1	17	29.8	57	2.2	0.27
14–55	5.4	4.6	0.1	1	5.6	10.6	5.3	0.9	0.1	16.9	27.7	61	0.7	0.07
55–90	6	4.7	0.1	1	4.4	12	4.7	1.6	0.1	18.4	21.4	86	0.2	0.03
90–135	6.5	5.1	0	1	4.4	18.6	5.9	0.2	0.1	24.8	36.6	68	0.2	0.02
135–200	6.6	5.4	0.1		4.4	14.1	5.4	0.1	0.1	19.7	38.2	52	0.1	0.01

Table 8.11 Soil physical characteristics of Haplic Nitisols at Adama

Depth (cm)	Texture (%)				Colour	Structure	Consistence	Stickiness and plasticity
	Sand	Silt	Clay	Textural class				
0–28	34	28	37	Clay	VDB (7.5YR2.5/2)	MOSTSAB	VFR	ST/PL
28–66	27	25	48	Clay	DRB (5YR3/3)	st Co AB	FR	ST/PL
66–110	23	22	55	Clay	DRB (5YR3/3)	st Co AB	FR	ST/PL
110–170	23	22	55	Clay	VDGB (10YR3/2)	st Co AB	FR	ST/PL
170–200				Clay	DYB (10YR3/4)	st Co AB	FR	ST/PL

8.4.5 Leptosols

Leptosols are very shallow soils limited in depth by continuous hard rock. They occur mainly on steep side slopes; mountains and hills are prominent in the Region. The soils are generally young and are limited by their topsoil horizon or directly over altered parent rocks from which they have developed. Leptosols are the most widely spread soils in the upper parts of Gilgel Abbay, Gumara, Rib, Megech and Infranz Catchments of Tana sub-basin. The thickness of the solum ranges from 10 to 50 cm. weathered bedrock (saprolite) or rock fragments are generally just under the sequm. The lower part of the sequm typically has many coarse fragments which increase with depth, excessively drained, sandy loam to clay. They have dark reddish brown (5YR3/4) colour, silty clay texture, moderate to medium sub angular blocky structure, friable, to slightly firm, many fine and very fine roots, few fine pores, acidic and gradual smooth boundary. They are characterized by common rockiness and stoniness. Erosion is moderate to severe. These soils developed from variable parent materials including basic volcanic rocks, intrusive rocks, alluvium and colluvium deposits, coarse grained clastics, non carbonate meta-sediments, meta volcanic rocks and laterite.

The major limitation of these soils is a shallow effective soil depth. The shallow depth does not allow sufficient moisture storage and root penetration. This renders the soils being prone to drought. Because they dry up easily, they are very likely to be truncated and washed away during heavy rains. They cannot also support sufficient vegetative growth to protect them from severe erosion. Crops do not perform well in these soils because of their in ability to conserve enough moisture and lack of space for root proliferation. The unfortunate situation in the steepest slope of the Tana sub-basin is these soils are intensively cultivated. In most cases the plow reaches the underlying hard rock making land preparation difficult. It is distributes in the Tana sub-basin at the highest elevation. Rock outcrops are typical of Leptosol landscapes. Trees must be shallow rooting or develop where the soil is a little

Table 8.12 Soil chemical characteristics Haplic Nitisols at Adama

Depth (cm)	pH 1:2.5		ECe (dS m^{-1})	ESP (%)	Olsen P mg kg^{-1}	Exchangeable cations, (cmolc kg^{-1})					CEC (cmolc kg^{-1})	BS (%)	OC	TN
	Water	KCl				Ca	Mg	K	Na	Sum			(%)	
0–28	5.7	4.8	0.1	2	13.7	10.8	14.3	0.5	0.1	25.7	31	84	1.7	0.09
28–66	5.6	4.6	0	1	6.5	9	10.8	0.3	0.1	20.1	26	54	1.3	0.08
66–110	5.4	4.7	0.1	1	27	11.6	6.3	0.3	0.1	18.3	25	46	0.8	0.06
110–170	5.6	4.8	0	1	13.8	9.9	8.1	0.4	0.1	18.4	26	47	0.6	0.05

Table 8.13 Soil physical characteristics of Luvic Nitisols at Sekela Woreda

Depth (cm)	Texture (%)				Colour	Structure	Consistence	Stickiness and plasticity
	Sand	Silt	Clay	Textural class				
0–13	7	31	62	Clay	VDB (7.5YR2.5/2)	MOSTSAB	VFR	ST/PL
13–55	1	13	86	Clay	VRB (7.5YR2.5/2)	st Co AB	SVFR	ST/PL
55–126	2	18	81	Clay	DR (2.5YR3/6)	st Co AB	VFR	ST/PL
126–200	6	15	79	Clay	DR (2.5YR3/6)	st Co AB	VFR	ST/PL

deeper and where impeded drainage can lead to higher water retention. Soil development is slow. Only limited extensive grazing is possible.

8.5 Management Implications of the Soil Properties

Soil fertility degradation due to deforestation, soil erosion, overgrazing, and inappropriate cultivation practices and overpopulation is a major problem in the Tana sub-basin. Areas that need urgent attention in soil conservation activities are interfluvial Ridges, upper Plateau, hill side slopes, and mountains. Water is the main cause of erosion in the Tana sub basin. Severe land degradation is observed in many parts of the highlands particularly in Rib, Gumara, Infranz, Megech and upper part of Gilgel Abbay. Suggested management options could be closing off degraded lands for re-vegetation, moving people from steep slopes and settling on flat lands, area closure and afforestation of degraded steep slopesing with the use of cut and carry feeding system.

Vertisols at Fogera, Dembia and Alefa-Takusa plains has drainage problem as it is often inundated during the rainy season. Soils like Vertisols at Fogera, Dembia and Alefa-Takusa plains have drainage problem as it is often inundated during the rainy season. There is need to monitor the water table and develop good irrigation schedule in order to allow the growth of water loving crops and vegetables. Therefore, a proper drainage ditch network will need careful design will be integral to the agricultural development planning. Drainage should ensure that surface water from heavy rainfall (1:5 year return period or similar) is removed from the fields within 24 h for dry foot crops and within 3 days for any rice crop. The priority is to remove surface water following rainfall and from excessive irrigation. Constructing dams upstream can control flooding from the rivers.

On the lower landscape areas, several chemical elements required for plant growth, most of which are supplied from the soil, and all of these are taken up in water soluble forms. The use of fertilizers is established as an important aspect of

Table 8.14 Soil chemical characteristic of Rhodic Luvic Nitisols at Sekela Woreda

Depth (cm)	pH 1:2.5		ECe (dS m^{-1})	ESP (%)	Olsen P mg kg^{-1}	Exchangeable cations (cmolc kg^{-1})					CEC (cmolc kg^{-1})	BS (%)	OC	TN
	Water	KCl				Ca	Mg	K	Na	Sum			(%)	
0–13	4.9	3.9	0	1	3.2	7.2	6.3	0.3	0.1	13.9	31	45	1.5	0.07
13–55	5.3	4.4	0	2	1.8	14.5	12.7	0.8	0.1	28.1	46	61	2.3	0.14
55–126	5.7	5	0	1	1.8	7.2	7.2	0.2	0.1	14.7	27	54	0.8	0.06
126–200	5	4.1	0	1	1.8	7.2	6.3	0.2	0.1	13.8	27	51	0.9	0.05

crop production and advice is required on the type of fertilizer to use, how much, when and how to apply them. Nutrients are obtained from the soil in a soluble form. If the soil is dry for prolonged time, nutrient intake is inadequate to meet the plants, needs and reduced growth may result.

As shown in the survey result soils found within the Tana sub-basin have greater than 50% base saturation, hence very high. Exchangeable calcium is mostly high to very high (>15 me/100 g soil) in most of the soils of the study area followed by magnesium have medium to very high in most of the soils of the study areas.

Soil pH is found within the normal range for most soils except for some Nitisols and Luvisols where the soil pH is below 5.5. These soils may suffer from aluminum toxicity. The low pH may also induce nutrient uptake imbalance. Lime is required to raise the pH to an acceptable level of above 5.5. There is a clear indication that Vertisols and Leptosols may require phosphorus fertilizers as far as the soils of the basin are concerned.

The level of total nitrogen is very low to medium in the soils of the study area. Although the availability of nitrogen depends on several aspects of the physical environment, the soils of the Tana sub basin are relatively deficient in nitrogen. This is also closely related to low level of organic matter in most soils of the study area in general.

The low fertility status of the soils of Tana sub-basin can be brought to better use for agriculture by increasing the organic matter level through incorporation of organic residues such as farmyard manure, plant residues, and household refuse. For sustainable crop production, there is need for guided inorganic fertilizer use and improved management practices in the area that will effectively minimize erosion and enhance and maintain soil quality and productivity.

8.6 Conclusion

Characterization of soils of Tana sub-basin reveal that the soils were predominantly of colluvio-alluvial origin and low in inherent fertility as evidenced by low organic matter, base saturation and low CEC. The fertility status of the soil at different landscapes (slopes and altitude) of the watershed showed higher differences, the lower being at the steep slopes which might be associated with old erosion histories. The soils are neutral to slightly alkaline. The study has brought out clearly relevant soil information that can guide decision on the use and management of soils of the area on a sustainable basis.

The low fertility status of the soils of Tana sub-basin can be brought to better use for agriculture by increasing the organic matter level through incorporation of organic residues such as farmyard manure, plant residues, and household refuse. For sustainable crop production, there is need for guided inorganic fertilizer use and improved management practices in the area that will effectively minimize erosion and enhance and maintain soil quality and productivity.

Levels of soil organic carbon are everywhere low. A good organic content takes time to build-up and is crucial to promote good structure, permeability, water-holding and nutrient-retention. Organic matter conservation is the single most important factor to improve soil fertility. Soil analyses indicate that, currently at least, levels of K are adequate but it is likely that most crops will respond to P fertiliser. Levels of N are low and regular (non-acidifying) nitrogenous fertiliser will be essential for all crops.

Vertisols at Fogera, Dembia and Alefa-Takusa plains has drainage problem as it is often inundated during the rainy season. There is need to monitor the water table and develop good irrigation schedule in order to allow the growth of water loving crops and vegetables.

References

Asadu CL, Nnaji GU, Ezeaku PI (2012) Conceptual issues in pedology. University of Nigerian Press Limited, University of Nigeria, Nsukka, Nigeria, pp 34–57

Esu IE (2005) Soil characterization and mapping for food security. A Keynote Address at the 29th annual conf. of SSSN held at University of Nigeria, Abeokuta, from 6th to 10th Dec. 2004

Esu IE, Akpan-Idiok AU, Eyong MO (2008) Characterization and classification of soils along a typical Hill slope in Afikpo area of Ebonyi state, Nigeria. Nigerian J Soil Environ 8:1–6

Eswaran H (1977) Soil Analysis for Soil Surveys. In: Soil Resource Inventories. A proceedings of a workshop held at Cornell University, Ithaca, USA, pp:315–324

FAO (1988) Soil map of the world. Revised legend, by FAO–UNESCO–ISRIC. World Soil Resources Report No. 60. Rome

FAO (1994) World Reference Base for Soil Resources, by ISSS–ISRIC–FAO. Draft. Rome/Wageningen, Netherlands

FAO (1998) World Reference Base for Soil Resources, by ISSS–ISRIC–FAO. World Soil Resources Report No. 84. Rome

FAO (2006) Guidelines for soil description. 4th edition. Rome

Kalivas DP, Triantakonstantis DP, Kollias VJ (2002) Spatial prediction of two soil properties using topographic information. Global Nest: the Int. J. 4(1):41–49

Kamara CS, Haque I (1988) Characteristics of Vertisols at ILCA research and outreach sites in Ethiopia. Revised version. Plant Science Division Working Document B5. ILCA (International Livestock Centre for Africa), Addis Ababa, Ethiopia

Mekonnen G (2011) Feasibility study of the soil survey on Megech and Dirma (Kola diba-Dembia) Irrigation projects, Unpublished

Mesfin A (1998) Nature and management of Ethiopian soils. Alemaya University of Agriculture, Alemaya, Ethiopia, p 272

Metson AJ (1961) Methods of chemical analysis for soil survey samples. Soil Bureau Bulletin No. 12, New Zealand Department of Scientific and Industrial Research. Government Printer, Wellington, New Zealand, pp 168–175

Mitiku H (1987) Genesis, characteristics and classification of soils of the central highlands of Ethiopia. PhD thesis, State University of Ghent, Belgium

Mitiku H (1991) Need for soil resources information for agro-technology transfer. Paper presented at the Second African Soil Science Society Conference, Cairo, Egypt, 4–10 November 1991.

Ogunkunle AO (2005) Soil Survey and Sustainable Land Management. Invited paper at the 29th annual conf. of SSSN held at University of Nigeria, Abeokuta, from 6th to 10th Dec. 2004

Probert ME, Fergus IF, Bridge BJ et al (1987) The properties and management of Vertisols CAB International, Wallingford, UK

Sims JT (2000) Soil fertility evaluation. In: M.E. Sumner (ed) handbook of soil science. CRC Press, Florida, pp. D-113–D-153

Tekalign M, Haque I (1991) Phosphorus status of some Ethiopian soils. III. evaluation of soil test methods for av. phosphorus. Tropical Agriculture

WRB (2014) Food and agriculture organization of the United Nations Rome 2014

Yihenew G (2002) Selected chemical and physical characteristics of soils of Adet research centre and its testing sites in Northwestern Ethiopia. Ethiop J Nat Res 4(2):199–215

Yihenew G (2004) Modeling of Nitrogen and Phosphorus Fertilizer Recommendations for Maize (*Zea Mays* L.) Grown on Alfisols of North-western Ethiopia. PhD Dissertation, Kasetsart University

Chapter 9
Hydrology of Lake Tana Basin

Elias Sime Leggesse and Biazenlegn S. Beyene

Abstract Lake Tana Basin located in the northeast Ethiopia is one of the tributaries of the Blue Nile. The rainfall over the Upper Blue Nile in general, originates from moist air coming from the Atlantic and Indian oceans following the north-south movement of the Inter Tropical Convergence Zone. The area has one main rainy season between June and September, receiving about 70–90% of the annual rainfall. The southern portion of the Tana basin receives the highest rainfall ($\sim$1600 mm) with a decreasing trend from south to north ($\sim$1200 mm). The dry season occurs between November and April while the wet season occurs mostly between May and October. The basin has four main rivers: the Gilgel Abay, Gumara, Rib and Megech. The Gilgel Abay River alone contributes 60% of the flow into Lake Tana. Lake Tana is the largest lake in the country but is relatively shallow; historically, lake levels have fluctuated within the range of 1784.26–1787.81 m.a.s.l. The hydrological process of the basin is poorly understood and only 42% of the basin is gauged. Recently, efforts have been made to identify the processes by different scientists.

Keywords Basin · Annual rainfall · Trend · Hydrological processes · Gauged · Erosion · Sedimentation · Water resources utilization

E.S. Leggesse (✉)
Blue Nile Water Institute and Civil and Water Resource Engineering,
Bahir Dar University, Bahir Dar, Ethiopia
e-mail: 2003sime@gmail.com

B.S. Beyene
Blue Nile Water Institute and College of Natural Sciences, Bahir Dar University,
Bahir Dar, Ethiopia
e-mail: biazenlegns@bdu.edu.et

K. Stave et al. (eds.), *Social and Ecological System Dynamics*,
AESS Interdisciplinary Environmental Studies and Sciences Series,
DOI 10.1007/978-3-319-45755-0_9

9.1 Introduction

Lake Tana Basin is of significant importance to the region and to the country concerning water resources aspects and the ecological balance of the area. It is also a basin of international significance because of the biodiversity and bird species it is endowed. As one of the growth corridor of the area, currently the basin and the lake water is subjected to huge water resources utilization projects which are under planning and some under implementation. However, the sustainable utilization of the lake basin and the lake resources requires a thorough understanding of the hydrological system of the basin, and how the development activities would adversely affect the environment. Such knowledge includes the basin hydrology which comprises the runoff processes, the soil erosion processes and sedimentation transport and deposition mechanisms in the basin. Other includes the various water resources utilizations demands and scenarios and how the hydrology would also be affected by climate change.

Despite efforts that have been made so far on the understanding of the hydrology especially on the quantification of the lake water components a major research still remains in the ungauged catchments hydrology, erosion processes, sediment transport and deposition mechanisms, effective management practices as to how to reduce sedimentation of water bodies etc. This paper tries to give the review of the some of the works done in the basin and gives insight to major gaps need to be done in the future.

9.2 Lake Characteristics

Lake Tana is located in the basaltic plateau in the north-western range of the Ethiopian highlands between 12° 00′N and 37° 15′E at 1786 m.a.s.l (meter above seas level). The topography of the Sub-Bain is variable (Fig. 9.1). It is an oligo-mesotrophic fresh water lake with weak seasonal stratification (Wood and Talling 1988; Wudneh 1998). It comprises of abundant suspended sediment, giving it a reddish-brown color (Fig. 9.2).

It forms a wide, but shallow depression believed to have been formed during the Miocene era (Chorowiz et al. 1998) followed by the damming of a 50 km-long and 100 m deep Quaternary basalt flow blocking the exit channel (Mohr 1962). Subsequently, the lake basin filled creating the present lake that is over 3060 km^2.

Lake Tana is the largest lake in Ethiopia and the third largest lake in Nile Basin. The only surface outflow from the lake is the Blue Nile River, with a mean annual discharge that contributes about 8% of Blue Nile flow (Shahin 1988; Conway 1997).

It is a shallow lake with a maximum depth of only 15 m. It is approximately 84 km long (North-South ends) and, 66 km wide (East-West ends) with a shoreline perimeter of 385 km.

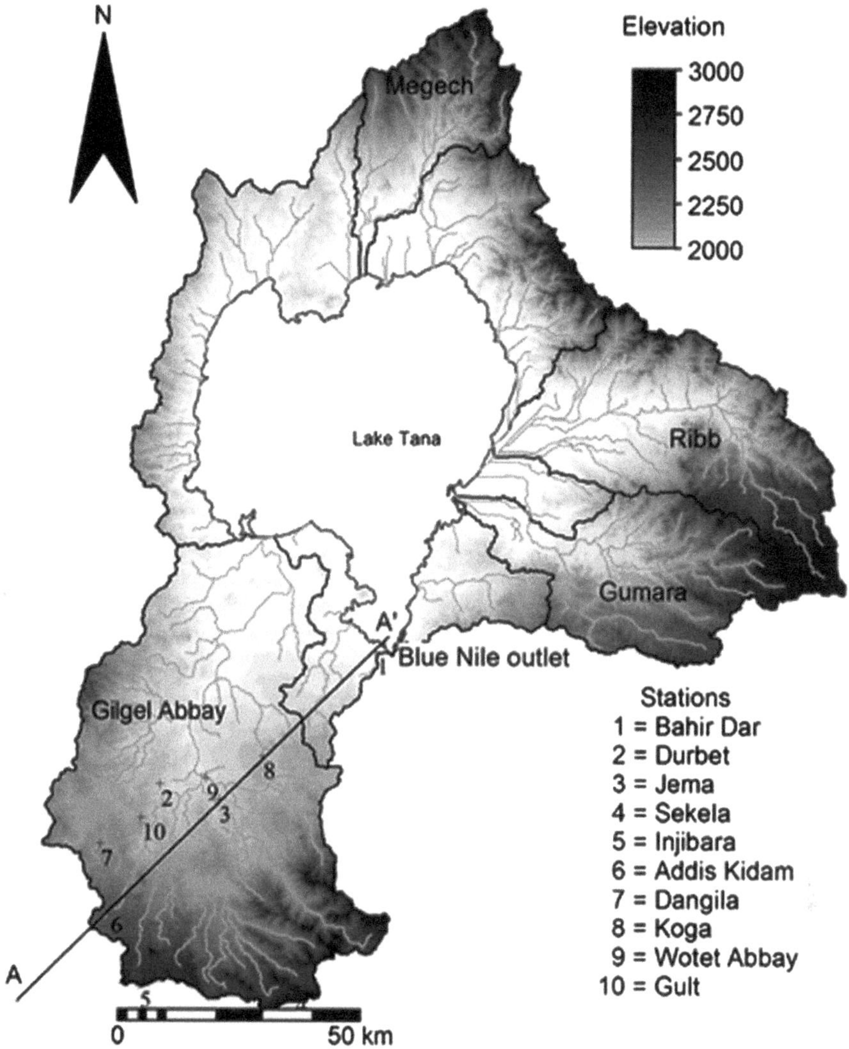

Fig. 9.1 DEM of Lake Tana basin with watershed boundaries of the major rivers and location and names of rain gauges as indicated by numbers (Haile et al. 2009)

9.3 Inlet Rivers and Streams Characteristics

Lake Tana is the source of the one of the longest rivers in the world, Blue Nile River. It is fed by four perennial rivers and many other intermittent streams. The rivers that contribute the majority of the inflow are Gilgel Abay from the South, Rib and Gumara Rivers from the East, and the Megech River from the North. Different

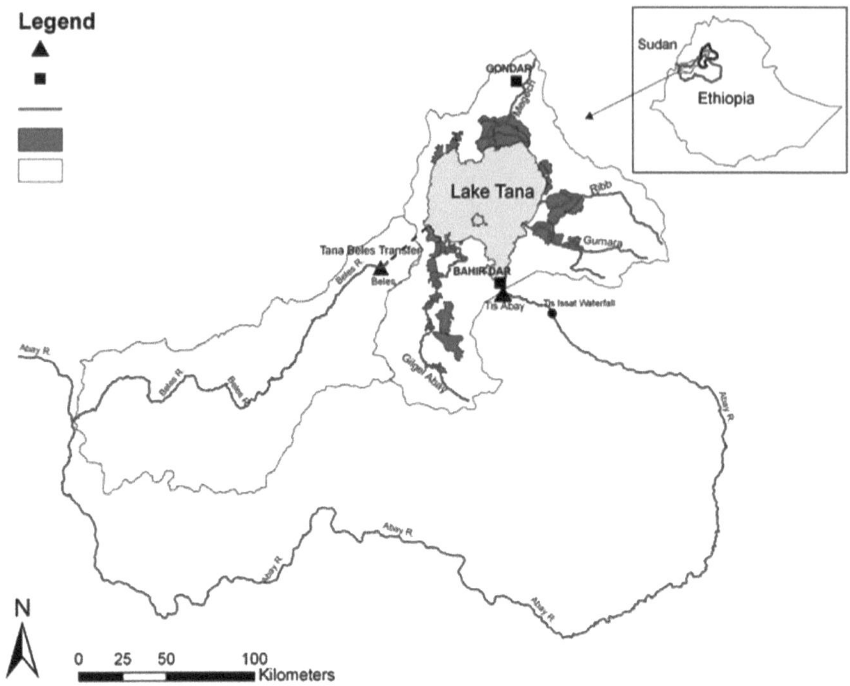

Fig. 9.2 Map showing the Lake Tana and Beles River catchments and future irrigation and hydropower development—(McCartney et al. 2010)

studies showed varying amounts of inflow to the lake from these rivers. For example, Chebud and Assefa (2009) and Kebede et al. (2006) indicated that these rivers account for 93% of the inflow into the Basin. Ligdi et al. (2010) and Lamb et al. (2007) stated that more than 95% of the inflow is contributed by these rivers.

Ligdi et al. (2010) stated that due to its slow response to peak and low flows because of large storage capacity, the lake stabilizes the outflow through its single outlet and produces a particularly smooth hydrograph. They stated that the lake receives an estimated inflow of $10.3 \times 10^9\ m^3yr^{-1}$ from 61 water courses, from which $3.7 \times 10^9\ m^3$ is lost as surface outflow and the remaining through evaporation (Ligdi et al. 2010). These values indicate that nearly two thirds of the lake water is lost through evaporation.

The annual outflow ranges from a minimum of 1075 Mm^3 (in 1984) to a maximum of 6181 Mm^3 (in 1964) (Kebede et al. 2006) and for the period 1976–2006 the average outflow is estimated at 3732 mm^3.

A total of an estimated 59 rivers drain water from Tana Basin into the lake. The major rivers are perennial rivers while smaller rivers/streams are ephemeral. Approximately 42% of Lake Tana basin is gauged (SMEC 2007) primarily in the upper part of the major four big rivers: Gilgel Abay at Merawi; Gumara at Bahir Dar; Rib at Addis Zemen; and Megech at Azezo. Most of the hydrological analysis

Table 9.1 Major sub catchments of 17 gauged sub-catchments of LTSB with respective gauged and ungauged areas (Wale et al. 2009)

Sub catchments	Gauging station	Gauged area in Km^2	Total sub catchments area in km^2
Gilgel Abay	Gilgel Abay near Merawi	1656.2	4557.8
	Kelti near Delgi	606.7	
	Koga at Merawii	299.8	
Rib	Near Addis Zemen	1302.6	2013.8
Gumara	Near Bahir Dar	1283.4	1768.3
Megech	Near Azezo	513.5	999.4
Gumara	Near Maksegnit	164.9	547.7
Garno	Near Infranz	98.1	359.0
Gelda	Near Ambessame	26.8	400.8

Table 9.2 Long term average annual river flow of the major tributaries (Wale et al. 2009)

Gauged rivers	Gauged area in km^2	Mean annual river flow in MCM
Gilgel Abay	1656.2	1753.5
Gumara	1283.4	1229.5
Rib	1302.6	510.4
Megech	513.5	195.2
Kelti	606.6	283.8

and estimates are derived from this 42% gauged sections in the LTSB (Lake Tana Sub-Basin) (Tables 9.1 and 9.2).

The hydrology of the ungauged catchment is poorly understood. Studies vary in their estimation of the inflow from ungauged catchments. For example, Kebede et al. (2006) indicated only 7% of the lake inflow is from ungauged catchments. SMEC (2007) indicates that some 29% of the lake inflow is from ungauged catchments while Wale et al. (2009) indicates that 42% of the lake inflow is from ungauged systems.

9.4 Lake Storage and Outflow

The Blue Nile is the only natural surface water outlet of the lake. The lake is regulated by "cherechera" weir constructed in 1996, a low height weir at the outlet of the lake. The Lake has a slow response time to precipitation fluctuations, due to its large surface area (>3000 km^2) and, thus, its large storage capacity and restrictions at its outlet. For example, the lake outflow is maximum in September, two months after the maximum rainfall amount in July (Conway 1997).

The lake balance is calculated from hydrometeorological variables such as rainfall, evaporation, runoff from gauged and ungauged catchments and lake outflow. There are a number of studies that investigated the water balance and hydrology of Tana Basin. Among them are Setegn et al. (2010), who showed that more than 60% the total inflow to the Basin is lost through evapotranspiration. They also indicated that base flow contributes to the discharge more than surface runoff.

9.5 Historical Lake Level

The elevation of the spillway was fixed at elevation of 1987 m a.s.l. The minimum operating of the weir is 1784 m a.s.l. The lake level varies by 1.6 m annually with lake levels averaging 1786.3 between the period 1976–2006. Since the operation of the weir lake level dropped to the historical minimum water level of 1784.46 m a.s.l on 6/30/2003.

Due to its large storage capacity, the response of lake level fluctuations with rainfall variability is relatively slow. The deviation of the lake level between the wet and dry seasons approximates only 1.6 m (Wale et al. 2009). Inter-annual lake level fluctuates primarily by year-to-year differences in the major rainy and dry seasons. The lake level usually reaches its maximum in September and minimum in June.

9.6 Hydrologic Budget

The water balance of Lake Tana is largely unknown because of insufficient gauging stations. Also, these devices are not regularly inspected and, therefore, regular data is often missing. Recently, Kebede et al. (2006) and SMEC (2007) investigated the contribution of runoff from ungauged catchments. Despite the limitation of data, shortage and adequacy of data the authors tried to analyse the water balance of the lake. They used different data quality management, methodologies, including simple Excel sheets of water balance, NHP, and SWAT models used in the analysis and at different time steps from daily to monthly.

9.7 Erosion and Sedimentation

Soil erosion is the removal of surface material by wind and water. Soil erosion by water is caused by detachment and transport action of rainfall and runoff. Soil erosion has both onsite and offsite damages. It reduces productivity of crops, exacerbate flood hazards, and decreases life expectancy of water storage structures.

The processes involved in sediment supply, transport and deposition are generally non-linearly. Rather, from small scale to river basin scale there will be a cycle of supply–transport–deposition which will repeat in both space and time. Thus, sediments deposited at a point in a river basin may become part of the supply chain for sites further downstream or may remain in situ for long time periods.

Despite major efforts by the government and other organizations, soil erosion and associated sedimentation of lake and reservoirs continues to affect the livelihoods of the local populations and compromise the major water infrastructures constructed throughout the country, and especially in the basin. The lack of long term and representative measured data has added to this long standing problem. With all the limitations research related to soil erosion are being conducted in the basin and show a high degree of vulnerability of the basin. Setegn et al. (2009) identified that 18.5% of the area has a high erosion potential (sediment yield >30 ton/ha). The study also indicated that the Rib and Gumara sub watersheds are the highest contributors of sediment to the lake. Awulachew (2010) found the sediment yield from Gumara sub-watershed of the basin ranges from 11 to 22 t/ha/yr. Assegahegn and Zemadim (2013) measurement in Mizewa watershed located in the Rib subwatershed indicated an average soil loss rate of 40.9 t/ha.

The studies addressing sediment yield for a small watershed in the lake basin by Assegahegn et al. (2013), and a study outside the lake basin by Guzman et al. (2013) show that the sediment in the basin strongly varies with early onset of rainfall. The hydrographs of the sediment yields shows that the peak of sediment precedes the peak of rainfall and subsequent runoff. Furthermore, the sediment cumulative curve shows that most of the sediment volume passes in the river in the first three months of the rainy season. Analysis with rainfall and runoff shows that the peak of sediment concentration and sediment load comes before rainfall and runoff. The annual sediment concentration (sediment weight per volume of water) measured in mg/l shows sediment load distribution is concentrated in June to September, with the highest peak in July.

Research also shows the extent of sedimentation in the flood areas around the lake and in the lake. Marye et al. (2011, Tourism Destinations Networking Development Plan for Lake Tana, unpublished) between 1986 and 2010 the Gilgel Abay Delta increased by 586.1 ha (49.8%) and the Gumara River Delta by 101.6 ha (218.4%) respectively. Zegeye et al. (2010) examined the contribution of sediment to the flood plains and showed that the lake's sediment may not all come from upland agricultural erosion. Comparing the gully and upland erosion rates, the soil loss rate of the upland plots (rill erosion) is approximately 20 times less than that transported due to gully erosion. His work agrees with recent studies done elsewhere that has shown that river banks and roads also play an important role in supplying sediment to the river (Walling and Woodward 1995; Walling et al. 1999; Carter et al. 2003; Gruszowski et al. 2003), and in some cases this input dominates over other sediment sources. Most other works in the basin were based on the assumption of stability in the driving processes. This may not be a reasonable assumption in the basin, where climate change is widely accepted to be influencing river flows and hydrological behaviour of river basins. Given the major changes in land use that has been seen in the lake basin over the course of time, such an assumption may become completely unrealistic and unsupportable. Setegn et al. (2009) Using multi-criteria evaluation (MCE) Showed how varied the estimation of sediment yield could be in the basin which suggests rigorous research needs in the future.

9.8 Current and Future Water Resources Utilization of the Lake Basin

Lake Tana, including the surrounding wetlands, provides huge socioeconomic and ecosystem benefits for the area. Lake Tana is important for domestic water supply, hydropower, fisheries, grazing and water for livestock, as well as papyrus reeds for boat construction. In addition, the lake is important for water transport and as a tourist destination. Due to the increase demand of water for agriculture, industries, domestic, and power generation, various water resources developments are under construction requiring additional feasibility studies to optimize usage of the water resources potential of the lake.

The Tana-Beles hydropower project which is currently under operation has a capacity of 460 MW (Mega Watt) project. It involves transfer of approximately 2 985 Mm^3 of water from Lake Tana to the Beles River through 12 km long, 7.1 m diameter tunnel. The planned irrigation development shows the number of irrigation schemes (up to approximately 60,000 ha) are under construction and feasibility study (Table 9.3).

Utilization of the water resources of the basin has recently accelerated. Such utilization of the water resources may have negative consequences as occurred in 2003, when navigation in the lake was affected when water levels dropped below 1785 m a.s.l the minimum level at which the ships can operate. In addition, large areas of papyrus reed were destroyed, there was significant encroachment of agriculture on the exposed lake bed and there was a decrease in fisheries production (EPLAUA 2004, water level reduction of Lake Tana and its environmental impacts, unpublished).

Table 9.3 Planned irrigation development in the Lake Tana catchment (Getachew et al. 2013)

Irrigation scheme	Irrigable area (ha)	Large dam storage (Mm^3)	Stage of development
Gilgel Abay	12,852	563	Feasibility studies
Gumara A	14,000	59.7	Feasibility studies
Rib	19,925	233.7	Under construction
Megech	7300	181.9	Under construction
Koga	6000	78.5	Construction completed
NE lake tana	5745	Withdrawals from the lake	Identification
NW lake tana	6720	Withdrawals from the lake	Identification
SW lake tana	5132	Withdrawals from the lake	

The planning and implementation such a large projects in the basin requires careful evaluation of the environmental and competing demands for water. One study revealed that a development and operation of all the planned projects could lower the mean annual water level of the lake by 0.44 m with a subsequent decrease in the average surface area of the lake by 30 km^2. In addition, the scale of the adverse ecological impacts and implications for shipping and the livelihoods of many local people could be significantly affected.

9.9 The Most Important Research Gaps

The estimated flow from the ungauged catchment highly varies across studies. It is due to the limitation of the data, the applicability of the methods employed or the underestimation of diversification of the basin in terms of climate, geomorphology, and geology of the basin are main ones. This suggests that further research is needed to understand the hydrology of the ungauged catchment.

Current research suggests how critical the soil erosion and sedimentation processes are in the basin. In addition, due to a number of projects planned in the basin the knowledge of soil erosion and sedimentation has become more important than ever. Hence, further research needs to be done to understand the sediment transportation and deposition, the main factor driving the soil erosion (soil, land use/cover, slope, climate), the effect of scale (watershed to basin) in soil erosion modeling, the effect of soil and water management efforts being implemented and the critical management areas in terms of maintaining cultivated land productivity and sedimentation of structures,

The lake water serves many functions apart from the ecosystem services it provides. Among these are; navigation, fishing, irrigation, hydropower etc. Currently, the basin water is subjected to different water uses apart from the upstream and downstream water use rights which needs a thorough understanding and resolution. Further research must ultimately be conducted so that the lake basin water is best utilized with a minimal negative effect from development scenarios.

References

Assegahegn MA, Zemadim B (2013) Erosion modelling in the upper Blue Nile basin: The case of Mizewa watershed in Ethiopia. In: Wolde Mekuria (ed) Rainwater management for resilient livelihoods in Ethiopia: Proceedings of the Nile Basin Development Challenge science meeting, Addis Ababa, 9–10 July 2013. NBDC Technical Report 5. International Livestock Research Institute Nairobi

Awulachew SB (2010) Improved water and land management in the Ethiopian highlands and its impact on downstream stakeholders dependent on the Blue Nile, CPWF Project Number 19, CGIAR Challenge Program on Water and Food Project Report series, www.waterandfood.org. Accessed 6 July 2011

Carter J, Owens PN, Walling DE, Leeks GJ (2003) Fingerprinting suspended sediment sources in a large urban river system. Sci total Env 314:513–534
Chebud YA, Assefa MM (2009) Modelling lake stage and water balance of Lake Tana, Ethiopia. Hydrol Processes 23(25):3534–3544
Chorowiz J, Collet B, Bonavia F et al (1998) The Tana basin, Ethiopia. Intra-plateau uplift, rifting and subsidence. Tectonophysics 295:351–367
Conway D (1997) A water balance model of the Upper Blue Nile in Ethiopia. Hydro Sci J 42: 265–286
Getachew T, Dereje H et al (2013) Lake Tana Reservoir Water Balance Model. Int J Appl Innovation Eng Manag (IJAIEM) 2(3):2319–4847
Gruszowski KE, Foster ID, Lees JA, Charlesworth SM (2003) Sediment sources and transport pathways in a rural catchment, Herefordshire. UK Hydrol Processes 17(13):2665–2681
Guzman CD, Tilahun SA, Zegeye AD et al (2013) Suspended sediment concentration–discharge relationships in the (sub-) humid Ethiopian highlands: Hydrol Earth. Syst Sci 17:1067–1077. doi:10.5194/hess-17-1067-2013
Haile AT, Rientjes T, Gieske A et al (2009) Rainfall Variability over mountainous and adjacent lake areas: the case of Lake Tana basin at the source of the Blue Nile River. J App Meteor Clima 48(8):1696–1717
Kebede S, Travi Y, Alemayehu T, Marc V (2006) Water balance of Lake Tana and its sensitivity to fluctuations in rainfall, Blue Nile basin. Ethiopia J Hydrol 316(1–4):233–247
Lamb HF, Bates CR, Coombes PV et al (2007) Late Pleistocene desiccation of Lake Tana, source of the Blue Nile. Quat Sci Revi 26(3):287–299
Ligdi EE, Kahloun M, Meire P (2010) Ecohydrological status of Lake Tana—a shallow highland lake in the Blue Nile: review. Ecohydro Hydrobio 10(2–4):109–122
McCartney M, Alemayehu T, Shiferaw A, Awulachew SB (2010) Evaluation of current and future water resources development in the Lake Tana Basin, Ethiopia. International Water Management Institute, Colombo, SriLanka 39p. (IWMI Research Report 134). doi:10.3910/2010.204
Mohr PA (1962) The geology of ethiopia. University-College Press, Addis Ababa
Setegn SG, Ragahavan S, Bijan D (2009) Hydrological modeling in the Lake Tana Basin, Ethiopia using SWAT model, Open Hydrol.Process. Published online in Wiley InterScience. www.interscience.wiley.com. doi:10.1002/hyp.7476
Setegn SG, Bijan D, Ragahavan S et al (2010) Modeling of Sediment Yield From Anjeni-Gauged Watershed, Ethiopia Using SWAT Model. J Am Water Res Assoc (JAWRA) 46(3):514–526. doi:10.1111/j.1752-1688.2010.00431.x
Shahin M (1988) Hydrology of the Nile basin. Elsevier, International Institute for Hydraulic and Environmental Engineering
SMEC (2007) Hydrological Study of the Tana-Beles sub-basins, main report. Snowy Mountains Engineering Corporation, Australia
Wale A, Rientjes THM, Gieske ASM, Getachew HA (2009) Ungauged catchment contributions to Lake Tana's water balance. Hydrol Process 23(26):3682–3693
Walling DE, Woodward JC (1995) Tracing sources of suspended sediment in river basins: a case study of the River Culm, Devon, UK. Marine and Freshwater Research 46(1):327–336
Walling DE, Owens PN, Leeks GJL (1999) Fingerprinting suspended sediment sources in the catchment of the River Ouse, Yorkshire, UK. Hydrol Process 13(7):955–975
Wood RB, Talling JF (1988) Chemical and algal relationships in a salinity series of Ethiopian inland waters. Hydrobiol 158:29–67
Wudneh T (1998) Biology and management of fish stocks in Bahir Dar Gulf, Lake Tana, Ethiopia. Dissertation, Wageningen University
Zegeye AD, Steenhuis TS, Blake RW et al (2010) Assessment of soil erosion processes and farmer perception of land conservation in Debre Mewi watershed near Lake Tana. Ethiopia Ecohydrology Hydrobiology 10(2):297–306

Chapter 10
Water Quality of Lake Tana Basin, Upper Blue Nile, Ethiopia. A Review of Available Data

Goraw Goshu, A.A. Koelmans and J.J.M. de Klein

Abstract Water is at the forefront of the economic agenda of Ethiopian government and Tana basin has been identified as a major economic corridor because of the basin's immense water resource potential for socioeconomic development. For effective and sustainable utilization of water resources in the basin, it is essential to assess the water quality in spatial and temporal dimensions. Nonetheless, scientific information on Ethiopian water bodies is rare and the available ones are of expeditious origin. In Tana basin, there is no detailed and systematic characterization of water quality based on long term and spatially representative data, due to the absence of a sustainable monitoring program. Despite the fact that data and studies are fragmentary, the available information on Lake Tana indicates that the trophic status of the lake has gradually changed to mesotrophic and eutrophic in some places due to nutrient loads. Sedimentation is a threat to Lake Tana. Fecal pollution and toxigenic cyanobacteria are detected in the lake especially in the shores and river mouths. Although the current dataset on water and Lake bottom sediment characteristics and quality impacts is spatially and temporally limited, the available information indicates occurrence of potential anthropogenic pollution mainly on river mouths and shore areas. Impairment of water quality has been going on for years, which has significantly affected the basin's potential for agriculture, industry, hydroelectric, ecosystem, water supply and recreation sectors. Sedimentation, eutrophication, fecal pollution, wetland encroachment and hydrological alterations have been identified as main issues of water quality

G. Goshu (✉)
Blue Nile Water Institute and College of Agriculture and Environmental Sciences, Bahir Dar University, P.O. Box 1701(private) or 79, Bahir Dar, Ethiopia
e-mail: gorawha@gmail.com; goraw.yemer@wur.nl

G. Goshu · A.A. Koelmans
Aquatic Ecology and Water Quality Management, Water Systems and Global Change, Wageningen UR, Wageningen, The Netherlands
e-mail: bart.koelmans@wur.nl

J.J.M. de Klein
Aquatic Ecology and Water Quality Management Group, Wageningen University, Wageningen, The Netherlands
e-mail: jeroen.deklein@wur.nl

K. Stave et al. (eds.), *Social and Ecological System Dynamics*,
AESS Interdisciplinary Environmental Studies and Sciences Series,
DOI 10.1007/978-3-319-45755-0_10

management in the Lake Tana basin. Efficient monitoring programs based on a practical selection of robust water quality indicators are recommended for the basin.

Keywords Water characteristics · Eutrophication · Sedimentation · Fecal pollution · Bacteriological characteristics · Chemical characteristics · Tana basin

10.1 Introduction

The surface water potential of Ethiopia is estimated to encompass about 7400 km^2 of lake area and a total river length of about 7000 km (Wood and Talling 1988). The storage capacity of the lakes is in excess of 88 billion cubic meters with Lake Tana basin comprising more than 52% of the total surface area or 33% of the total volume. Tana basin is the second largest sub-basin of the Blue Nile.

Lake Tana has many unique characteristics. It is the source of the Blue Nile, a home of unique fish species flock, feeding and breeding ground for birds and it is the home of century old island monasteries dating back to the 13th century with a unique cultural heritage.

Given the importance of Lake Tana, its tributary rivers and ground water resources to the region's and nation's economy, ecosystems, human health and livelihoods, understanding its water quality across spatiotemporal scales for selected indicators is very critical for sustainable development.

Water quality is a dynamic balance of physical, chemical, biological and hydrological characteristics and processes occurring in an aquatic system that can be defined in terms of certain user functions (Koelmans et al. 2001). Water quality indicators are useful tools for monitoring the ecosystem integrity and as an indicator for the appropriateness of the water for the intended uses.

Scientific information about water quality of Ethiopian lakes including Lake Tana is generally rare and the available ones are of expedition origin. Chemical and related biological features of Ethiopian water bodies have been reported in the literature over more than fifty years but the studies have overall not been systematic nor sustained. Relatively, a suite of water quality studies has been conducted in South Ethiopia where the physico-chemical characteristics of the Rift Valley Lakes have been studied systematically and in much detail (e.g. Kebede et al. 1992; Zinabu et al. 2002). The physical, chemical and biological characteristics of most water bodies in Lake Tana basin however, are not characterized systematically.

Most of our current understanding of water characteristics in the Lake Tana basin comes from irregular monitoring and short term biological and hydrological expeditions conducted in the recent decade. Therefore, it is not possible to unambiguously assess whether the magnitude and direction of changes in the water bodies are due to natural or anthropogenic sources. Consequently, it is difficult to predict the possible consequences.

These water resources of Lake Tana are not utilized sustainably for the socioeconomic development of the region in particular and the nation at large.

The development intervention outweighs the environmental protection and management, which leads to resource degradation (Teshale et al. 2002; Tadesse et al. 2010). The sustainable management of Lake Tana basin water quality is challenged by easily perceivable extensive catchment degradation, sediment load, competing development needs, soil fertility losses and declining land productivity, disappearance of wetlands, drying of streams in the upper slopes and seasonal flooding of the plains. These problems are exacerbated by population pressure in the basin.

Lake Tana used to be an oligotrophic lake (Wondie et al. 2007; Teshale et al. 2002; Wudneh 1998; Nagelkerke 1997) but its trophic status has changed gradually. Especially river mouths have experienced seasonal eutrophication (Goraw 2012). This is caused by increased concentrations of phosphorous and nitrogen from hotels, recreation centers, and nonpoint sources with negative consequences on public and ecosystem health as well as on various sectors of the economy. Microbial contamination is also a factor for public and ecosystem health risks and the recurrent water related disease outbreaks in the basin are most likely to be caused by such problems.

This review is the first to consolidate different data and information related to water quality and to bring forward relevant development, research and policy intervention ideas for a sound management of water quality and water resources in the Lake Tana basin.

10.2 Morphometric Characteristics of Lake Tana Basin

Lake Tana is the largest water body (ca. 3094 km^2) in Ethiopia (Table 10.1). It is located at 12°N, 37°15′E, and 1830 m altitude in the north-western highlands of Ethiopia. The Lake Tana catchment area covers 16,111 km^2 (Table 10.1). The distance from north to south is approximately 84 km and east to west is 66 km. It is a shallow lake with a mean depth of 8 m and a maximum depth of 14 m (Table 10.1). More than seven large, permanent rivers and approximately 40 small seasonal rivers feed the lake. The main tributaries to the lake are the Gilgel Abbay

Table 10.1 Morphometric characteristics of Lake Tana and or Tana Basin (after various sources)

Morphometric characteristics	Values	Reference
Maximum depth (m)	14	Vijverberg et al. (2009)
Mean depth (m)	9	
Latitude Longitude	12°N, 37°15′E	"
Altitude (m)	1800	"
Lake area (AL) (km^2)	3111	Wubneh and Goraw (2013)
Catchment area (Ac) (km^2)	15,114	"
Lake volume (VL) (km^3)	28.4	Seifu et al. (2005)
Water residence time (years)	3	"
Runoff coefficient (k)	0.22	"

(Little Nile River), Megech River, Gumara River and the Rib River. Together they contribute more than 95% of the total annual inflow (Lamb et al. 2007). The Blue Nile is the only out flowing river.

Lake Tana's bottom substrate is volcanic basalt mostly, covered with a muddy substratum with a low organic matter content of 1% in 1994 (Howell and Allan 1994) and reached to 14% in 2011 (Goraw 2011a). The lake has high silt concentrations with a loading rate of 8.96–14.84 M tons of soil per year (Yitaferu 2007) and the trophic status is oligotrophic to mesotrophic (Ilona et al. 2011; Wondie et al. 2007; Teshale et al. 2002; Wudneh 1998; Nagelkerke 1997). The Lake Tana area has warm temperatures and the mean annual rainfall is about 1564 mm, of which 59% falls in the months of July and August, when the rainfall can be 444–483 mm per month. The seasonal rains cause the lake level to fluctuate regularly with an average difference between the minimum, in May–June, and maximum in September–October of about 1.5 m before the Tana beles hydroelectric power plant starts operation.

The land cover of the Lake Tana catchment in 2013 includes farmland (32.23%), waterbody (19.57%), built-up area (18.6%), grassland (11.3%), forestland (7.3%), wetland (4.26%), shrub land (3.66%), woodland (2.1%) and plantation forest with natural trees (0.87%) (Wubneh and Goraw 2013 Land use/cover change in Lake Tana basin, unpublished).

The lake is believed to be created by basalt outflow in the Pleistocene cutting of the basin at the southern extreme of the lake at Chara-Chara in Bahir Dar. In other words it is thought to have originated by volcanic blocking of the Blue Nile River two million years ago (Mohr 1962 in Vijverberg et al. 2009). Geologic evidence shows that these volcanic activities resulted from tectonic movements. Lake Tana is formed by damming of Abbay River by 50 km long quaternary volcanic (Aden Volcanoes) basalt flow (Chorowicz et al. 1998).

The soils of islands, peninsulas and surrounding wetlands and dry uplands of the lake are thought to be dominated by Nitosols, Luvisols and Vertisols, but require further investigation (Mekonnen 2011).

The permanent and seasonal rivers and streams flow down in different geologic landscapes contributing to different habitat structures. All these structures support diverse riverine organisms, especially serving as the endemic fish labeobarbs spawning grounds. The rivers are also essential for the development of agriculture, transportation, and fishing, especially from the perspective of tourism, as they are rich in different tourist attraction resources and they are essential for various tourism activities.

10.3 Water Characteristics

10.3.1 Physico–Chemical Characteristics of Major Tributary Rivers

The physico-chemical characteristics of major tributaries of Lake Tana are summarized in Table 10.2.

The physico-chemical characteristics of major tributaries of Lake Tana are summarized in Table 10.2.

There is limited data on spatially and temporally variability of the physico-chemical water characteristics, hence it is not possible to describe this in detail. Nonetheless, the available pooled data showed that there is significant pH, conductivity, total dissolved solids, dissolved oxygen and temperature differences among the Lake Tana basin rivers ($P < 0.05$, $n = 3$–119, Kruskal-Wallis H test). The above differences are more attributed to the differences in the background geology. The pH, electrical conductivity, total dissolved solids, dissolved oxygen and temperature ranges from 6.43–8.93;10 to 1000 μS cm^{-1};5 to 490 mgL^{-1};1.1 to 8.6 mgL^{-1} and 11.7 to 28.6 °C respectively (Table 10.2). The pH of most natural waters can range from 6.0 to 8.5, although lower values can occur in dilute waters high in organic content, and higher values in eutrophic waters, groundwater brines and salt lakes (APHA 1995). The measured pH values in the rivers are in the normal range but it does not mean that the rivers are free from contaminants. Concentrations of especially nitrogen and phosphorous are high and the turbidity is also very high. The inherent self-purification capacity of rivers should be also taken into account.

10.3.2 Physico-Chemical Characteristics of Hand Dug Wells and Protected Pumps

The physico-chemical water characteristics of traditional hand dug wells and protected hand pumps in Lake Tana basin is presented in Table 10.3. The examined water bodies are significantly different in temperature, conductivity, pH and total hardness ($P < 0.05$, $n = 40$, Kruskal-Wallis H test). The conductivity and total hardness of two water points F3 and F4 were significantly different from other sampling sites ($P < 0.05$, $n = 40$, Kruskal-Wallis H test). The pH of F4 was significantly different from the traditional hand dug well sites except H2. There was also significant difference among traditional hand dug wells in terms of conductivity, pH, total hardness and total dissolved solids. Within protected hand pumps, sites were significantly different from each other in terms of conductivity, pH, total suspended solids, total hardness and total dissolved solids ($P < 0.05$, $n = 39$, Kruskal-Wallis H test).

Generally, all traditional hand dug wells and F3, F4 water points from protected hand pumps showed significantly distinct patterns in terms of total hardness and conductivity and pH to a lower extent. This pattern was caused by different levels of anthropogenic influence on the sampling sites, which correlated excellently with the different kinds of usage and pollution of the different water sampling sites. It was possible to see a range of differing habitats, land use and pollution pattern in the investigated water systems.

Table 10.2 Physico-chemical characteristics of major tributary rivers of Lake Tana, analysis from different water samples collected during various times in the year 2010–2015

	G/Abay				Gumara				Rib			
	Min	Max	SD	N	Min	Max	SD	N	Min	Max	SD	N
NO_3–N (mgL^{-1})	0.027	9.0	2.32	14	0.0	1.45	0.55	10	0.08	1.55	0.56	8
NO_2–N (mgL^{-1})	0.0	0.9	0.26	18	0.0	11	3.01	13	0.00	11	3.42	10
NH_3–N (mgL^{-1})	0.04	6.60	2.34	13	0.02	3.3	1.38	6	0.02	1.40	0.77	3
PO_4–P (mgL^{-1})	0.080	20.6	7.26	19	0.03	12	3.46	13	0.06	9.70	3.44	10
Sulphate (mgL^{-1})	5	320.0	83.06	18	5.00	170	48.59	13	5.00	300	93.85	10
Sulphide (mgL^{-1})	0.0	22.0	6.27	12	0.02	12	4.18	8	0.02	7	2.76	6
Fe^{2+} (mgL^{-1})	0.0	11.0	4.35	17	0.00	6.5	2.00	10	0.00	5.80	2.08	7
Total hardness (mgL^{-1})	60	1080.0	286.27	19	22.00	500	182.41	13	47.00	1350	391.67	10
Alkalinity (mgL^{-1} Ca CO_3)	35	1080.0	250.89	18	47.00	520	162.12	12	73.00	675	209.22	9
Temperature (°C)	13.0	25.8	3.23	35	11.70	27.6	4.87	18	12.10	28.60	5.61	13
pH	6.63	8.8	0.59	35	7.40	8.83	0.39	18	7.12	8.87	0.49	13
Dissolved Oxygen (mgL^{-1})	1.1	5.1	1.31	21	3.11	7.90	1.58	11	3.26	8.60	2.05	7
Electrical Conductivity (μScm^{-1})	12	284.0	72.15	34	10.0	280	75.97	20	11	490	153.92	14
Total Dissolved Solids (mgL^{-1})	0	180.0	46.50	39	5.0	180	46.83	20	6	240	75.83	14
Salinity (ppt)	0.0	0.13	0.04	15	0.01	0.1	0.03	6	0.01	0.09	0.04	3
Turbidity (NTU)	5	1002.0	280.50	31	8.45	993	355.36	17	3.06	869.00	347.83	11

	Dirma				Megech				Arno				Garno			
	Min	Max	SD	N	Min	Max	SD	N	Min	Max	SD	N	Min	Max	SD	N
NO_3–N (mgL^{-1})	0.15	2.420	0.72	8	0.01	1.4	0.68	5	0.25	2.8	1.80	3	0.09	2.1	1.42	3
NO_2–N (mgL^{-1})	0.01	1.01	0.42	10	0.02	1.0	0.45	6	0.0	0.01	0.0	3	0.00	0.02	0.01	3
NH_3–N (mgL^{-1})	0.03	5.5	3.13	3	0.10	2.7	1.50	3	0.09	1.3	0.7	3	0.02	0.5	0.25	3
PO_4–P (mgL^{-1})	0.04	13.8	4.55	10	0.07	9.4	3.89	8	0.36	3.2	1.47	3	0.36	2.40	1.06	3

(continued)

Table 10.2 (continued)

	Dirma				Megech				Arno				Garno			
	Min	Max	SD	N	Min	Max	SD	N	Min	Max	SD	N	Min	Max	SD	N
Sulphate (mgL^{-1})	3	380	119.05	10	5.00	560	193.65	8	0.0	5	2.89	3	0.0	7	4.04	3
Sulphide (mgL^{-1})	0.04	5	1.97	6	0.02	8.0	3.33	5	0.0	0.7	0.49	3	0.0	0.7	0.49	3
Fe^{2+} (mgL^{-1})	0	18	7.95	7	0.00	35.2	13.98	6	0.0	1.52	0.82	3	0.0	1.05	0.54	3
Total hardness (mgL^{-1})	80	1080	377.84	10	110	600	177.23	8	102	1080	562.92	3	108	380	153.16	3
Alkalinity (mgL^{-1} Ca CO_3)	90	1730	620.42	9	132	2050	649.1	8	112	1300	679.35	3	153	800	363.86	3
Temperature (°C)	12.40	24.8	3.39	14	17.70	25.4	2.92	11	19.4	27.4	2.86	7	19.7	25.7	2.18	7
pH	6.40	8.68	0.57	14	7.20	8.93	0.57	11	7.88	8.86	0.44	7	7.81	8.51	0.28	7
Dissolved Oxygen (mgL^{-1})	2.17	7.2	1.84	7	3.73	5.9	1.16	4	3.01	4.0	0.57	4	2.96	4.1	0.55	4
Electrical Conductivity (μScm^{-1})	100	1000	240.74	14	110.00	380	100.92	10	76.0	271	61.42	7	180	633	163.35	7
Total Dissolved Solids (mgL^{-1})	40	490	118.56	14	50.00	190	51.19	11	49.0	198	45.42	7	100	411	110.17	7
Salinity (ppt)	0.05	0.18	0.08	3	0.05	0.14	0.06	3	0.08	0.1	0.01	3	0.09	0.31	0.13	3
Turbidity (NTU)	2.96	1002	411.14	11	3.89	962	286.61	11	2.45	1002	397.89	6	2.13	1002	394.85	6

Abbreviations Min Minimum, *Max* Maximum values, *SD* Standard deviation and *N* refers to number of samples

Table 10.3 Physico chemical characteristics of the examined water habitats

Water resources						
Parameters	THDS (n = 20)			PHPs (n = 20)		
	Range	MD		Range	MD	
T (°C)*	21	23.6	22.1	22.5	23.9	23.5
DO (mgL^{-1})	1.3	4.0	2.6	1.3	5.0	3.3
Con (μS/cm)*	50	347*	166	50	1516*	386
pH (–)*	5.8	6.7*	6.3	6.0	7.6*	6.7
TSS (mgL^{-1})	1	32	2.5	1	10*	4
CI (mgL^{-1})	0	50	19	0	40	10
TH (mgL^{-1})*	0	180*	80	10	700*	255
NO_3 (mgL^{-1})	0.18	7.66	2.02	0.08	8.89	1.45
NO_2 (mgL^{-1})	0.00	0.36	0.02	0.00	0.63	0.01
NH_3 (mgL^{-1})	0.0	12.00	0.42	0.0	12.0	0.54
SAK_{254} (m^{-1})	9.9	35.2	19.0	13.7	21.8	18.3
TDS (mgL^{-1})*	0.02	0.16*	0.09	0.02	0.35*	0.20

Abbreviations THD Traditional hand dug wells, *PHPS* Protected hand pumps, *MD* Median, *To* Temperature, *DO* Dissolved oxygen, *Con* Conductivity, *TSS* Total suspended solids, *Cl* Chloride, *TH* Total hardness, NO_3 Nitrate, NO_2 Nitrite, NH_3 Ammonia, *TDS* Total dissolved solids, SAK_{254} Specific absorption at 254 nm

*Significant difference ($P < 0.05$, n = 20–40, Kruskal—Wallis H test)

Source Goraw (2007)

10.3.3 Physico-Chemical Water Characteristics of Lake Tana

The physico - chemical characteristics of Lake Tana are summarized in Table 10.4. Lake Tana has relatively low water temperatures, varying only within small limits (range: 20.8–28.6 °C) (Table 10.4).The conductivity of Lake Tana ranges from 100–1000 μS cm^{-1}. Higher values of electrical conductivity were reported in river mouths and shore areas in adjacent towns. The conductivity of most freshwaters ranges from 10–1000 μS cm^{-1}(APHA 1995), but may exceed 1000 μS cm^{-1}, especially in polluted waters, or those receiving large quantities of land run-off (APHA 1995). The pH of Lake Tana ranges from 6.8–8.3, which is common for most natural waters. The pH of most natural waters is between 6.0 and 8.5, although lower values can occur in dilute waters high in organic content, and higher values in eutrophic waters, groundwater brines and salt lakes (APHA 1995). The dissolved oxygen concentration in Lake Tana ranges from 3–7.6 mg L^{-1}.

The trophic status of Lake Tana is described as oligotrophic (2.6–8.5 mg m^{-3} chl-a) having low nutrient concentrations (Vijverberg et al. 2009; Wondie et al. 2007; Dejen et al. 2004; Degraaf 2003). However, in rivers, river mouths and shore areas of the lake, increasing concentrations of chlorophyll-a (maximum of 50.46 μg L^{-1}) were measured from time to time, indicating an increase of trophic

Table 10.4 Some physico-chemical features of Lake Tana (after various sources)

Parameter	Mean/range	Reference
Rainfall (mm yr^{-1})	1326	SMEC (2008)
Annual evaporation (mm)	1675	"
Temperature (°C)	20.8–28.6	Goraw (2011a, b)
Turbidity (NTU)	11.2–125	"
Conductivity (μS cm^{-1})	100–1000	"
Total dissolved solids (mgL^{-1})	148–178	"
pH	6.8–8.3	"
Dissolved Oxygen (mgL^{-1})	3–7.6	"
Ca^{2+}(mgL^{-1})	14–15	"
Mg^{2+} (mgL^{-1})	12–17	"
Fe^{2+}(mgL^{-1})	2.2	"
Total hardness (mgL^{-1})	22–390	"
Alkalinity (mgL^{-1}Ca CO_3)	35–440	"
Color (pt COU)	64–1140	"
NO_3–N (mgL^{-1})	n.d.–3.66	"
NO_2–N (mgL^{-1})	n.d.–0.366	"
NH_3–N (mgL^{-1})	n.d.–12	"
SAK $_{254}$ (m^{-1})	8.2–48	"
PO_4-P (mgL-1)	0.1–9.1	"
Transparency/secchi disk (cm)	31–182	Nagelkerke (1997)
Na^{+1} (mgL^{-1})	7–9	Rzóska (1976)
Sulphide (mgL^{-1})	0.3	"
Sulphate (mgL^{-1})	16.3	"
Silicate (mgL^{-1})	0.93	"
Chlorophyll a (μg L^{-1})	2.6–50.46	Wondie et al. (2007), Dejen et al. (2004), Ilona et al. (2011)
BOD_5 20 (mgL^{-1})	8.5–226.3	Tenagne (2009)
TN (mgL^{-1})	41	"
TP (mgL^{-1})	1	"
Pb (ngL^{-1})	0.04–42.6	Hirut (2014)
Cd (ng l^{-1})	2.0–19.8	"
Cr (ng l^{-1})	11–18	"
Microcystins (μg L^{-1})	n.d.–2.65	Ilona et al. (2011)*
Cyanobacteria biomass (mgL^{-1})	188.18	"

*The cyanobacteria biomass and microcystin concentration measurements were taken only in post rainy season in November and was only from limited samples

status (Ilona et al. 2011; Goraw, 2011b; Vijverberg et al. 2009). The increased algal biomass in the major tributary rivers and river mouths was clearly the result of increasing inputs of nitrate and phosphate from the catchment. A similar pattern was also observed in Rift Valley lakes (Zinabu et al. 2002). In Lake Tana, the nitrate concentration ranges from n.d. to 3.66 mg L^{-1} and reactive phosphorous from 0.1–9 mg L^{-1}. The transparency of Lake Tana is low due to the high silt load of the inflowing rivers during the rainy seasons (May to October), and due to daily re-suspension of sediments in the inshore zone.

Although our data is spatially limited to the Gulf of Lake Tana, Biological Oxygen Demand on which incubation was done for five days at 20 °C (BOD_5 20) and Specific absorbance coefficient at 254 nm wave length ($SAK_{254\ nm}$) measurements are indicative for organic pollution in the gulf. BOD_5 20 ranges from 8.5–226.3 mg $L^{-1}$and SAK_{254nm} from 8.2–48 (m^{-1}). Data on heavy metal concentrations was based on graphite furnace atomic absorption spectrometry (AAS) and total aqueous Pb concentrations ranged from 0.04–42.6 (ng L^{-1}); Cd from 2–19.8 (ng L^{-1}) and Cr from 11–18 (ng L^{-1}) (Hirut 2014).

The concentration of toxigenic bacteria (i.e. microcystins) is available only from Ilona et al. (2011) and the concentration ranged from n.d. (nondetected) to 2.65 (µg L^{-1}) whereas cyanobacteria biomass density was assessed as 188.18 mg L^{-1}.

10.3.4 Bacteriological Characteristics of Lake Tana

The bacteriological characteristics of Lake Tana are summarized in Table 10.5 and Fig. 10.1. The bacteriological parameters that have been researched include total coliforms, faecal coliforms, and *E. coli*. Goraw (2011a, b) and Yemenu (2005) reported faecal pollution levels that were significantly increased and clearly discernible in the Bahir Dar Gulf locations, as compared to presumptively anthropogenic uninfluenced reference locations near the outlet of the Blue Nile River of Lake Tana (Fig. 10.1).

The minimum, maximum, median and percentage occurrence of microbial parameters at the investigated sites for pooled data are shown in Table 10.5. Total coliforms (TC), fecal coliforms (FC), *Escherichia coli* (EC) and *Clostridium perfringens* (CP) were detected in 100, 86, 82 and 90% of all sampling sites analyzed throughout the sampling period, respectively. The pooled median value of Standard heterotrophic plate count (HPC) was log 2.3 CFU per ml (Fig. 10.1).

Table 10.5 Levels of fecal indicators (log CFU/ 100 ml), HPC (log CFU/ 1 ml)

Statistic	TC	FC	*E. coli*	CP	HPC
Maximum	6.3	6.2	6.1	4	4
Minimum	2.4	n.d.	n.d.	n.d.	1.1
Median	3.1	1.4	1.3	1.6	2.3
% occurrence	100	86	82	90	100

n.d. refers to non-detectable (n = 22). *TC* Total coliforms, *FC* Fecal coliforms, *EC* Escherichia coli, *CP* Clostridium perfringens spores in log (CFU/100 ml) and *HPC* standard heterotrophic plate count in log (CFU/1 ml) for sampling sites

Table 10.6 GPS reading of water sampling sites with a description of locations for bacteriological characterization

Site	Northing 00° 00′ 00″	Easting 00° 00′ 00″	Relative location
L1	11°37.1′66″	37°21.8′ 34″	Near Medhalelem Orphanage
L2	11°36′37.1″	37°22′24.7″	Felege Hiwot Referral Hospital
L3	11°36′37.3″	37°22′40.2″	Fish Production & Marketing Enterprise
L4	11°35′56.4″	37°23′4″	Hidar 11 Recreational Center
L5	11°35′47.7″	37°2322.5″	Near Mango Recreational Center
L6	11°36′21.6″	37°24′5.2″	Near Regional Prison
L7	11°37′28.7″	37°24′5.2″	Near Blue Nile River outlet
L8	11°38′05″	37°24′25.5″	Near Blue Nile river outlet

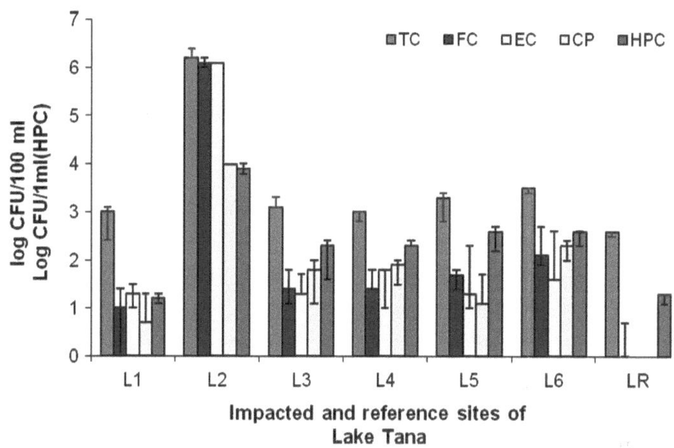

Fig. 10.1 Abundance of total coliforms (TC), fecal coliforms (FC), Escherichia coli (EC), Clostridium perfringens spores (CP) in log (CFU/100 ml) and standard heterotrophic plate count (HPC) in log (CFU/1 ml) for sampling sites. The error bars indicate the maximum, minimum and median concentration. For description of abbreviation of sampling sites please refer Table 10.6. *Source* Goraw et al. (2010)

10.3.5 Physico–Chemical Characteristics of Lake Tana Bottom Sediment

The physico-chemical characteristics of Lake Tana sediment sampled in the period 2010/2011 are summarized in Table 10.7. The physico-chemical characteristics of Lake Tana bottom sediment were described from 97 sediment samples collected from 33 sites all over Lake Tana at different times during 2010, 2011, 2013. The sites are distributed over different stations representing the lake shore (adjacent to cites/towns including anthropogenic influenced ones), pelagic and river mouths.

Table 10.7 The Physico—chemical characteristics of the Lake Tana Sediment sampled from March 2010–2013, number of sample, minimum, maximum, mean and standard error of the mean

Parameter	N	Min	Max	Mean	Std. error
pH	97	5.74	7.96	6.81	0.051
Conductivity ($\mu S\ cm^{-1}$)	97	0.04	1.45	0.22	0.022
TN (%)	96	0.05	3.51	0.38	0.060
OM (%)	97	0.07	14.05	3.87	0.310
AVP (mg Kg^{-1})	97	1.07	80.64	20.98	1.45
OC (%)	40	0.65	4.87	2.05	0.16

TN refers total nitrogen, *OM* Organic matter, *OC* Organic carbon and *Av* p-refers available phosphorous

The pH of the Lake Tana bottom sediment ranges from slightly acidic to slightly alkaline. The minimum was noted in Bahir Dar gulf of Lake Tana where the influence from anthropogenic activities should been very high. Especially, a sewerage line from Kebele 03 all the way down to Lake Tana is inflowing into the Bahir Dar gulf area and Tana Fish production and marketing enterprise also disposes a lot of fish offal into the Lake. The maximum pH was noted in Delgi area.

The electrical conductivity of Lake Tana bottom sediment ranges from 40 to 1450 ($\mu S\ cm^{-1}$). The minimum was noted in Delgi and the maximum was noted in Bahir Dar gulf area.

The total nitrogen content of Lake Tana bottom sediment ranges from 0.02 to 3.5% and the minimum was noted in the pelagic part of Lake tana and the maximum was noted in Gorgora shore area.

Low values of organic matter was noted in Korata area (0.07%) and maximum value of organic matter was noted in Bahir Dar gulf close to the dense stands of wetland vegetation (14.05%). The minimum values of Lake bottom sediment available phosphorous was noted in Bahir Dar gulf near mango recreational center (1.07 ppm). On the other hand the maximum available phosphorous (80.64 ppm) was recorded in Bahir Dar gulf where fish offal has been dumbed. The minimum % organic carbon was noted in Delgi area (0.65%) and the maximum was noted in Bahir Dar gulf area (4.87%).

The physico-chemical characteristics of the Lake Tana bottom sediment clearly indicate Lake Tana has been influenced significantly by anthropogenic activities of adjacent cities.

10.4 Conclusions

The water quality indicators are important to assess the suitability of the water for the intended water uses. Understanding the water characteristics is also vital input for water resources development and environmental protection plans and management. Nevertheless, to date there is no clear and comprehensive water quality assessment available for Lake Tana and its environs. The data is not systematically organized. The water characteristics in the Lake basin are affected by anthropogenic and natural processes but mainly by human activities from point and diffuse sources.

Although only a limited data set in terms of time and space has been established, it is clear that anthropogenic activities are significantly influencing water quality in and around Lake Tana. This was evidently shown by a set of multiple fecal indicators, organic pollution by saprobity determination using HPC and SAK_{254nm} measurement and microcystin concentration measurement and measurement of other chemical and physical water quality and pollution indicators.

Sedimentation, eutrophication, fecal pollution, wetland encroachment and hydrological alterations have been identified as main issues of water quality management of the Lake Tana basin.

Present results urgently call for further research and continuous data generation concerning the whole lake area and the basin. As a first immediate step in establishing good knowledge on the general situation of the Lake and basin water characteristics, simple monitoring programs based on a good and practical selection of robust physical, biological, hydrological indictors and basic chemical parameters are urgently needed.

10.5 Recommendations

As basin water resources are of crucial importance—serving multiple purposes and being the largest fresh water body in Ethiopia—the following recommendations are given for further work and sustainable development:

- Quantify contaminants load to Lake Tana from the catchment and bottom sediment
- Quantify and characterize organic contaminants from large scale floriculture and agriculture farms
- More data on physical, chemical, biological and microbiological characteristics of the surface and ground water
- Identifying point and diffuse pollution sources and erosion hotspots
- Development of decision support tools for lake ecosystem management
- Determination of phosphorous and nitrogen critical loads
- Evaluation of technical and socioeconomic competitiveness of different waste water treatment and quality management technologies
- Conduct microbial risk assessment
- Conduct pesticide risk assessment
- As a first and immediate step in establishing good knowledge on the general situation of the water quality in the basin, simple monitoring programs based on a good and practical selection of robust physical, bacteriological, biological indictors and basic chemical parameters are urgently needed. The programs should account for the spatial and temporal variations.

References

American Public Health Association (1995) Standard Methods for the Examination of Water and Wastewater, 19th edn. Am Publ Health Assoc, Washington

Chorowicz J, Collet B, Bonavia F et al (1998) The Tana basin, Ethiopia: intra-plateau uplift, rifting and subsidence. Tectonophysics 295:351–367

Degraaf M (2003) Lake Tana's Piscivorous Barbs (Cyprinid, Ethiopia). PhD thesis, Wageningen University

Dejen E, Vijverberg K, Nagelkerke L et al (2004) Temporal and spatial distribution of micro crustacean zooplankton in relation to turbidity and other environmental factors in a large tropical lake (L. Tana, Ethiopia). Hydrobiologia 513:39–49

Goraw G (2007) Evaluation of microbial fecal Indicators and quantifying the respective level of pollution in ground and surface water of Bahir Dar and peri-urban areas, Ethiopia. MSc thesis, UNESCO-IHE

Goraw G et al (2010) A pilot study on anthropogenic faecal pollution impact in Bahir Dar Gulf of Lake Tana, Northern Ethiopia. Ecohydrol Hydrobiol 10(2):271–279

Goraw G (2011a) Annual report on Lake Tana water quality base line monitoring. Tana Beles Integrated Water Resources Development Project, Bahir Dar

Goraw G (2011b) Challenges and opportunities of water quality management in tana sub basin. In: Goraw G, Muluken L (eds).1st National awareness creation workshop, Bahir Dar, October 2011. Anthropogenic fecal and chemical pollution impact on ecosystem and public health, Bahir Dar Gulf of Lake Tana, pp 1–15

Goraw G (2012) Wetlands for sustainable development and climate change mitigation. In: Seyoum M, Tadesse F (eds) 2nd National workshop, Bahir Dar, January 2012. Water quality deterioration as potential enabling environment for Proliferation of floating water hyacinth (*E. crassipes*) in NW-river mouths of Lake Tana, pp 162–190

Hirut G (2014) Proximate analysis & Determination of heavy metals from water & common fish species in Lake Tana. MSc thesis, Addis Ababa University

Howell PP, Allan P (1994) The nile: sharing a scarce resource. Cambridge University Press,

Ilona G, Goraw G, Katarzyna k et al (2011) Impacts of climate change and population on tropical aquatic resources. In: Brook L, and Abebe G (eds). The 3rd ethiopian fisheries and aquatic science conference, Haramya University, Haramya, 5–6 February 2011. Detection of toxigenic cyanobacteria in Bahir Dar Gulf of Lake Tana"—pilot study, pp 271–284

Kebede E, GetachewT Taylor WD et al (1992) Eutrophication of Lake Hayk in the Ethiopian highlands. J Plankton Res 4:1473–1482

Koelmans AA, Van der Heijde A, Knijff L et al (2001) Modelling feedbacks between Eutrophication and organic contaminant fate & effects in aquatic ecosystems. Rev Water Res 35:3517–3536

Lamb HF, Bates CR, Coombes PV et al (2007) Late Pleistocene desiccation of Lake Tana, source of the Blue Nile. Quat Res 26:287–299

Mekonnen G (2011) Challenges and opportunities of water quality management in tana sub basin. In: Goraw G, Muluken L (eds) 1st National awareness creation workshop, Bahir Dar, October 2011. Evaluation of Chemical Residues on Vegetable Crops through Wastewater Irrigation, Amhara Region (Bahir Dar & Kombolcha), pp 32–47

Nagelkerke L (1997) The barbus of Lake Tana, Ethiopia. Morphological diversity and its implications for taxonomy, trophic resources portioning, and fisheries. PhD thesis, University of Wageningen

Rzóska J (1976) Lake Tana, headwaters of the Blue Nile. In: Rzóska J (ed) The Nile, Biology of an Ancient River. Junk, The Hague, pp 223–232

Seifu K, Travia Y, Tenalem A et al (2005) Water balance of Lake Tana and its sensitivity to fluctuations in rainfall, Blue Nile basin. Ethiopia J Hydrol 316(2006):233–247

SMEC (2008) Hydrological study of the Tana-Beles sub basins, main report. Ministry of Water Resource, Addis Ababa

Tadesse A, Matthew M, Seifu K (2010) The water resource implications of planned development in the Lake Tana catchment. Ethiop Eco-hydrology Hydrobiology 10(2–4):211–222
Tenagne A (2009) The impact of urban storm water runoff and domestic waste effluent on water quality of lake tana and local groundwater near the city of Bahir Dar, Ethiopia. MSc thesis, Cornell University
Teshale B, Lee C, Girma Z (2002) Development Initiatives and Challenges for Sustainable Resource Management and Livelihood in the Lake Tana Region of Northern Ethiopia International. J Technol Manage Sustain Dev 1(2):111–124
Vijverberg K, Sibbing FA, Dejen E (2009) Lake Tana: Source of the Blue Nile. Chapter In: Dumont HJ (ed), The Nile, Origin, Environments, Limnology and Human Use, Springer, pp 163–191
Wondie A, Mengistu S, Vijverberg K, Dejen E (2007) Seasonal variation in primary production of a large high altitude tropical lake (Lake Tana, Ethiopia): effects of nutrient availability and water transparency. Aquatic Ecol 41:195–207
Wood RB, Talling JF (1988) Chemical and algal relationships in a salinity series of Ethiopian inland waters. Hydrobiologia 158:29–67
Wudneh T (1998) Biology and Management of Fish stocks in Bahir Dar gulf, Lake Tana, Ethiopia. PhD thesis, Wageningen University
Yemenu A (2005) Characterization of domestic waste water disposal as point source pollution in southern Gulf of Lake Tana, North-western Ethiopia. MSc thesis, Addis Ababa University
Yitaferu B (2007) Land Degradation and Options for Sustainable Land Management in the Lake Tana Basin (LTB), Amhara Region, Ethiopia. PhD thesis, University of Bern
Zinabu GM, Kebede E, Desta Z (2002) Long-term changes in chemical features of waters of seven Ethiopian Rift-valley lakes: chemical characteristics along a salinity-alkalinity gradient. Hydrobiologia 288:1–12

Chapter 11
Plankton of Lake Tana

Ayalew Wondie and Seyoum Mengistu

Abstract This review paper presents the compositions, distribution, abundance, biomass and production of plankton communities in Lake Tana from various research works. A total of 85 phytoplankton species, consisting of 15 blue greens, 37 diatoms, 28 greens, and 5 other minor groups are recorded. The phytoplankton biomass showed a certain decline when turbidity increased during the rainy season. *Microcystis* sp. was found to dominate the phytoplankton community of Lake Tana throughout the rainy and post-rainy seasons. The chlorophyll *a* concentration in the water column of the lake varied between 0.03 and 13.44 μg L^{-1} and showed marked seasonal variations. The highest chlorophyll concentrations were observed in the post-rainy season. The average gross primary production in Lake Tana (mean: 1.8 g O_2 m^{-2} d^{-1}) is relatively lower compared to African lakes. A number of studies on zooplankton in Lake Tana recorded a total of 26 zooplankton species, consisting of 3 copepods, 11 cladocerans and 12 rotifers. Copepods showed usually higher densities than cladocerans and rotifers. In the case of Lake Tana, the major factors responsible for temporal and spatial change of the plankton are resource limitations and a high sediment concentration. During the productive season (PORS), the phytoplankton is dominated by large species (e.g. *Melosira* and *Microcystis* spp.), which are difficult to be consumed by the herbivorous zooplankton.

Keywords African lake · Lake Tana · Phytoplankton · Zooplankton · Seasonal dynamics

---

A. Wondie (✉)
College of Natural Science, Bahir Dar University, Bahir Dar, Ethiopia
e-mail: ayaleww2001@yahoo.com

S. Mengistu
College of Natural Sciences, Addis Ababa University, Addis Ababa, Ethiopia

K. Stave et al. (eds.), *Social and Ecological System Dynamics*,
AESS Interdisciplinary Environmental Studies and Sciences Series,
DOI 10.1007/978-3-319-45755-0_11

11.1 Introduction

The study of plankton production in lakes is fundamental to understand both water quality and fisheries. Plankton plays a central role in transformation of energy to the higher trophic levels. An increased phytoplankton production generally results in higher zooplankton and higher fish production.

Although most of the aquatic information on Ethiopian lakes, originates from studies of central rift Valley, A number of limnological investigations (Dejen 2003; Wondie 2006; 2012; Goshu 2010) had been carried out in Lake Tana almost for the last 10 years.

In some large lakes and reservoirs in Africa and S.E. Asia, small pelagic zooplanktivores fish play a key role in the food web (Dejen et al. 2009) and contribute considerably to artisanal fisheries. Therefore zooplankton can constitute an important link in the food chain between the primary producers and zooplanktivores. Only a few studies were carried out on plankton diversity, spatial and temporal distribution, biomass and productivity (Wood and Talling 1988; Wudneh 1998; Dejen et al. 2004, Wondie et al. 2007).

Although Lake Tana is an oligotrophic lake from analysis of nutrient concentration and phytoplankton productivity, their zooplankton biomass is relatively high. Turbidity is the most regulating factor for structuring and functioning of the plankton community in Lake Tana (Wondie et al. 2007). Previous studies observed that the phytoplankton is dominated by diatoms and blue green algae. The most dominant blue green algae are usually *Microcystis* and the usual most frequent green algae are *Pediastrum and Straurastrum* species (Wood and Talling 1988).

Secchi-disk depth ranged from 5 to 80 cm in the inshore zone and from 10 to 100 cm in the open water zone. The highest values were recorded in February (dry season) and the lowest in July (main-rainy season). Water transparency was significantly higher in the open water zone.

Seasonally low Secchi-disc depths were recorded during the main-rainy season because of high flushing rate, inorganic silt turbidity, and high loading of dissolved organic matter from the inflowing rivers, which is like any other tropical lake. The concentrations of silicate, soluble reactive phosphate and nitrate showed pronounced seasonal variations, but the variation patterns were different. Silicate concentrations were low in the dry season, but high during the rest of the year. Phosphate concentrations were very low during early dry season (Dec.–Feb.), whereas nitrate showed a maximum during the post-rainy season and low concentration during the rest of the year.

11.2 Phytoplankton

The list of phytoplankton in Lake Tana are summarized in Table 11.1. A total of 85 phytoplankton species; consisting of 15 blue greens, 37 diatoms, 28 greens, and 5 other minor groups are recorded. Although a number of phytoplankton species

Table 11.1 Lists of Major phytoplankton species in Lake Tana

Blue green algae	Diatoms	Green algae
Aphanocapsa sp.	*Cyclotella* sp.	*Ankistrodesmus* sp.
Aphanothece sp.	*Melosira agassizi*	*Pediastrum* sp.
Microcystis sp.	*Thalassiosira fauril*	*Planctonema* sp.
Cyanodictylon sp.	*Nitzschia amphibia*	*Tetrastrum* sp.
Peudoanabaena sp.	*Synedra* sp.	*Closterium* sp.
Apanizomenon sp.	*Cylinderotheca* sp.	*Ankyra* sp.
Lyngbya sp.	*Cymbella* sp.	*Coelastrum* sp.
Mallomonas sp.	*Frgilaria virescens*	*Dictyosphaerium*
	Tabellaria fenestrata	*Volvox* sp.
	Surirella sp.	*Oocystis* sp.
	Pinnularia major	*Sphaerocystis* sp.
	Rhizosolenia eriensis	*Staurastrum* sp.
		Monoraphdium sp.
		Cryptomonas sp.
		Scendesmus sp.
		Schreuderia sp.

observed in most tropical lakes were absent here, all the species observed in Lake Tana are common to both tropical and temperate lakes (Wondie 2006).

The lake phytoplankton showed marked seasonality. During the rainy and dry seasons characterized by low phytobiomass; diatoms dominated the community until a shift in dominance from diatoms to blue greens. Dense phytoplankton population developed immediately after the rainy season, followed by succession of blue greens and diatoms (Wondie 2006). The first peak was usually attained by the expansion of species of Blue green algae. Then after the renewal of the lake water and end of the rainy season, a series of populations developed in which *Microcystis* was prominent early and *Melosira* late, represented also by a variety of green algae (e.g. *Coelastrum, Volvox, Pediastrum* spp.).

The three taxonomic groups, blue greens, diatoms and greens were dominant in the sense that they represented more than 90% of the total net phytoplankton as biovolume (Wondie 2006). The contribution of green algae never exceeded more than 20% the total phytoplankton biovolume (Fig. 11.1).

The most important contributors to the total blue greens biovolume were *Microcystis* and *Aphanothece* species (>90% in the post-rainy and late dry seasons); *Melosira* and *Synedra* species (>80% in the rainy and during the early post-rainy and dry seasons) among diatoms; and *Staustrum* and *Ankisterodesmus* species (>70% on months of dry season) among green algae (Wondie 2006).

The phytoplankton biomass showed a certain decline when turbidity increased during the rainy season. Large phytoplankton usually decline disproportionately following periodic lowering of the euphotic zone and the algae spend increasing amounts of time in the dark. The dominant influence of turbidity during the rainy season was compensated by the increased nutrient input through river inflow and lake water dilution, as well as delay in mixing in most of the open water of the lake. Similar conditions have been observed in other lakes. The onset of summer calm condition in the temperate zone has major implications for the phytoplankton

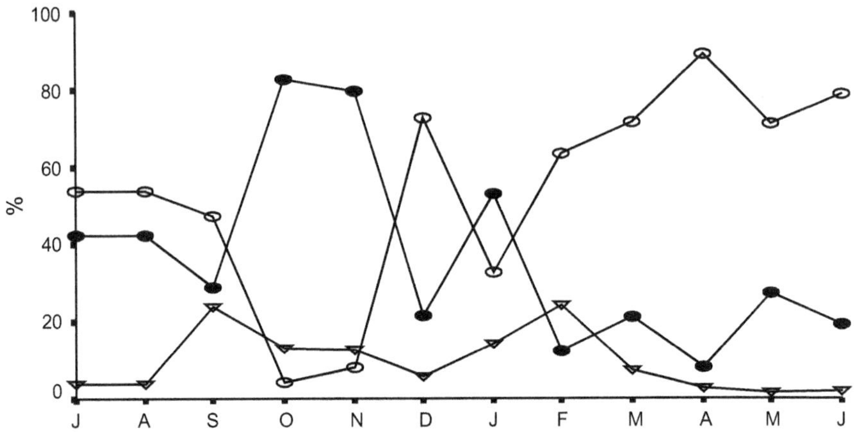

Fig. 11.1 Temporal changes in percentage contribution of blue greens (*filled circle*), diatoms (*open circle*) and greens algae (*down triangle*) to the total phytoplankton biovolume in Lake Tana (July 2003–June 2004) (Wondie 2006)

community. In addition, external inputs from catchments increase through major and minor river discharges (Lewis 1979, 1987).

The predominance of diatoms as observed in L. Tana, is characteristic for tropical oligotrophic lakes (Kalff 1978). As the dilution of the lake (increase in water level) and nutrient concentration reached their seasonal peaks, all taxa were favoured, but mainly dominated by cyanobacteria. Furthermore, the cause of post-rainy season maximum of phytoplankton is the regulation in top-down control of the algae, resulting from increased predation pressure exerted on the zooplankton, by rapidly growing juvenile fish and zooplanktivores fish (de Graaf 2003; Dejen 2003), and later both trophic groups (phytoplankton and zooplankton) drastically declined due to exhausted nutrient supply and lowered temperature in December.

Among the four periods of the seasonal cycle of the phytoplankton community in Lake Tana, diatoms dominate three. In the present study *Microcystis* sp. was found to dominate the phytoplankton community of Lake Tana throughout the rainy and post-rainy seasons—period of lake water dilution. Similar pattern were observed in three other large African lakes (Turkana, Chad and Volta), a strong hydrological influence on phytoplankton seasonality was also established. The two lakes, Turkana and Chad are enclosed basins, which are chemically polarized by inputs of nutrient-rich floodwater. *Microcystis aeruginosa* became seasonally abundant in Lake Turkana, whereas several diatoms rose to ascendancy in Lake Chad, after initial dilution (Compere 1983).

The presence of dense phytoplankton with *Microcystis* and *Melosira* spp., in both the proportion and absolute abundance increase during periods of vertical mixing and higher chlorophyll a concentrations. Unlike Lake Tanganyika where there is no well-established *Melosira* component in the plankton and Lake Malawi (Hecky and Kling 1981) where no co-dominance of diatoms and blue greens is seen, Lake Tana showed positive relationship between total diatom and total blue green algae abundance, particularly during the post-rainy season. However, the predominance of diatoms generally occurred in periods of stronger vertical mixing and that of blue green algae under relatively less mixed water column (post-rainy season). In addition, the fact that the lake is shallow and lacks marked thermal stratification has enabled it to have its own pattern of seasonality compared with other large African lakes.

The biovolume of the phytoplankton population during post-rainy season in Lake Tana is in excess of 3 orders of magnitude (1order = 10:1 ratio) between the dry and rainy seasons. Very high relative amplitudes in excess of 3 orders of magnitude are probably common in hydrologically dominated seasonality with strong washout (Rzoska et al. 1955; Reynolds 1990). Whereas, hydrological dominated seasonality with nutrient input from floodwater were observed (Wondie 2006). High nutrient (especially nitrate) content of the silty floodwater was also reported from the Blue Nile River (Rzoska and Talling 1967).

In cycles not dominated hydrologically, a wide range of amplitudes can occur (e.g. offshore water of Lake Victoria) (Talling 1987). In Lake George where the normal amplitude of phytoplankton changes within a year is small, there is an opportunity for long term persistence of similar levels of total biomass and many of its component species (Burgis 1971; Ganf 1974). In Lake Chad, inter-annual variations of phytoplankton population are large because of hydrological instability (water balance).

The chlorophyll *a* concentration in the water column of the lake varied between 0.03 and 13.44 $\mu g\ L^{-1}$ and showed marked seasonal variations. The highest chlorophyll concentrations were observed in the post-rainy season. Gross primary production and the gross primary production per unit of chlorophyll over the euphotic zone were substantially higher in the inshore zone than in the open water. This is a clear indication that nutrients are less limiting for algal growth in the inshore zone than in the offshore region and this is obviously not related to underwater light climate, since Secchi-disk depth is lower in the inshore zone. The increased gross primary production in the inshore zone is the combined effect of higher chlorophyll levels and higher phytoplankton growth rates. The first is mainly the result of the higher chlorophyll levels in the inshore zone. Nitrate availability is also higher in the inshore zone due to increased mixing and resuspension of bottom sediments.

The average, chlorophyll-specific gross primary production over the euphotic zone in Lake Tana (0.52 mg $O_2\ m^{-2}\ d^{-1}$ per mg chlorophyll m^{-2}) lies in the same range as in 30 other tropical lakes and reservoirs. However, the average gross primary production in Lake Tana (mean: 1.8 g $O_2\ m^{-2}\ d^{-1}$) is relatively lower value (Mean 10.0; range 2.6–23 g $O_2\ m^{-2}\ d^{-1}$). Since the primary production per unit of chlorophyll is in the same range as in other tropical lakes it is very unlikely that for a tropical lake relatively moderate temperatures, were limiting primary production. In general, low annual primary production rates and chlorophyll concentration in Lake Tana are primarily attributable to a combination of severe nutrient limitation (nutrient-poor catchment) and low irradiance as a result of clay and silt turbidity (Table 11.2).

Table 11.2 Time series data on zooplankton taxonomy in Lake Tana by different researchers

Brunelli and Cannicci (1940 in Wudneh 1998)	Frequency	Tesfaye Wudneh (1998)	(1nd/lit)	Eshete Dejen (2003)	Frequency (%)
Copepoda (3 species)	Dominant	**Copepoda** (3 species)	4.1	**Copepoda**	4.2
Thermodiaptomu.galebi	Rare	*Mesocyclops* sp.	11.7	*Mesocyclops aequatorialis*	15.2
Mesocyclops leuckarti	Non-planktonic	*Thermocyclops* sp.	14.5	*similis*	31.2
Cyclops albidu	Abundant	*T. galebi*	7.3	*Thermocyclops ethiopiensis*	v. rare
Cladocera (11 species)	Abundant	**Cladocera** (7 species)	5.5	*Thermodiaptomus galebi*	rare
Bosmina.longirostris	v. rare	*Diaphanosoma*	0.7	*lacustris*	13.9
Diaphanosoma excisum	frequent	*excisum*	1.8	*Microcyclops varicans*	0.3
Ceriodaphnia biscuspidata	not abundant	*B.longirostris*	0.9	**Cladocera**	1.4
C. cornuta	not abundant	*Daphnia longispina*	0.7	*Bosmina.longirostris*	v. rare
Daphnia longispina	not abundant	*D. similis*	1.9	*Ceriodaphnia cornuta*	9.1
D. lumholtzi	———	*Ceriodaphnia cornuta*	7.3	*C. dubia*	8.1
Moina dubia		*C. dubia*	2.7	*Chydorus sphaericus*	3.3
Rotifera (12 species as cited by Wudneh		*Moina* sp.	2.6	*Daphnia hyalina*	10.8
1998)		**Rotifera** (7 species)	1.5	*D. lumholtzi*	2.5
		Keratella quadrata	2.8	*Diaphanosoma. excisum*	
		Keratella crassa	1.1	*D. sarsi*	
		Brachionus falcatus	1.3	*Moina. micrura*	
		Brachionus caudatus			
		Filinia terminalis			
		Lecane sp.			
		Trichocera sp.			

11.3 Zooplankton

The major aquatic invertebrates in Lake Tana include rotifers, molluscs, crustaceans and insects. Benthic life in abundance and diversity is poor. The heterotrophic bacterioplankton and benthic community of Lake Tana have not been systematically studied so far.

A number of studies on zooplankton in Lake Tana recorded a total of 26 zooplankton species, consisting of 3 copepods, 11 cladocerans and 12 rotifers (Brunelli and Cannicci 1940 in Wudneh 1998; Dejen 2003; Wondie 2006).

Numerically *Thermocyclops* is higher from cyclopoid species. The mean density was 20.4 (range 8.04–55.2 ind L^{-1}) for thermocyclops, and 17 (range 3.0–22.9 ind L^{-1}) for cyclopoid spp. In general, copepods increased during pre-rainy season and reached peaks in post-rainy season (av. 90.64 ind L^{-1}in November). The densities decreased during the main rainy season (av. 19.0 ind L^{-1}in September) and early dry season (21.0 ind L^{-1}in February) (Fig. 11.2).

Calanoids and cyclopoid copepods are disproportionately abundant in oligotrophic waters at all latitudes (Kalff 1978). The zooplankton community in Lake Tana has shown a shift from dominance by cyclopoids (Wudneh 1998) to calanoids in recent years (Wondie 2006).

Dejen et al. (2004) noted that the size of zooplankton in Lake Tana is relatively large for a tropical lake, which are generally assumed to be unfavourable for small zooplanktivores. This condition results in low food chain efficiency and poor utilization of larger copepods, like calanoida. For example, although calanoids represent >50% of the total zooplankton density and biomass in Lake Tana, they represent only 2.7–9.6% (by biovolume) of the diet of zooplanktivores small barbs (Dejen 2003). This may be because catching calanoids is difficult as they are fast swimmers-even against water currents.

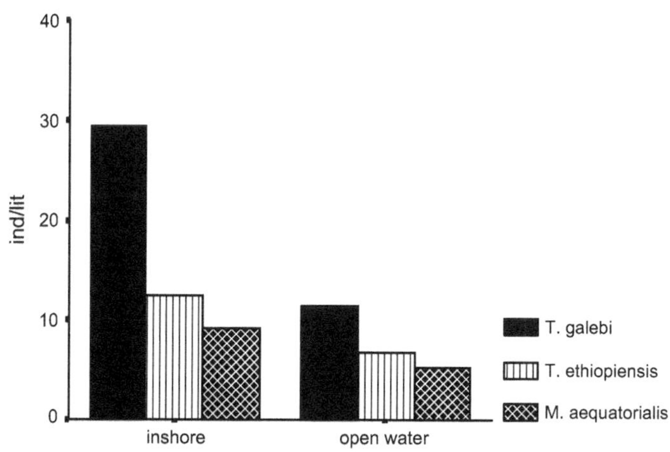

Fig. 11.2 Spatial variations of copepod species between sampling zones (Wondie 2006)

11.4 Cladoceran Species

Of the total cladocerans recorded, 35% of these were *Diaphanosoma* species followed by 32% of *Bosmina* species, 20% of *Daphnia* species, 7% of *Miona* species and 6% of *Ceriodaphnia* species. All the cladocerans together contributed about 11% of the zooplankton abundance in the lake.

Cladocerans increased during the post-rainy season at 11.13 ind L^{-1} and pre-rainy season at 19.0 ind L^{-1} except in August (at 8.8 ind L^{-1}) when *Bosmina* species is dominant. The densities decreased during the rainy-season at 3.26 ind L^{-1} in September and early dry season at 4.07 ind L^{-1} in January. *Bosmina* species were higher in August and May, at 4.07 ind L^{-1} and 5.5 ind L^{-1}, respectively. *Diaphanosoma* have peaks in May as 8.1 ind L^{-1} and *Daphnia* as 3.4 ind L^{-1} in May. The remaining species, *Moina* and *Ceriodaphnia* reached peak in October with average value of 0.57 ind L^{-1} and 0.48 ind L^{-1}, respectively.

The three dominant cladocerans (*Diaphanosoma, Daphnia* and *Bosmina* species) tend to co-occur in relatively clear water with each other (peak in May), except, *Bosmina,* which shows higher density in August.

Diaphanosoma, Daphnia and *Bosmina* genera were the most frequently recorded and most abundant (>80%) cladoceran species. *Daphnia* are considered rare in tropical regions, and detail information on species distribution and community ecology is scarce. High incidence of *Daphnia* species in this lake shows that *Daphnia* can be equally widespread in tropical regions.o

The rotifer population was dominated by two species (80%) of which 73.6% was composed of *Keratella* sp. (av. 19.5 ind L^{-1}, range 6.24–52.58 ind L^{-1}) and 26.4% of *Brachionus* sp. (av. 7.02 ind L^{-1}, range 2.15–14.3 ind L^{-1}). Rotifers contributed on average about 36.8% (av 26.52 ind L^{-1}, range 8.4–66.9 ind L^{-1}) of the zooplankton community in the lake (Fig. 11.3).

Rotifer species increased beginning from rainy season (June—av. 33.5 ind L^{-1}) up to post-rainy season (Nov—av. 66.9 ind L^{-1}). The densities decreased during the dry season (March—April—av. 9.8 in April). Total rotifers reached a peak in November, where the highest density (ca. 266.7 ind L^{-1}) was recorded at Gumietirs inshore station.

Copepods showed usually higher densities than cladocerans and rotifers. The calanoid, *T. galebi* largely determined this dominance and occasionally reached 50% or more of the total zooplankton abundance. In general, cladocerans accounted for 11% of the total zooplankton abundance in Lake Tana while copepods and rotifers contributed 52 and 37%, respectively.

11.5 Role of Phyto- and Zooplankton in the Food Web

In the case of Lake Tana, the major factors responsible for temporal and spatial change of the plankton are resource limitations and a high sediment concentration (Scheffer 1998; Wondie 2006). During the productive season (PORS), the

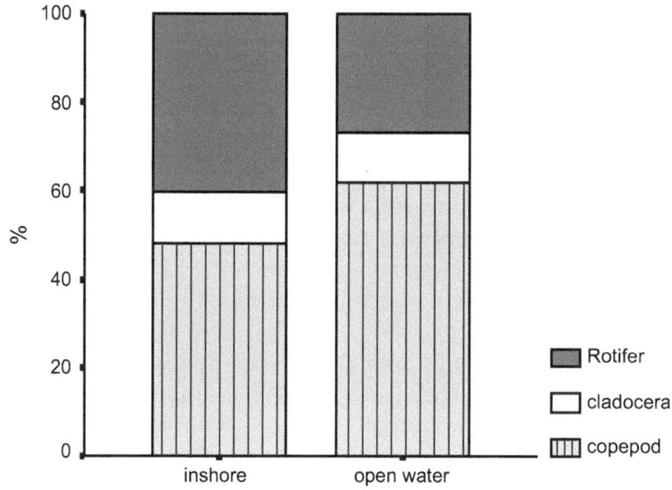

Fig. 11.3 The % contribution in mean density of the major zooplankton—Copepoda, Cadocera and Rotifera in Lake Tana during July 2003–June 2005 (Wondie 2006)

phytoplankton is dominated by large species (e.g. *Melosira* and *Microcystis* spp.), which are difficult to be consumed by the herbivorous zooplankton. However, this situation is improved during the second productive season (PRS) where small-sized diatoms and high chlorophyll-a concentration is observed. Generally, zooplankton herbivores in Lake Tana do not efficiently utilize the phytoplankton community (Wondie 2006).

Secondary production of zooplankton in Lake Tana is primarily, contributed by copepods, especially calanoids. A zooplankton community permanently dominated by copepods, either cyclopoids or calanoids, is characteristic for almost all tropical and subtropical lakes and reservoirs (e.g. Burgis 1971; Lewis 1979; Vareschi and Vareschi 1984; Hart 1987; Mavuti 1990; Amaraseinghe et al. 1997).

Seasonal dynamics of zooplankton coincided with phytoplankton abundance, water level, turbidity and temperature. Food (quality and quantity) limitation plays an important role in regulating zooplankton dynamics in the lake. In post-and pre-rainy seasons intensified production and decomposition following nutrient release and sedimentation stimulate secondary production. Although temperature did not vary significantly between seasons, there was significant reduction in the productivity during lower-temperature months. There was also effect of turbidity. Whereas zooplankton abundance was not negatively correlated with turbidity, strong positive ($r^2 = 0.58$) and significant ($P < 0.001$) relationship between phytoplankton and zooplankton production rates were observed.

As zooplankton dynamics was influenced by primary productivity, phytoplankton production rates in turn were associated mainly with nutrient concentration and availability, wind-driven turbulence, light condition, temperature and grazing pressure. Since seasonal pattern of pH, conductivity and dissolved oxygen showed relatively little variation, phytoplankton seasonality was dominated by hydrological and hydrographic changes. A combination of low temperature and increased grazing pressure from higher densities of zooplankton and fish communities of the preceding season may have accounted for the generally reduced densities of phytoplankton during the dry season.

One of the effects of turbidity, which is to limit euphotic zone, is compensated by nutrient availability and improved autotrophic and allochotonous food sources. The allochotonous dissolved nutrients washed into the lake by streams and rivers during the rainy season induce higher phytoplankton production, which in turn supports higher zooplankton densities. Such seasonal increases in zooplankton densities during the rainy periods were also reported in Lake Naivasha, Kenya and Lake George, Uganda (Mavuti and Litterick 1981), and in Lake Awassa. However, the strong inorganic turbidity at the start of the rainy season (July) and beginning of lake water dilution in the river mouths limits primary production during the rainy season. Primary production in most seasons was enhanced by wind-driven turbulence, which resulted in algal cells being subjected to different light conditions for varying lengths of time and consequent reduction of light limitation.

The post-rainy season phytoplankton peak in Lake Tana was coupled with high zooplankton densities (Fig. 11.4). Similarly, the spring phytoplankton increase in temperate lakes is followed by zooplankton development rate that greatly lags the phytoplankton growth rate during periods of low water temperature (Reynolds 1990). In Lake Tana, ciliate protozoans, rotifers and micro-crustacean zooplankton are abundant during periods of low temperature (late post-rainy and early dry seasons), but they were only capable of removing the smallest algae, thus losses of large phytoplankton due to herbivores zooplankton tend to be small in this period.

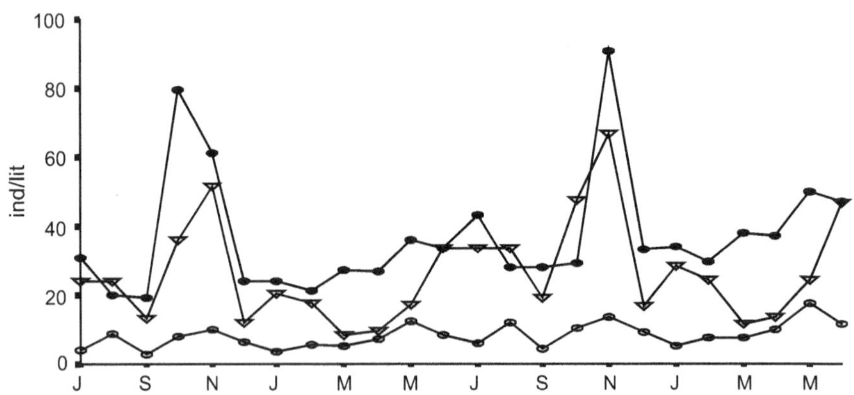

Fig. 11.4 Temporal variations in the total density of Copepoda (*filled circle*), Cladocera (*open circle*) and Rotifera (*down triangle*) in Lake Tana during July 2003–June 2005 (Wondie and Mengistu 2006)

There is no clear-cut seasonal succession of phytoplankton and zooplankton taxa, but a consistent seasonal pattern was seen for the different functional groups. Most taxa co-occur and reach their peak at the same season (PORS) which indicates that their density, biomass and production is controlled not by competition or predation but by food availability. In general, plankton productivity in Lake Tana seems to be slightly lower than other African lakes. Littoral periphyton and macrophytes, together with allochotonous sources may be contributors of additional organic carbon to the secondary producers in Lake Tana. For example, Hecky and Kling (1981) noted that bacteria and protozoa biomasses are potential sources of fixed carbon other than primary production by algae in Lake Tanganyika. They found that within the euphotic zone, bacterial production is about half the mean rate of algal production.

11.6 Recommendations for Further Plankton Research

- Detailed studies on zooplankton feeding behaviour. Partitioning of food items such as the role of phytoplankton, detritus, and sediment-associated microorganisms etc. have not been done. There is need to conduct grazing studies to distinguish the quantity and quality of the food of the various prey groups by using different feeding techniques.
- The effect of sediment on growth of zooplankton. Sediment resuspension is generally accepted as having an important role in controlling the structure and functioning of shallow lake ecosystems (Scheffer 1998). Sediment is characterized by its size and by the nature and relative proportions of the chemical and mineral components. The role of sediment may vary among zooplankton species, depending on the quantity, particle size, organic content and grain composition of the sediment encountered.
- The role of detritus in the feeding of zooplankton. Detritus contains a large amount of energy and nutrients. Wondiie (2006) concluded that phytoplankton is not efficiently utilized by the higher trophic level groups in Lake Tana. Therefore, detritus both from external sources and internal recycling probably plays an important role.
- Usually littoral zone of a lake is more productive when compared to open water zone. However in the case of Lake Tana littoral zone shows low primary productivity due to siltation and absence of buffer stability. Therefore further studies will be required on the impact of siltation in primary productivity

References

Admasu D (1998) Age and growth determination of tilapia, Oreochromis niloticus L. (Pisces: Cichlidae). In: some lakes in Ethiopia. PhD thesis, Addis Ababa University

Amarasinghe PB, Vijverberg K (2002) Primary production in a tropical reservoir in SriLanka. Hydrobiologia 487:85–93

Amarasinghe PB, Vijverberg J Boersma M (1997) Production biology of copepods and cladocerans in three south-east Sri Lankan low-land reservoirs and its comparison to other tropical freshwater bodies. Hydrobiologia 350:145–162

Belay A, Wood RB (1984) Primary productivity of five Ethiopian rift valley lakes. Verh Internat Verein Limnol 22:1187–1192

Burgis MJ (1971) An ecological study of zooplankton of Lake George, Uganda. PhD thesis, University of London

Compere P, Iltis A (1983) The phytoplankton. 6. In Carmouze J-P, Durand J-R, Leveque C (eds) Lake Chad. Ecology and production of shallow tropical ecosystem. Monogr Boil (Junk) 53:145–197

Dadeboo E (2002) Reproductive biology and habit of some fish species in Lake Chamo, Ethiopia. PhD thesis, Addis Ababa University

De Graaf M (2003) Lake Tana's piscivorous Barbus (Cyprinidae, Ethiopia): Ecology, Evolution, Exploitation. PhD thesis, Wageningen University

Defaye D (1988) Contribution a la connaissance des Crustaces copepodes d Ethiopie. Hydrobiologia 164:103–147

Dejen E (2003) Ecology and potential for fishery of the small barbs (Cyprinidae, Teleostei) of Lake Tana, Ethiopia. PhD thesis, Wageningen University, Wageningen, The Netherlands

Dejen E, Vijverberg k, Nagelkerke LA et al (2009) Growth, biomass, and production of two small barbs (Barbus humilis and B. tanapelagius, Cyprinidae) and their role in the food web of Lake Tana (Ethiopia)ation. Hydrobiologia 636:89–100

Dejen E, Vijverberg k, Nagelkerke LA et al (2004) Temporal and spatial distribution of microcrustacean zooplankton in relation to turbidity and other environmental factors in a large tropical lake (L. Tana, Ethiopia). Hydrobiologia 513:39–49

Fernando CH (1994) Zooplankton, fish and fisheries in tropical fresh waters. Hydrobiologia 272:105–123

Fetahi T (2005) Trophic analysis of Lake Awassa using Mass-balance Ecopath model. MSc thesis, Addis Ababa University

Ganf GG, (1974) Phytoplankton biomass and distribution in a shallow eutrophic lake (Lake George, Uganda). Oecologia 16:9–29

Goshu G, Byamukama D, Manafi M et al (2010) A pilot study on anthropogenic fecal pollution impact in Bahir Dar Gulf of Lake Tana, Northern Ethiopia. Eco Hydrol Hydrobiology 10(2–4): 271–280

Hart RC (1987) Population dynamics and production of five crustacean zooplankters in a subtropical reservoir during years of contrasting turbidity. Freshwat Biol 18:287–318

Hecky RE, Kling HJ (1981) The phytoplankton and protozooplankton of the eutrophic zone of Lake Tanganyika: species composition, biomass, chlorophyll content and spatio-temporal distribution. Limnol Oceanogr 26:548–564

Kalff J, Knoechel R (1978) phytoplankton and their dynamics in oligotrophic and eutrophic lakes. Ann Rev Ecol Syst 9:475–495

Kebede E, Belay A (1994) Species composition and phytoplankton biomass in a tropical African lake (Lake Awassa, Ethiopia). Hydrobiologia 284:13–32

Kifle D (1985) Variation in phytoplankton primary production in relation to light and nutrients in Lake Awassa. MSc. thesis, Addis Ababa University

Lewis WM Jr (1979) Zooplankton community analysis studies on tropical systems. Springer, New York, p.163

Lewis WM Jr (1987) Tropical limnology. Ann Rev Ecol Syst 18:159–184

Mavuti KM, Litterick MR (1981) Species composition and distribution of zooplankton in a tropical lake. Lake Naivasha Kenya Arch Hydrobiol 93:52–58

Mavuti KM (1990) Ecology and role of zooplankton in the fishery of Lake Naivasha. Hydrobiologia 208:131–140

Mehari AK, Wondie A, Mingist M et al (2014) Spatial and seasonal variation in the macro-invertebrates and physico-chemical parameters of the Enfranz River. Lake Tana sub-basin (Ethiopia) Ecohydrology & Hydrobiology 14:304–312

Melack JM (1976) Limnology and dynamics of phytoplankton in equatorial African lakes. PhD thesis, Duke University
Mengestou S, Fernando CH (1991) Seasonality and abundance of some dominant crustacean zooplankton in Lake Awassa: a tropical rift valley lake in Ethiopia. Hydrobiologia 226: 132–152
Reynolds CS (1990) Temporal scales of variability in pelagic environments and the response of phytoplankton. Freshwat Biol 23:25–53
Rzoska J, Brook AJ, Prowse GA (1955) Seasonal plankton development in the White and Blue Nile near Khartoum. Verh Int Ver Limnol 12:327–334
Rzoska J, Talling JF (1967) The development of plankton in relation to hydrological regime in the Blue Nile. J Ecol
Scheffer M (1998) Ecology of sallow lakes. Population and Community Biology Series 22. Chapman and Hall, London
Talling JF (1986) The seasonality of phytoplankton in African lakes. Hydrobiologia 138:139–160
Talling JF (1987) The phytoplankton of Lake Victoria (East Africa). Arch Hydrobiol Bieh 25:229–256
Talling JF, Rzóska J (1967) The development of plankton in relation to hydrological regime in the Blue Nile. J Ecol 53:637–662
Teshita A, Wondie A (2014) The impact of impoundment on downstream macro invertebrate communities at Koga Irrigation Dam, West Gojjam, Ethiopia. Int J Sci Res 3:6
Vareschi E, Vareschi A (1984) The ecology of Lake Nakuru (Kenya) IV. Biomass Distrib Consum Org Oecologia 62:78–98
Vijverberg K, Richter AF (1982) Population dynamics and production of Daphnia hyalina Leydig and Daphnia cucullata Sars in Tjeukemeer. Hydrobiologia 95:235–259
Wondie A (1996) Primary production and chemical characteristics of concentrated Ponds in Lake Abijata, Ethiopia. MSc. thesis, Addis Ababa University
Wondie A (2006) Dynamics of the major phytoplankton and zooplankton communities and its role in the food web of Lake Tana, Ethiopia. PhD thesis, Addis Ababa University
Wondie A (2009) Current land use practices and possible management strategies in shore area wetland ecosystem of Lake Tana: towards improving livelihoods, productivity, and biodiversity conservation Journal of Ethiopian Fishery and other aquatic sciences Associations. Addis Ababa, Ethiopia
Wondie A (2010) Improving management of shoreline and riparian wetland ecosystems: the case of Lake Tana catchment. Ecohydrology and Hydrobiology 6:229–235
Wondie A, Mengistu S (2006) Duration of development, biomass and rate of production of the dominant copepods (calanoida and cyclopoida) in Lake Tana, Ethiopia. SINET: Ethiop J Sci 29(2):107–122
Wondie A, Mengistu S (2014) Seasonal variability of secondary production of cladocerans and rotifers, and their trophic role in Lake Tana, Ethiopia, a large, turbid, tropical highland lake. Afr J Aquat Sci 39(4):403–416
Wondie A, Mengistu S, Fetahi S (2012) Trophic interactions in lake tana, a large turbid highland lake in Ethiopia in: Jordan, F, Jorgensen, SE (eds) Models of the ecological hierarchy; from molecules to the ecosphere. Elsevier B.V., 217–235
Wondie A, Mengistu S, Vijverberg J et al (2007) Seasonal variation in primary production of a large high altitude tropical lake (Lake Tana, Ethiopia): effects of nutrient availability and water transparency. Aquat Ecol 41:195–207
Wondie A, Osondu A (2008) Seasonal variation of phytoplankton biomass in Lake Tana (Ethiopia). Trop Freshw Biol 17(2):21–31
Wondie A, Seyoum M (2006) Duration of development, biomass and rate of production of the dominant copepods (calanoida and cyclopoida) in Lake Tana. Ethiopia SINET: Ethiop J Sci 29(2):107–122
Wondimagegne T, Wondie A, Mengist M et al (2012) Seasonality in abundance, biomass and production of the phytoplankton of welala and sheher wetlands: Lake Tana Sub-basin (Ethiopia). J Water Resour Prot 4:877–884

Wood RB, Talling JF (1988) Chemical and algal relationships in a salinity series of Ethiopian inland waters. Hydrobiologia 158:29–67
Wosenie A, Wondie A (2014a) Bahir Dar tannery effluent characterization and its impact on the head of Blue Nile River. Afr J Environ Sci Technol 8(6):312–318
Wosenie A, Wondie A (2014) Assessment of downstream impact of Bahir Dar tannery effluent on the head of Blue Nile River using macroinvertebrate as bioindicators. Int J Biodivers Conserv 6(4)
Wudneh T (1998) Biology and management of fish stocks in Bahir Dar Gulf, Lake Tana, Ethiopia. PhD thesis, Wageningen Agricultural University

Chapter 12
The Fish and the Fisheries of Lake Tana

Abebe Ameha Mengistu, Chalachew Aragaw, Minwyelet Mengist and Goraw Goshu

Abstract The fishes of Lake Tana show impressive diversity and unprecedented uniqueness. The lake harbors the only species flock of large cyprinids in the world, and of the 28 known species, approximately 68% are endemic to the lake. Most of the habitats harboring the ichthyofaunal diversity of this lake are relatively intact, and still naturally attractive. Most of the fisheries development activities are characterized by the traditional reed-raft gillnet fishery and some motorized boat gillnet fishery. Fish production from capture fisheries including Nile tilapia, Large African barbs and African catfish from the lake and its influent and effluent rivers has not yet been able to satisfy the ever-growing local demand for fish. No alternative aquaculture production system has been developed. Current and past research findings depict the importance of the fish and the fisheries in Lake Tana, as well as the challenges that these valuable resources are facing. Fast increase in human population of the Lake Tana Sub-basin and associated human-induced impacts are degrading the natural habitats and negatively affecting the fish and the fisheries. The impact on the fishes occurs indiscriminately and does not appear taxon-specific. However, endemic taxa such as the Large African barbs of the lake can be more prone than other widely distributed species. Further research, development and

A.A. Mengistu (✉)
Program of Applied Biology, Adama Science and Technology University,
Nazareth, Ethiopia
e-mail: abbefish@yahoo.com

C. Aragaw
Amhara Region Livestock Resources Development Promotion Agency,
Bahir Dar, Ethiopia
e-mail: chalachew71@hotmail.com

M. Mengist
College of Agriculture and Environmental Sciences, Bahir Dar University,
Bahir Dar, Ethiopia

G. Goshu
Blue Nile Water Institute and College of Agriculture and Environmental Sciences,
Bahir Dar University, Bahir Dar, Ethiopia
e-mail: gorawha@gmail.com; gorawg@bdu.edu.et

K. Stave et al. (eds.), *Social and Ecological System Dynamics*,
AESS Interdisciplinary Environmental Studies and Sciences Series,
DOI 10.1007/978-3-319-45755-0_12

conservation activities need to consider proper planning and implementation to preserve the fishes and the fisheries in the Lake Tana Sub-basin.

Keywords Fisheries management · Freshwater ecosystem · Highland lake · Labeobarbus · Tana Subbasin

12.1 Fish Habitats in the Lake Tana Sub-basin

12.1.1 Lake Tana

Lake Tana is the largest lake in Ethiopia which accounts to 50% of the freshwater resource of the country. The south-north maximum length is ca. 90 km and east west width of ca 65 km. The maximum depth of Lake Tana is 14 m, with mean depth of 9 m. The lake is located at an elevation of 1800 m a.s.l., and is fed by many streams and a few large rivers.

12.1.2 Lake Tana Catchment

Tana Sub-basin which covers an area of 15,100 km^2 is the second largest sub-basin of the Abbay (Blue Nile) Basin. The highest elevation is 4100 m a.s.l., with mean elevation of 2025 m a.s.l. (meter above sea level), and at Abbay River out flow 1785 m a.s.l. The average annual rainfall at Lake Tana is 1248 mm/year, 7% lower than the surrounding catchments. Atmospheric temperature decreases 0.7 °C per 100 m and ranges from 13 to 22. The main landforms in and around Lake Tana comprise plains, hills, mountains, mountain cliffs and depressions. The flat plains and depressions are generally wetlands. The land use in the Lake Tana sub-basin is predominantly cultivable land (71%), grazing (9%), infrastructure (6%), forest (3%) and others. The major type of land cover includes farmland, water bodies, wetlands, forest, woodland, shrubs, rangeland, grassland and settlements.

The soils of islands, peninsulas and surrounding wetlands and dry uplands of the lake are dominated by Nitosols, Luvisols and Vertisols. Soil analysis taken from Zegie peninsula of the lake showed a pH value of 5.5–6.7 (Alemnew 2001). Flood plains of Fogera, Libo Kemkem and Dembia and the river mouth delta of Stumit, Kristos Semera, Nabega, Angara and Dirma are dominated by alluvial deposits making shallow water depth which is cultivated in the dry season following the retreating lake water. Most of the transported high sediment loads are deposited and silted down in the water body of the lake. Thus, it can be said that Lake Tana is both a natural water reservoir and silt refinery for the Abbay River, but detrimental to the long term existence of the lake to function as habitats for aquatic organisms.

12.1.3 Fish Habitats

12.1.3.1 Tributary Rivers

Lake Tana is fed by over 60 rivers and streams draining from the watershed and forms complicated hydrologic networks (Fig. 12.1). These temporary and permanent rivers and streams flow down in different geologic landscapes contributing to have different habitat structures. All these structures provide habitats for various

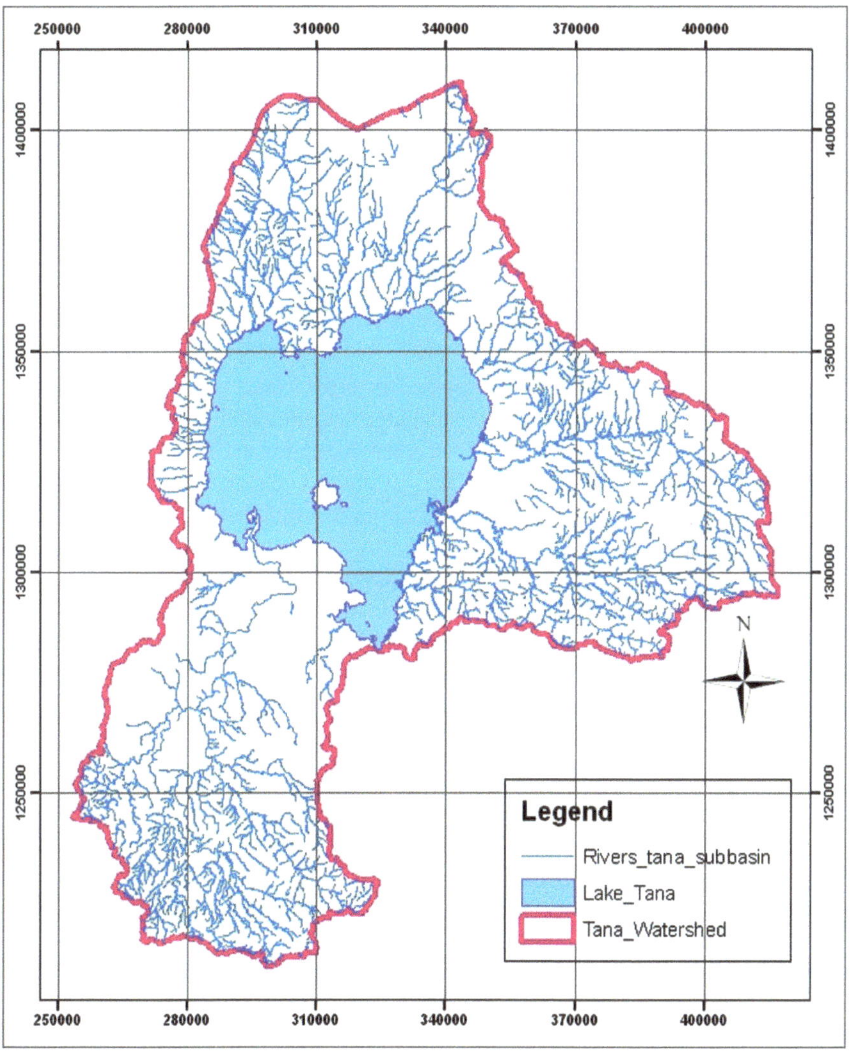

Fig. 12.1 Drainage map of Lake Tana and its environs

riverine organisms, especially serving as spawning grounds of the endemic fish, African large barbs (*Labeobarbus*). They are also very essential for the development of agriculture, transportation, and fishing; especially from the perspective of tourism, they are rich in different tourist attraction resourses and are essential for various tourism activities.

There are more than five major tributaries of Lake Tana, including Gilgel Abbay, Ribb, Gumara, Megech, Dirma and Arno-Garno. Gilgel Abbay is located at southwestern part of Lake Tana. Gelda River is located at the southeastern part of Lake Tana while Gumara and Rib Rivers are situated on the eastern shore of the lake. Megech and Dirma Rivers are located at the northern part. The only outflow from Lake Tana is Abbay (Blue Nile) River which flows from the southeast direction.

12.1.3.2 Vegetation

In general, there are four types of indigenous vegetation in the Tana sub-basin. These are, Dry Evergreen Afromontane Forest, Combretum-Terminalia, Evergreen Scrub (shrubs and small trees) and Riverine and Swamp Vegetation (*Ficus* sp.).

12.1.3.3 Wetlands and Flood Plains

Different forms of wetlands are represented in Lake Tana. Wetlands/swamp areas make up 1.6% of the Lake Tana watershed area. The potential natural vegetation of the Lake Tana area consists of Dry Evergreen Afromontane Forest and Grassland complexes, Lake Tana as freshwater lake, freshwater marshes and swamps, floodplains and lake shore vegetation and Combretum-Terminalia woodland and wooded grassland. Emergent macrophytes fringe the flat swampy parts of the shoreline, with the dominant species being *Cyperus papyrus, Echinochloa pyramidalis, Echinochloa stagnina, Polygonum barbatum, Polygonum senegalense* and *Typha domingensis*. Floating-leaved species include *Nymphaea caerulea, Nymphaea lotus and Pistia stratiotes. Ceratophyllum demersum and Vallisneria spiralis* are noted as the most important submergent plants occurring in the lake area.

12.2 Diversity and Distribution of Fishes

12.2.1 History of Exploration

Scientific studies on the fishes of Lake Tana and its rivers dates back to 1829 when Rüppell described several cyprinid species during his travel for trade and exploration (Rüppell 1829, 1836). There have been occasional collection and identification works done for some species of fishes in the second half of the 19th Century (e.g., Vinciguerra 1883; Boulenger 1902). Detailed investigation on the fishes of

Lake Tana was done in the 1930s (Bini 1940; Brunelli 1940a, b; Brunelli and Cannicci 1940). There was an unsuccessful effort to introduce the Mosquito Fish (*Gambusia* sp.) and the European Pike (*Esox lucius*) into Lake Tana (Tedla and Haile Meskel 1981). The most recent studies on the taxonomy, ecology and other biological aspects of the fishes of Lake Tana started in the 1980s through fisheries resources development projects of the Ministry of Agriculture (MoA). The Lake Tana Fisheries Resources Development Programme (LTFRDP) was initiated in collaboration with the Ethiopian Orthodox Church and Interchurch Foundation Ethiopia (ISE) from Urk, The Netherlands. Following this, the MoA and European Commission (EC) launched a country-wide Programme named the Lake Fisheries Development Project (LFDP) within which the Lake Tana system was one of the major units. The Bahir Dar Fisheries Research and Training Center was established by LFDP with a sub-center at Gorgora. Research and development programs were integrated in LFDP highlighting the status of fish stocks in rivers (Ameha and Assefa 2002) and the biology of commercially important species of fishes (e.g., Ameha et al. 2006). Graduate students from Wageningen University(WU, the Netherlands) and researchers from the Joint Ethio-Soviet Biological Expedition (JESBE, later the Russian JERBE) team worked with Ethiopian experts. The LFDP, JERBE and WU programs were running on Lake Tana side by side for a few years. Subsequent studies and ongoing research by MSc students of Bahir Dar University (e.g., Mohammed et al. 2013), Addis Ababa University (e.g., Abdissa 2009), and the Bahir Dar Fisheries and other Aquatic Life Research Center are based on the foundation laid by the abovementioned programs.

12.2.2 Species Diversity

Lake Tana is not regarded as species-rich as compared to other tropical lakes in Africa; this could be attributed to the relative placement of the lake on a highland area at about 1800 m above sea level. Generally, with increasing altitude, species diversity is expected to decrease while endemicity increases; the same principle applies to the Ethiopian Highlands where there is high proportion of endemicity of terrestrial fauna. However, as compared to other Ethiopian lakes, Lake Tana harbors many species of fish due to its relatively large surface area, its geographic isolation from other such water bodies and varied microhabitats within the lake and its rivers (Vijverberg et al. 2012).

In Lake Tana, there are a total of 28 described species of fishes grouped under four families (Cyprinidae, Balitoridae, Clariidae and Cichlidae) and seven genera (*Labeobarbus*, *Barbus*, *Garra*, *Varicorhinus*, *Afronemacheilus*, *Clarias* and *Oreochromis*) (Getahun and Dejen 2012; Habteselassie 2012). The first three genera are relatively species-rich, whereas each of the last four genera is represented in Lake Tana by a single species. Figure 12.2 shows some of the species of fishes in Lake Tana. The taxonomic distinction of the species in the genera *Labeobarbus*, *Barbus* and *Garra* is based on external morphology, osteology and

Fig. 12.2 Some fish species of Lake Tana

feeding ecology. Some effort has been made to show distinctions of species of *Labeobarbus* using molecular evidence (e.g., Kruiswijk et al. 2005; de Graaf et al. 2010) but the genetic variations among the studied species appear to be too small to prove long time evolutionary diversification.

Cyprinid fishes dominate the species composition of fishes in Lake Tana by having 25 species (89%) as compared to the other families, each of which are represented by a single genus and species. A complete listing of the currently known species of fishes of Lake Tana is presented in Table 12.1.

12.2.3 Ecological Importance

The fishes of Lake Tana are part of complex communities and habitats within an aquatic-wetland-terrestrial system. Some populations of fishes are strict lake or river inhabitants, spending their whole life in the lake or rivers, while others wander between the lake, rivers and surrounding wetlands for feeding and reproduction (e.g., Anteneh et al. 2006, 2012). The different species of fishes in the area have important contribution in the food web of this complex system. The adults of some species such as *Labeobarbus surkis*, *Varicorhinus beso* and *Oreochromis niloticus* feed on aquatic vegetation (macrophytes or algae); other species such as *L. crassibarbis* specialize on specific food items as molluscs and other invertebrates. Still others are planktivores (e.g., *L. brevicephalus*) or are piscivorous (e.g., *Clarias gariepinus, L. acutirostris*). In addition, the young and adult fishes are prey to other animals, including birds (e.g., Pelicans, African darter), mammals (e.g., freshwater otter) and reptiles (e.g., Nile monitor). Fish consumption by humans should not be ignored here. The dead body of fishes and fish offal are also consumed by invertebrate animals (e.g., freshwater crabs) and decomposed by bacteria into inorganic nutrients that in turn are inputs for primary production.

12.2.4 Economically Important Species

The commercially most important fishes of Lake Tana and its rivers include species of the Large African barbs, Nile tilapia, African catfish and bezo. Details of the fishing and marketing of these fishes are presented below in the "fisheries" section

Table 12.1 List of known species of fishes from Lake Tana, Ethiopia (Extracted from: Getahun and Dejen 2012; Habteselassie 2012)

Family	Scientific name	Common name	Vernacular name
Balitoridae	*Afronemacheilus abyssinicus*		
Cichlidae	*Oreochromis niloticus tana*	Nile tilapia	Qereso
Clariidae	*Clarias gariepinus*	African catfish	Ambaza/Sorz
Cyprinidae	*Barbus humilis*	Small barbs	
	Barbus paludinosus		
	Barbus tanapelagius		
	Garra dembecha		
	Garra dembeensis		
	Garra regressus		
	Garra tana		
	Labeobarbus acutirostris	Long snout barb	Bilcha, Modmuade
	Labeobarbus brevicephalus	Short-head barb	Bilcha, Shasha
	Labeobarbus crassibarbis	Barbel barb	Bilcha, Bajete
	Labeobarbus dainellii		Bilcha
	Labeobarbus degeni		Bilcha, Motmuate
	Labeobarbus gorgorensis		Bilcha
	Labeobarbus gorguari	Black hunch barb	Bilcha, Asa-baria
	Labeobarbus intermedius		Bilcha, Afe-qey
	Labeobarbus longissimus	Big mouth-mini eye barb	Bilcha
	Labeobarbus macrophthalmus	Big mouth-small eye barb	Bilcha, Fetate
	Labeobarbus megastoma	Big mouth-big eye barb	Bilcha, Afe-qefo
	Labeobarbus nedgia	Flap lip barb	Bilcha, Qibe-bemuday
	Labeobarbus ossensis		Bilcha
	Labeobarbus platydorsus	White hunch barb	Bilcha, Gobate
	Labeobarbus surkis		Bilcha, Yechibaasa
	Labeobarbus truttiformis	Trout-like barb	Bilcha, Tereza
	Labeobarbus tsanensis		Bilcha
	Varicorhinus beso		Bezo

of this chapter. Other than fishing for food (nutrition) for humans and domestic animals (dogs and cats), there are no well-established economic values of the fishes of the sub-basin. Some traditionally established uses of these fishes in rural areas include the use of fish body fats as fuel for lamps (replacing kerosene), and the use of bile for medicinal purpose.

Species of the small-sized genera *Barbus* and *Garra* are also seen in the fish markets; these fishes are commercially harvested from the tributary rivers of Lake

Tana where fishes are seasonally collected, whole sun-dried and grounded into flour with mortar and pestle.

12.3 The Fisheries

12.3.1 Types of Fisheries of Lake Tana

There are two major types of fisheries in Lake Tana, the motorized boat gill net fishery and the reed raft gill net fishery, the later being the dominant type. Over 80% of the fishers use reed boats (Fig. 12.3) that are made of locally available papyrus, which grows around the south-western part of the lake. The riverine fisheries upstream at major tributaries can be regarded as a third type of fishery in the Lake Tana system; locally made scoop nets, traps, hook and line as well as blockage of rivers with fences and toxic plants are used for fishing in the rivers.

The carrying capacity, lower durability and smaller size makes the reed boats inefficient as seen and compared to the physical condition of the lake. The number of crew per trip in the reed boat is limited to be one. Due to these facts, all reed boat fishers do not go for fishing far from the shore of the lake. The average reed boat has a dimension of 3.2 m length and 0.60 m width, and the average service life is only two to three months.

The Bureau of Agriculture (Amhara Region) reports that the number of motorized boats in Lake Tana is not greater than forty and these are currently operating at southern part of the lake around Bahir Dar. Another report indicates

Fig. 12.3 Fisherman rowing a reed boat; and papyrus vegetation in the south western part of Lake Tana. *Photo credit* Friedrich zur Heide

that this figure reached 75 in 2010 (Mohammed et al. 2013). The motorized boats have a crew of two or three persons. This fishery operates more offshore where the reed boats can't reach. The size of the steel boats ranges from 5 to 8 m with engine 15–25 Hp.

Currently, the most frequently used fishing gear by both types of fisheries are gill nets. Other fishing gear such as cast nets, scoop nets, traps, and hooks are also occasionally used. Most of the motorized boat and reed boat fisheries use monofilament gillnets that are illegally imported. The current government law prohibits monofilament gill nets for fishing, though currently it appears difficult to enforce this law.

12.3.2 Major Benefits of the Fisheries of Lake Tana

Socio-economic benefit: In terms of employment and income generation, the fisheries of Lake Tana supports large numbers of fishers, fish traders, processors, and other segments of the society. Currently, there are about 3514 fishers, with an average six dependents per fisher, and in total 21,084 beneficiaries directly dependent on the fishing activities. The primary beneficiaries would be estimated above 40,000 if the number of fish processors, traders, gillnet makers, boat builders, etc. were included. The lake borders 54 kebeles/localities under ten administrative Woredas/districts and three administrative zones.

Contribution for regional income: Exporting fish has already started to the Sudan. Within five years' time alone (2008–2012), USD 1,967,735 (ETB 37,878,892) was generated from export fish market; and an income of ETB 59 million was generated in 2009 from domestic market.

Nutritional benefits and food security: In areas around the lake and its rivers, fish is the major and important source of animal protein especially for the poor who can't afford buying other animal protein sources. Animal protein is the most deficient nutrient in the diet of most rural communities, because it is expensive. The market price of a kilogram of beef at Bahir Dar town is ETB 100–130 while that of filleted tilapia is ETB 60–70, and ETB 25–40 for filleted catfish and African large barbs. Therefore, the current market price of fish is nearly two times lower than that of beef.

Fish meal: The offal of fish can be processed and used for animal feed. More than 40% of the fish body is being wasted as offal. But there is high level of animal feed shortage in the region. These days, there is an increased demand of fishmeal from poultry farmers. There is only one established fish meal processing unit at Bahir Dar town. However, the production is very limited and the drying system of the offal is traditional and needs improvement. This practice should be improved and should be scaled up in the future to reduce pollution and to produce more processed animal feed.

12.3.3 Number of Fishers and Fish Production

According to a census conducted in 2012, there were 3514 fishers around the lake. Fishermen census takes place every three years. Ten Woredas/districts surround Lake Tana (Fig. 12.4), and there are 54 major landing sites and the majority of them are located at the northern part of the lake (Table 12.2). Nearly all of the landing sites have no facilities.

Recorded fish production increased from 756 t in 2000 to 9980 t in 2013. Nile tilapia contributed to more than 50% by volume of this production. The yield has been increasing steadily in the last few years (Fig. 12.5). The number of fishers also increased from 1700 in 2010 to 3514 in 2012 (Aragaw 2012). This was due to large coverage of data collection system at most of the landing sites of the lake, and the increasing demand and price of fish in the main towns including Addis Ababa.

The Maximum Sustainable Yield (MSY) of the fisheris of Lake Tana has previously been estimated based on empirical models. The estimates are based on empirically found relationship between fish yield and other parameters such as lake area in km^2, the average depth of the lake in meters, shore line development, Morpho-Edaphic-Index (MEI), and Total Dissolved Solids (TDS mg/l) of the lake. Based on these different models (Table 12.3), the average (MSY) of the lake was estimated at about 15,000 t per year. Currently, there is no regular fish stock assessment programme designed to estimate the biomass and yield of Lake Tana using more reliable analytical models.

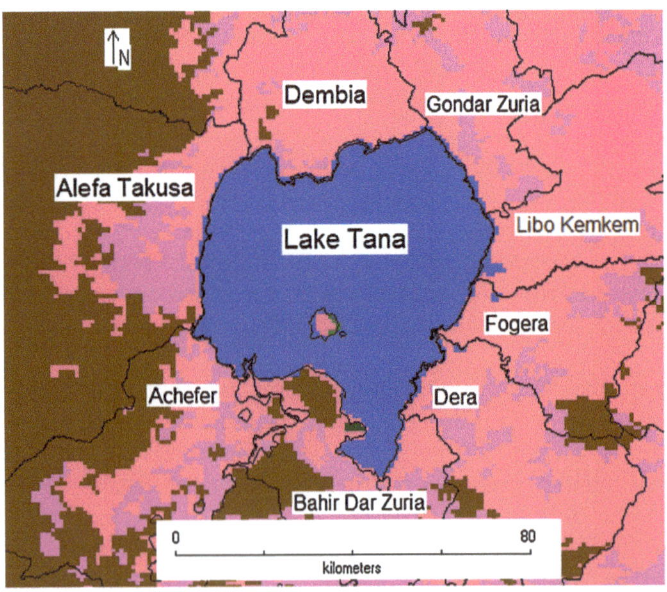

Fig. 12.4 Land cover map of Lake Tana and the surrounding fish producing Woredas/districts. Note the cultivated and managed area (*pink*) that dominates the surrounding area as contrasted with the Shrubland (*brown*) that lies outside the Lake Tana Sub-basin. (Map modified by the first author from spatial data in DIVA-GIS 2015)

Table 12.2 The Current number of fishers at Lake Tana in the main fish producing Woredas

Administrative zone	Area of production (woreda/district)	Number of main landing sites	Number of fishers	% out of total number of fishers
North Gondar	Alefa	3	149	4.24
	Takusa	6	188	5.35
	Dembia	10	605	17.22
	Gondar Zuria	5	328	9.33
	Zone total	24	1270	36.14
South Gondar	Libo	3	372	10.59
	Fogera	2	479	13.63
	Dera	3	439	12.49
	Zone total	8	1290	36.71
West Gojam	Bahir Dar town	10	341	9.70
	Bahir Dar Zuria	7	371	10.56
	North Achefer	5	242	6.89
	Zone total	22	954	27.15
Lake total		54	3514	100

Source Aragaw (2012)

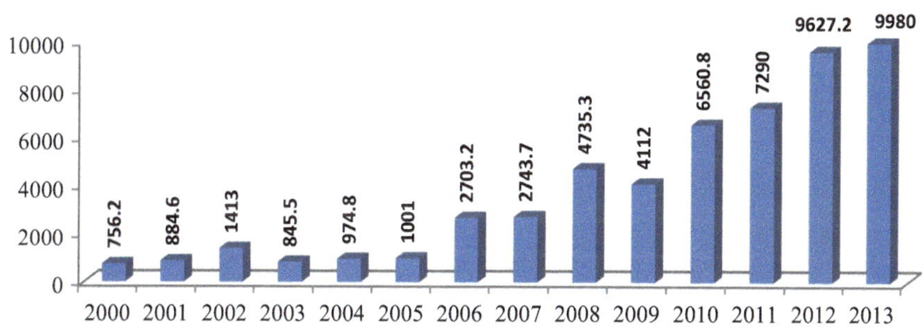

Fig. 12.5 Trends of fish catch (in tons) in Lake Tana. *Source* LRDPA (2013)

12.3.4 *Fish Processing and Marketing*

Processing of fish involves primarily the application of preservation techniques in order to retain quality and increase shelf life. It may also mean adding value to produce a wide variety of products. Fish processing is also concerned with proper waste management and with adding value to fish products. The processing of fish at Lake Tana takes place both in a traditional system as well as in better (modern) fish processing facilities.

Table 12.3 Estimated potential yield (tons/year) of the fisheries of Lake Tana based on empirical models

Model used	Estimated productivity kg/ha/year	Estimated potential yield in tons/year
Marshal (1984)	50	17,536
Marshal (1984)	27	9450
Crul (1992)	43	15,159
Schlesinger and Regier (1982)	18	6300
Henderson and Welcome (1974)	63	22,050
Toews and Griffith (1979)	57	19,950
Schlesinger and Regier (1982)	42	14,700

Source Aragaw (1998)

Fig. 12.6 Processing of fish on bare ground at Delgi landing site, northwestern Lake Tana. *Photo credit* Herwig Waidbacher

A. Traditional fish processing and marketing

The processing of fish at Lake Tana widely takes place in a traditional system, mainly under non-hygienic conditions on bare ground for gutting, filleting or drying with little to no regard for safety (Fig. 12.6).

After gutting and cleaning, the fish are carried by shoulder and taken to local markets, restaurants and hotels in the surrounding town. Some local traders transport whole, gutted or filleted fish to Addis Ababa, which is about 565 km from Bahir Dar, without cold chain transport facilities. At the southern part of the lake, there are more than 100 women who directly buy fish from the fishers and sell the fish by walking house to house. These women are the most disadvantaged group of the fisheries community in the area.

Fig. 12.7 Split fish drying on plastic sheet on the ground (*left*) and heap of dry fish ready for transportation (*right*) in Chuahit town, Dembia district. *Photo credit* Herwig Waidbacher

On the distant northern parts of the lake (Dembia, Takusa and Alefa), due to inaccessibility of landing sites, problem of transport network and limited market for fresh fish, huge amount of sun-dried or wet fish is exported to the Sudan. At the landing sites, fish are gutted and split, then hung up to dry on strings. Fishes are also sun-dried on rocks and stones. Dried fish are then packed in sacks for storage and transportation to the Sudan. Typically, sun-dried fish are prepared in poor hygienic conditions (Fig. 12.7).

B. Processing and marketing of fish in better (modern) facilities

Fish processing with improved (modern) facilities takes place around Bahir Dar town by the Fish Production and Marketing Enterprise (FPME) and to some extent by the Lake Tana No. 1 Fishers' Cooperatives. Fish are filleted, packed and chilled/frozen and then sold at domestic market in fish shops. FPME has a considerable number of processing plants, freezing facilities and cold storage at production points, cold storage in Addis Ababa, refrigerated transport and other vehicles, and 18 retail outlets with multiple freezers. In general, FPME buys fish through contractual arrangements with fishing cooperatives and individual fishers operating on the lake. Fish are gutted and filleted near the landing site and quickly frozen. The cold chain remains intact throughout transport and retailing. Fish are sold through FPME's own retail outlets in the capital and in production areas: 14 in Addis Ababa and one each in Debre Zeit, Ziway, Bahir Dar and Gondar (Gordon et al. 2007). The fish processing system used by both the FPME and the Cooperatives is close to a traditional approach; that is, it does not match any of the criteria required to meet hazard analysis critical point control system compliant with the requirements.

Licenses for trading of fish are issued to individual or corporate traders by the Regional Bureau of Trade and Transport (BoTT) after consulting the Bureau of Agriculture (BoA) but now legally given to Livestock Resources Development Promotion Agency (LRDPA). Fish trade license are valid for one year and renewed every year.

Domestic market: The domestic markets are for fresh whole tilapia, catfish and large barbs, or traditionally processed fillets of these fish. Most of the fish are sold at fish shops and some at the landing sites. The domestic fish market outlets are not as such very developed. In each fish marketing link in the value chain, there are

Table 12.4 Fish prices (ETB/kg) of producers', wholesalers and retailers by type of fish and type of product at Bahir Dar town, Lake Tana in 2014

Type of fish	Producers' price (ETB/kg)	Wholesale price (ETB/kg)	Retail Price (ETB/kg)	Filleted (whole and retail) price (ETB/kg)
Nile tilapia	19.00	22.50	24.00	62.00
African catfish	8.50	10.00	13.00	37.00
African large barbs	7.50	9.00	11.00	23.00
Average	11.67	13.83	16.00	40.67

Source Lake Tana No. 1 Fishers' Cooperatives 2014

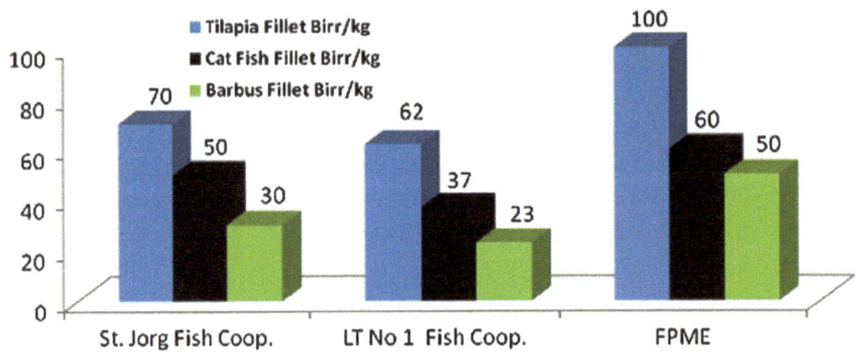

Fig. 12.8 Prices (in ETB) of filleted fish at wholesale/retail markets by St. George Fishers' Cooperative, Lake Tana No. 1 Fishers' Cooperative and FPME in Bahir Dar in April 2014

noticeable differences in the prices of fish (Table 12.4). For example, the fishers' cooperative buys whole fish on average for ETB 11.67 per kg from its members and sells for ETB 13.83 and 16.00 per kg for the wholesale and retail market, respectively. The farm gate prices are increasing from year to year following reduced level of fish catches and increased market demad. Nile tilapia is the most preferred fish by the consumers and have better market price.

The price of fish varies not only by type of fish, type of product and place of market, but also between the cooperatives and FPME. For example, because of 15% VAT, the FPME sells fish relatively at higher prices than that of the cooperatives (Fig. 12.8).

Except FPME and the Lake Tana No. 1 Fishers' Cooperative who sell fish by weight, the majority of fishermen sell fish by numbers and size of fish. This marketing is mostly done by women who sell fish by walking from house to house carrying fish on their shoulder. In this trade, the buyer has high bargaining power since the fish has to be sold within a few hours, or else it spoils, and it is common to find fish prices going down late in the afternoon.

Export market for wet or partially dried fish: The market for dried fish has been established in 1993 at the northern part of Lake Tana, especially at Delgi and Gorgora. At that time, there was no organized fish market system for fresh fish. The BoA then introduced fish drying technology package, especially the solar drier tent.

Table 12.5 Trends in the quantity and value of fish exported from Lake Tana to the Sudan

Year	Quantity exported (tons)	Value (USD)	Equivalent in ETB
2008	443.89	863,761	16,627,412
2009	365.20	261,023	5,024,693
2010	621.20	254,000	4,889,500
2011	531.60	265,800	5,116,650
2012	642.60	323,150	6,220,638
Total	2604.488	1,967,735	37,878,892

Source LRDPA (2013)

Exports of fish are dominated by the export of wet/partially dried fish to the Sudan. Nearly all wet fish exported was from northern part of the lake, especially from Dembia district (Gorgora area) and Takusa district around Delgi. At the northern part of the lake, the majority of the fish catch is split wet fish or sundried in a very poor processing and storage system. Table 12.5 shows trends of export quantity and value. As is shown in the table, the wet fish/dried fish generated considerable amount of income from export market.

12.4 Conservation and Management

It is shown above that the Lake Tana fishery makes important contributions for the livelihoods of many local communities living around the lake. To sustain these benefits for the generations to come, it is vital to use them in responsible and sustainable way. The fish populations are poorly managed. Currently, all three main commercial fish species/taxa are over-exploited, and in particular, several migrating *Labeobarbus* species are frequently exposed to several threats.

12.4.1 Existing Challenges

Currently, the fisheries in Lake Tana is under immense pressure from a number of challenges more than ever. Unless proper actions are taken immediately, the fisheries may collapse soon. The major challenges on the fish and fisheries of Lake Tana include the following.

(i) Population pressure

The ever-increasing human population around Lake Tana demands for agricultural and industrial expansion to meet demands of food and energy. The Lake Tana sub-basin is considered as a development corridor whereby more infrastructure are to be built at the expense of the natural resources. This will have a direct impact on the natural resources including the fisheries. The impacts are clearly manifested in

Fig. 12.9 Anthropogenic activities around Lake Tana. **a** Recession agriculture. **b** Grazing land. **c** Massive poison fishing in rivers. **d** Illegal monofilament. *Photo credits* **a**, **b** and **d**

habitat degradation, deforestation, overexploitation, pressure on biodiversity and pollution.

(ii) Water level fluctuation

Fluctuation of Lake Tana's water level remained regular from 1960 to 1996, but showed a negative slope afterwards (Kebede et al. 2006; Sewnet and Rao 2011). The construction of Chara Chara Weir at the outflow of Abbay (Blue Nile) River reduced the seasonal (but increased the yearly) variation in water level. Huge amount of water is also pumped out of Lake Tana for irrigation projects. These conditions could result in expansion of recession agriculture (ca. 6000 ha; Fig. 12.9a) and grazing land (Fig. 12.9b); these in turn brought in destruction of shore wetlands (EPLAUA 2004) and accumulation of much more sediment. All these had serious impact on the existing fish stocks and other living resources of the lake.

(iii) Water quality deterioration

The water quality of Lake Tana is affected by urban waste discharges and rural agricultural activities as well as sediment loads due to soil erosion, especially in the flooded and exposed areas of the lake. This condition favored algal blooms, changing the color and smell of the lake's water, reducing the aesthetic value and storage capacity of the lake, and affecting the productivity of the aquatic resources. A few years ago, there was change in the taste of fish and the color of water in the north-eastern and eastern parts of the lake. In addition, deterioration of the water quality favored the growth of the highly invasive weed water hyacinth (*Echhornia crassipes*). Currently, water hyacinth is spreading in the north-eastern and eastern parts of the lake′s shores including the rivers and river-mouths, blocking sunlight from reaching to native aquatic phytoplankton and hampering oxygen supply to the water body. In addition, the weed interrupts regular activities of local fishers by blocking access to the shores and affecting feeding and breeding grounds of fishes. Ongoing efforts to control spread of the weed have proven highly challenging and costly.

(iv) Overfishing and illegal fishing

According to monitoring studies conducted on the motorized boat fishery for small parts of Lake Tana during 1991–1993, 2001 and 2010, the CpUE of *O. niloticus, C. gariepinus* and *Labeobarbus* species decreased by 10, 88 and 90%, respectively (Wudneh 1998; de Graaf et al. 2006; Mohammed 2013). In 1991–1993 and 2001, populations of juvenile *Labeobarbus* in the littoral area (length range 5–18 cm) decreased by more than 85% (Wudneh 1998; de Graaf et al. 2004, 2006). The danger cast by the riverine fishery on spawning *Labeobarbus* has been shown using continued decrease both in CpUE and average weight of the fish caught (Ameha and Assefa 2002).

The collapse of fish stocks of Lake Tana, especially of *Labeobarbus* species is attributed to the following factors.

First, fishing pressure during the peak spawning season especially from June to October in the littoral region and inflowing river mouths as well as along the inflowing rivers (Fig. 12.9c) and their upstream spawning grounds;

Second, fishing using illegal narrow-meshed monofilament gillnets (5–7 cm stretched mesh sizes; Fig. 12.9d), fabricated in Egypt and imported from the Sudan since 2008 (use of gillnets with ≤ 8 cm stretched mesh size, beach seine and cast nets are common practices in the Lake);

Third, targeting of *Labeobarbus* species during spawning aggregations in the river mouths and upstream in the rivers, by blocking the fish passages and poisoning (with crushed seeds of the "Birbira" tree *Milletia ferruginea*) the spawning stocks (Nagelkerke and Sibbing 1996; Ameha 2004; de Graaf et al. 2006; Getahun et al. 2008);

Fourth, destruction of migratory routes and spawning grounds by irrigation and sand mining activities; and construction of dams in the major inflowing rivers and diversion structures in small tributary rivers is the graveyard for the migratory *Labeobarbus* species and riverine fishes.

(v) Lack of enforcement of the existing fisheries legislation and regulation

The Amhara National Regional State has issued the region's fisheries development, protection and utilization proclamation (Proclamation No. 92/2003) and its regulation (Regulation No. 50/2006). However, neither the proclamation nor the legislation is fully enforced at the grass root level, mainly due to lack of concern or commitment. Thus, the fish resources are at the verge of collapse.

12.4.2 Conservation Efforts and Future Needs

Some of the fishery management measures that can be recommended to conserve the fish resources in Lake Tana are:

(i) Closed seasons and areas

The issue of closed fishing seasons at the inflowing rivers of Lake Tana and their river mouths, flooded wetlands like Welala and Shesher, littoral vegetation and associated wetlands is clearly stated in the region's fishery legislation. Some Woredas are implementing these conservation measures. It is known that most of the cyprinids, especially *Labeobarbus* spp and *Clarias gariepinus*, reproduce during the rainy season (June–October) at the littoral shores, flooded areas and upstream areas in the rivers. These periods and areas should be off-fishing activities if recruitment overfishing is to be minimized. Thus, it is important to take strict monitoring activities all over the lake in these sensitive areas to protect the spawning adults and their early life stages (eggs and juveniles).

(ii) Prohibition of destructive fishing methods

The regional fisheries proclamation and regulation clearly identifies legal and illegal fishing gear. But these days, almost all fishers are using the destructive monofilament net. In addition, use of plant extracts such as birbira, toxic chemicals such as DDT, explosives, use of beach seine and fencing are practiced seldom. All these fishing gear and practices potentially kill all sizes of fishes and other organisms in the water. These destructive techniques should be banned.

(iii) Mesh size regulations

The regional fisheries proclamation and regulation specifies gillnets with mesh sizes of 8 cm and lower as illegal; these mesh sizes do not allow immature fishes to pass through to safety. However, fishers are observed using gillnets with mesh sizes even below 8 cm.

(iv) Restocking

The worst can yet come for *Labeobarbus* species of Lake Tana when dams and river diversions are constructed in all the inflowing rivers of Lake Tana (Getahun and Dejen 2012). Under such conditions, replenishment of the depleted stocks is mandatory. Restocking programs should be based on good scientific evidence, therefore, the Bahir Dar Fish and other Aquatic Life Research Center should play an important role in this regard.

(v) Environmental impact assessment reports

Any development activity (construction of dams, recreation centers, houses, etc.) on or near aquatic ecosystems may likely have a negative impact on the fisheries. In this regard, some of the irrigation development activities have made fisheries impact assessment studies and came up with some mitigation measures (example, the Ribb Irrigation Project; Getahun et al. 2008). However, almost all small and large dams and irrigation projects, never considered the negative effect of these structures on migratory routes of fishes and spawning grounds. This calls for the active involvement of multidisciplinary team members to better protect the aquatic resources and the environment.

(vi) Training and awareness creation

Training and awareness creation are important tools for improving people's know-how. In the case of Lake Tana, although there has been training given to different stakeholders on various aspects of the fisheries in the area, the majority of the fishers and local communities still remain unaware of the major problems of Lake Tana fisheries and associated regulations in the region. Thus, timely and frequent training and awareness creation programs should be designed for different stakeholders based on their knowledge gaps.

12.5 Conclusion

Lake Tana, along with its surrounding catchment area, is geo-climatically suitable to harbor common as well as unique diversity of fish. The species diversity is high as compared to other Ethiopian lakes, and the high proportion of endemicity of the fishes is impressing. It harbors the only endemic species flock of large barbs in the world. These geographical and biological attractions are also accompanied by severe challenges of human population pressure, land and water degradation, pollution and gaps in knowledge, infrastructure and enforcement of policies. There are good opportunities and prospects in research, skilled manpower, infrastructure and policy aspects that can be considered to ensure sustainable utilization of these resources. The establishment of commercial aquaculture in the area could be a strong helping hand to supplement fish markets and to conserve the fish stocks in Lake Tana and its rivers. Concerted effort of all stakeholders is mandatory for practical improvement in fisheries development, conservation of the fish resources and societal welfare.

References

Abdissa B (2009) Effect of different temperature regimes on ontogenic development of *Labeobarbus intermedius* of Lake Tana, Ethiopia. Unpublished MSc Thesis, Addis Ababa University, Ethiopia

Alemnew A (2001) Diversity and socio-economic importance of woody plants on the Peninsula of Zegie, north-western Ethiopia: Implication for their sustained utilization

Ameha A (2004) The effect of Birbira, *Milletia ferruginea* (Hochst.) Baker on some *Barbus* spp. (Cyprinidae, Teleostei) in Gumara River (Lake Tana). MSc Thesis, Addis Ababa University

Ameha A, Abdissa B, Mekonnen T (2006) Abundance, length-weight relationship and breeding season of *Clarias gariepinus* in Lake Tana, Ethiopia. SINET: Ethiop J Sci 29(2):171–176

Ameha A, Assefa A (2002) The fate of the barbs of Gumara River, Ethiopia. SINET, Ethiop J Sci 25(1):1–18

Anteneh W, Getahun A, Dejen et al (2012) Spawning migration of the endemic *Labeobarbus* (Cyprinidae, Teleostei) species of Lake Tana, Ethiopia, and its role in their population decline. J Fish Biol 81:750–765

Anteneh W, Getahun A, Dejen E (2006) Reproductive biology of *Labeobarbus intermedius* in the Dirma and Megech tributary rivers of Lake Tana, Ethiopia. Ethiop. J Sci Tech 4(2):13–16

Aragaw C (2012) Lake Tana fisheries co-management plan and working manual. Livestock resources development promotion agency, Bahir Dar, Amhara Region

Aragaw C (1998) Development and management approaches for inland fisheries with special reference to Lake Tana, Ethiopia. MSc Thesis. University of Portsmouth, England

Bini G (1940) I pescidel Lago Tana. Missione di Studio al Lago Tana 3(2):137–179

Boulenger GA (1902) Description of new fishes from collection made by Mr E Degen in Abyssinia. Ann Mag Nat Hist (Ser. 7) 10(60):421–437

Brunelli G (1940a) Considerazioni sullapesca e la pescosita del lago Tana modo di migliorarla. Missione di Studio al Lago Tana 3(2):215–229

Brunelli G (1940b) Le mutazioni del genere Barbus del lago Tana. Missione di Studio al Lago Tana 3(2):207–213

Brunelli G, Canniccii G (1940) Le caratteri stiche biologiche del lagoTana. Missione di Studio al Lago Tana 3(2):69–116

de Graaf M, Machiels MAM, Wudneh et al (2004) Declining stocks of Lake Tana's endemic *Barbus* species flock (Pisces; Cyprinidae): natural variation or human impact? Biol Conserv 116:277–287

de Graaf M, Nagelkerke LAJ, Palstra et al (2010) Experimental evidence for the biological species status in Lake Tana's *Labeobarbus* flock (Cyprinidae). Anim Biol 60:183–193

de Graaf M, van Zwieten et al (2006) Vulnerability to a small-scale commercial fishery of Lake Tana's (Ethiopia) endemic *Labeobarbus* compared with African catfish and Nile tilapia: an example of recruitment-overfishing? Fish Res 82:304–318

EPLAUA (2004) Water level reduction of Lake Tana and its environmental impacts. Environmental protection land administration and use authority, Bahir Dar, Ethiopia

Getahun A, Dejen E (2012) Fishes of Lake Tana, a guide book. Addis Ababa University Press, Addis Ababa

Getahun A, Dejen E, Anteneh W (2008) Fishery studies of Rib River, Lake Tana Basin, Ethiopia. A report submitted to World Bank. Vol. 2, E1573

Gordon A, Demissie S, Tadesse M (2007) Marketing systems for fish from Lake Tana, Ethiopia: Opportunities for improved marketing and livelihoods. IPMS (Improving Productivity and Market Success) of Ethiopian Farmers Project Working Paper 2. ILRI (International Livestock Research Institute), Nairobi, Kenya. 49 p

Habteselassie R (2012) Fishes of Ethiopia: annotated checklist with pictorial identification guide. EFASA, Addis Ababa 250p

Kebede S, Travi Y, Alemayehu et al (2006) Water balance of Lake Tana and its sensitivity to fluctuations in rainfall, Blue Nile basin, Ethiopia. J Hydrol 316:233–247

Kruiswijk CP, Hermsen T, van Heerwaarden et al (2005) Major histocompatibility genes in the Lake Tana African large barb species flock: evidence for complete partitioning of class II B, but not class I, genes among different species. Immunogenetics 56:894–908

LRDPA (2013) Annual Report of Livestock Resources Development Promotion Agency (LRDPA), Amhara Region Bureau of Agriculture, Bahir Dar

Mohammed B, de Graaf M, Nagelkerke et al (2013) Assessment of motorized commercial gillnet fishery of the three commercially important fishes in Lake Tana, Ethiopia. In: Proceedings of the fourth Annual Conference of the Ethiopian Fisheries and Aquatic Sciences Association (EFASA) p 75–106

Nagelkerke LAJ, Sibbing FA (1996) Reproductive segregation among the large barbs (*Barbus intermedius* complex) of Lake Tana, Ethiopia. An example of intralacustrine speciation? J Fish Biol 49:1244–1266

Rüppell E (1829) Atlas zur der Raise imnördlichen Afrika, Zoologie. Fische des Rothen Meeres. Part 2. Frankfurt-am-Main

Rüppell E (1836) NeuerNachtrag von Beschreibungen und Abbildungen neuer Fische im Nil entdeckt. Museum Senckenbergianum, Abhandlungen aus dem Gebiete der beschreibenden Naturgeschichte. Johann David Sauerländer, Frankfurt-am-Main

Sewnet A, Rao KK (2011) Hydrological dynamics and human impact on ecosystems of Lake Tana, northwestern Ethiopia. Ethiop J Env Stud Manag 4:56–63

Tedla S, Haile Meskel F (1981) Introduction and transplantation of freshwater fish species in Ethiopia. SINET: Ethiop J Sci 4:69–72

Vijverberg J, Dejen E, Getahun et al (2012) The composition of fish communities of nine Ethiopian lakes along a north-south gradient: threats and possible solutions. Anim Biol 62:315–335

Vinciguerra D (1883) Spedizion eItaliananell' Africa equatorial, Risulati zoologici; Pesci d'aqua dolce. Annali di Museo Civico di Storia Naturale "Giacomo Doria". Genova 18:691–703

Wudneh T (1998) Biology and management of fish stock in Bahir Dar gulf, Lake Tana, Ethiopia. PhD. Thesis, Wageningen Agricultural University

Chapter 13
Birds of Lake Tana Sub-basin

Shimelis Aynalem

Abstract Birds play a vital role in keeping the balance of nature. Lake Tana is the largest lake in Ethiopia, with a surface area of 3200 km^2 and a watershed of 15,100 km^2 located at 1800 m a.s.l. Most of the wetlands situated along the rivers and Lake Tana shore support several thousands of birds. Taking previous work in the area into consideration, the Lake Tana sub-basin possesses six globally threatened species, 35 highland biome species and four Sudan–Guinea Savanna biome species. The sub-basin area is particularly important for wetland and water birds. The total population of birds counted during the winter season exceeds 150,000 seasonally, in Lake Tana area alone. Shesher-wallala area contributes the largest winter population. The Lake Tana sub-basin consists of 78 families of 437 birds; the most dominant is Accipitridae with 39 species. The distribution of birds in the area is changeable. However, the Lake Tana area and associated wetlands hold the largest bird populations compared with forest, farmland and bush land habitats. The Fogera and Dembia flood plain provides suitable feeding and roosting sites for winter birds, and is also known as a congregation site. Both Palearctic (112) and Afrotropical migrants (7) occur. In addition, the Lake Tana sub-basin holds 15 species of endemic birds, some of which are shared with Eritrea. Degradation of the wetlands is underway, with overgrazing, vegetation removal, cultivation, deforestation, occurrence of invasive weeds, wetland drainage, flooding, continuing sedimentation, major water resources development in the basin are the main threats.

Keywords Avifauna · Endemic birds · Migratory birds · Lake Tana Sub-basin birds

---

S. Aynalem (✉)
Blue Nile Water Institute and College of Agriculture and Environmental Sciences, Bahir Dar University, P.O. Box 2205, Bahir Dar, Ethiopia
e-mail: shimelisay@yahoo.co.in; shimelis.aynalem@gmail.com

S. Aynalem
College of Natural Science, Addis Ababa University, Addis Ababa, Ethiopia

S. Aynalem
Research Associate of International Crane Foundation, Baraboo, USA

K. Stave et al. (eds.), *Social and Ecological System Dynamics*,
AESS Interdisciplinary Environmental Studies and Sciences Series,
DOI 10.1007/978-3-319-45755-0_13

13.1 Introduction

Lake Tana is the largest lake in Ethiopia, with a surface area of 3200 km^2 and a watershed of 15,100 km^2 located at 1800 m a.s.l (meter above sea level). It forms the head water of the Blue Nile River, which carries about 80% of the total volume of the Nile River at Khartoum, Sudan. The lake is fed by 61 small streams, but the Gilgel Abay, Rib, Gumara and Megech Rivers are the major ones. The volume of water they carry is variable. Gilgel Abay is considered to be the source of the Blue Nile, because it is the longest of all. This river carries the overspill of the lake from its southern extremity. Most of the wetland areas are situated along the rivers and Lake Tana shore. These wetlands support many thousands of birds.

Birds are an important component of biotic community of wetland ecosystems (Hillman 1993). They respond quickly to changes in habitat (Krebs et al. 1999; Sauer and Link 2002). Thus, birds are good bio-indicators of wetland habitat quality, productivity, and stability. Birds are often strongly linked with specific habitats (Jones 1998) and have also been used as surrogates for assessing the impact of habitat changes (Jones 2009). Monitoring the species abundance, habitat preference, and correlation between species abundance and habitat provides basic information for determining factors causing population fluctuation of bird species (Weller 1999). Subsequently, the information helps in conservation and management of threatened and endangered species (Ministry of Health 2005). Long-term avian monitoring can identify bird species in decline due to habitat loss or degradation.

Birds play a vital role in keeping the balance of nature; for example, fruit-eating birds help spread seeds; hummingbirds pollinate plants (Mitsch 2005). Many kinds of birds help farmers by consuming weed seeds, harmful insects and other agricultural pests. Some birds are especially helpful in keeping certain kinds of insects under control. Robins and sparrows, for example, are highly effective against cabbage worms, tomato worms, and leaf beetles. Rats and mice can cause huge losses on farms by eating stored grain. Hawks and owls prey on these mammals, control their population and limit such losses (Mitsch 2013).

An inventory of species is the most basic element of study of the avifauna of a site, because it summarises the diversity. It also shows the presence of rare species, if any, in the area. The number of rare and endemic species and the diversity of the species present at the site can be used as indicator of the importance of different sites or habitats for bird conservation. Birds are among the best known parts of the earth's biodiversity.

In Ethiopia, different ecosystems of high biological importance are threatened and there is a need for strong conservation action that should be supported by legislation. The threats include habitat destruction, human population growth, and encroachment by undesirable plant species following flooding, shrinkage of lakes and wetlands due to industrial and agricultural developments.

At Lake Tana sub-basin, a number of studies have been carried out on the birds of Lake Tana area; Francis and Shimelis (2007 the birds of Lake Tana report, unpublished), Shimelis and Afework (2008 and 2009), Shimelis et al. (2008), Nega and Afework (2008), Shimelis et al. (2011), Nowald et al. (2010), Nowald et al. (2007 First Survey of Eurasian Cranes Grus grus in Ethiopia, unpublished), and Shimelis (2013). However, comprehensive baseline information on the avifauna resource, distribution, habitats, and their threats is lacking. This chapter therefore will fill this gap by summarizing previous literature and by adding to the existing information of the area.

13.2 Avifaunal Characteristics of the Lake Tana Sub Basin

Lake Tana qualifies as an Important Bird Area (IBA) (Tilahun et al. 1996). It possesses the globally threatened species such as wattled crane, pallid harrier, greater spotted eagle, Egyptian vulture, lesser kestrel, and sociable lapwing. It is estimated that the population of water birds around Lake Tana is likely to exceed 150,000 seasonally (Shimelis et al. 2011). Previously the Lake Tana area was considered to hold 19 highland biome species (Tilahun and Edwards 1996). However, the Lake Tana sub-basin has 35 highland biome species, out of 56 in the country as a whole (Ash and Atkins 2009). These species are: blue-winged goose, Erckel's francolin, wattled ibis, spot-breasted lapwing, black-winged lovebird, yellow-fronted parrot, white-cheeked turaco, Abyssinian (African long-eared) owl, Nyanza swift, banded barbet, Abyssinian woodpecker, Abyssinian oriole, thick-billed raven, white-backed black-tit, brown woodland-warbler, Abyssinian catbird, parisoma (brown warbler), montane (broad-ringed) white-eye, Abyssinian slaty-flycatcher, Rueppell's robin-chat, little rock-thrush, Rueppell's chat, white-winged cliff-chat, moorland chat, slender-billed starling, white-billed starling, Tacazze sunbird, Abyssinian longclaw, African citril, yellow-rumped serin, streaky seedeater, brown-rumped seedeater, Swainson's sparrow, Baglafecht weaver, and yellow-bellied waxbill.

In addition, four Sudan–Guinea Savanna biome species are found: green-backed eremomela, white-fronted black-chat, red-billed pytilia and bar-breasted firefinch. Other species of interest include bronze sunbird, which has been reported on a number of occasions, and bar-breasted firefinch, which is fairly common in Gorgora area. Silvery-cheeked hornbill nests in the large figs around the lake, including in hotels along lake shore in Bahir Dar, and both brown-backed woodpecker and black-backed cisticola are known from the area to the west of Lake Tana.

The sub-basin area particularly is important for wetland and water birds. Some of which occur in large numbers. For example, the total population of birds counted during the winter season exceeds 150,000 seasonally (Shimelis 2013) and more than 100,000 bird were counted in *Shesher-wallala* alone in 2009 (Shimelis et al. 2011).

13.3 Bird Species Diversity and Distribution

13.3.1 Birds Species in Lake Tana Area and Its Sub-basin

The Lake Tana sub-basin holds of 78 families of 437 bird species, which is more than 50% of the total bird population of the country. The most numerous family is Accipitridae with 39 bird species; Muscicapidae, Anatidae, Ploceidae and Scolopacidae have 27, 20, 15 bird species, respectively. Conversely, Anhingidae, Buphagidae, Coliidae, Dicruridae, Heliornithidae, Leiothrichidae, Macrosphenidae, Monarchidae, Numididae, Otididae, Pandionidae, Paridae, Phoenicopteridae, Pteroclidae, Pycnonotidae, Sagittariidae, Scopidae, Trogonidae, Tytonidae and Upupidae have only one species each. Details of species counts are given in Annex 11.1.

13.3.2 Distribution

The distribution of birds in the area varies. However, the Lake Tana area and associated wetlands hold the largest number of birds compared to forest, farmland and bush land. The Fogera and Dembia flood plain with the specific sites Shesher-wallala and the Megech and Dirma River delta provides suitable feeding and roosting sites for winter birds. However, the occurrence and population of birds depend on the presence of habitat suitable for them. The highland biome species are mostly distributed in Zegie peninsula, gulf of Lake Tana-Bahir Dar area, and islands. The yellow-fronted parrot in particular is strongly restricted to Zegie forest, but occasionaly can be found in Bahir Dar area.

13.4 Habitats

The major habitats around Lake Tana sub-basin are farmland, grassland, forest, rocky areas, wetlands, and the lake itself. Water retention is high, making the area prone to inundation, particularly the *Fogera* and *Dembia* plain (eastern and northeastern). The marshes support a variety of grasses, sedges and climbers, which also support a number of birds. One of the most striking features of Lake Tana is the extensive papyrus beds found along the *Gilgel* Abay River (Chimba area) and the Yiganda wetland area northwestern of Zegie Peninsula. The dominant macrophytes and trees are Typha, Echinochloa spp. grasses and Polygonum. Several aquatic plants, including *Nymphaea coerulea*, are noticeable (Shimelis 2013).

Farmlands: the Lake Tana sub-basin/catchment supports more than three million people. Ninety per cent of the area's rapidly growing population depends on subsistence agriculture for its livelihood, which crop-livestock farming is main one. The productivity and sustainability of mixed farming practices however, depend on ecosystem goods and services, which rely on the functional integrity of the watershed's ecosystems—rivers, wetlands, lake, forests, pastures and soils (Dessalegn et al. 2011). As a result most of the open lands are used for farming. The farmlands favour several seedeating birds because they can utilize left over crops after cultivation, as post harvest technology is not developed yet. Therefore the opportunities for utilizing farmlands for foraging is great; however, some birds such as cranes (common cranes, mostly) conflict with the local farmers if they arrive before crop harvesting time. In addition, since farmers use sequential farming, cranes will not have opportunity to utilize fallen grain and then shift to early growing crops such as legumes.

Farmers have a long tradition of retaining trees in their farmland for the following reasons: shade, animal fodder and for the construction of fencing. Common trees in the area are: scattered fig trees, *Croton macrostachys*, *Cordia africana*, *Vernonia amaygidalina*, *Mangifera indica*, *Sapium elliptium*, *Ficus thoningii* and *Euphorbia abyssinica*. The birds use these trees for shelter, resting to look for prey, and to get food, particularly frugivorous birds during fruiting time, such as like barbets, starlings, pigeons and weavers.

Forest/trees: some remnant indigenous trees are located on various farmland areas, islands, churches and the peninsula. In particular, Alemsaga forest (Debre Tabor area), Taragedam (Addis Zemen area), the Zegie peninsula (Bahir Dar area), the islands of Daga Estifanos, Dek and Kibran, and some smaller islands in the northern have a variety of dense tree stands. These areas harbour monasteries, churches and residences as well. On these sites significant numbers of bird species and highland biome species live. The common indigenous tree species are: *Cordia africana*, *Juniperus procera*, *Millettia ferruginea*, *Apodytus dimidiata var. acutifolia*, *Ficus sur*, *Ficus sycomorus*, *Ficus vasta*, *Sizygium guineense*, *Podocarpus falcatus*, *Rothmannia urcelliformis*, *Mimusops kummel*, *Celtis afaricana* (Shimelis 2013; Shimelis and Afework 2009).

Grass and bush lands: these are limited in extent. However, there are still volcanic mountains in the area. These harbor different types of small mammals, ground dwelling animals, insects, reptiles and raptors. The herbaceous and grass land areas along the Gilgel Abay River in particular commonly hold the black-bellied bustard and several lark species, some cisticolas, finches and pipits are abundant.

Riverine: this habitat supports many birds and functions as a roosting and feeding area. Birds found in the area are high canopy snatchers and gleaners in addition to those that prefer the low understory. The edge of the Kibran-Gebriel and Entos-Iyesus islands, the Zegie peninsula, Abay, Megech, Dirma, Gumara and Infranz hold riverine habitats known to harbor woodland bird species, especially raptors that prefer these sites for nesting, roosting and resting. This area also supports frugivorous and highland biome birds.

Modified habitats: mixed/modified types of vegetation are located along the coastal areas of the lake. They include the recreation areas, such as Gorgora, and the residential area in Bahir Dar. Because of the diversified vegetation of these areas, they attract various types of bird species, including nesting and roosting by eagles, herons, stork and vultures. However, due to advancement of construction and development, these trees are being removed and cut at alarming rate. The Abdim's stork in particular constructs nests on fig trees in Bahir Dar City.

Water bodies: the lake itself and some of the temporary water reservoirs or ponds formed during the wet season are included under this category. Depending on the season, the type and number of water fowl is variable. For example, during October–March, migratory species are dominant and abundant. However, residents like great white pelicans, Egyptian and spur-winged geese, black and yellow billed ducks, cormorants, some gulls, some terns, grebes and shore birds like rails, crake, gallinules, coots, herons, egrets and others are also common.

Mudflats and Wetlands: following the shore of the lake and rivers, several natural ponds are found. Their distribution and size varies; however, Shehser-wallala is a site known to harbour wintering birds. In addition, Gudo-Bahir (in Bahir Dar city) harbours thousands of geese and ducks as well as the Abay River delta (south west of Lake Tana). During the dry season and post rainy season, both of these areas hold very large numbers of birds. Most of these wetlands are situated around the lake shore and are underwater during the rainy season. The water level stays high until the middle of the dry season (January–February); however it decreases as the dry season advances. Some of the extensive wetlands are Chimba, Yiganda, Gerima (near Bahir Dar city), Infranz (behind Bahir Dar Airport), Abay inflow delta, Kunzila area, Delghi and Dirma. These areas are known as both breeding and feeding sites for birds. A few fragmented and pocket wetlands still serve for feeding so long as the wet surface exists, but most of them are at risk for encroachment. The lake shore is predominantly rocky and has undulating marshes and swamps. The most common and dominant macrophyte in wetlands are *Andropogon gayana, Cheilanthes species, Cyperus rotundus, Cyperus papyrus, Echinochloa stagina, Hyperrhenia rufa, Ipomoea aquatic, Leersia hexandra, Ludwigia stolonifera, Nymphaea noucgali, Oryza longistaminata,* and *Typha latifolia*. This vegetation supports large number of birds, but is under heavy grazing pressure and invasive weed infestation.

13.5 Migratory, Congregators and Endemic Species

13.5.1 Migratory

Migrants include representatives from a wide range of avian families, such as seabirds, freshwater species, raptors and waders. The largest proportion involves birds breeding in the Palearctic region in the north, which migrate south to Africa in the autumn and return north in the spring. At the opposite extreme are birds which breed in the southern Afrotropical region or in areas even further south, which migrate north after breeding (Ash and Atkins 2009). In the Lake Tana area both Palearctic and Afrotropical migrants occur. The sub-basin holds about 112 Palearctic migrants and seven known Afro-tropical migrants.

A number of species, partly or entirely breeding south of Ethiopia, move north in the non-breeding season, including comb duck, African openbill, Abdim's stork, woolly-necked stork, black-necked eared grebe, marabou stork, and pink-backed pelican. However, the Abdim's stork is a breeding winter visitor around Lake Tana unlike the others. It usually stays from March–June. It prefers old fig trees located in the urban area to establish nests. The mere presence of these indigenous trees in the city attracts this bird to breed. However, the future of the Abdim's stork as a breeding bird in the area is questionable since the construction of houses; hotels and other infrastructure are threatening indigenous trees.

The most abundant Palearctic migrants are: avocet (1209), common crane (21,358), common teal (15,000), godwit (28,000), greater flamingo (830), gull-billed tern (300), northern shoveler (8000), pintail (3800), and ruff (15,000) based on only one count in 2009 at Shesher-wallala flood plain (Shimelis et al. 2011). The Lake Tana basin holds the largest population of wintering common crane in Ethiopia if we sum counts made in different areas around the Lake Tana area (Nowald et al. 2007, first survey of Eurasian cranes *Grus grus* in Ethiopia, unpublished; Nowald et al. 2010; Shimelis et al. 2011).

13.5.2 Congregators

Congregatory behaviour is a feature of many bird families and often occurs at particular stages of the life- cycle. While some species congregate to breed, others congregate during the non-breeding season. Congregation is particularly common in waterbirds including many pelicans, herons, egrets, ducks, geese and swans (not found in Ethiopia), storks, shorebirds and other waders, gulls and terns. Lesser flamingo (*Phoenicopterus minor*) in Abijjata Shalla Lake relies on a handful of large soda lakes and pans, mainly within the Great Rift Valley, for feeding and nesting. Millions of birds may congregate at a single site, moving on to another when local conditions change. Conservation of each of these sites is vital for the species' continued survival. Some migratory waterbirds are heavily reliant on a small number of suitable stopover sites, at which most of the population may congregate over a short time period to refuel and rest while on migration. Congregation offers advantages, including providing 'safety in numbers' from natural predators, but can also increase vulnerability to site-related threats. Destruction or degradation of key sites can have serious impacts on congregatory birds at the population level (BirdLife International 2008).

Lake Tana area is an IBA (Important Bird Area) site based on the criteria of "Category A4" That is congregation criteria III—sites known or thought to hold on a regular basis $\geq$ 20,000 water birds or $\geq$ 10,000 pairs of sea birds of one or more species (Tilahun and Edwards 1996). In Lake Tana basin there are sites that are important for congregatory birds that are already mentioned in Sect. 13.5.1. The *Shesher-wallal* flood plain area in the east, the Seraba-Megech flood plain area in the north, the Chimba wetlands and Abay River inflow delta in the south are very important places for congregatory birds. These places may support 25,000–100,000 birds at a time, during the winter season.

13.5.3 Endemics

An Endemic Bird Area (EBA) is defined as an area that encompasses the overlapping breeding ranges of two or more restricted-range landbirds, such that the complete ranges of at least two species fall entirely within the boundary of the EBA. A total of 218 EBAs has been identified across the world, covering the ranges of 93% of restricted-range birds (2451 species). The majority of EBAs (77%) are in the tropics and subtropics. There are approximately equal numbers of island EBAs (105) and mainland EBAs (113). Of the island EBAs, 70% are on oceanic islands, 30% on continental-shelf islands, while for the mainland ones, 42% are located largely in montane areas, and 35% in lowland areas and 24% span both. The predominant natural habitat in most EBAs (c. 80%) is forest, especially tropical lowland and montane moist forest (Bird Life International 2004).

In the Lake Tana basin there are 15 species of birds endemic to Ethiopia, some of which are shared with Eritrea. They are: blue-winged goose, spot-breasted lapwing, black-winged lovebird, yellow-fronted parrot, banded barbet, Abyssinian woodpecker, Abyssinian oriole, Erlanger's lark, white-backed black-tit, Abyssinian catbird, Rueppell's chat, white-winged cliff-chat, Abyssinian longclaw, yellow-rumped serin, and red-billed pytilia. No detailed study has been conducted on these endemic birds. However, the status and distribution of these species are described the IUCN Red List of Threatened Species Version 2014.3 (IUCN 2014).

13.6 Human Impacts on Birds and Their Habitats

Together with natural succession, the impacts of which are slow, human factors are the most significant threat to birds and their habitats. In Lake Tana sub-basin, degradation of wetlands due to overgrazing, vegegation removal, sedimentation, cultivation, deforestation, occurrence of invasive weeds, wetland drainage, flooding, construction and industrial and domestic wastes are the main threats.

Cultivation: there is no demarcated boundary of private farm lands; as a result farmers encroach to the wetland or communal grazing areas. The flood plain are being drained, which shortens the duration of the flood to before winter birds return back north to their breeding grounds. Sequential farming in particular affects cranes that are dependent on fallen grain.

Continuing sedimentation because of poor watershed management practice, water resources development in the basin, continuing land use change, planned and current practice of irrigation schemes in the tributaries, and high nutrient load to the lake are also threatening the wetlands.

Overgrazing: due to human pressure most of agricultural lands are cultivated year to year without fallow periods. Private grazing land is not common; however, most of the wetlands (Yiganda, Dirma, Chimba and others) are grazing places for enormous number of cattle. Since wetlands are communally owned, they are free for anyone to exploit the resource. Thus livestock are a major factor in the degradation of wetlands and associated breeding and feeding sites of birds too.

Vegetation removal: Papyrus is removed for commercial purposes. The stems are cut, taken and transported to *Bahir Dar*. During this process, disturbance occurs to those birds dependent on the habitat for shelter, nesting and resting. Cut and carry is common, but there is no appropriate management practice. However, birds dependent on the presence of stems for nest building are particularly affected. Village and Baglafecht weaver are the main species concerned, and also birds that forage on the tubers of sedges and papyrus.

Invasive weeds: the invasive species Water hyacinth (*Eichornia crassipes*) is the most striking feature in *Dirma* area, with the weed *Hygrophila schulli* also. Water hyacinth infestation of the area is very extensive. Movement of patches of the weed leads to it being highly dispersed in the study area. Even though a weed eradication campaign began in 2012, the management techniques were ineffective in controlling growth during the next wet season. The waste weeds accumulated at the edge of wetlands and regrew where there was enough moisture, though weed that was exposed at rocks did not regrow. In addition, use of fertilizer at the lake shore aggravates the situation, since it is an input for the growth and multiplication of the weed. Invasive species continue to dominate the center of Dirma wetland. However, the weed is not yet recorded in the southern and eastern part of the Lake. Those birds that depend on the area for feeding and nesting such as cranes, geese, crakes and herons are being affected (Shimelis 2015, effect of water hyacinth (*Eichhornia crassipes* (Mart.) Solms) on waterfowls of Megech, Dirma and Rib River mouths at Lake Tana, Ethiopia, unpublished).

13.7 Research Gaps

- Little research has been conducted on the biology and ecology of wetland, forest and water birds; therefore, birds that are globally threatened, endemic and restricted range species must be prioritized.
- The impact of environmental pollution (domestic waste and effluents) in Bahir Dar and nearby area should be assessed.
- The impacts of irrigation, fertilizer application, and farming practice on the degradation of bird habitats and birds themselves should be evaluated.
- Indigenous trees, fig in particular, are very important for resting, feeding and nesting sites of several birds; therefore GIS-based mapping of trees in Bahir Dar should be carried out for future protection of these old indigenous trees, allowing future urban planning to co-exist with the environment.
- Lake Tana area is rich in terms of avifauna. Hence further study on the distribution, density, status, behavior of each species relative to existing conditions such as human impact and habitat preference should be carried out.

13.8 Recommendations

Dirma, Gelda, Yiganda and Chimba wetlands are very important breeding places of the globally threatened bird species black-crowned crane. Chimba wetland is also the only breeding site for wattled crane and black-crowned crane as well. Several species including geese, storks, ducks and rails and crake also nest here. The flood plain vegetation of the area provides suitable habitat for them. The area not only supports the resident wetland-dependent resident birds but also provides winter grounds for many migratory birds. These sites are the most important represention of the wetland ecosystem of Lake Tana area. Therefore, it should be protected and conserved for the next generation; since this should help guarantee hydrological functions can sustain many organisms.

Shesher-wallala is a flood plain that is very important as a roosting site for common crane and also feeding place for several thousands of winter birds. In order to protect and manage for future generation the following recommendations are made:

- The hydrology and wetland interaction of *Chimba* area should be studied, since the presence of this wetland is crucial as a breeding area for wattled and black-crowned cranes.
- Annual data collection on the avifauna of the area should be undertaken, as birds are ecological indicators.
- The specific breeding sites of globally threatened birds must be protected.
- The feeding ecology of birds in relation to habitat and human interaction should be studied.
- The unique feature of each wetland in terms of hydrology, ecological characteristics, human intervention, should be studied. This could give insights allowing lost wetlands to be rehabilitated.
- Wise utilization of the wetland resource should be given great emphasis.

13.9 Conclusion

The Lake Tana sub-basin forms the head water of the Blue Nile River, with a 15,100 km^2 catchment area. The presence of varied habitats—woodland to wooded grassland, cultivated land to natural forest and mixed forest area, water body to open wetland, together with low direct persecution on birds by humans has given rise to a diverse birdcommunity. Four hundred and thirty seven species totalling more than one hundred thousand birds occur in the area. In addition, the area is important for several species of globally threatened, highland biome, winter migrant and endemic birds. The distribution of each species of birds differs according to the availability of good breeding grounds, foraging areas, and roosting places. The eastern and north eastern part of Lake Tana area supports the largest numbers of birds as compared to other sites. The presence of ample food and roosting sites, particularly for winter migrants, in the Fogera and Dembia plain helps to explain this. But, highland biome species are situated mostly where there are indigenous forest patches in riverine, island, monastery and inland church forest areas in the southern and eastern parts of Lake Tana area. A wide array of threats is recognized. Human-induced factors such as wetland conversion to agricultural lands, deforestation, sedimentation, pesticides and fertilizer application, and recent occurrence of water hyacinth are the main ones. One positive point is that hunting of birds for food is unlikely to happen, since it is a taboo, except for francolin and guineafowl species in some cases.

Hence the Lake Tana sub-basin area should be properly managed, conserved and protected for sustainable use of resources. Degradation and pollution of the area could disrupt the ecosystem of the sub-basin where birds are biodiversity components of great importance for the ecological functioning of the area. Comprehensive studies and investigation of the avifauna resource are still lacking; therefore, the research gaps mentioned above should be given due attention for future conservation and utilization of the resource. This would be helped by promoting eco-tourism and bird watching activities to support the national economy in general, and the Lake Tana area in particular.

Annex 11.1 Checklists of the Birds of Lake Tana Sub-basin, Ethiopia (Shimelis A 2014 Own Survey Unpublished)

Number of Families: 78, Number of species: 437, Number of endemics: 15, Number of globally threatened species: 6, Number of near threatened species: 11, Date last reviewed: 2012-02-11, Source: ©Denis Lepage 2012: http://avibase.bsc-eoc.org/checklist.jsp?region=etandlist=clements; see also: http://www.africanbirdclub.org/countries/checklists/ accessed 15/06/2012.

Family name	Common name	Scientific name	Status/Habits
Anatidae	White-faced Whistling-Duck	*Dendrocygna viduata*	Resident
	Fulvous Whistling-Duck	*Dendrocygna bicolor*	Resident
	White-backed Duck	*Thalassornis leuconotus*	Resident
	Blue-winged Goose	*Cyanochen cyanoptera*	Endemic Vulnerable
	Comb Duck	*Sarkidiornis melanotos*	Afrotropical migrant Resident
	Egyptian Goose	*Alopochen aegyptiaca*	Resident
	Spur-winged Goose	*Plectropterus gambensis*	Resident

(continued)

(continued)

Family name	Common name	Scientific name	Status/Habits
	African Pygmy-Goose	*Nettapus auritus*	Resident
	African Black Duck	*Anas sparsa*	Resident
	Gadwall	*Anas strepera*	Palearctic migrant
	Eurasian Wigeon	*Anas penelope*	Palearctic migrant
	Yellow-billed Duck	*Anas undulata*	Palearctic migrant
	Northern Shoveler	*Anas clypeata*	Palearctic migrant
	Red-billed Duck	*Anas erythrorhyncha*	Palearctic migrant
	Northern Pintail	*Anas acuta*	Palearctic migrant
	Garganey	*Anas querquedula*	Palearctic migrant
	Eurasinan (Green-winged) Teal	*Anas crecca*	Palearctic migrant
	Southern Pochard	*Netta erythrophthalma*	Palearctic migrant
	Common Pochard	*Aythya ferina*	Palearctic migrant
	Ferruginous Duck	*Aythya nyroca*	Near-threatened rare—to Lake Tana area
Numididae	Helmeted Guineafowl	*Numida meleagris*	Resident
Phasianidae	Erckel's Francolin	*Francolinus erckelii*	Resident
	Clapperton's Francolin	*Francolinus clappertoni*	Resident
	Harlequin Quail	*Coturnix delegorguei*	Resident
	Stone Partridge	*Ptilopachus petrosus*	Resident
Podicipedidae	Little Grebe	*Tachybaptus ruficollis*	Resident
	Great Crested Grebe	*Podiceps cristatus*	Resident
	Black-necked Eared Grebe	*Podiceps nigricollis*	Resident (Interafrica migrant)
Phoenicopteridae	Greater Flamingo	*Phoenicopterus roseus*	Palearctic migrant
Ciconiidae	African Openbill	*Anastomus lamelligerus*	Resident (Interafrica migrant)
	Black Stork	*Ciconia nigra*	Palearctic migrant
	Abdim's Stork	*Ciconia abdimii*	Breeding visitor (Interafrica migrant)
	Woolly-necked Stork	*Ciconia episcopus*	Resident (Interafrica migrant)
	White Stork	*Ciconia ciconia*	Palearctic migrant
	Saddle-billed Stork	*Ephippiorhynchus senegalensis*	Resident
	Marabou Stork	*Leptoptilos crumeniferus*	Resident (Interafrica migrant)
	Yellow-billed Stork	*Mycteria ibis*	Resident

(continued)

(continued)

Family name	Common name	Scientific name	Status/Habits
Phalacrocoracidae	Great Cormorant	*Phalacrocorax carbo*	Resident
	Long-tailed Cormorant	*Phalacrocorax africanus*	Resident
Anhingidae	African Darter	*Anhinga rufa*	Resident
Pelecanidae	Great White Pelican	*Pelecanus onocrotalus*	Resident
	Pink-backed Pelican	*Pelecanus rufescens*	Resident (Interafrica migrant)
Scopidae	Hamerkop	*Scopus umbretta*	Resident
Ardeidae	Eurasian (Great) Bittern	*Botaurus stellaris*	Rare/Accidental Palearctic migrant
	Little Bittern	*Ixobrychus minutus*	Resident
	Gray Heron	*Ardea cinerea*	Palearctic migrant
	Black-headed Heron	*Ardea melanocephala*	Resident
	Goliath Heron	*Ardea goliath*	Resident
	Purple Heron	*Ardea purpurea*	Resident
	Great Egret	*Ardea alba*	Resident
	Intermediate Egret	*Mesophoyx intermedia*	Resident
	Little Egret	*Egretta garzetta*	Resident
	Western Reef-Heron	*Egretta gularis*	Rare/Accidental
	Black Heron	*Egretta ardesiaca*	Resident??
	Cattle Egret	*Bubulcus ibis*	Resident
	Squacco Heron	*Ardeola ralloides*	Resident
	Striated Heron	*Butorides striata*	Resident
	Black-crowned Night-Heron	*Nycticorax nycticorax*	Resident
	White-backed Night-Heron	*Gorsachius leuconotus*	Resident-rare
Threskiornithidae	Glossy Ibis	*Plegadis falcinellus*	Palearctic migrant
	Sacred Ibis	*Threskiornis aethiopicus*	Resident
	Hadada Ibis	*Bostrychia hagedash*	Resident
	Wattled Ibis	*Bostrychia carunculata*	Endemic-resident
	African Spoonbill	*Platalea alba*	Resident
Pandionidae	Osprey	*Pandion haliaetus*	Resident
Accipitridae	Black-shouldered Kite	*Elanus caeruleus*	Resident
	African Swallow (Scissor)-tailed Kite	*Chelictinia riocourii*	Palarecric migrant
	Black Kite	*Milvus migrans*	Palearctic migrant

(continued)

(continued)

Family name	Common name	Scientific name	Status/Habits
	African Fish-Eagle	*Haliaeetus vocifer*	Resident
	Hooded Vulture	*Necrosyrtes monachus*	Resident
	Lammergeier	*Gypaetus barbatus*	Resident, but rare to the area
	Egyptian Vulture	*Neophron percnopterus*	Endangered Palearctic migrant, rare to Lake Tana
	White-backed Vulture	*Gyps africanus*	Near-threatened-Resident
	Rueppell's Griffon	*Gyps rueppellii*	Near-threatened Palearctic migant
	Yellow billed Kite	*Milvus aegyptius*	Resident
	Lappet-faced Vulture	*Torgos tracheliotus*	Resident-migrant
	White-headed Vulture	*Trigonoceps occipitalis*	Vulnerable-Resident migrant
	Black-chested (breasted) Snake-Eagle	*Circaetus pectoralis*	Resident
	Brown Snake-Eagle	*Circaetus cinereus*	R-migrant
	Western-banded Snake-Eagle	*Circaetus cinerascens*	Resident
	Bateleur	*Terathopius ecaudatus*	Resident
	Eurasian Marsh-Harrier	*Circus aeruginosus*	Palearctic migrant
	Pallid Harrier	*Circus macrourus*	Near-threatened Palearctic migrant
	Montagu's Harrier	*Circus pygargus*	Palearctic migrant
	African Harrier-Hawk	*Polyboroides typus*	Resident
	Dark Chanting-Goshawk	*Melierax metabates*	Resident
	Gabar Goshawk	*Micronisus gabar*	Resident
	African Goshawk	*Accipiter tachiro*	Resident
	Shikra	*Accipiter badius*	Resident
	Little Sparrowhawk	*Accipiter minullus*	Resident
	Great-sparrow Hawk (Black Goshawk)	*Accipiter melanoleucus*	Resident
	Common Buzzard	*Buteo buteo*	Palearctic migrant
	Long-legged Buzzard	*Buteo rufinus*	Palearctic migrant
	Augur Buzzard	*Buteo augur*	Resident
	Greater Spotted Eagle	*Aquila clanga*	Vulnerable Palearctic migrant
	Tawny Eagle	*Aquila rapax*	Resident
	Steppe Eagle	*Aquila nipalensis*	Palearctic migrant
	Verreaux's Eagle	*Aquila verreauxii*	Resident
	African Hawk-Eagle	*Aquila spilogaster*	Resident
	Wahlberg's Eagle	*Hieraaetus wahlbergi*	Palearcic migrant
	Booted Eagle	*Hieraaetus pennatus*	Palearctic migrant
	Ayres's Hawk-Eagle	*Hieraaetus ayresii*	Resident

(continued)

(continued)

Family name	Common name	Scientific name	Status/Habits
	Martial Eagle	*Polemaetus bellicosus*	Resident
	Long-crested Eagle	*Lophaetus occipitalis*	Resident
Sagittariidae	Secretary-bird	*Sagittarius serpentarius*	Resident migrant
Falconidae	Lesser Kestrel	*Falco naumanni*	Vulnerable Palearctic migrant
	Common (Eurasian) Kestrel	*Falco tinnunculus*	Palearctic migrant
	Greater Kestrel	*Falco rupicoloides*	Resident-migrant
	Gray Kestrel	*Falco ardosiaceus*	Resident
	Red-necked Falcon	*Falco chicquera*	Palearcti migrant
	Red-footed Falcon	*Falco vespertinus*	Near-threatened
	Eurasian Hobby	*Falco subbuteo*	Palearctic migrant
	African Hobby	*Falco cuvierii*	Resident
	Lanner Falcon	*Falco biarmicus*	Resident
	Peregrine Falcon	*Falco peregrinus*	Palearctic migrant
Otididae	Black-bellied Bustard	*Lissotis melanogaster*	Resident
Rallidae	Corn Crake	*Crex crex*	Near-threatened Pm
	African Rail	*Rallus caerulescens*	Resident-migrant
	Black Crake	*Amaurornis flavirostra*	Resident
	Little Crake	*Porzana parva*	Palearctic migrant
	Spotted Crake	*Porzana porzana*	Palearctic migrant
	Allen's Gallinule	*Porphyrio alleni*	Resident-migrant
	Common (Eurasian) Moorhen	*Gallinula chloropus*	Resident
	Red-knobbed Coot	*Fulica cristata*	Resident
Heliornithidae	African Finfoot	*Podica senegalensis*	Resident
Gruidae	Black Crowned-Crane	*Balearica pavonina*	Near-threatened Resident
	Wattled Crane	*Bugeranus carunculatus*	Vulnerable-Resident
	Common Crane	*Grus grus*	Palearctic migrant
Burhinidae	Senegal Thick-knee	*Burhinus senegalensis*	Resident
	Spotted Thick-knee	*Burhinus capensis*	Resident
Charadriidae	Spur-winged Plover	*Vanellus spinosus*	Resident
	Black-headed Lapwing	*Vanellus tectus*	Resident
	Black-winged Lapwing	*Vanellus melanopterus*	Resident
	Wattled Lapwing		Resident

(continued)

(continued)

Family name	Common name	Scientific name	Status/Habits
		Vanellus senegallus	
	Spot-breasted Lapwing	*Vanellus melanocephalus*	Endemic
	Sociable Lapwing	*Vanellus gregarius*	Critically endangered
	White-tailed Lapwing	*Vanellus leucurus*	Rare/Accidental
	Gray (Black-bellied) Plover	*Pluvialis squatarola*	Palearctic migrant
	Caspian Plover	*Charadrius asiaticus*	Palearctic migrant
	Kittlitz's Plover	*Charadrius pecuarius*	Palearctic migrant
	Common Ringed Plover	*Charadrius hiaticula*	Palearctic migrant
	Little Ringed Plover	*Charadrius dubius*	Palearctic migrant
	Three-banded Plover	*Charadrius tricollaris*	Resident
Recurvirostridae	Black-winged Stilt	*Himantopus himantopus*	Palearctic migrant
	Pied Avocet	*Recurvirostra avosetta*	Palearctic migrant
Jacanidae	Lesser Jacana	*Microparra capensis*	Rare/Accidental Resident
	African Jacana	*Actophilornis africanus*	Resident
Scolopacidae	Common Sandpiper	*Actitis hypoleucos*	Palearctic migrant
	Green Sandpiper	*Tringa ochropus*	Palearctic migrant
	Spotted Redshank	*Tringa erythropus*	Palearctic migrant
	Common Greenshank	*Tringa nebularia*	Palearctic migrant
	Marsh Sandpiper	*Tringa stagnatilis*	Palearctic migrant
	Wood Sandpiper	*Tringa glareola*	Palearctic migrant
	Common Redshank	*Tringa totanus*	Palearctic migrant
	Eurasian Curlew	*Numenius arquata*	Near-threatened—Palearctic migrant
	Black-tailed Godwit	*Limosa limosa*	Near-threatened—Palearctic migrant
	Little Stint	*Calidris minuta*	Palearctic migrant
	Temminck's Stint	*Calidris temminckii*	Palearctic migrant
	Curlew Sandpiper	*Calidris ferruginea*	Palearctic migrant
	Ruff	*Philomachus pugnax*	Palearctic migrant
	Common Snipe	*Gallinago gallinago*	Palearctic migrant
	African Snipe	*Gallinago nigripennis*	Resident

(continued)

(continued)

Family name	Common name	Scientific name	Status/Habits
Glareolidae	Temminck's Courser	*Cursorius temminckii*	Resident
	Collared Pratincole	*Glareola pratincola*	Palearctic migrant
	Rock Pratincole	*Glareola nuchalis*	Resident-migrant
Laridae	Black-headed Gull	*Chroicocephalus ridibundus*	Palearctic migrant
	Pallas's Gull	*Ichthyaetus ichthyaetus*	Palearctic migrant
	Lesser Black-backed Gull	*Larus fuscus*	Palearctic migrant
	Gull-billed Tern	*Gelochelidon nilotica*	Palearctic migrant
	White-winged Tern	*Chlidonias leucopterus*	Palearctic migrant
	Whiskered Tern	*Chlidonias hybrida*	Palearctic migrant
	Slender-billed Gull	*Larus genei*	Palearctic migrant
Pteroclidae	Four-banded Sandgrouse	*Pterocles quadricinctus*	Resident
Columbidae	Speckled Pigeon	*Columba guinea*	Resident
	White-collared Pigeon	*Columba albitorques*	Resident
	Lemon Dove	*Columba larvata*	Resident
	Red-eyed Dove	*Streptopelia semitorquata*	Resident
	Vinaceous Dove	*Streptopelia vinacea*	Resident
	Laughing Dove	*Streptopelia senegalensis*	Resident
	Blue-spotted Wood-Dove	*Turtur afer*	Resident
	Namaqua Dove	*Oena capensis*	Resident
	Bruce's Green-Pigeon	*Treron waalia*	Resident
Psittacidae	Black-winged Lovebird	*Agapornis taranta*	Endemic (country)
	Yellow-fronted Parrot	*Poicephalus flavifrons*	Endemic
Musophagidae	White-cheeked Turaco	*Tauraco leucotis*	Resident
	White-bellied Go-away-bird	*Corythaixoides leucogaster*	Resident
	Eastern Plantain-eater	*Crinifer zonurus*	Resident
Cuculidae	Pied Cuckoo	*Clamator jacobinus*	Resident
	Levaillant's Cuckoo	*Clamator levaillantii*	Resident
	Great Spotted Cuckoo	*Clamator glandarius*	Resident

(continued)

(continued)

Family name	Common name	Scientific name	Status/Habits
	Common Cuckoo	*Cuculus canorus*	Resident
	African Cuckoo	*Cuculus gularis*	Resident
	Red-chested Cuckoo	*Cuculus solitaries*	Resident-migrant
	Klaas's Cuckoo	*Chrysococcyx klaas*	Resident
	African Emerald Cuckoo	*Chrysococcyx cupreus*	Resident
	Blue-headed Coucal	*Centropus monachus*	Resident
	Senegal Coucal	*Centropus senegalensis*	Resident
	White-browed Coucal	*Centropus superciliosus*	Resident
Tytonidae	Barn Owl	*Tyto alba*	Resident
Strigidae	African Scops-Owl	*Otus senegalensis*	Resident
	European Scops-Owl	*Otus scops*	Palearctic migrant
	Grayish Eagle-Owl	*Bubo cinerascens*	Resident
	Verreaux's Eagle-Owl	*Bubo lacteus*	Resident
	Pearl-spotted Owlet	*Glaucidium perlatum*	Resident
	African Wood-Owl	*Strix woodfordii*	Resident
	Abyssinian (African Long-eared) Owl	*Asio abyssinicus*	Resident
	Marsh Owl	*Asio capensis*	Resident
Caprimulgidae	Mountane (Abyssinian) Nightjar	*Caprimulgus poliocephalus*	Resident
	Standard-winged Nightjar	*Macrodipteryx longipennis*	Resident
Apodidae	Alpine Swift	*Apus melba*	Resident-migrant
	Mottled Swift	*Apus aequatorialis*	Resident-migrant
	Common Swift	*Apus apus*	Palearctic migrant
	Nyanza Swift	*Apus niansae*	Resident
	Little Swift	*Apus affinis*	Resident
	Horus Swift	*Apus horus*	Resident
	White-rumped Swift	*Apus caffer*	Resident-breeding
	African Palm-Swift	*Cypsiurus parvus*	Resident
Coliidae	Speckled Mousebird	*Colius striatus*	Resident
Trogonidae	Narina Trogon	*Apaloderma narina*	Resident
Alcedinidae	Malachite Kingfisher	*Corythornis cristatus*	Resident
	African Pygmy-Kingfisher	*Ispidina picta*	Resident
	Gray-headed Kingfisher	*Halcyon leucocephala*	Resident

(continued)

(continued)

Family name	Common name	Scientific name	Status/Habits
	Woodland Kingfisher	*Halcyon senegalensis*	Resident
	Striped Kingfisher	*Halcyon chelicuti*	Resident
Meropidae	Red-throated Bee-eater	*Merops bulocki*	Resident
	Little Bee-eater	*Merops pusillus*	Resident
	Blue-breasted Bee-eater	*Merops variegatus*	Resident
	Northern Carmine Bee-eater	*Merops nubicus*	Resident-migrant
Coraciidae	European Roller	*Coracias garrulus*	Near-threatened Palearctic migrant
	Abyssinian Roller	*Coracias abyssinicus*	Resident-migrant
Upupidae	Eurasian Hoopoe	*Upupa epops*	Palearctic migrant
Phoeniculidae	Black-billed Woodhoopoe	*Phoeniculus somaliensis*	Resident
	Black Scimitar-bill	*Rhinopomastus aterrimus*	Resident
Bucerotidae	Hemprich's Hornbill	*Tockus hemprichii*	Resident
	African Gray Hornbill	*Tockus nasutus*	Resident
	Silvery-cheeked Hornbill	*Ceratogymna brevis*	Resident
	Abyssinian Ground-Hornbill	*Bucorvus abyssinicus*	Resident
Lybiidae	Red-fronted Tinkerbird	*Pogoniulus pusillus*	Resident
	Yellow-fronted Tinkerbird	*Pogoniulus chrysoconus*	Resident
	Banded Barbet	*Lybius undatus*	Endemic (country)
	Black-billed Barbet	*Lybius guifsobalito*	Resident
	Double-toothed Barbet	*Lybius bidentatus*	Resident
Indicatoridae	Wahlberg's Honeyguide	*Prodotiscus regulus*	Resident
	Lesser Honeyguide	*Indicator minor*	Resident
	Greater Honeyguide	*Indicator indicator*	Resident
Picidae	Eurasian Wryneck	*Jynx torquilla*	Palearctic migrant
	Red -throated (Rufous-necked) Wryneck	*Jynx ruficollis*	
	Nubian Woodpecker	*Campethera nubica*	
	Abyssinian Woodpecker	*Dendropicos abyssinicus*	Endemic
	Cardinal Woodpecker	*Dendropicos fuscescens*	Resident
	Bearded Woodpecker	*Dendropicos namaquus*	Resident
	Gray Woodpecker	*Dendropicos goertae*	Resident
			Resident

(continued)

(continued)

Family name	Common name	Scientific name	Status/Habits
	Gray-headed Woodpecker	*Dendropicos spodocephalus*	
	Brown-backed Woodpecker	*Dendropicos obsoletus*	Resident
Platysteiridae	Brown-throated Wattle-eye	*Platysteira cyanea*	Resident
	Black-headed Batis	*Batis minor*	Resident
Malaconotidae	Brubru	*Nilaus afer*	Resident-migrant
	Northern Puffback	*Dryoscopus gambensis*	Resident
	Black-crowned Tchagra	*Tchagra senegalus*	Resident
	Ethiopian (Tropical) Boubou	*Laniarius aethiopicus*	Resident
	Sulphur-breasted Bushshrike	*Telophorus sulfureopectus*	Resident
	Gray-headed Bushshrike	*Malaconotus blanchoti*	Resident
Campephagidae	Gray Cuckoo-shrike	*Coracina caesia*	Resident
	Red-shouldered Cuckoo-shrike	*Campephaga phoenicea*	Resident
Laniidae	Red-backed Shrike	*Lanius collurio*	Palearctic migrant
	Rufous-tailed Shrike	*Lanius isabellinus*	Palearctic migrant
	Southern Gray Shrike	*Lanius meridionalis*	Palearctic migrant
	Lesser Gray Shrike	*Lanius minor*	Resident
	Gray-backed Fiscal	*Lanius excubitoroides*	Resident
	Common Fiscal	*Lanius collaris*	Resident
	Masked Shrike	*Lanius nubicus*	Palearctic migrant
	Woodchat Shrike	*Lanius senator*	Palearctic migrant
Oriolidae	Eurasian Golden Oriole	*Oriolus oriolus*	Palearctic migrant
	African Golden Oriole	*Oriolus auratus*	Resident
	Abyssinian Oriole	*Oriolus monacha*	Endemic (country)
Dicruridae	Fork-tailed Drongo	*Dicrurus adsimilis*	Resident
Monarchidae	African Paradise-Flycatcher	*Terpsiphone viridis*	Resident
Corvidae	Cape Crow	*Corvus capensis*	Resident
	Pied Crow	*Corvus albus*	Resident
	Fan-tailed Raven	*Corvus rhipidurus*	Resident
	Thick-billed Raven	*Corvus crassirostris*	Resident
Alaudidae	Flappet Lark	*Mirafra rufocinnamomea*	Resident
	Chestnut-backed Sparrow-Lark	*Eremopterix leucotis*	Resident

(continued)

(continued)

Family name	Common name	Scientific name	Status/Habits
	Erlanger's Lark	*Calandrella erlangeri*	Endemic
	Thekla Lark	*Galerida theklae*	Resident
Hirundinidae	Plain Martin	*Riparia paludicola*	Resident
	Sand Martin (Bank Swallow)	*Riparia riparia*	Palearctic migrant
	Banded Martin	*Riparia cincta*	Palearctic migrant
	Rock Martin	*Ptyonoprogne fuligula*	Resident
	Barn Swallow	*Hirundo rustica*	Palearctic migrant
	Ethiopian Swallow	*Hirundo aethiopica*	Resident
	Wire-tailed Swallow	*Hirundo smithii*	Resident
	Red-rumped Swallow	*Cecropis daurica*	Resident
	Lesser Striped-Swallow	*Cecropis abyssinica*	Resident
	Mosque Swallow	*Cecropis senegalensis*	Resident
	Common House-Martin	*Delichon urbicum*	Palearctic migrant
	Black Sawwing	*Psalidoprocne pristoptera*	Resident
	Gray-rumped Swallow	*Pseudhirundo griseopyga*	Resident
Paridae	White-backed Black-Tit	*Melaniparus leuconotus*	Endemic
Pycnonotidae	Common Bulbul	*Pycnonotus barbatus*	Resident
Macrosphenidae	Northern Crombec	*Sylvietta brachyura*	Resident
Phylloscopidae	Brown Woodland-Warbler	*Phylloscopus umbrovirens*	Palearctic migrant
	Willow Warbler	*Phylloscopus trochilus*	Palearctic migrant
	Common Chiffchaff	*Phylloscopus collybita*	Palearctic migrant
Acrocephalidae	Eastern Olivaceous Warbler	*Hippolais pallida*	Palearctic migrant
	Sedge Warbler	*Acrocephalus schoenobaenus*	Palearctic migrant
	Eurasian Reed-Warbler	*Acrocephalus scirpaceus*	Palearctic migrant
	Marsh Warbler	*Acrocephalus palustris*	Palearctic migrant
	Great Reed-Warbler	*Acrocephalus arundinaceus*	Palearctic migrant
	Lesser Swamp-Warbler	*Acrocephalus gracilirostris*	Palearctic migrant

(continued)

(continued)

Family name	Common name	Scientific name	Status/Habits
Locustellidae	Little Rush-Warbler	*Bradypterus baboecala*	Resident
	Cinnamon Bracken-Warbler	*Bradypterus cinnamomeus*	Resident
	Common Grasshopper-Warbler	*Locustella naevia*	Rare/Accidental Palearctic migrant
	Savi's Warbler	*Locustella luscinioides*	Palearctic migrant??
Cisticolidae	Grey-backed Camaroptera	*Camaroptera brachyura*	Resident
	Buff-bellied Warbler	*Phyllolais pulchella*	Resident
	Red-faced Cisticola	*Cisticola erythrops*	Resident
	Singing Cisticola	*Cisticola cantans*	Resident
	Ethiopian (Winding) Cisticola	*Cisticola galactotes*	Resident
	Stout Cisticola	*Cisticola robustus*	Resident
	Croaking Cisticola	*Cisticola natalensis*	Resident
	Siffling Cisticola	*Cisticola brachypterus*	Resident
	Pectoral-patch Cisticola	*Cisticola brunnescens*	Resident
	black-backed cisticola	*Cisticola eximius*	Resident
	Tawny-flanked Prinia	*Prinia subflava*	Resident
	Yellow-bellied Eremomela	*Eremomela icteropygialis*	Resident
	Green-backed Eremomela	*Eremomela canescens*	Resident
Sylviidae	Blackcap	*Sylvia atricapilla*	Palearctic migrant
	Garden Warbler	*Sylvia borin*	Palearctic migrant
	Barred Warbler	*Sylvia nisoria*	Palearctic migrant
	Common (Greater) Whitethroat	*Sylvia communis*	Palearctic migrant
	Lesser Whitethroat	*Sylvia curruca*	Palearctic migrant
	Abyssinian catbird	*Parophasma galinieri*	Endemic
	Parisoma (Brown Warbler)	*Parisoma lugens*	Resident
Zosteropidae	African Yellow White-eye	*Zosterops senegalensis*	Resident
	Montane (Broad-ringed) White-eye	*Zosterops poliogastrus*	Resident
	White-breasted White-eye	*Zosterops abyssinicus*	Resident

(continued)

(continued)

Family name	Common name	Scientific name	Status/Habits
Leiothrichidae	White-rumped Babbler	*Turdoides leucopygia*	Resident
Muscicapidae	Pale Flycatcher	*Bradornis pallidus*	Resident
	Abyssinian Slaty-Flycatcher	*Melaenornis chocolatinus*	Resident
	Northern Black-Flycatcher	*Melaenornis edolioides*	Resident
	Spotted Flycatcher	*Muscicapa striata*	Palearctic migrant
	African Dusky Flycatcher	*Muscicapa adusta*	Resident
	Rufous-tailed Scrub-Robin	*Cercotrichas galactotes*	Palearctic migrant
	Rueppell's Robin-Chat	*Cossypha semirufa*	Resident
	Thrush Nightingale	*Luscinia luscinia*	Resident
	Common Nightingale	*Luscinia megarhynchos*	Resident
	Bluethroat	*Luscinia svecica*	Palearctic migrant
	Common Redstart	*Phoenicurus phoenicurus*	Palearctic migrant
	Little Rock-Thrush	*Monticola rufocinereus*	Resident
	Common (Rufous-tailed) Rock-Thrush	*Monticola saxatilis*	Palearctic migrant
	Blue Rock-Thrush	*Monticola solitarius*	Palearctic migrant
	Whinchat	*Saxicola rubetra*	Palearctic migrant
	African Stonechat	*Saxicola torquatus*	
	Rueppell's Chat	*Myrmecocichla melaena*	Endemic (country)
	White-fronted Black-Chat	*Myrmecocichla albifrons*	Palearctic migrant
	Mocking Cliff-Chat	*Thamnolaea cinnamomeiventris*	Resident
	White-winged Cliff-Chat	*Thamnolaea semirufa*	Endemic (country)
	Familiar Chat	*Cercomela familiaris*	Resident
	Moorland Chat	*Cercomela sordida*	Resident
	Northern Wheatear	*Oenanthe oenanthe*	Palearctic migrant
	Pied Wheatear	*Oenanthe pleschanka*	Palearctic migrant
	Black-eared Wheatear	*Oenanthe hispanica*	Palearctic migrant
	Isabelline Wheatear	*Oenanthe isabellina*	Palearctic migrant
	Red-breasted Wheatear	*Oenanthe bottae*	Resident

(continued)

(continued)

Family name	Common name	Scientific name	Status/Habits
Turdidae	Abyssinian Ground-Thrush	*Zoothera piaggiae*	Resident
	Groundscraper Thrush	*Psophocichla litsipsirupa*	Resident
	Abyssinian/mountain Thrush	*Turdus (Olivaceous) abyssinicus*	Resident
	African Thrush	*Turdus pelios*	Resident
Sturnidae	Wattled Starling	*Creatophora cinerea*	Resident
	Greater Blue-eared Glossy-Starling	*Lamprotornis chalybaeus*	Resident
	Lesser Blue-eared Glossy-Starling	*Lamprotornis chloropterus*	Resident
	Violet-backed Starling	*Cinnyricinclus leucogaster*	Resident
	Fischer's Starling	*Spreo fischeri*	Resident
	White-crowned Starling	*Spreo albicapillus*	Resident
	Red-winged Starling	*Onychognathus morio*	Resident
	Slender-billed Starling	*Onychognathus tenuirostris*	Resident
	White-billed Starling	*Onychognathus albirostris*	Resident
Buphagidae	Red-billed Oxpecker	*Buphagus erythrorhynchus*	Resident
Nectariniidae	Kenya Violet-backed Sunbird	*Anthreptes orientalis*	Resident
	Scarlet-chested Sunbird	*Chalcomitra senegalensis*	Resident
	Tacazze Sunbird	*Nectarinia tacazze*	Resident
	Variable Sunbird	*Cinnyris venustus*	Resident
	Copper Sunbird	*Cinnyris cupreus*	Resident
Motacillidae	Western Yellow Wagtail	*Motacilla flava*	Palearctic migrant
	Gray Wagtail	*Motacilla cinerea*	Palearctic migrant
	Mountain Wagtail	*Motacilla clara*	Palearctic migrant
	White Wagtail	*Motacilla alba*	Palearctic migrant
	African Pied Wagtail	*Motacilla aguimp*	Resident
	Grassland African Pipit	*Anthus cinnamomeus*	Resident
	Tawny Pipit	*Anthus campestris*	Palearctic migrant
	Plain-backed Pipit	*Anthus leucophrys*	Resident
	Tree Pipit	*Anthus trivialis*	Palearctic migrant
	Red-throated Pipit	*Anthus cervinus*	Palearctic migrant
	Abyssinian Longclaw	*Macronyx flavicollis*	Endemic Near-threatened

(continued)

(continued)

Family name	Common name	Scientific name	Status/Habits
Emberizidae	Ortolan Bunting	*Emberiza hortulana*	Palearctic migrant
	Cinnamon-breasted Bunting	*Emberiza tahapisi*	Resident
Fringillidae	Yellow-crowned Canary	*Serinus flavivertex*	Resident
	Yellow-fronted Canary	*Serinus mozambicus*	Resident
	African Citril	*Serinus citrinelloides*	Resident
	Yellow-rumped Serin	*Serinus xanthopygius*	Endemic (country)
	Streaky Seedeater	*Serinus striolatus*	Resident
	Brown-rumped Seedeater	*Serinus tristriatus*	Resident
Passeridae	Swainson's Sparrow	*Passer swainsonii*	Resident
	Bush Petronia	*Petronia dentata*	Resident
Ploceidae	Speckle-fronted Weaver	*Sporopipes frontalis*	Resident
	Chestnut-crowned Sparrow-Weaver	*Plocepasser superciliosus*	Resident
	Baglafecht Weaver	*Ploceus baglafecht*	Resident
	Little Weaver	*Ploceus luteolus*	Resident
	Spectacled Weaver	*Ploceus ocularis*	Resident
	Village Weaver	*Ploceus cucullatus*	Resident
	Red-billed Quelea	*Quelea quelea*	Resident
	Orange Bishop	*Euplectes franciscanus*	Resident
	Black-winged Bishop	*Euplectes hordeaceus*	Resident
	Yellow-crowned Bishop	*Euplectes afer*	Resident
	Yellow Bishop	*Euplectes capensis*	Resident
	White-winged Widowbird	*Euplectes albonotatus*	Resident
	Yellow-shouldered Widowbird	*Euplectes macroura*	Resident
	Red-collared Widowbird	*Euplectes ardens*	Resident
	Fan-tailed Widowbird	*Euplectes axillaris*	Resident
Estrildidae	Yellow-bellied Waxbill	*Coccopygia quartinia*	Resident
	Common Waxbill	*Estrilda astrild*	Resident
	Red-cheeked Cordonbleu	*Uraeginthus bengalus*	Resident
	Red-billed Pytilia	*Pytilia lineata*	Endemic
	Red-billed Firefinch	*Lagonosticta senegala*	Resident

(continued)

(continued)

Family name	Common name	Scientific name	Status/Habits
	Bar-breasted Firefinch	*Lagonosticta rufopicta*	Resident
	African Firefinch	*Lagonosticta rubricata*	Resident
	Cut-throat	*Amadina fasciata*	Resident
	Zebra Waxbill	*Sporaeginthus subflavus*	Resident
	African Quailfinch	*Ortygospiza fuscocrissa*	Resident
	Bronze Mannikin	*Spermestes cucullatus*	Resident
	African Silverbill	*Euodice cantans*	Resident
Viduidae	Pin-tailed Whydah	*Vidua macroura*	Resident
	Long-tailed Paradise-Whydah	*Vidua interjecta*	Resident
	Eastern Paradise-Whydah	*Vidua paradisaea*	Resident
	Village Indigobird	*Vidua chalybeata*	Resident
	Cuckoo Finch (Parasitic Weaver)	*Anomalospiza imberbis*	Resident

References

Ash J, Atkins J (2009) Birds of Ethiopia and Eritrea: an atlas of distribution. Christopher Helm, London

BirdLife International (2004) Most endemic bird areas are in the tropics and important for other biodiversity too. http://www.birdlife.org/datazone/sowb/casestudy/61. Accessed 19 Aug 2014

BirdLife International (2008) Congregation at particular sites is a common behaviour in many bird species. http://www.birdlife.org/datazone/sowb/casestudy/58. Accessed 19 Aug 2014

Dessalegn MK, Chlosom N, Enright P (2011) Putting research knowledge into Action: the missing link for sustainability of Lake Tana ecosystem, Ethiopia. Ee-JRIF 3(2):4–19

Hillman JC (1993) Ethiopia: compendium of wildlife conservation information. Vol 1 and 2. NYZS - The wildlife conservation society international, New York and Ethiopian Wildlife Conservation Organisation, Addis Ababa

Jones M (1998) Study design. In: Bibby C, Jones M, Marseden S (eds) Expedition field techniques: bird surveys. Royal Geographical Society, London, pp 15–34

Jones K, Lanthier Y et al (2009) Monitoring and assessment of wetlands using Earth Observation: the Glob Wetland project. J Envir Manag 90(7):2154–2169

Krebs JR, Wilson JD et al (1999) The second silent spring? Nature 400(6745):611–612

Ministry of Health: National hygiene and sanitation strategy. Addis Ababa, Ethiopia (2005) http://www.wsp.org/sites/wsp.org/files/publications/622200751450_EthiopiaNationalHygieneAndSanitationStrategyAF.pdf. Accessed 8 August 2004

Mitsch WJ (2005) Wetland creation, restoration, and conservation: a wetland invitational at the Olentangy River Wetland Research Park. Ecol Engin 24(4):243–251
Mitsch WJ (2013) Wetland Utilization in the World: Protecting Sustainable Use http://www.globalwetlands.org/Conference. Accessed 6 Nov 2013
Nega T, Afework B (2008) Diversity and habitat association of birds of Dembia plain wetlands, Lake Tana. Ethiopia. SINET: Ethiopi. J. Sci 31(1):1–10
Nowald G, Schröder W, Günther V et al (2010) Kraniche Grus grus in Äthiopien—Common Cranes (Grus grus) in Ethiopia. Vogelwelt 131:1–13
Sauer JR, Link WA (2002) Hierarchical modeling of population stability and species group attributes from survey data. Ecology 83(6):1743–1751
Shimelis A, Afework B (2009) Species composition, relative abundance and habitat association of the bird fauna of montane forest of Zegie peninsula and nearby islands, lake tana, ethiopia. SINET: Ethiop. J. Sci 32(1):45–56
Shimelis A Z (2013) Birds of Lake Tana area, Ethiopia. A photographic field guide. View Graphics and Printers, Addis Ababa
Shimelis A, Afework B (2008) Species composition, relative abundance and distribution of bird fauna of riverine and wetland habitats of Infranz and Yiganda at southern tip of Lake Tana, Ethiopia. Trop Ecol 49(2):199–209
Shimelis A, Afework B, Abebe G (2008) Species diversity, distribution, relative abundance and habitat association of the Avian Fauna of modified habitat of Bahir Dar and Debre Mariam Island, Lake Tana, Ethiopia. Int J Ecol Env Sci 34(3):259–267
Shimelis A, Nowald G, Schroder W (2011) Observation on the biology and ecology of cranes: wattled cranes (grus carunculatus), black-crowned cranes (balearica pavonina), and eurasian cranes (grus grus) at Lake Tana, Ethiopia. INDWA. J Afr Crane Res Conservationist 7:1–12
IUCN (2014). The IUCN Red List of Threatened Species. Version 2014.3. <www.iucnredlist.org>. Accessed 28 May 2015
Tilahun S, Edwards S (eds.) (1996) Important bird areas of Ethiopia: a first inventory. Ethiopian Wildlife and Natural History Society, Semayata press, Addis Ababa
Weller MW (1999) Wetland birds: habitat resources and conservation implications. Cambridge University Press, Cambridge

Chapter 14
Herpetofauna and Mammals

Shimelis Aynalem and Abebe Ameha Mengistu

Abstract Of the 64 known amphibian species of Ethiopia, the Lake Tana Sub-basin is believed to contain 19 species of amphibian belonging to 12 genera and 9 families. All of these species are grouped under Order Anura (tail-less frogs). Most of the species found in the area appear to be those that are tolerant to changed habitats. Fourteen species of lizards, 20 species of snakes and one species of freshwater turtle, a total of 35 reptile species occur in the area. However, the herpetofauna taxonomic status and evolutionary relationships of many of the species has not been confirmed using molecular and biogeographic studies or updated using modern phylogeographic approaches. In addition a total of 27 species of mammals that belongs to 16 families are listed. But there is little information on Order-Rodentia (rodents), Lagomorphs (lagomorphs), Erinaceomorpha (hedgehogs and gymnures), Soricomorpha (shrews, moles, and soledons) and Chiroptera (bats). Depending on the land use/land cover history of the area, the wildlife habitats of the area can be categorized in many ways, and each habitat harbors a number of amphibians, reptiles, and mammals. The wildlife values can be expressed in terms of their economic importance, nutritional value, ecological role and socio-cultural significance. Still Ethiopia is not benefiting as there is little attention for the resource. The existence of proclamations specifically dealing with biodiversity does not guarantee conservation unless they are put into action. The main challenge faced in conservation of herpetofauna and mammals is habitat degradation and fragmentation.

S. Aynalem (✉)
Blue Nile Water Institute and College of Agriculture and Environmental Sciences,
Bahir Dar University, P.O. Box 2205, Bahir Dar, Ethiopia
e-mail: shimelisay@yahoo.co.in; shimelis.aynalem@gmail.com

S. Aynalem
College of Natural Science, Addis Ababa University, Addis Ababa, Ethiopia

S. Aynalem
Research Associate of International Crane Foundation, Baraboo, USA

A.A. Mengistu
Department of Biology, Adama University, Nazareth, Ethiopia
e-mail: abbefish@yahoo.com

K. Stave et al. (eds.), *Social and Ecological System Dynamics*,
AESS Interdisciplinary Environmental Studies and Sciences Series,
DOI 10.1007/978-3-319-45755-0_14

keywords Amphibian and reptiles · Mammals · Lake Tana Sub-Basin · Values

14.1 Introduction

Amphibians and reptiles are found on all continents except the Antarctic. They play a very important role in the efforts to conserve biological diversity world-wide. They account for about one quarter of all known vertebrate species. They are important in terms of aesthetic and cultural interest, as well as their significance to science. They also have a great economic value. Unfortunately, in conserving the herpetofauna, very significant obstacles may be faced. In particular, many species do not have a very positive image in the mind of the general public (Masood and Asiry 2012).

Africa's tropical savannas support the world's most spectacular mammalian fauna (East 1999). They play key roles within the ecosystems in which they live, and are critical in maintaining the functions and services provided by ecosystems. They achieve this through their roles as grazers, predators, pollinators and seed dispersers. Mammals also provide numerous benefits to humans, both directly and indirectly; they are an important food source for many cultures, and are used in recreation and improving livelihoods around the world. Indirectly, mammals are very important in maintaining balanced ecosystems and the services they provide to mankind.

Despite their economic value, most mammalian species are declining in distribution and numbers as human populations continue their rapid growth and expansion. This decline is caused by destruction of natural habitats, competition with livestock and excessive off take by meat hunters, in many cases through the actions of poor rural communities who lack alternative development options (East 1999). Careful evaluation of the role of small mammals and their relationships with their environment is necessary to fully appreciate the impact of control programs on the ecosystem. Particularly, small mammals regarded as "pests" should not be viewed separately from other components in the ecosystem. Rather, small mammals must be viewed in terms of their interrelationships with other components. Alteration of small mammal communities through control programs could influence other components and ultimately the whole system (Carolyn 1987). However, at least 25% of the world's mammal species are at risk of extinction, according to the first assessment of their status for a decade (IUCN 2008).

Mammal characteristics include numerous adaptations that enable them to survive in a wide range of environments. The Lake Tana Sub-basin consists of broadly homogenous, but locally highly varied micro-habitats. The uniformity of the habitats comes from the geographic position and small size of the sub-basin. The watershed area has surrounded by flood plains, natural ponds, rivers, and chains of mountains. Much of the area is occupied by agricultural land and human settlement. These features characterize the sub-basin as a tropical highland environment consisting of a shallow oligotrophic lake, perennial rivers, wetlands and terrestrial habitats.

The aquatic, wetland and terrestrial systems of Lake Tana Sub-basin are interconnected both spatially (physically and biologically) and temporally (human history and evolution). This linkage is on one hand vital to sustain the ecosystem for the wild animals and on the other catastrophic when normal functioning is disrupted at some point. Within the remaining natural habitats and changed ecosystems, various types of micro-habitats provide food, water, shelter and breeding sites. The micro-habitats contained within this environment are home to amphibians, reptiles, mammals and birds. Scientific information on the mammalian and herpetofauna of the Lake Tana Sub-basin is scarce, and existing data have not been compiled systematically. And hence this chapter tries to briefly outline and discuss existing knowledge, potential threats and future prospects in research, valuation and conservation of the herpetofauna and mammals in the Lake Tana Sub-basin.

14.2 Lake Tana Area and the Catchment

The Lake Tana basin comprises an area of 15,096 km^2 including the lake area. The mean annual rainfall of the catchment area is about 1280 mm. The annual mean actual evapotranspiration and water yield of the catchment area is estimated to be 773 and 392 mm, respectively. The climate of the region is 'tropical highland monsoon' with main rainy season between June and September. The annual average air temperature is 20 °C. The mean annual relative humidity (1961–2004) at Bahr Dar gauge station is 0.65 (Shimelis et al. 2008). The basin has significance national importance due to its high potentials for irrigation, hydroelectric power development, high value crops and livestock production, and ecotourism.

Lake Tana, the main source of the Blue Nile River, is the largest lake in Ethiopia. It is approximately 84 km long, 66 km wide and is located in the country's north-west highlands (Latitude 12° 0' North, Longitude 37° 15' East). The lake is a natural freshwater lake which covers 3000–3600 km^2 area at an elevation of 1800 m above sea level (m a.s.l). The lake is shallow with a maximum depth of 15 m.

The Lake Tana sub-basin is found in the north-western highlands of the Amhara Region, Ethiopia. The basin forms the upper course of the Blue Nile Basin in Ethiopia and its catchment is at the southern tip of the lake where the Bahr Dar City is situated, and the basin is the source of the Blue Nile River (Biru 2007).

The plains around Lake Tana are land areas often flooded in the rainy season and form extensive wetlands such as the *Fogera* plain in the East, the *Dembiya* plain in the North, the *Kunzila* plain in the southwest, and marshlands at the peripheries of the lake bordering the city of Bahr Dar in the South (Shimelis et al. 2008).

14.3 Wildlife Habitats in the Lake Tana Sub-basin

Wildlife habitats are diverse. Depending on the land use/land cover history of the area, the wildlife habitats of the area can be categorized as: forest/shrubland, cultivated land, grass land/pasture, wetlands/swampy, water body (the Lake and Rivers) and settlement areas. Of course this classification is most widely used globally, published by FAO's (Food and Agricultural Organization) Production Yearbooks. The proportion and distribution of the habitat-land use/cover with respect to the total area of the Lake Tana sub-basin is variable. However, the habitat description of the area is summarized below.

Forest/shrub land: the term "forest" has a variety of meanings that have not yet been standardized in global change studies; however, the FAO's tropical forest inventories use a figure of 10% canopy cover to separate forested from deforested area (Lanly et al. 1991). The case of Lake Tana area, there are a few remnant forest/tree patch areas that mostly concentrated around monasteries, churches and some locally protected areas. The overall forest cover of the basin is not more than 2% (Biru 2007).This habitat harbors different range of wild animals, but little information on smaller mammals such as: Order-Rodentia (rodents), Lagomorphs (lagomorphs), Erinaceomorpha (hedgehogs and gymnures), Soricomorpha (shrews, moles, and soledons) and Chiroptera (bats). Based on the land use/cover study of Biru (2007), shrubs that dominate the foot-slopes and riverine landscapes and whose vegetation cover is assumed to be >65% as well as those lands with 15–65% cover and usually degraded from other land uses, such as grazing, cultivation and deforestation are all put into shrublands. But forestland in Lake Tana watershed includes some closed natural forests usually under closure, forests at cultural and religious centers, and plantation forests that are dominantly of eucalyptus species where settlements are common.

Cultivated land: it is defined as those lands that are regularly used to grow domesticated plants, ranging from long-fallow, land rotational systems to permanent, multi-cropping systems. The expansion of cultivation land is different from countries to countries. The same is true in Lake Tana Sub-basin, as the population increases several wetlands, forest and grasslands are put for cultivation and settlement. According to Biru (2007), cultivated land includes areas that are plain and slightly undulating landscapes, where flood is not a problem, are usually intensively cultivated (>75%). Foot-slopes and undulating landscapes are usually under moderately cultivated condition (50–75%). Still this habitat supports a number of burrowing animals such as porcupines, aardvark, mongoose and arvicanthis species, but usually they are considered pests (Carolyn 1987).

Grass land/pasture: this land cover includes short time flooded flat lands that are usually used for intensive grazing. It also includes degraded mountains and foot-slopes that are usually sparsely covered by shrubs and forbs (<15%) but usually used for grazing purposes. It is main habitat for rodents like grass rat and some antelopes.

Wetlands/swampland: includes seasonally flooded lands where the water impedes cropping in the main season. Unless otherwise some drainage activity is done, this land cover type usually stays under water cover or is at least remaining wet for most times of the growing season.

The water body (aquatic habitat) includes the Lake Tana itself, rivers, streams and ponds. However, the major aquatic habitats in the sub-basin are the lake (Tana), inflowing rivers (Gelda, Gumara, Rib, Megech, Dirma, Arno-Garno, Gilgel Abay), the out flowing Abay River and ponds (Shesher and Wallala). Several small rivers and streams arising from springs in the surrounding highlands feed the major rivers and the lake. In general, the rivers are relatively fast-flowing at the rocky upstream reaches due to the steep slope of the landscapes. At the interface between the mountains and the plains, there could be waterfalls, sometimes as big and impressing as the Duriba waterfall on Gumara River. At the lower reaches, the rivers flow slowly winding through the vast sandy or muddy plains to join the lake. During the main rainy season, the rivers overflow and flood the plains, and also feed the ponds. The only effluent river, Abay, leaves Lake Tana to run through the southern plains until it drops into the gorge at Tiss Issat Fall. The standing waters of the lake and the ponds are very shallow and get their water mainly from the inflowing rivers, and substantial amount from direct rainfall. The lake's substratum and shoreline is covered by sand, mud or rocks. Turbulent and muddy rivers bring eroded mud into the lake along with a lot of nutrients. Primary production and lower levels of the lake's food chain are constituted by phytoplankton, macrophytes, zooplankton, benthos, and larvae and adults of macro-invertebrates and lower vertebrates. Besides to fishes, the aquatic habitat of Lake Tana area in particular is home of Hippos and several Monitor lizards (Nile Monitor). A few crocodiles are also inhabits between the Abay Fall and Lake Tana outlet delta, Bahir Dar area.

The proportion of each habitats in the watershed can be summarized as: Water body consists of 307,020 ha (20.37%), Swampland, 32,157 ha (2.13%), Forest, 15773 ha (1.05%), Shrubland 75,700 ha (5.02%), Grassland 301,799 ha (20.03%), Cropland 765,677 ha (50.81%), Urban area 6107 ha (0.41%), others unclassified areas consists of about 2791 ha (0.19%) (Biru 2007).

14.4 Wildlife Values

It has taken time for the international community to realize the value of wildlife. The World Charter for Nature, adopted and solemnly proclaimed by the General Assembly of the United Nations in 1982, addressed the concern of wildlife conservation without referring to the concept of wildlife value. It was only in 1992 at the International Convention on Biodiversity in Rio de Janeiro that a clear declaration of intent to secure the 'value' of the biodiversity of the Earth was made (Chardonnet et al. 2002).

Wildlife values can be classified majorly into two: direct and indirect values. When we say direct values it includes—consumptive use value that is non-market value of firewood, game, etc., and—productive use value such as commercial value of timber, fish, etc. whereas indirect values includes—non-consumptive use value: scientific research, bird watching, etc.,—option value: value of maintaining options available for the future and existence value: value of ethical feelings of existence of wildlife.

These values carry different weights, which vary according to the respective interests of the stakeholders involved. Although important, virtual values, such as the ethical value, are not as powerful in terms of justification for conserving wildlife as pragmatic ones, such as economic values. Be it relevant or not, financial profitability, economic yield and environmental sustainability are often dominant values for high-level decision makers as well as for grass-root level individuals who live in close proximity to wildlife (Bojö 1996). For this reason, the classification adopted here rather relies on a pragmatic approach differentiating between the following: the economic importance of wildlife, the nutritional value of wildlife, the ecological role of wildlife, and the socio-cultural significance of wildlife. All the mentioned values are positive. Wildlife, however, may be seen as sometimes presenting negative or adverse values when they become causalities to humans, pests for agricultural products and predator of domestic animals.

To appraise the economic importance of wildlife is as difficult in developing countries as is a classic academic exercise in developed countries. In the world, some of the wildlife values cannot or can hardly be quantified as aesthetic, educational, ecological or ethical values. The rationale of the economic approach is therefore limited to some aspects of the entire issue. The classic categories of wildlife economics comprise the consumptive uses of wildlife, i.e. a number of activities whereby the wildlife resource is exploited by removing a certain quota of either live or dead animals, and the non-consumptive uses of wildlife, i.e. the activity of giving value to wildlife without removing the resource.

The entire range of wildlife activities produces revenues and brings added value which contributes to the gross national product (GNP), however the amount of income generated varies from high levels of US$131.7 million in Zimbabwe for example to low levels, such as US$30 million in the Central African Republic. However, in Ethiopia taking all tourism activities, the contribution of tourism to Ethiopian GDP is very low it was about its 0.5% GDP in 1996 (Adem 2008).

The non-consumptive use of wildlife is mostly based on the aesthetic value of wildlife. Wildlife becomes the support of the tourism industry, as beaches are the support of the seaside tourism industry. This category of tourism is essentially based on wildlife viewing and is almost entirely part of the service sector. A Persian word, adopted in Swahili, 'safari', has become a world-wide term used for journeys through the African national parks with the purpose of observing wildlife, landscape and local atmosphere. Several nations in Africa, mainly in the eastern and southern parts of the continent, earn substantial income through wildlife tourism (Chardonnet et al. 2002).

Consumptive use of wildlife is an ancient practice, as old as humankind and is responsible for the development of the human brain, having been the support of livelihood for most ancient civilizations and enabled survival for many, e.g. the hunter-gatherers, trappers, reindeer (*Rangifer tarandus*) herders, etc. The modern man progressively distanced himself from using wild animals as dependence on domesticated animals increased. However, wild animal production remains important to many developing countries and for many developed countries provides an opportunity to diversify crowded domestic animal production, or sometimes even becomes a replacement activity (Scandinavia).

Sustainable use of wildlife is fully recognized as legitimate by all international institutions and conventions. During the last World Conservation Union (IUCN) Congress held in Amman in 2000, sustainable use of wildlife was again officially reconfirmed as a way in which biodiversity could be protected and the development of rural communities could be assisted.

Identification and recognition of the ecosystem and economic values of the rich diversity of amphibians and reptiles in Ethiopia, including those in the Lake Tana Sub-basin, has not been done systematically before. Elsewhere in other parts of the world, these animals are known to provide economic benefits to humans as sources of food, for medicinal use, biological control of agricultural pests, production of leather and ornaments, or in tourism in the wild, in zoos or in museums (East 1999; Pough et al. 2003).

The use of amphibians and reptiles in Ethiopia for economic and social use is mainly exercised in traditional systems instead of modern and scientific approaches (Mengistu et al. 2013). In most parts of Ethiopia, religious and traditional norms do not allow the use of amphibians and reptiles for food. The body size of the species of frogs in Ethiopia also appears too small to produce considerable amount of leg muscles for food. However, there are practices of utilizing amphibians and reptiles, in particular dermal secretions of some frogs and the venom of snakes in traditional medicine (Mengistu 2012). The predatory nature of amphibians and reptiles helps to control insect and rodent pests that cause damage on agricultural crops as well as vectors of diseases (Pough et al. 2003). The only known commercial production of reptiles in Ethiopia is the Arba Minch Crocodile Farm located in the southern part of the country in the Rift Valley. Stuffed, fluid-preserved and skins of amphibians and reptiles are displayed for visitors and stored for researchers at the Zoological Natural History Museum at Addis Ababa University (Getahun and Mengistu 2006). The African Rock Python and tortoises are sometimes help in captivity in different parts of the country.

The ecosystem value of amphibians and reptiles comes mainly from their role in the food web. Amphibians and reptiles are mainly predators, the former mainly feeding on invertebrates (Pough et al. 2003). Different species of reptiles feed on invertebrates, the eggs, young and adults of fishes, amphibians, other reptiles, birds and mammals. Some species such as tortoises and turtles are herbivores (Largen and Spawls 2010). Amphibians and reptiles in turn are also prey mainly to birds, mammals and bigger reptiles. This implies that they are important elements of aquatic, wetland and terrestrial ecosystems. Some burrowing or fossorial species such as typhlopid snakes and puddle frogs can be useful in soil fertility and aeration. The reproductive process of most frogs includes aquatic eggs and larvae, and terrestrial adults; this is indicative of the vital role of amphibians at the interface between water and land. In addition, the moist skin of frogs is used for breathing, and this special feature enables researchers and conservationists to use amphibians as very good indicators of environmental health (Mengistu et al. 2013).

14.5 Diversity and Distribution of Species

14.5.1 Herpetofauna

The herpetofauna—amphibians and reptiles—of the Lake Tana Sub-basin have not been separately documented before. Previous documentations focused either on country-wide assessments or on specific taxa and localities of interest (such as protected areas). The following information summarizes the recorded and documented listing of amphibians and reptiles from the Lake Tana Sub-basin. Revision of the taxonomy and evolutionary history of some amphibian genera of Ethiopia has been started in the past few years (Mengistu et al. 2013). Traditional morphological taxonomy Largen (1977, 2001) is being supplemented by genetic and biogeographic studies (Evans et al. 2011; Zimkus 2008; Mengistu et al. 2013). The taxonomic names used are updated ones (Mengistu et al. 2013). Taxonomic studies on the reptiles of Ethiopia, however, require update using modern molecular studies.

Amphibians

Of the 64 known amphibian species of Ethiopia, Frost and AMNH (2010), Mengistu et al. (2013) the Lake Tana Sub-basin is believed to contain 19 species belonging to 12 genera and 9 families. All of these species are grouped under Order Anura (tail-less, frogs). There is no record of species belonging to Order Apoda (leg-less, caecilians) in this sub-basin (Wilkinson et al. 2011; Mengistu et al. 2013). Order Caudata (tailed, salamanders) are not found in Sub-Saharan Africa (Poynton 1999; Largen 2001; Largen and Spawls 2010). A list of the known species of amphibians from the Lake Tana Sub-basin is described in this section (Annex 14.1). The amphibian species composition of the Lake Tana Sub-basin is a clear indication of the type of habitats available for different types of frogs. That is, the diversity of tree frogs is relatively low due to absence of wet forests of indigenous trees. Most of the species found in the area are tolerant to changed habitats (Mengistu et al. 2013). Recent phylogeographic investigation revealed that some of the Ethiopian endemics, *Leptopelis yaldeni*, *Ptychadena wadei* and an undescribed species of *Ptychadena* from Debre Markos area are local endemics to the Lake Tana Sub-basin (Mengistu 2012). In addition, *Ptychadena erlangeri* and *P. neumannii*, which were previously suggested to have been found in this sub-basin, are now morphologically, genetically and biogeographically confirmed non-existent in this area.

The taxonomic status of many of the species (such as conspecificity of the closely related *P. porosissima* and *P. wadei*) has not been confirmed using molecular and biogeographic studies. One species (*L. yaldeni*) has a Near Threatened conservation status, while *Ptychadena* sp. found in Debre Markos and *P. wadei* are Data Deficient; all other amphibians of the sub-basin have a Least Concern status (Mengistu et al. 2013; IUCN 2013).

Reptiles

There are 35 species of reptiles known from the Lake Tana Sub-basin; these make up around 16% of all Ethiopian species of reptiles (Largen and Rasmussen 1993; Mengistu et al. 2013). The reptiles in the Lake Tana Sub-basin includes 14 species of lizards (grouped under eight genera and five families), 20 species of snakes (grouped under 17 genera and five families) and one species of freshwater turtle. Of these, only three species of lizards (*Acanthocerus zonurus*, *Chamaeleo affinis* and *Panaspis tancredii*) and two species of snakes (*Lamprophis erlangeri* and *Letheobia somalica*) are Ethiopian endemics (Largen and Spawls 2010). To our knowledge, there is no scientific report on the natural occurrence of tortoises and crocodiles in Lake Tana and its tributary rivers. The Nile crocodile (*Crocodilus niloticus*) is readily abundant in parts of the Abay River below the Tisissat Falls mainly. A list of the known species of reptiles from the Lake Tana Sub-basin is described (Annex 14.2).

The taxonomy and evolutionary relationships of Ethiopian reptiles has not been updated using modern phylogeographic approaches (Largen and Rasmussen 1993). The very limited knowledge that we have on Ethiopian reptiles is in part attributed to the bad reputation that most local people have towards reptiles in general and snakes in particular (Abebe A. Mengistu, unpublished data). Indiscriminate killing of snakes when in sight is a common practice in many parts of the country. Among the snakes in the Lake Tana Sub-basin, two species (*Dendroaspis polylepis* and *Causus rhombeatus*) are venomous and highly dangerous, and one species African Rock Python (*Python sebae*) (Fig. 14.1) is a non-venomous dangerous snake due to its big size and muscular body (Spawls and Branch 1995; Spawls et al. 2002; Largen and Spawls 2010; Gower et al. 2012b; Abebe A. Mengistu, unpublished data). Four other species (*Philothamnus battersbyi*, *Psammophis angolensis*, *Psammophis sibilans* and *Psammophylax variabilis*) also need to be handled with care.

14.5.2 Mammals

The distribution and occurrence of mammals in a certain area are determined on the geographic distribution of species and the local distribution of individuals. The home ranges, territories and microhabitats are indicators of the distribution of individuals within an area of convenient habitats; because these are governed by

Fig. 14.1 African rock python (*Python sebae*) in Chimba wetland edge (©Photo: Shimelis A. 2009)

access to important resources such as food, living space and availability of mates (Smith 1992).

Of the 4000–4500 living species of mammals worldwide, approximately 25%, or some 1100, occur on the African continent. There is no absolute certainty as to how many species there are, particularly in the case of the bats, shrews and smaller rodents. Taxonomists are constantly revising and assessing the scientific status of many mammals, and with the development of more sophisticated methods of establishing genetic diversity, this can be expected to accelerate (Stuart and Stuart 2006). With regard to the faunal diversity, 284 species of terrestrial mammals, 201 reptiles, 60 amphibians, and 154 fishes are known to occur in Ethiopia. Among these, 31 (11%), 9, 24 and 49 are endemic to the country respectively (Yalden et al. 1996; Hillman 1993).

As far as mammals and their habitats are concerned, different wild animals are living in Lake Tana sub basin. The data were organized from different reports of the area done by the authors, field observation during various times in the area, published articles and personal communications of different professionals. This manuscript tries to review some species of mammals that were actually recorded and seen at different times. However major animals encountered during field work are only listed in this chapter. A total of 27 species of mammals that belongs to 16 families are enumerated (Annex 14.3) (Shimelis 2011 unpublished).

Order Primates: Cercopithrcidae (baboons) and Chlorocebus (grivet monkey). Savanna (Common) Baboon (*Papio cynocephalus*). Even though there are different species of baboons in Ethiopia, in Lake Tana Basin the most common ones is the Savanna (common) Baboon, and the Girvet Monkey (*Chlorocebus aethiops*). The Common Baboon is the fourth largest African primate after the gorilla, chimpanzee and bonobo. It is widely distributed in the basin where there is savanna associations, forest edges, montane areas. Common to see around Lake Tana southern part mostly, Guna Mountain area, and Alemsaga forest area.

Grivet Monkey, Grivet Tota (*Chlorocebus aethiops*): formerly Grivet and Vervet Monkey were included with other guenons in *Cercopithecus*, but are now separated as *Chlorocebus*. They are common in the area. Particularly Zegie peninsula holds the largest population. They are known for pests of crops such as vegetables and fruits. However, they were located in the upper course of the forest and in the riverine habitat of the area. The presence of such species could also indicate the presence of potential predators such as serval cats; but the status of this species requires further study whether their number is increasing or decreasing or stable.

Order Whippomorpha: Hippopotamuses. Common Hippopotamus (*Hippopotamus amphibius*). The distribution of Hippos particularly in the Amhara region is so restricted to Lake Tana, where adequate vegetation, protection cover and food are located. Hippo is found in the southernmost tip of the lake from the Abay delta in the south west, Gorgora-Achera area, Geldaw southeastern and Bahir Dar Abay River below Abay Bridge. Even though, Hippo requires a large area and wide range of habitat, food and sufficient water to allow for complete submergence is a requirement, and a preference is shown for permanent waters with sandy substrates. Access to adequate grazing is also essential but these animals will move several kilometers away from water-bodies to reach suitable feeding area. Due to its rarity of the species and existence of typical habitats describing as a charismatic species for Lake Tana may be paramount importance. A total of 8–11 families consisting of 5–7 members are supposed to remain at Lake Tana (personal observation 2009–2013).

Order Suiformes: Family Suidae (Pigs and Hogs). Common Warthog (*Phacochoerus africanus*). It is most frequently observed wild pig where there is open protected vegetation. It is common to see in *Alemsaga-Debre* Tabor area in particular, but rare to see in Derbanata area southwestern part of Lake Tana. However, the Bushpig (*Potamochoerus larvatus*) is widely distributed throughout much of its range although habitat modification and hunting have fragmented population in some regions. It is considered as pest of crops though hunting is not common at Lake Tana area. However, the species is exterminated in Zegie Peninsula (Getachew and Afework 2009).

Order Ruminatia: under this order there are four families such as Girfafidae, Cervidae, Tragulidae, and Bovidae; however, the Family Bovidae has numerous sub-families, which we only see on three subfamilies that are found in the area. These are Subfamily Tragealphinae: Kudu and Bush Buck; Subfamily Reduncinae: Waterbuck, Reedbuck; Subfamily Antilopinae: Gazelles and Dawarf Antelope; Subfamily Cephalophinae: Duiker.

Subfamily Tragealphinae: no Kudu has been recorded, but greater Kudo could be found in the Lake Tana watershed where there is suitable habitat such as Alemsaga forest area. Bushbuck (*Traglaphus scriptus*), the species is solitary, but pairs and small, loosely knit groups of ewes and lambs are commonly observed. Whether or not there is some seasonal variation on home range largely depends on the availability of water. It is seen in Chimba wetlands—Dehana Mariam in the thick vegetation in 2012. Similarly, it was observed in Alemsaga-Debretabor. However, the occurrence of the species is decreasing due to hunting and habitat loss.

Subfamily Antilopinae: Gazelles and Dwarf Antelope, no detail information is found. Subfamily Cephalophinae: Duiker. Common (Grey) Duiker (*Sylvicapra grimmia*), even though the population of duiker is getting low due to illegal hunting and habitat destruction, some droppings of the species was found in several places of the basin. The secretive nature of the species makes real observation very difficult; however, the local people information was justified by the observation of fresh and late droppings.

Order Carnivora: Family Canidae includes Foxes, Jackals. Golden (common) Jackal (*Canis aureus*). It is known to distribute throughout Ethiopia. It is omnivorous, taking a wide range of food items including small mammals, birds, reptiles, invertebrates, carrion, as well as wild fruits and berries. It is commonly distributed throughout the Lake Tana sub-basin. However, the family felidae are rare to distribute, but the serval cat commonly occurring and leopard are found in some parts of the watershed.

Family Mustelidae: Badger. Honey Badger (Ratel) (*Mellivora capensis*): no detail information is found for this species, but it said that Honey badger is still widely distributed in the area, and some farmers still complain about the pest nature of the species if no protection is made for their local beehives.

Family Viverridae, Nandiniidae, Herpestidae: Civets, Genets and Mongoose. Under this family we have only Civets and Mongoose group. Taking the case of African Civet, Civet cat is believed to be distributed in some parts of the Amhara region. Though this species is known to produce an economic important secretion known as Civetrine 'Zibad' in Amharic, the status, distribution and abundance have never and ever been documented. The presence of the species is merely identified simply by identifying the feacal type and the common latrine site. Unless they are being ecologically disturbed, Civets tend to use common latrine site. Fresh and late feacal were accumulated and found along the road to Dengors Elementary school-Alemsaga natural forest around Debre Tabor, *Bezawit Mariam* - Bahir Dar, *Gorgora*, *Infranz* River, *Zegie* and *Kunzila* area. Most probably the Civet species that are found in the area is *Civettictis civetta*. Look the real picture of latrine site found during the survey and the species as well (Fig. 14.2).

Family Hayenidae and Protelidae: Hyaena. Spotted hyaena (*Crocuta crocuta*): Best known of the hyaenas, with heavily built forequarters standing higher than the rump. In Africa, its status is still common in a number of savannah areas but generally reduced, even local extinct, in others. It is widely distributed in Lake Tana sub-basin.

Order Hyracoidea: Hyraxes (Dassies) (Family Procaviidae). Commonly seen in Zegie peninsula where there are churches for instance in Ura Kidane-Mihiret. Also has been seen in the rocky cliffs of the eastern shore of Lake Tana area. The *Derbanta* hills in the southwestern part of Lake Tana, and *Alemsaga* area are potential sites for this species to occur, though further investigation shall be done in the future.

Fig. 14.2 Civet latrine site Bezawit area (*L*), Civet cat dead, Kunzila area (*R*) (©Photo: Shimelis A. 2011)

Order Tubulidentata: Family Orycteropodidae—Aardvark (*Orycteropus afer*). Aardvark is nocturnal and is a solitary creature that feeds almost exclusively on ants and termites (formicivore). An aardvark emerges from its burrow in the late afternoon or shortly after sunset, and forages over a considerable home range encompassing 10–30 km, swinging its long nose from side to side to pick up the scent of food. Its claws enable it to dig through the extremely hard crust of a termite or ant mound quickly, avoiding the dust by sealing the nostrils. When successful, the aardvark's long (as long as 30 cm) tongue licks up the insects; the termites' biting, or the ants' stinging attacks are rendered futile by the tough skin (Stuart and Stuart 2006).

Aside from digging out ants and termites, the aardvark also excavates burrows in which to live: temporary sites are scattered around the home range as refuges, and a main burrow is used for breeding. Main burrows can be deep and extensive, have several entrances and can be as long as 13 m. The aardvark changes the layout of its home burrow regularly, and from time to time moves on and makes a new one. Only mothers and young share burrows. If attacked in the tunnel, it will seal the tunnel off behind itself or turn around and attack with its claws. It is common to find in Bahir Dar-international Air Port compound and in the cultivation field around *Zegie, Chimba* and *Kunzilla* area where reddish/brown light soil is dominant.

Order Lagomorpha: Family Leporidae—Hares and Rabbits. Bush Hare (*Lepus fagani*). It is a species of intermediate altitude (500–2500 m) and scrubby or forest edge habitats in the area. It is distinguished by a coarser fur yellowish-brown fur than *Lepus habessinicus*, shorter ears and a yellow edged tail. It is widespread and common in the area where there is adequate cover.

Order Rodentia: Family Sciuridae—Squirrels. Arboreal squirrel (*Heliosciurus gambianus*): The most common small mammal ever recorded in *Zegie Peninsula*, Lake Tana area. Particularly they are also common at Infranz Riverine habitat-Bahir Dar, Abay Riverine Habitat, Bezawit forest. They prefer forest and riverine habitat where there is fruit tree. They are recognized as pests of fruits as complained by the local people. Genetically characterization of the species should be carried out as most of the species are isolated from others (Fig. 14.3). However, the family Muridae: the grass rats in particular, Arvicanthis are commonly occurring in the

Fig. 14.3 Sun squirrel common to see at Zegie Peninsula. (©Photo: Shimelis A. 2011)

Lake Tana Sub-basin where agriculture is practicing, pasture land and grass land. They are diurnal terrestrial rodents, characteristically pests of agricultural products. They live in loose colonies and create well-marked runs from their burrows to their feeding areas. In addition, the genus is widespread in the savanna zones of Africa. (Afework and Yalden 2013). There are still several species occurring in the area such as *Arvicanthis dembeensis* again pests of agricultural crops and grass, *Mus musculus,* and *Desmonys harringtoni* (Afework and Yalden 2013).

Family Hystricidae: Porcupines. African porcupines—It is distributed in most part of the park. It is only this species (*Hystrix africaeaustralis*) that distributed in Ethiopia. Although foraging is primarily a solitary activity, several animal, may use same burrows system. Several spikes were found in the southern part of Lake Tana, Bezawit-Bahir Dar, North Gondar-Dembiya and Gorgora, and western part of Lake Tana as well. They are commonly found, but peoples are complaining when the cropping season comes, because porcupines are known pest of potatoes and maize.

14.6 Conservation and Management

In Ethiopia, different ecosystems of high biological importance are threatened. The threats include habitat destruction, human population growth, and encroachment by undesirable plant species following flooding, shrinkage of lakes and wetlands due to industrial and agricultural developments (Pol 2006). The existing conservation and management practices are only restricted to protected areas such as National Parks, Wildlife Reserves etc.; however, there are a number of areas that has to be protected because of their potential biological resources. Several proclamations, policies at National and Regional level have been declared since a century, but the action and implementation are so reluctant to conserve and manage the wildlife resources.

14.6.1 Policy and Legal Issues

One of the good opportunities for conservation of amphibians and reptiles in Ethiopia is the existence of proclamations specifically dealing with biodiversity, wildlife, protected areas, forestry, fisheries and water resources (IBC 2005). The legal framework potentially enables to conserve Ethiopia's herpetofauna in nature along with other biodiversity, as well as to protect illegal transfer of live animals and genetic resources away from- and into the country. But practically, these issues face a lot of implementation problems. The Ethiopian Wildlife Conservation Authority (EWCA) and regional bureaus for wildlife and national parks are trying their best to implement national and regional policies in protected areas. However, still this issue remains one of the problems to safeguard the wildlife resources of the country.

14.6.2 Conservation Efforts

So far, there are no specific conservation endeavors known to target species of amphibians, reptiles and mammals in the Lake Tana Sub-basin. This could have resulted mainly from lack of data on the status of populations and their specific ecological requirements. However, habitat based conservation at Lake Tana area is being initiated, which will make Lake Tana area as one of Biosphere Reserves in Ethiopia. The Regional Government in collaboration of the Ethiopian Science and Technology with the support of German Based NGO–NABU (Nature and Biodiversity Conservation Union) is working for the designation by UNESCO as a Biosphere area.

14.7 Threats and Challenges

The main challenge faced in conservation of amphibians and reptiles is habitat degradation and fragmentation through urbanization, deforestation, Beebee (1996), Cheng et al. (1998), Stuart et al. (2004) large scale modern agriculture and chemical pollution Largen (2001), Gower et al. (2012b), Mengistu et al. (2013). The Lake Tana area has been subject to large scale habitat change by humans for hundreds of years. Recent activities are simply happening at a faster rate than before and aggravating the situation. Another global threat to the diversity of amphibians is the enigmatic amphibian chytrid fungus (*Batrachochytrium dendrobatidis*) (Rödder et al. 2009; Kielgast et al. 2010; Lötters et al. 2010). This fungus is considered to be one of the main causes of recent extinction of several species of amphibians worldwide. A recent study on some highland areas in southern and southwestern Ethiopia has shown that about half of the studied specimens were positive for presence of the fungus (Gower et al. 2012a).

Another important conservation challenge for reptiles in Ethiopia is indiscriminate killing of snakes and other big sized reptiles when in sight. Many people have the wrong perception of considering all snakes and most reptiles as poisonous and dangerous; this is a result of lack of basic knowledge on snakes and other reptiles (Mengistu 2012).

Many large animal species have a high risk of extinction. This is usually thought to result simply from the way that species traits associated with vulnerability, such as low reproductive rates, scale with body size (Cardillo et al. 2005; East 1999). Most of the diurnal large mammals in Lake Tana warershed are exterminated, but still poaching is being practiced for these rare animals. This day, killing nocturnal animals such as civet, mongoose, and aardvark have becoming a common practice as the habitats are being encroached for agriculture. Extermination of grevet monkeys in Zegie peninsula had been tried in the past two-three years since monkeys are considered to be pest of agriculture as their natural habitats is being fragmented. But, still some monkeys are surviving in the area. Different study shows leopard, grivet monkey, bushpig, duiker and common bushbuck are the most illegally hunted wild animals in Zegie area, but warthog has been eradicated from Zegie as a result of illegal hunting (Getachew and Afework 2009). Destruction of habitats: burrow site destruction of porcupines, bush pigs, jackals; smoking, watering, poisoning and poaching becomes a common phenomenon in the area.

The last but very important challenge is lack of local professional herpetologists in Ethiopia. This slowed not only conducting research, but also prioritizing several populations and species of amphibians and reptiles for conservation. In addition, the absence of technical skill, materials, equipments and generally professionals in the field of zoology could be accounted for having insufficient research outputs in the field of herpetology and mammalogy as well.

14.8 Research Gaps

- Little information is existed particularly on small mammals such as: Rodentia, Lagomorphs, Erinaceomorpha, Soricomorpha and Chiroptera. On known species of mammals, the diversity, distribution, population status generally on the ecology these species should be worked out.

- There is little research in the area of herpetofauna in general.
- The status of the amphibian chytrid fungus in the Lake Tana Sub-basin has to be studied.
- Even though there are small families of Hippos occur in Lake Tana, no comprehensive research is ever conducted on the distribution, population and genetic information on the species.
- There are Rock Pythons in wetlands of the area, however, no detail information on the species.

14.9 Recommendations and Management Options

Recommendations

- Strengthen graduate researches in the field of zoology particularly in amphibians, reptiles and to do research on selected mammalian species.
- Develop awareness of the community on the values and ecological functions of herpetofauna and mammals.
- Establish collaboration with international institution, and sharing of expertise in the field of herpetology and mammalogy.
- Establish protected area of core places used for breeding of amphibians, reptiles and mammals.
- Capacity building in the field of herpetology and mammalogy.
- Inspection water pollution and recommend safe drainage and water treatment practices at sites where domestic and industrial wastes are deposited.
- Strengthen coordination among the major stakeholders.
- Undertake Wildlife disease surveillance.

Management options

- Wildlife animals inventory and monitoring: with a certain time of interval, at least annual monitoring of the resource has to be carried out.
- Practice wetland and riparian resource management action plan.
- Conservation of priority ecosystem components by detail study of the area.
- Controlling encroachment and ecological degradation of the natural habitat of amphibians, reptiles and mammals.
- Establish and implement watershed based habitats conservation and restoration practice by participating the community and GO and NGO stakeholders.
- Develop ecotourism promotion and regulation strategy.

14.10 Conclusion

The faunal diversity spans from the minute and cryptic puddle- and tree-frogs, to the medium sized Nile monitor and African rock python, and to the large-sized Nile crocodile, leopard and hippopotamus. The occurrence of 19 amphibians, 35 reptiles and 28 mammalian species could be associated with presence of diverse habitats. The woodland to wooded grassland, cultivated land to natural forest and mixed forest area, water body to open wetland provides good habitat for the fauna. However, the total number of species listed could be underestimating. The absence of qualified professional particularly in herpetology and the insufficient research in the field of mammalogy could be accounted for this problem. But the list of species could provide highlight information for future research.

The species diversity, richness, distribution and the status of each species should be worked out. Currently, the wetland and forest habitats are being degraded, encroached and converted into agriculture. The high population growth along with the industrial wastes let into the lake, the use of pesticides and fertilizers in the catchment area is becoming a threat for the faunal diversity as pollution increases. Proper management practice must be implemented and practiced to balance the integrity of the Lake Tana sub-basin ecosystem and to sustainably use the resources. Capacity building in the field of Herpetology, community awareness on the value and function of wildlife, and a comprehensive study on the subject matter should be encouraged.

Annex 14. 1

Amphibian species of Lake Tana Sub-basin (Mengistu 2012 and Mengistu et al. 2013)

Family	Scientific name and authority	Common name	Endemicity to Ethiopia
Arthroleptidae Mivart 1869	*Leptopelis bocagii* (Günther 1865)	–	Non-endemic
	Leptopelis yaldeni (Largen 1977)	Grassland forest tree frog	Endemic
Bufonidae Gray 1825	*Amietophrynus kerinyagae* (Keith 1968)	Keith's toad	Non-endemic
	Amietophrynus regularis (Reuss 1833)	African common toad	Non-endemic
Hemisotidae Cope 1867	*Hemisus marmoratus* (Peters 1854)	Marbled snout-burrower	Non-endemic
	Hemisus microscaphus (Laurent 1972)	Ethiopian snout-burrower	Endemic

(continued)

(continued)

Family	Scientific name and authority	Common name	Endemicity to Ethiopia
Hyperoliidae Laurent 1943	*Hyperolius acuticeps* (Ahl 1931)	–	Non-endemic
	Kassina senegalensis (Duméril & Bibron 1841)	Senegal kassina	Non-endemic
	Paracassina obscura (Boulenger 1894)	Ethiopian striped frog	Endemic
Petropedetidae Noble 1931	*Conraua beccarii* (Boulenger 1911)	Beccari's giant frog	Non-endemic
Phrynobatrachidae Laurent 1941	*Phrynobatrachus minutus* (Boulenger 1895)	Ethiopian dwarf puddle frogs	Endemic
	Phrynobatrachus natalensis (Smith 1849)	Natal dwarf puddle frog	Non-endemic
Pipidae Gray 1825	*Xenopus clivii* (Peracca 1898)	Peracca's clawed frog	Non-endemic
Ptychadenidae Dubois 1987	*Ptychadena porosissima* (Steindachner 1867)	–	Non-endemic
	Ptychadena pumilio(Boulenger 1920)	–	Non-endemic
	Ptychadena sp. (Debre Markos)	–	Endemic
	Ptychadena wadei (Largen 2000)	Tisisat grassland frog	Endemic
Pyxicephalidae Bonaparte 1850	*Amietia angolensis* (Bocage 1866)	Angola river frog	Non-endemic
	Cacosternum boettgeri (Boulenger 1882)	Boettger's dainty frog	Non-endemic

Annex 14. 2

Reptile species of Lake Tana Sub-basin (Largen and Spawls 2010 and Abebe A. Mengistu, unpublished data)

Order	Family	Scientific name	Common name
Sauria (Lizards)	Agamidae	*Acanthocerus cyanogaster*	Blue-bellied Ridgeback Agama
		Acanthocerus zonurus	Ethiopian Ridgeback Agama
		Agama doriae	Doria's Agama
	Chamaeleonidae	*Chamaeleo affinis*	Ethiopian Highland Chameleon
		Chamaeleo africanus	Sahel Chameleon
		Chamaeleo gracilis	Gracile Chameleon
	Gekkonidae	*Hemidactylus laticaudatus*	Andersson's Gecko

(continued)

(continued)

Order	Family	Scientific name	Common name
	Scincidae	*Chalcides ragazzii*	Ragazzi's Bronze Skink
		Mabuya isselii	Issel's Skink
		Mabuya maculilabris	Speckle-lipped Skink
		Mabuya quinquetaeniata	Five-lined Skink
		Mabuya wingatii	Wingate's Skink
		Panaspis tancredii	Ethiopian Snake-eyed Skink
	Varanidae	*Varanus niloticus*	Nile Monitor
Serpentes (Snakes)	Typhlopidae	*Afrotyphlops blanfordii*	Blanford's Blind Snake
		Letheobia somalica	Ethiopian Blind Snake
	Pythonidae	*Python sebae*	African Rock Python
	Lamprophiidae	*Lamprophis erlangeri*	Erlanger's House Snake
		Lamprophis fuliginosus	Brown House Snake
		Pseudoboodon lemniscatus	Striped Ethiopian Snake
		Lycophidion capense	Cape Wolf Snake
		Psammophis angolensis	Dwarf Sand Snake
		Psammophis sibilans	Hissing Sand Snake
		Psammophylax variabilis	Grey-billed Skaapsteker
	Colubridae	*Dasypeltis scabra*	Common Egg-eater
		Dasypeltis atra	Montane Egg-eater
		Meizodon regularis	Eastern Crowned Smooth Snake
		Platyceps florulentus	Geoffroy's Racer
		Philothamnus battersbyi	Battersby's Green Snake
		Scaphiophis raffreyi	Ethiopian Hook-nosed Snake
		Crotaphopeltis hotamboeia	White-lipped Herald Snake
		Natriciteres olivacea	Olive Marsh Snake
	Elapidae	*Dendroaspis polylepis*	Rhombic Night Adder
	Viperidae	*Causus rhombeatus*	Black Mamba
Testudines (Tortoises and Turtles)	Trionychidae	*Trionyx triunguis*	African Softshell Turtle

Annex 14. 3

List of Mammals in the Area (Shimelis a 2006–2013, Unpublished Data)

Order	Family name	Scientific name and authority	Common name	Conservation status
Primate	Cercopithecidae	Papio cynocephalus Anubis (Linnaeus 1766)	Savanna (common) Baboon	LC
		Cercopithecus pygerythrus (F. Cuvier 1821)	Vervet (Green) Monkey	LC
	Suiformes	Phacochoerus africanus (Gmelin 1788)	Common Warthog (Kerkero)	LC
Ruminatia	Bovidae	Sylvicapra grimmia (Linnaeus 1758)	Bush Duiker (Midaqua)	LC
		Tragelaphus scriptus (Pallas 1766)	Bushbuck* (Dikula)	LC
		Oreotragus oreotragus (Zimmerman 1783)	Klipspringer (Sessa)	LC
		Ourebia ourebi (Zimmermann 1782)	Oribi (Feko)	LC
		Kobus ellipsiprymnus (Ogilby 1833)	Waterbuck (Difarsa)	LC
Carnivora	Canidae	Canis aureus (Linnaeus 1758)	Golden (common) Jackal	LC
		Mellivora capensis (Storr 1780)	Honey Badger (ratel)	LC
	Viverridae	Civettictis civetta (Schreber 1776)	African Civet	LC
	Herpestidae	Galerella sanguine Rüppell 1836	Slender mongoose	LC
	Hyaenidae	Crocuta crocuta (Erxleben 1777)	Spotted hyaena	LC
	Felidae	Leptailurus serval (Schreber 1776)	Serval cat	LC
		Panthera pardus (Linnaeus 1758)	Leopard	NT
Hyracoidea	Procaviidae	Procavia capensis (Pallas 1766)	Rock hyrax	LC
Tubulidentata	Orycteropodidae	Orycteropus afer (Pallas 1766)	Aardvark	LC
Lagomorpha	Leporidae	Lepus fagani (Thomas 1903)	Bush hare	LC

(continued)

(continued)

Order	Family name	Scientific name and authority	Common name	Conservation status
Rodentia	Sciuridae	Heliosciurus gambianus (Ogilby 1835)	Sun squirrels	LC
	Hysticidae	Hystrix cristata (Linnaeus 1758)	African porcupines	LC
	Muridae	Arvicanthis abyssinicus (Rüppell 1842)	Grass rats	LC
		Arvicanthis dembeensis (Rüppell 1842)	–	LC
		Desmomys harringtoni (Thomas 1902)	Harrington's rat	LC
		Mus musculus (Linnaeus 1758)	House mouse	LC
Suiformes	Suidae	*Potamochoerus larvatus*	Bushpig*	LC
		Phacochoerus africanus	Common Warthog	LC
Whippomorpha	Hippopotamidae	*Hippopotamus amphibius*	Hippopotamus	LC

LC Least concern *NT* Near Threatened (IUCN 3.1)

References

Adem G (2008) Assessment of ecotourism potentials for sustainable natural resources management in and around Abijata-Shala Lakes National Park in the Central Ethiopian Rift Valley. Thesis, Addis Ababa University

Afework B, Yalden DW (2013) The mammals of Ethiopia and Eritrea. Addis Ababa University Press, Addis Ababa

Beebee TJC (1996) Ecology and conservation of amphibians. Chapman and Hall, London

Biru Y (2007) Land degradation and options for sustainable land management in the Lake Tana Basin (LTB), Amhara Region, Ethiopia. Dissertation, University of Bern

Bojö J (1996) The economics of wildlife: case studies from Ghana, Kenya, Namibia and Zimbabwe. AFTES Working Paper No. 19, The World Bank, Washington DC

Cardillo M, Mace GM, Jones KE et al (2005) Multiple causes of high extinction risk in large mammal species. Sc 309:1239–1241

Carolyn HS (1987) Small mammals: Pests or vital components of the ecosystem (1987). Great plains wildlife damage control. In: Paper presented at the 8th wildlife damage control workshop, in Rapid City, South-Dakota, April 28–30, 1987

Chardonnet P, Clers B, des Fischer J et al (2002) The value of wildlife. Rev sci tech Off Int Epiz 21(1):15–51

Cheng S, Hiwatashi Y, Imai H et al (1998) Deforestation and degradation of natural resources in Ethiopia: forest management implications from a case study in the Belete-Gera Forest. J Res 3:199–204

East, R (1999). African antelope database 1998. IUCN/SSC Antelope Specialist Group

Evans B, Bliss SM, Mendel SA et al (2011) The Rift Valley is a major barrier to dispersal of African clawed frogs (Xenopus) in Ethiopia. Molec Ecol 20(20):4216–4230. doi:10.1111/j.1365-294X.2011.05262.x

Frost DR, AMNH (2010) Amphibian species of the world: an Online Reference. Version 5.3. Electronic Database accessible at http://research.amnh.org/herpetology/amphibia/ American Museum of Natural History, New York, USA. Accessed 12 February 2009

Getachew G, Afework B (2009) Human-wildlife conflict in Zegie peninsula (Ethiopia) with emphasis on grivet monkey (Cercopithecus aethiops aethiops). SINET: Ethiop J Sci 32(2): 99–108

Getahun A, Mengistu AM (eds) (2006) Field and laboratory techniques for zoological museum collections: manual on methods of collection, preservation and management. The Zoological Natural history museum, Addis Ababa University, Addis Ababa

Gower D, Doherty-Bone TM, Aberra RK et al (2012a) High prevalence of the amphibian chytrid fungus (*Batrachochytrium dendrobatidis*) across multiple taxa and localities in the highlands of Ethiopia. Herpeto J 22:225–233

Gower D, Garrett K, Stafford P (2012b) Snakes. The Natural History Museum, London

Hillman JC (1993) Compendium of wildlife conservation information. Vol. 1: Wildlife conservation in Ethiopia. Ethiopian Wildlife Conservation Organization, Addis Ababa

IBC (2005) National biodiversity strategy and action plan. Institute of Biodiversity Conservation, Government of the Federal Democratic Republic of Ethiopia, Addis Ababa

IUCN (2008) IUCN red list of threatened species. Version 2014.2. <www.iucnredlist.org>. Accessed 16 October 2014

IUCN (2013) IUCN red list of threatened species. Version 2010.4. <www.iucnredlist.org>. Downloaded on 20 February 2013 IUCN, Gland, and Cambridge

Kielgast J, Rödder E, Veith M et al (2010) Widespread occurrence of the amphibian chytrid fungus in Kenya. Anim Conserv 13(Suppl 1):1–8

Lanly JP, Singh KD, Janz K (1991) FAO's 1990 reassessment of tropical forest cover. Nat Resour 27(2):21–26

Largen MJ (1977) The status of the genus Leptopelis (Amphibia, Anura, Hyperoliidae) in Ethiopia, including descriptions of two new species. Monit Zool Ital NS Suppl 9:85–136

Largen MJ (2001) Catalogue of amphibians of Ethiopia, including a key for their identification. Trop Zool 14:307–402

Largen MJ, Spawls S (2010) Amphibians and Reptiles of Ethiopia and Eritrea. Edition Chimaira. Frankfurt am Main

Largen MJ, Rasmussen JB (1993) Catalogue of the Snakes of Ethiopia (*Reptilia Serpentes*) including identification keys. Trop Zool 6:313–434

Lötters S, Kielgast J, Bielby J et al (2010) The link between rapid enigmatic amphibian decline and the globally emerging chytrid fungus. Eco Health. I Asso Ecol Heal 6(3):358–372. doi:10.1007/s10393-010-0281-6

Masood MF, Asiry AA (2012) Ecological studies on diversity of Herpetofauna in Asir region, Kingdom of Saudi Arabia. Egypt Acad J Biolog Sci 4(1):143–163

Mengistu AA (2012) Amphibian diversity, distribution and conservation in the Ethiopian highlands: morphological, molecular and biogeographic investigation on Leptopelis and Ptychadena (Anura). Dissertation, University of Basel

Mengistu AM, Nagel P, Getahun A et al (2013) Updated review of amphibian diversity, distribution and conservation in Ethiopia. Ethiopian J Biol Sci 12:81–116

Pol JLV (2006) A guide to endemic birds of Ethiopia and Eritrea, 2nd edn. Shama Books, Addis Ababa

Pough FH, Andrews RM, Cadle JE et al (2003) Herpetology, 3rd edn. Benjamin Cummings, San Francisco

Poynton JC (1999) Distribution of amphibians in Sub-Saharan Africa, Madagascar and Seychelles. In: Duellman WE (ed) Patterns of distribution of amphibians: a global perspective. The John Hopkins University Press, Baltimore, pp 483–539
Rödder D, Kielgast J, Bielby J et al (2009) Global amphibian extinction risk assessment for the panzootic chytrid fungus. Diversity 1:52–66
Shimelis GS, Srinivasan R, Dargahi B (2008) Hydrological Modelling in the Lake Tana Basin, Ethiopia Using SWAT Model. Open Hydro J 2:49–62
Smith RL (1992) Elements of ecology, 3rd edn. Harper Collins Publishers, New York
Spawls S, Branch WR (1995) The dangerous snakes of Africa. Southren Book Publishers, Halfway House, South Africa
Spawls S, Howell K, Drewes R et al (2002) A field guide to the reptiles of East Africa. Academic Press, San Diego, San Francisco
Stuart C, Stuart T (2006) Field guide to the larger mammals of Africa. Struik Publishers, South Africa
Stuart SN, Chanson JS, Cox NA et al (2004) Status and trends of amphibian declines and extinctions worldwide. Sci 306:1783–1786
Wilkinson M, San Mauro D, Sherratt et al (2011) A nine-family classification of caecilians (Amphibia: Gymnophiona). Zootaxa 2874:41–64
Yalden DW, Largen MJ, Kock D et al (1996) Catalogue of the mammals of Ethiopia and Eritrea. 7. Revised Checklists. Zoogeography and Conservation. Trop Zoo 9:73–160
Zimkus B (2008) Phrynobatrachus bullans: geographic distribution (Tanzania, Kenya, and Ethiopia). Herpet Rev 39(2):235

Further Reading

Largen M, Spawls S (2010) The amphibians and reptiles of Ethiopia and Eritrea. Edition Chimaira. Andreas S. Brahm publisher

Chapter 15
Forest Resources in Amhara: Brief Description, Distribution and Status

Alemayehu Wassie

Abstract The Amhara National Regional State (ANRS), with a total area of approximately 170,052 km^2, is situated in the north western and north central part of Ethiopia. The forest resource of the Amhara Region that may qualify as 'forests' following the definition of FAO (2001) is estimated to be around 6% of the total area of the region, which includes high forest (0.48% of the total area); woodlands (4.2%); bamboo and plantation forests (1.23%). These vegetation types further can be classified as: Afroalpine and subafroalpine, Montane dry forest and scrub, Combretum-Terminalia woodland, Acacia-Comiphora woodland, High and Lowland bamboo, and Plantations. The overall forest resources of the region have shown slight increase in area coverage for the last 15 years. The major plus comes from Eucalyptus plantation and rehabilitation efforts in Eastern and recently in Western highlands of Amhara. On the other hand, the Afroalpine and Subafroalpine vegetation, the high Dry afromontane forests, Bamboo forests and the Woodlands faced remarkable shrink not only in area coverage but also degradation in their structural and species composition.

Keywords Amhara · Forest resource · Forest description · Forest distribution · Biodiversity

15.1 Introduction

The Amhara National Regional State (ANRS), with a total area of approximately 170,052 km^2, is situated in the north western and north central part of Ethiopia (EPRDF 2014). The Region has a common boundary with Tigray in the north, Oromia in the south, Benshangul Gumz to the west and southwest and Afar in the east. It shares an international boundary with the neighboring country, the Sudan, in the west. The vast majority of the region is rugged landscape with 34% of the land

A. Wassie (✉)
College of Agriculture and Environmental Sciences, Bahir Dar University, Bahir Dar, Ethiopia
e-mail: alewas2008@yahoo.com

K. Stave et al. (eds.), *Social and Ecological System Dynamics*,
AESS Interdisciplinary Environmental Studies and Sciences Series,
DOI 10.1007/978-3-319-45755-0_15

in the region estimated to have a slope of over 35% (EPLAUA 2004 Environmental Protection Authority report, unpublished). The region is also endowed with plains like the Fogera, Dembia, Bechena and Dawachefa areas.

The Region can be roughly divided into a set of highland blocks separated by deep river valleys and the Eastern and Western Escarpments and their associated Lowlands:

- The Highlands of East and West Gojam and Awi Zones that peak on Mount Choke and are bounded in the west by the Western Escarpment and in the south and east by the Abay Gorge,
- The South Gonder Highlands peaking on Mount Guna, bounded in the north by the Tekeze, the south and west by the Abay river, the east and south by the Beshlo river, and the east by the Eastern Escarpment,
- The North Gonder Highlands peaking on Mount Dashen of Simien Mountains bounded in the east and north by the Tekeze River, and the west by the Western Escarpment.
- The Highlands of North Wellò and Wag Hemra Zones peaking on Mount Abune Yosef bounded in the west and south by the Tekeze river, and the east by the Eastern Escarpment,
- The Highlands and high plateau of South Wello, Oromiya and North Shewa Zones peaking at Mount Tarmaber together with hills to the east of the Borkena Valley and bounded in the west by the Abay river, the north by the Beshlo river, and to the east by the Eastern Escarpment,

The Lowlands are found in the west and the east:

- The Gonder Lowlands are found between the Western Escarpment and the border with Sudan
- The eastern Lowlands are found in the Borkena Valley and the foothills of the Eastern Escarpment adjoining the Afar Plains.

Altitude governs the temperature and overall biodiversity distribution (Alemayehu 2007). Temperature is inversely related to altitude, with mean annual temperatures of 22–27 °C in the Lowlands and between 10 and 22 °C in the Highlands up to 3000 m above sea level (m.a.s.l). Frost is a phenomenon during the dry season in valley bottoms of areas above 2600 m.a.s.l with increasing frequency with altitude. Two broad rainfall pattern zones and Length of Growing Period (LGP) are observed in the Region:

- Areas with a single rainy season of inconsistent length between May and October, located in the western part of the Region (North and South Gonder, East and West Gojam, and Awi Zones)
- Areas with a short rainy season (Belg) of inconsistent length and intensity between February and April, and a main rainy season (Mehr) of variable length between July and October (Wag Hemra, North and South Wello, Oromiya and North Shewa Zones) (ANRS 2004).

There is no generally agreed definition of forest. The UN Food and Agriculture Organization (FAO) has given various definitions including the following: FAO (2006) defined a forest as a minimum land area of 0.05–1 ha with tree crown cover more than 10–30% and tree height of 2–5 m at maturity; FAO (2001) defined a forest as "land with a tree crown cover (or equivalent stocking level) of more than 10% and an area of more than 0.5 hectare; the trees should be able to reach a minimum height of 5 m at maturity in situ".

Although a definitive survey of forests has not been conducted, the forest resource of the Amhara region that qualifies as 'forest' following the definition of FAO (2001) is estimated to be around 6% (1,007,761 ha) of the total area of the region. These forest resources include: high forest (0.48% of the total area or 81,047 ha); woodlands (4.2% of the total area or 716,915 ha); bamboo and plantation forests (1.23% of the total area or 209,799 ha) (Woreta 2007). These vegetation types further can be classified as: Afroalpine and subafroalpine, Montane dry forest and scrub, Combretum-Terminalia woodland, Acacia-Comiphora woodland, High and Lowland bamboo, and Plantations (IBC 2005).

In addition to forests that meet FAO definitions, a considerable number of trees are grown in agroforestry forms. Agroforestry is a farming system that integrates trees into the farmland. In the Amhara Region different traditional farm forestry practices exist in different agro-ecological zones. The main types of agroforestry practices are:

- Homestead trees: trees such as Eucalyptus species, *Cordia africana*, *Croton macrostachus* and other valuable trees like *Rhamnus prinoides* are planted around the home.
- Dispersed trees on crop and grazing land: indigenous trees are either retained or planted on cultivated field for the purpose of fodder, fuel wood and poles. Prominent retained tree species are *Cordia Africana, Croton macrostachus, Mellitia ferrugenia, Ficus species, Ziziphus species* and *Balanityes species.*

15.2 Forest Resource Types of the Amhara Region

15.2.1 Afroalpine and Sub-afroalpine Forests

Areas which are on average higher than 3200 m.a.s.l are generally referred to as the Afroalpine and Subafroalpine (Hedberg 1957). These areas include chains of mountains, mountain slopes and tops of highest mountains in Amhara Region. The highest peak in Ethiopia, Ras Dashen (4533 m.a.s.l), the tips of the chain of mountains including Mount Guna in South Gondar, chain including Mount Tarmaber in North Shoa, and the chain including Mount Abune Yosef of South Wello and Mount Choke in East Gojjam are the major ecosystems where Afroalpine and Sub afroalpine forests exist in Amhara Region. The vegetation occurs in a combination of giant herbs and shrubs, succulents and grasses, with patches of *Erica arborea*.

These forests used to be intact and wild due to temperatures too cold for human settlement and agriculture which as a result made them home for many endemic plants and animal species. However the last 40 years seems they have not been immune. Surprisingly the human settlements with sedentary agriculture practice now reach up to 3800 m.a.s.l, which has resulted in the clearance of the original vegetation and serious soil erosion and land degradation. Uncontrolled agriculture and overgrazing are becoming serious threats that are already showing their effects. The fact that these mountain blocks and their ecosystems are the source of several major streams and rivers means their degradation will have serious consequences in the hydrology of the vast majority of the region.

15.2.2 Dry Evergreen Montane Forest and Evergreen Scrub

The Amhara Region highlands constitute more than 60% of the land area with Afromontane vegetation, of which dry montane forests form the largest part (EPLAUA 2004, Environment Protection Authority report, unpublished). Evergreen scrubland vegetation occurs in the highlands of Amhara either in association with the remnant dry evergreen montane forest or as the rehabilitated condition after deforestation of the dry evergreen montane forests.

Dry Evergreen Montane Forests are occur in an altitudinal range of 1500–2700 m, with average annual temperature and rainfall of 14–25 °C and 700–1100 mm, respectively (Friis 1992). As it is found in wide range of altitude it is designated as a very complex vegetation type. These types of forests develop in areas of relatively high humidity, but not much rain, and where there is a prolonged dry season. As a result, tropical dry forests are noted for their seasonality with respect to rainfall in contrast with the rain forests where the environment is stable throughout the year. Dry evergreen montane forests experience long dry seasons (4–8 months) and inconsistent rainy periods. During the dry season, both moisture stress and temperature increase. Daytime humidity drops and watercourses either dry up or have greatly diminished flows (Demel 1996).

In the wetter western parts of the region, the forests comprise mixed *Podocarpus falcatus–Juniperus procera* forest with a well-developed understory often comprising *Croton macrostachyus, Ficus* spp., *Olea europaea subsp. cuspidata, Trema orientalisand, Maesa lanceolata*. On the shores and islands of Lake Tana a more humid variant of this forest occurs. In the drier eastern parts of the Region, the forest remnants comprise mixed Juniperus and Olea woodland (ANRS 2004).

The altitudinal range of 1500–2700 m.a.s.l has been inhabited by the majority of the Ethiopian population practicing sedentary, cereal-based, mixed agriculture for centuries. As a result these forests are among the most exploited forest ecosystems of the world and have been increasingly fragmented. In Amhara Region these forests have been seriously diminished due to agricultural expansion and replaced by bush lands in most areas. There are no extensive continuous dry afromontane forests in Amhara Region currently; rather they exist as remnant forests scattered in the landscape. Under the increasing pressure by the human population, via e.g. crop land expansion, grazing activities or tree harvesting, natural regeneration may be hampered and, as a result, the persistence of the remnant forest patches and their indigenous species in many areas are threatened.

The remnant dry afromontaine exists in two forms in the region:

Protected State Forests: designated as priority areas. Although there are 13 regional forest priority areas, the prominent dry afromontane forests of the region are:

- *Wof-Washa Forest*:
  The forest of Wof-Washa is located at N 9° 35' E 39° 45' in the south east highlands of the region in Northern Shoa Zone. It spreads over two woredas of Ankober and Baso and Worana. The total size of the forest area is 3600 ha with altitudinal range of 2000–3730 m.a.s.l (Fekadu 2011). The sides of the mountains are so steep that the forest is one of the most inaccessible in the country. Inaccessibility seems to be the only reason for its survival to the present day. The altitude ranges from 2000 m to 3730 m.a.s.l. Wof–Washa vegetation type is dry montane mixed broad-leaved/conifer forest. Recent studies show that the forest has an estimated 252 species of plants of which 33 are trees and 18 are shrubs. Dominant woody trees include *Juniperus procera, Podocarpus falcatus, Olea europaea subsp. cuspidata, Hagenia abyssinica,* and *Allophylus abyssinica*. At higher altitudes commonly found plants include *Erica arborea, Hypericum revolutum, Lobelia* spp. (Demel and Tamirat 1995; Tamirat 1994).
- *Yegof Forest*:
  The forest of Yegof is found in South Wello on a mountain ridge on western side of the high way from Dessie to Kombolcha. Its area formerly was estimated to be around 18,000 ha but as the result of several threats it faced, the current total size of the forest area is 1530 ha (Amhara Region Bureau of Agriculture 2008 unpublished report). It has an altitude range between 2000 and 3002 m.a.s.l. The natural forest, which once covered Mt Yegof is classified as dry evergreen mixed and broad-leaved type. Dominant woody species at higher altitudes include *Juniperus procera, Olea europaea subsp. cuspidata, Erica arborea, Hypericum revolutum and H. quartinianum. At lower altitudes Bersama abyssinica, Croton macrostachyus, Syzygium guieense, Rhus vulgaris, Cordia africana, Euphorbia* spp., *Albizia* spp. and a few *Acacia* species are found (Tilahun et al. 1996; Fekadu 2011).
- *Denkoro Forest*:
  Denkoro Forest is found in Amhara Saynt and Debresina District of South Wello situated approximately between 10° 47'–10° 53'N and 38° 35'–38°48'E. The land area of Denkoro forest is 5500 ha. The area is generally characterized by rough topography with mountains, deeply incised valleys, which ranges from 1500 to 3500 m.a.s.l. The natural vegetation of the forest is classified into six clusters as: (1) *Erica arborea-Hypericum revolutum,* (2) *Myrsine melanophloeos–Dombeya torrida,* (3) *Myrsine africana–Maesa lanceolata–Prunus*

africana, (4) *Olinia rochetiana–Olea europaea,* (5) *Olinia rochetiana–Allophylus abyssinicus–Apodytes dimidiata,* and (6) *Maytenus gracilipes–Teclea nobilis*. Eleven endemic species, which are included in the IUCN Red List categories, have been recorded (Abate et al. 2006).

- *Tara–Gedam Forest*:
  The forest of Tara-Gedam is located at N 12° 8′ 42″ E 37° 44′ 43″ east of Lake Tana along the high way from Bahidar to Gondar in Libokemekem woreda of South Gondar. The total size of the forest area is 515.6 ha (Dagnachew 2001), of which 66.89 ha are church forest, the other parts are state forests which have been owned and managed by the state since 1979 (Zegeye et al. 2011). The most frequent tree species were *Albizia gummifera*, followed by *Carissa spinarum, Calpurnia aurea* and *Olea europaea subsp. cuspidata* (Dagnachew 2001; Renée 2013; Zegeye et al. 2011).
- *Ethiopian Orthodox Church Forests*:
  In the northern highlands of Ethiopia, patchy remnants of old-aged Afromontane forests can be found mainly around the Ethiopian Orthodox Tewahido Churches (EOTC). Hence, when a traveler sees a patch of indigenous old-aged trees in the northern highlands of Ethiopia, he/she can be sure that there is an Orthodox Church in the middle. They are visible from a great distance, with a majestic appearance, usually built on small hills "overlooking" the surrounding villages. The local people call these churches with the surrounding trees as "debr" or "geddam". "Debr" or "Geddam" is seen by the followers as the most holy place religiously as well as a respected and powerful institution socially. Although the main purpose of churches is as places for worship, burials, meditating and religious festivals, they also provide valuable, often unique and secured habitats for plants and animals, and green spaces for people. The Christian population in Northern Ethiopia inhabits mostly the highlands within the range of 1500–3500 m.a.s.l. Churches of the EOTC are distributed following this settlement pattern. This churches that are located in the Zegie peninsula and island monasteries of Lake Tana are endowed with rich biodiversity. The tradition of the community to protect and conserve church forests in particular has contributed a lot for that.

Some studies suggest that church forests are relics or blueprints of ancient and largely lost forest ecosystems, and as such, that they are hotspots of biodiversity for indigenous species (Alemayehu 2002; Alemayehu et al. 2005a, b; Alemayehu 2007; Bingelli et al. 2003; Bongers et al. 2006). The most dominant species in the church forests are *Juniperus* and *Olea,* which were species mainly composing the predominating diverse vegetation of dry afromontane forests of Northern Ethiopia since 1400–1700 AD (Alemayehu 2007; Darbyshire et al. 2003). Not all church forests, however, are remnants of the original forest of the area. In churches established since the 1960's after the original forest had already vanished, forests are mostly rehabilitated vegetation types and plantations of exotic species.

From vegetation sampling done on 28 of these churches in South Gondar of Amhara Region (Northern Ethiopia), it was found that the size of forests that surround the church range in area from 1.6 to 100 ha. A total of 168 woody species representing 69 families were recorded in forests covering only 500 ha. Out of the 23 threatened indigenous tree species listed as national priority and commercially important species, 15 were found in these churche forests. Moreover two indigenous tree species (*Juniperus procera* and *Prunus africana*) currently included in the IUCN red list of threatened species are common to abundant in church forests. The total number of woody species in each of the 28 churches ranged from 15 to 78. Mean basal area ranged from 4.8 to 111.5 m^2/ha. The average maximum height ranged between 9.1 and 22.9 m. Mean canopy cover ranged from 18 to 86% (Alemayehu 2007). Church forests possess diverse woody species and accommodate good wood biomass compared to some of the biggest continuous forest in the central highlands of Ethiopia. The overall species richness of these remnant forests was higher than that of Wof-Washa forest (51 woody species) (Demel and Tamirat 1995), which is one of the biggest continuous forest in central highlands of Ethiopia. The maximum basal area/ha (111 m^2/ha) recorded in those forest samples is comparable to that of Wof-Washa forest (100.3 m^2/ha) (Tamirat 1994). The fact that the study surveyed only 28 churches out of over 10,000 churches that exist in Amhara Region strongly suggests that more species will be found if more church forests are assessed.

Church forests, embedded in a heavily degraded landscape matrix, serve a critical role in the landscape ecology of the region as in situ and ex situ conservation sites of indigenous biodiversity. These sites, can serve as sources of seeds/propagules/germplasm for future development of indigenous forests (Alemayehu et al. 2005a, b). They can provide knowledge and a point of orientation to determine habitats of individual species and synecologies—associations or assemblages of species—for communities (Alemayehu et al. 2005b; Bongers et al. 2006; van Andel and Aronson 2006).

Although the contribution of church forests to the conservation of biodiversity in the region and to the country is significant, they have been undermined and overlooked by policy makers, extension workers, NGOs and researchers to the extent that they have not been counted as part of the region's resources. As a result many of these forests are threatened by the increasing cattle population and increasing demand for wood products and farm lands, in spite of the fact that the local people generally respect the integrity of the church forests (Alemayehu 2002). The church has been using wood from these forests for church construction and mass services in a reasonably sustainable way for generations. However, logging in some of the churches is becoming much more intensive and in the interest of individuals instead of the church community. Undefined forest borders and irregular forest edges surrounded by individual land holdings have led to continuous but 'hidden' encroachments of agricultural fields into the forest. The conversion of the forest land to other land forms occurs in steps of only a few meters every year. Given the small size of the forest, such changes might drastically reduce forest areas in the coming years and even lead to disappearance of the forests. As a result of both human and environment induced factors, many of the church forests have only very few seedlings of woody species. In fact the understory of the forests is completely dominated by shrubs and lianas.

15.2.3 Combretum-Terminalia Woodland

According to Ethiopian vegetation classification, Woodlands comprises Combretum-Terminalia broadleaved deciduous forests and Acacia-Commiphora narrow-leaved deciduous (IBC 2005). Combretum-Terminalia broadleaved deciduous forests are characterized by *Combretum* spp., *Terminalia* spp., *Oxytenanthera abyssinica, Boswellia papyrifera, Anogeissus lieocarpa, Sterospermem kuntianum, Pterocarpus lucens, Lonchocarpus laxiflorus, Lannea* spp. *Albizia malacophylla* and *Enatada africana.* These are small trees with fairly large deciduous leaves, which often occur with the lowland bamboo–*Oxytenanthera abyssinica.* The understory is a combination of herbs and grasses. The herbs include *Justecia* spp., *Barleria* spp., *Eulophia, chlorophytum, Hossolunda opposita* and *Ledeburia* spp. Usually the herbs dominate the ground layer at the beginning of the rainy season while grasses dominate toward the end of the rainy season. These forests have developed under the influence of recurring fire as a result the trees have evolved very thick bark while most herbs developed perennial bulbs (Mulugeta and Zenebe 2011 unpublished).

These forests occupy the altitudinal range between 500–1900 m.a.s.l. They are predominantly found in wetter areas of the western lowlands of Amhara Region (Jawi, Quara, Metema, Armachiho, Adiarkay) and on the vallies of big rivers like Abay, tekeze and Beshelo (ANRS 2004).

In Amhara Region predominantly in Western lowlands of Quara, Metema and Armachiho both the communities and companies have been involving for decades in tapping Frankincense and Gum-Arabic which ultimately made it one of the foreign currency generating sectors for Ethiopia.

The regional government has been promoting large agricultural investments to produce sesame and cotton in the western lowlands of the region for the last 15 years. Re-settlement programs of the government from the degraded highlands were the other land demanding programs in these areas. Traditional shifting cultivation with uncontrolled burning has led to severe destruction of these wood lands. The source for these investment and resettlement lands were clearing of these forests. In addition, traditional tapping of the resinous trees has brought severe mortality of trees (Abeje 2002). All these issue combined have resulted in massive destruction of woodlands which had previously served as a green belt protecting the highlands from expansion of Sahara Desert.

15.2.4 Acacia-Commiphora Woodland

The Acacia-Commiphora forests are known for their varying soils, topography, and diverse biotic and ecological elements. These plant species are with either small deciduous leaves or leathery persistent ones. The density of trees varies from 'high', in which they form a closed canopy to scattered individuals to none at all forming open grasslands (Mulugeta 2011 unpublished).

These forests are recognized in the drier lowlands of the eastern foothills (Kobo, Habru, Tehuwledre, Kalu, Dawa chefa, Efratana Gidim and Shewarobit areas) and the inner Tekeze valley of Amhara Region in areas of altitude above 1000 and below 1900 m.a.s.l (ANRS 2004).

Although these forests are known for their gum and resin producing species such as Acacica, Boswellia and Commiphora species (Mulugeta 2011 unpublished), in Amhara Region predominantly they serve as agroforestry component of the sedentary agriculture. They have been facing devastation primarily for charcoal making and harvestings for farm tools.

15.2.5 Bamboo Forests

Bamboos are perennial woody grasses belonging to the Poaceae (Gramineae) family and Bambuseae subfamily (Ohrnberger 1999). Since most bamboos have a tree morphology and attain tree size at maturity they are named tree-grasses (Kelecha 1980 unpublished). The main stem of the aboveground part of the plant is the culm, while the underground part constitutes the rhizome and root system. Ethiopia has two indigenous bamboo species which are endemic to Africa: the African Alpine Bamboo (Yushane alpine K. Shumann Lin; synonym: *Arundinaria alpina K. Schumann*) and the monotypic genus lowland bamboo *Oxytenanthera abyssinica (A. Richard) Munro* (Kasshun 2000). These two species of bamboo commonly called highland and lowland bamboo exist in various parts of Amhara Region (Seyoum 2011 unpublished). When mature the culm of the Ethiopian highland bamboo is hollow, while that of the lowland bamboo is solid (Kelecha 1980).

The highland bamboo is predominantly found in:

- Awi zone with better coverage: Banja, Guagusa, Shkudad, Injibara town adminstartion, Fagita Lekoma and Ankasha Woredas
- West Gojam–Sekela, Dega damot and Quarit Woredas
- East Gojam–Sinan and Bibugi Woredas
- South Gondar–Farta and Eastern Estie Woredas
- South Wello–Debresena woredas.

The lowland bamboo predominantly exists in:

- Awi zone–Jawi, Guangua and Ankesha Woredas
- North Gondar–Quara, Metema, West and Tach Armachiho and Adiarkay Woredas.

The lowland bamboos exist together with Combretum-Terminalia broadleaved deciduous forests in these areas.

Continued agricultural land expansion has almost eliminated the highland bamboo forests of Amhara Region. Nonetheless in Awi zone, considerable numbers of people depend on handcrafts made from bamboo for their livelihood. It is also an important source of construction material, fuel wood and joinery for construction purposes for local population (Tana Pulp and Paper SC 2008 unpublished).

In the lowlands, recurring uncontrolled fire used for shifting cultivation and clearance of vast areas of land for agricultural investments and settlements have devastated the lowland bamboo forests. As a result the lowland bamboo is in danger of being eliminated entirely.

15.2.6 Plantations

Plantation forestry in Ethiopia, which began in 1894, is almost solely practiced with exotic species (Pohjonen and Pukkala 1990). The reason for the popularity of the exotic species is that they are fast-growing pioneers, and are easy to plant and care for compared to the lesser-known indigenous trees. Moreover the absence of natural forests that can serve as a seed source for the plantation of the native woody species diversity is a major limiting factor particularly in the degraded central and northern highlands of Ethiopia. Most scholars argue relying on a few exotic species increases the ecological risks of plantations, and hence, increasing the share of indigenous tree species is becoming critically important (Pohjonen and Pukkala 1992).

Studies estimate that there are about 209,799 ha of plantations in Amhara National Regional State (Woreta 2007). These are subdivided according to ownership as 32,213 ha of state owned plantations, 10,261 ha of community and 167,325 ha of individual woodlots (Tana Pulp and Paper SC 2008 unpublished). The majority of the woodlots and plantations owned by the state, community or individuals are composed of Eucalyptus species, although there are plantations of *Cupressus lusitanica* and Pine species in few areas of Amhara Region. Despite the fact that Eucalyptus faces ecological criticism, its use has expanded dramatically in the region, particularly in the last 15 years, to the extent that fertile and irrigable crop lands such as in the Koga area of West Gojjam are continuously planted in Eucalyptus (Birru et al. 2003).

A great portion of the Amhara Region is covered already by Eucalyptus and the trend shows it will continue to expand (Tana Pulp and Paper SC 2008 unpublished). At this point, the expansion of Eucalyptus would be difficult to stop. Instead, efforts are being made to tap it as an economically viable forestry sector. However, these resources are currently being undervalued and are bought by neighboring countries at unfairly low prices.

Eucalyptus has more than 600 different species. In Ethiopia there are over 120 species but *Eucalyptus globules* and *Eucalyptus camaldulensis* are the most common species in Amhara region.

- *Eucalyptus globules*:
  In Amhara region almost all zones and woredas above the elevation of 2200 m. a.s.l are rich with *E. globules*, especially the following areas:
  - Debere Tabor-Geregera-Gashena and its vicinity
  - Dessie and its vicinity
  - Debre Brihan and its vicinity
  - Tarmaber-Menze and its vicinity
  - Debre Markos and its vicinity
  - Enjibara and its vicinity
  - Dega Damot, Sekela and Quarit vicinity.
- *Eucalyptus camaldulensis*
  This species is found in Amhara in mid-altitude areas (1700–2400 m.a.s.l). The following areas are predominantly covered by *E. camaldlunesis*:
  - Bahirdar–Dera–Fogera–Estie
  - Bahirdar–Marawi–Dangela
  - Awi(Adds–Alem)–Bure–Fenoteselam
  - Engibara–Chagni
  - Gondar town and its vicinity
  - North Wello

15.3 Summary

The overall area of forest resources in the Amhara Region has increased slightly over the last 15 years (Woreta 2007 and AFAP 1999 unpublished). The major source of the increase comes from Eucalyptus plantation and rehabilitation efforts in the Eastern highlands, and recently also in the Western highlands of Amhara. The rehabilitated forests have resulted in green scrub forests which are expected to develop into dry afromontane forests. However, the size and quality of Afroalpine and Subafroalpine vegetation, the high Dry afromontane forests, Bamboo forests and the Woodlands have decreased significantly in the same period. These ecosystems have suffered considerable degradation in structural and species composition. This has resulted in a serious loss of indigenous biodiversity and created opportunites for invasive species. The major problems devastating the region forest resources are uncontrolled grazing, cropland expansion and massive clearance of wood land for investments and resettlements.

References

Abate A, Tamrat B, Sebsebe D (2006) The undifferentiated afromontane forest of Denkoro in the central highland of Ethiopia: a floristic and structural analysis. Ethiop J Sci 29(1):45–56

Abeje E (2002) Regeneration status, soil seed banks and socio-economic importance of *Boswellia papyfifera* (Del.) Hochst. In two woredas of North Gondar Zone, northern Ethiopia. Thesis, Swedish University

Alemayehu W (2002) Opportunities and prospects of the Ethiopian Orthodox Tewahido Churches in conserving forest resources: the case of churches in South Gondar, Northern Ethiopia. Thesis, Swedish University

Alemayehu W (2007) Ethiopian church forests opportunities and challenges for restoration. Dissertation, Wgeningen University

Alemayehu W, Demel T, Powell N (2005a) Church forests provide clues to restoring ecosystems in the degraded highlands of Northern Ethiopia. J Ecological Restor 23:2

Alemayehu W, Demel T, Powell N (2005b) Church forests in North Gondar Administrative Zone, northern Ethiopia. Forests Trees Livelihoods 15:349–374

ANRS (2004) A strategic plan for the sustainable development, conservation, and management of the woody biomass resources. Final Report. Available Via Dialog. http://www.eap.gov.et/sites/default/files/ AFAR_STRAT_PLAN.pdf. Accessed 15 Oct 2014

Bingelli P, Desalegn D, Healey J et al (2003) Conservation of ethiopian sacred groves. Euro Trop For Res Netw Newslett 38:37–38

Birru Y, Anteneh A, Tadele A (2003) Expansion of eucalyptus woodlots in the fertile soils of the highlands of Ethiopia: could it be a treat on future cropland use? J Agri Scie 5:8

Bongers F, Alemayehu W, Sterck F et al (2006) Ecological restoration and church forests in northern Ethiopia. J drylands 1:35–45

Dagnachew G (2001) Sampling strategy for assessment of woody plant species diversity: a study in Tara Gedam natural forest, Ethiopia. Thesis, Swedish University

Darbyshire I, Lamb H, Umer M (2003) Forest clearance and regrowth in northern Ethiopia during the last 3000 years. Holocene 13:537–546

Demel T (1996) Seed ecology and regeneration in dry Afromontane forests of Ethiopia. Thesis, Swedish University

Demel T, Tamirat B (1995) Floristic composition of Wof-Washa natural forest, Central Ethiopia: Implications for the conservation of biodiversity, Feddes Repertorium (106):127-147

EPRDF (2014) Ethiopian Government Portal. http://www.ethiopia.gov.et/stateamhara. Accessed 23 Nov 2013

FAO (2001) Trees outside forests: Towards rural and urban integrated resources management. Rome, Italy. Available Via Dialog. http://www.fao.org/3/a-y1785e.pdf. Accessed 10 November 2014

FAO (2006) Choosing a forest definition for the clean development mechanism: Forest and climate change working paper 4, Rome, Italy. Available Via Dialog. http//: www. fao. org/forestry/11280-1-0.pdf. Accessed 10 November 2014

Fekadu G (2011) Dry evergreen montane forests of Ethiopia. In: Forest types in Ethiopia: status, potential contribution and challenges. Forum for Environment Occasional Report 2011. 7: 79–102

Friis I (1992) Forest and forest trees of northeast tropical Africa: their natural habitats and distribution pattern in Ethiopia, Djibouti and Somalia. Kew Bull Add Ser

Hedberg O (1957) Afroalpine vascular plants: a taxonomic revision Symb Bot Upsal (15) 411: 1–12

IBC (2005) Institute of biodiversity conservation. Ethiopian Biodiversity Strategy and Action Plan. http://www.cbd.int/doc/world/et/et-nbsap-01-en.pdf. Accessed 15 November 2014

Kasshun E (2000) The indigenous Bamboo forests of Ethiopia: an overview. Ambio 29:8

Ohrnberger D (1999) The Bamboos of the World. Elsevier, Amsterdam

Pohjonen Y, Pukkala T (1990) *Eucalyptus globulus* in Ethiopian forestry. For Ecol Manage 36: 19–31

Pohjonen Y, Pukkala T (1992) Juniperus procera Hocht. ex. Endl. in Ethiopian Forestry. For Ecol Manage 49(75–85):75

Renée M (2013) Remaining forests of the Lake Tana Region in Ethiopia. A participatory, social-ecological approach to rehabilitation. Thesis, University of Greifswald

Tamirat B (1994) Studies on remnant Afromontane forests on the Central Plateau of Shewa, Ethiopia. Disseretation, Unifersitatis Upsaliensis

Tilahun S, Edwards S, Tewolde-Berhan GE (eds) (1996) Important bird areas of Ethiopia: a first inventory. Ethiopian wildlife and natural history society, Semayata press, Addis Ababa

Van Andel A (2006) Restoration ecology. Blackwell Publishing, Massachusetts

Woreta A (2007) Amhara National Regional State's (ANRS) efforts towards forest cover increment. In: Bane, J, Sisay N (eds) Proceedings of Policies to increase forest cover in Ethiopia, Addis Ababa, 2007

Zegeye H, Demel T, Ensermu K (2011) Diversity and regeneration status of woody species in Tara Gedam and Abebaye forests, Northwestern Ethiopia. J Forestry Res 22(3):315–328

Chapter 16
Wetlands of the Lake Tana Watershed

Shimelis Aynalem, Goraw Goshu and Ayalew Wondie

Abstract The Lake Tana watershed lies within the Afro-tropical wetland system in Ethiopia. The total wetland area of the watershed is estimated to be 32,157 ha. These wetlands have different physical and hydrological, chemical and water quality, and biological and habitat functions. The Lake Tana watershed wetlands and its associated rivers holds about 28 species of fishes, of which fifteen are endemic, and there are over 300 species of birds recorded so far. The watershed is famous for having natural pasture land and is the home of the Fogera cattle. The wetlands in the Lake Tana watershed extend from the headwaters of Guna and Gishe-Abay to Fogera and the Dembia floodplains. The river mouths of the Gelda, Gumara, Arno-Garno, Megech, Dirma, Abagenen, Gilgel Abay and Infranz are the major delta wetland ecosystems. Yiganda and Amluk wetlands are of the lacustrine wetland type formed by the lake water while Chimba and Infranz wetlands are the common riverine wetlands. Several other wetlands are formed from cold and hot springs. There are also man-made wetlands such as the Koga Dam area and other constructed small water bodies and weirs as well. Lake Tana watershed wetlands provide habitat for globally threatened and endangered species. However, wetland loss is evident wherever major developments such as dams, irrigation schemes and conversion projects are present in the developing world. Humans usually and very dramatically accelerate natural processes. In the Lake Tana watershed, human

S. Aynalem (✉) · G. Goshu
Blue Nile Water Institute and College of Agriculture and Environmental Sciences, Bahir Dar University, P.O. Box 2205, Bahir Dar, Ethiopia
e-mail: shimelisay@yahoo.co.in; shimelis.aynalem@gmail.com

S. Aynalem
College of Natural Science, Addis Ababa University, Addis Ababa, Ethiopia

S. Aynalem
Research Associate of International Crane Foundation, Baraboo, USA

G. Goshu
e-mail: gorawha@gmail.com

A. Wondie
College of Natural Sciences, Bahir Dar University, Bahir Dar, Ethiopia
e-mail: ayaleww2001@yahoo.com

K. Stave et al. (eds.), *Social and Ecological System Dynamics*,
AESS Interdisciplinary Environmental Studies and Sciences Series,
DOI 10.1007/978-3-319-45755-0_16

induced threats such as conversion of wetlands into agriculture, draining of wetlands and lack of defined ownership of the resource is the main threats.

Keywords Wetland features of Lake Tana · Classification and functions of wetlands · Threats of wetlands

16.1 Wetland Characteristics of Lake Tana Watershed

The Lake Tana watershed lies in the Afro-tropical wetland system and serves as the prime water catchments and sources of its major rivers in Ethiopia. The Highlands receive an average annual rainfall of more than 2000 mm. with a bi-modal rainfall pattern. June to September is a long rainy season, while February to May is short season (Tilahun et al. 1996). However, the Lake Tana watershed has a unimodal rain fall pattern, with rains falling in the June to September period.

The Lake Tana watershed wetlands are distributed along the Lake Tana tributaries of Gilgel Abay, Rib, Gumara and Megech-Dirma Rivers. The floodplains of Dembia and Fogera areas also hold wetlands. The wetlands of Lake Tana area are home of globally threatened species while also world recognized migratory wintering sites. But these wetlands are still under threat by the community at large. Even though the Lake Tana areas are rich in terms of biodiversity resources, there is no comprehensive baseline information on the distribution and status of wetlands in the area. With this idea, therefore, this chapter provides general information on types, function and values of the Lake Tana wetlands, status and threats. And then identifies possible conservation actions and research gaps that should be addressed now and then.

In Amhara National Regional State, there are estimated to be 400,000–600, 000 hectares of wetlands. However, the total wetland area in Lake Tana watershed in particular is estimated to be 32,157 ha, which is 2.14% of 1,507,024 ha of the watershed area (Biru 2007).

Wetlands in Lake Tana watershed are known for their rich biodiversity and high ecological functions. The lake and its associated rivers hold about 28 species of fish, of these 15 are labeobarbs species endemic to Ethiopia. Over 300 species of birds have been recorded so far in the Lake Tana watershed, which is recognized as an international bird site by BirdLife International (BLI) (Shimelis 2013). The wetlands are famous for their natural pasture land and the home of the well known Fogera cattle. The lakeshore vegetation includes reeds which provide important fish breeding and nursery habitat, as well as multipurpose functions such as reed boat making, matting, and housing. In addition, sand is mined from the lake to provide construction materials to Bahir Dar and Gondar cities.

Wetlands in Lake Tana watershed are found from the headwaters of Guna and Gishe-Abay to the Fogera and Dembia floodplains. The river mouths of Gelda, Gumara, Arno-Garno, Megech, Dirma, Abagenen, Gilgel Abay and Infranz are the major delta wetland ecosystems. The Yiganda and Amluk areas are of the lacustrine wetland type formed by the lake water while Chimba and Infranz are the common

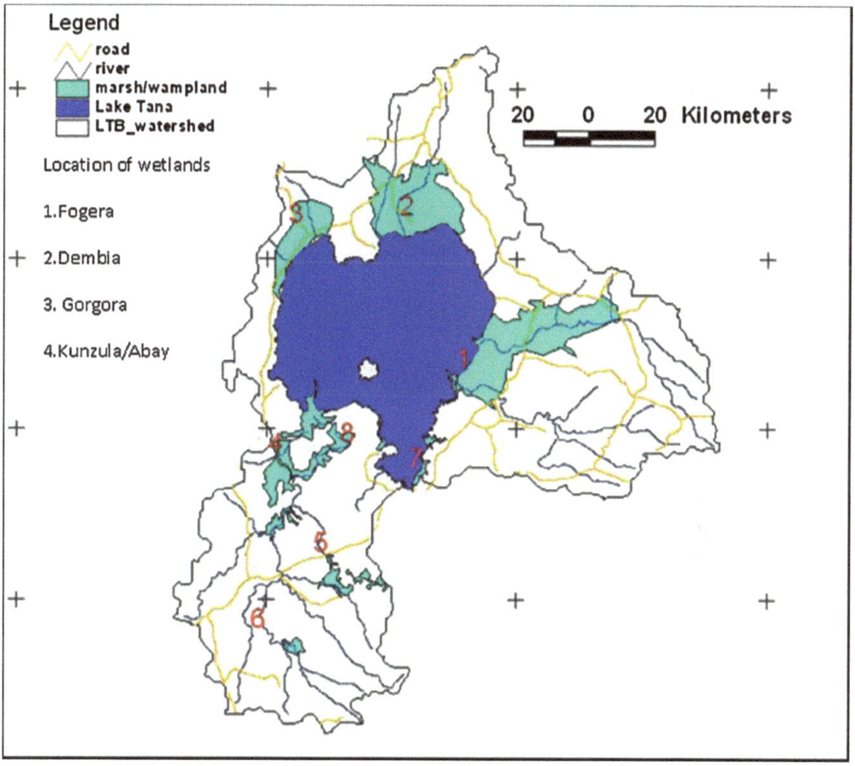

Fig. 16.1 Wetlands and flood plains of the LTB (Reproduced from Biru 2007, used by permission)

riverine wetlands. Several other wetlands are formed from cold and hot springs. Also included are the manmade wetlands such as the Koga Dam and other constructed small water bodies and weirs (Fig. 16.1).

16.2 Lake Tana Watershed Wetland Classification

The Canadian Wetland Classification System (National Wetlands Working Group 1997), the Stewart and Kantrud Wetland Classification System (Stewart and Kantrud 1971), and Cowardin Wetland Classification System (Cowardin et al. 1979) were developed based on the physiographic features and climatological conditions of Alberta, for example. However, this classification system is difficult to adopt for developing countries like Ethiopia in Africa, with very different climatic conditions, and where there is no sea or ocean-like ecosystem. Instead, we can think of wetlands by their position in the basin system. For example, the Gumara River sub-basin wetland includes headwater wetlands, riverine wetlands, floodplain wetlands like the

Table 16.1 Wetland types, distribution and area estimate in the Lake Tana Watershed

Major wetland areas in the Lake Tana watershed	Type	Source of water	Area estimated in hectare (ha)	Location
Chimba wetlands	Riverine	Gilgel Abay River	3230	Include seven "Kebele" administration: Chimba, Lata Amba, Legedia, Easetumete, Lejme, Dehena Mesenta and Addis Amba area
			1234	Only taking the case of Lata Amba Kebele area
Abay Inflow Delta	Riverine	The Lake and Gilgel Abay River	63	Easetumete and Lejme
Infranz	Riverine	Infranz River	391	Chenta Sostu and Yeloma Yeganfate
Gelda	Riverine	Gelda River	315.8	Korata Petros
Yiganda wetland	Lacustrine	Lake Tana and Mina stream	468	Wenejta
Dirma	Lacustrine	Lake Tana	71.5	Seraba Mariam
Debre Mariam-Bahir Dar	Lacustrine	Lake Tana	100–120	Gulf of Lake Tana (Deshete Goradema and Zenzelma)
Amluk	Lacustrine	Lake Tana	10	Weramit
Ambo Bahir	Lacustrine	Lake Tana	56	Debanta and Wenejta
Shesher and Wallala	Palustrine	Fogera floodplain	1354	Nabega Georgis, Shega Mariam
Kurt Bahir	Palustrine	Koga river	45	Merawi area
Dibanko	Palustrine	Dangila	25	Gerarege
Koga wetland	Artificial wetland	Koga dam (Oxbow lake)	1200	Sira Betegebar, Meder Genet and Berakate mainly

Shesher and Wallala, rivermouth and delta wetlands. Even though there are several wetlands located in isolation along the Lake shore, rivers, and inland flood plain areas, most wetlands in the watershed can be categorized riverine, lacustrine or palustrine, depending on where they are located in the landscape (Table 16.1).

The Chimba wetland system comprises the largest wetland area in the watershed. It is annually flooded by the Gilgel Abay River during the rainy season. The largest Papyrus bed is also found in this wetland area. The; Yiganda is formed due to the seasonal expansion of the Lake; the wetland is supported by a perennial small stream known as the Mina River. The vegetation is similar to that in the *Chimba* wetland (Fig. 16.2).

Fig. 16.2 The Yiganda wetland Lake Tana area during January (©*Photo* Shimelis A. 2014)

16.3 Services and Values of Wetlands in Lake Tana Watershed

The Millennium Ecosystem Assessment distinguishes four main categories of services that are provided by or derived from wetlands. These are: *provisioning services*, the most directly and easily visible services from which human populations benefit in wetlands; *regulating services*, which include the regulation of processes related to water, sediment and climate; *cultural services*, including spiritual and inspirational, recreational and educational services; and *supporting services*, including soil formation (accumulation of sediment and organic matter) and nutrient cycling (http://www.millenniumassessment.org/en/About.html. Accessed 7/6/2015).

Wetlands in the Lake Tana watershed are used for crop farming, grazing for livestock, water supply, fishing, fuel wood, medicinal plants such as *Hygrophila auriculata*, papyrus (*Cyprus papyrus*), and thatch grass. Based on different surveys carried out in the area, the greatest use of wetlands is to provide food for livestock. However, no quantified data is available to estimate the amount of fooder produced annually. But for example, at the Yiganda wetland about 15,000 of livestock are depending on the wetland pasture particularly during the dry season. Huge numbers of cattle also graze in Dirma, Chimba, Kurtbahir and other wetlands mentioned so far.

The contribution of these wetlands for the provision of water for livestock and people is very important. There are a number of springs and rivers from where large numbers of households collect their daily water requirements at and around the wetland area. Quarrying of boreholes is common in that every household has at least one. Fish are collected from the wetland and the Lake as well. Fishing is the main means of livelihood for significant number of households especially in the Zegie Peninsula. On average, a household generates about 3550 Ethiopian Birr per fisherman per annum in the Zegie peninsula area (Yiganda wetland management plan 2014, unpublished). Besides the Zegie community generates some amount of income from selling of wood, coffee beans, fruits and vegetables. However, low fish stocks, non-selective fishing activities and fishing net theft are major constraints of the local people. This situation has made them to great impact on deforestation of the peninsula and to migrate to Bhair Dar City.

Collection of edible plants, medicinal plants, papyrus, fire wood and thatch grasses is commonly practiced activities. About 84.2% of households collect edible fruits for home consumption. About 38.6% of households have harvested thatch grass. More than 45.6% and nearly 3/4th of households have collected medicinal plants and firewood, respectively. Firewood collection is for subsistence use and income generation. It has served as a source of income for large number of residents of Zegie Peninsula. Besides, 80.7% of interviewed households has harvested papyrus for construction purpose such as roofing materials and making ropes (Integrated Management Plan of Yiganda wetland, For People and Nature-Establishment of a UNESCO Biosphere Reserve at Lake Tana, Ethiopia 2014, unpublished). For these purposes Dengel (*Cyperus papyrus*), bulrush (*Typha latifolia)* are commonly used wetland plants. The Negede tribe around the Lake and the disadvantaged people in Bahir Dar town usually come to areas where there is papyrus vegetation. The Yiganda and Infranz wetlands and the Lakeshore around Bahir Dar are major sources of the papyrus vegetation.

In recent times, the Chimba, Yiganda and Dirma wetlands are becoming tourist destination sites because they provide many opportunities for recreational activities. They are valued for nature and wildlife observation, plant identification, photography, swimming, bird watching, cultural and aesthetic values. They also provide educational opportunities and outdoor classrooms for natural and social sciences and scientific research.

Wetlands, since they produce so much plant biomass and invertebrate life, estuaries and their coastal marshes serve as important nursery areas for fish and nesting sites of birds. The Yiganda and the Dirma Riverine wetlands in Lake Tana area are known for feeding and the breeding grounds of *Labeobarbus*, *Claris*, *Varihrominus* and *Oreochromis* fish species.

The Lake Tana watershed wetlands provide habitats for globally threatened and endangered species directly or indirectly for survival. Mammal such as Hippos (*Hippopotamus amphibious*), reptiles such as the Nile Monitor (*Varanus niloticus*), African rock python (*Python sebae*), and the Nile Crocodile (*Crocodylus niloticus*); the wetlands are also extremely important for several species of globally threatened wetland birds such as the Wattled Crane (*Bugeranus carunculatus*), Black-Crowned Crane (*Balearica pavonina*), Rouget's Rail (*Rougetius rougetii*), Lesser Jacana (*Microparra capensis*) and others such as the African Rail (*Rallus caerulescens*), Black Crake (*Amauromis flavirostis*), Lesser Gallinule (*Porphyrio alleni*), Common Moorhen (*Gallinula chloropus*), Lesser Moorhen (*Gallinula *angulata*), and Africa Jacana (*Actophilomis africanus*). The sub-basin again provides home for more than 100,000 migratory birds. Migratory waterfowl, including ducks, geese, and waders, use the flood plain and inland wetlands as resting, feeding, breeding, or nesting grounds for at least part of the year. For example, in the *Shesher–Wallala* flood plain area is a major wintering area for Eurasian Crane (*Grus grus*) (Shimelis et al. 2011).

16.4 Wetland Challenges and Threats

In developing world, dams, irrigation schemes and wetland conversion projects accounted for the loss of wetlands. Illegal extraction of wetland resources and miss use besides to natural dynamic system change has also contributed a lot. As a result, many wetlands are temporary features that disappear, reappear and re-create themselves over time (Barbier et al. 1996).

Mining is the most destructive activity if practiced in wetlands (Williams 1990). The human activities like agriculture, industry and urban development usually and very dramatically accelerate natural processes often unintentionally, which eventually, drainage and diverting water, to dredging and loading water sources with toxic chemicals. Humans contribute about 65% of wetland disturbances, while the remainders have natural origins. Of these, 73% of disturbances are thought to result from direct human actions, while the remaining 27% are believed to come from indirect sources (Dugan 1990).

The results of wetland loss are disastrous and far reaching. Humans and other life close to wetlands, and those who depend upon them, are the first to feel the impact of wetland loss. Dam construction can significantly impact the lives of people living downstream. It also affects water chemistry, flooding cycles, sediment behavior and fish migrations (Maltby 1986). All too often, wetland functions, including flood protection, nutrient retention, and erosion control or sediment retention will be compromised by well-meant development interventions.

Once a wetland has been destroyed, the services it previously provided now have to be paid for by tax payers. While industrialized countries can probably pay for most of these services from tax incomes, this is not so in developing countries. Apparently wetland destruction can have a very serious impact on the livelihoods of the rural poor (Dugan 1990).

Wetland destruction and alteration has been and is still seen as an advanced mode of development, even at government level, and their values remain little understood and their loss is increasingly becoming an environmental disaster. While rates of wetland loss are documented for the developed world, the limited study of these ecosystems in countries like Ethiopia leaves us with little to say.

In spite of multifaceted benefits, wetland have decreased in size and lost their natural characteristics. Based on the view of the local people interviewed at various times, indigenous trees, papyrus, different edible plants, medicinal plants and thatch grasses are declining (Yiganda Wetland Management Plan, Bahir Dar 2014, unpublished). The most important perceived threats for the last decades are overgrazing and unregulated harvesting of wetland resources. Eucalyptus trees and cultivation of chat, a narcotic drug on the edges of Lake Tana area wetlands in particular becoming a threat since most of the areas are situated along the road to Bahir Dar and Gondar. The most significant threats to wetlands in the Lake Tana sub-basin are as follows:

Overgrazing: due to human pressure most of the agricultural lands are cultivated year to year without fallowing. Since the wetland is not privatized, it is free for anyone to exploit. Livestock can have great impacts on the degradation of the wetland ecosystem though contributions to nutrient recycling and shortening of vegetation, which is incidentally good for birds. However, some wetlands such as the Chimba area and the Yiganda wetlands are still in good condition and intact compared to the Dirma (Northern part of Lake Tana), Infranz (southern part of Lake Tana), Kurt Bahir (Mecha area), Adrako (Dangila area) and the Gelda wetlands. Unmanaged utilization of the wetlands could destroy the vegetation and also the feeding places of many birds and breeding sites of Black-Crowned Crane and several types of fishes so far mentioned.

Vegetation removal: Papyrus is removed for commercial purposes. The spikes are removed and transported to Bahir Dar. During these operations disturbance occurs for those organisms that are dependent for shelter, nesting and resting in the Papyrus. Cut and carry is common, but there is no any appropriate management practice. However, birds that are depending on the presence of spikes to make nest are suffering. The Village Weavers (*Ploceus cucullatus*), Baglafecht Weaver (*Ploceus baglafecht*) are the main species affected. Birds that forage on the tubers of sedges and papyrus are also affected. In addition, Papyrus reeds are removed and taken to the market (Fig. 16.3).

Cultivation: the most striking feature of many wetlands is their conversion to agriculture by encroachment. The borders of wetlands are given to the local groups of young people. These people then cultivate are cultivating a narcotic drug known

Fig. 16.3 Papyrus Products Bahir Dar area (©*Photo* Shimelis A. 2007)

as 'Chat' (*Catha edulis*). Extensive cultivation of this plant is the predominant crop. There is also pesticide application, which this again has disastrous effect on the wetland.

Deforestation: it is fortunate that the Monzi-Medhanealem and Zegie Peninsula forest, Alemsaga natural forest, Lake Tana Islands (Kibran Gabriel, Entos Iyesus, Daga Estifanos, Mitsili Fasildes, Tana Kirkos, Quorata Welete Petros, etc.) and Taragedam–Addis Zemen area are in the ownership the Church which provides protection; however, there still is illegal removal of old and big trees at the Zegie Peninsula and Infranz riverine forests. Areas adjacent to the wetland had been covered by forest, but the settlement program in earlier times led to some forest removal the northern part in particular. There still exists cutting of indigenous trees without any substitution of planting.

Free land: most of the wetlands in the watershed are free, and this type of disowner shipment has resulted in the area being made unsecure for future sustainability.

Fire: people think that fire would initiate a new growth of grass, and of course it is true, but the negative impact is much greater. The fire can endure for a long time which has the effect of total destruction of the papyrus bed. So, with a loose or restricted management system, this is happening at the Yiganda wetland at various times.

16.5 Research Gaps

- There is no detailed classification and characterization of the wetlands of the sub-basin.
- There is no clear distribution map of wetlands and area covered relative to the total sub-basin area.
- The biodiversity resources (invertebrates, reptiles, fish, birds and mammals) and threats to all wetlands are not well documented.
- The hydrological function and interaction of wetlands with the respective watershed is not well documented.
- Small wetlands situated in the Lake Tana watershed areas should be explored and studied for further conservation and sustainable utilization of the resource.
- Energy flows in the ecosystem through lower trophic levels (phytoplankton dynamics, detritus, bacterio-plankton, etc.) is not studied.
- The effect of sediment load on the abiotic and biotic factors, its rate and possible measures to combat this problem, and community dynamics, e.g. multi-species interactions, predator-prey interactions, parasite-host relations should me studied.
- Significance of biodiversity for resilience (adaptability) and productivity of the ecosystem, including molecular and genetic approaches should also be known.
- Comparative study on the structure and functioning of the Lake Tana with other Ethiopian Lakes shall be addressed.

16.6 Possible Management Options

- Demarcation of communal wetlands and the promotion of community based wetland conservation and utilization management system by applying strong community bylaws.
- Implement public and government oriented invasive weed eradication campaign and program.
- Develop core areas for future protection and conservation of keystone species.
- Apply wetland inventory practice by higher institutions, Universities and NGO's.
- Promote and carry out soil and water conservation measures in the watershed areas.
- Demarcate and promote bird sanctuary areas particularly at the Shesher and Wallala wetlands.
- Apply legal enforcement where necessary.
- Implement wetland restoration practice where appropriate, based on studies.
- Encourage the Government of Ethiopia to ratify the Ramsar Convention as did AEWA.

16.7 Conclusion

In the Lake Tana watershed, wetlands cover a sizable area with functions of nutrient and sediment retention, biodiversity and hydrological and hydraulic functions. Under this topic the Lake Tana area watershed wetlands can provide some information where the major wetlands are distributed, the functions they are providing and the current challenges they face today.

The Lake Tana Watershed wetlands are mainly occurring around the Lake Tana shore and the along the major rivers of Abay River. The Chimba wetlands situated along the Gilgel Abay River, the Abay inflow delta, the Yiganda wetland, Shesher and Walla wetlands in the Fogera flood plain, and the Dirma wetlands in the northern part of Lake Tana are the main wetlands. Bio-physically important wetlands are situated in several parts of the watershed area. These wetlands support hundreds of thousands of people in various ways. However, these wetlands are under threat, the humans are the major factors for this case.

The current major problem of the Lake Tana area wetlands is the increasing occurrence of the water hyacinth (*E. crasipes*), which removes the focus from other problems. Comprehensive information on the water hyacinth is still lacking: particularly on the distribution, abundance, the threat to the ecosystem, the conservation status are a few of them.

The current conservation status of the wetlands is very poor, calling for urgent management intervention. The watershed area in general, the wetlands in particular should be urgently managed. Project based wetland conservation and utilization research have to be promoted. The biological resources of the wetlands, the hydrology and hydrogeology characteristics of the wetlands in the watershed should be characterized for future conservation and sustainable utilization. The biodiversity resources should be well studied and documented for further protection and rehabilitation of the fauna and flora.

16.8 Recommendations

- The current data set on the Chimba, Yiganda, Dirma wetlands and others in Lake Tana watershed area can only be considered spatially and temporally limited. However, the results impressively demonstrate the lack of data in all aspects for further conservation plans. Results urgently call for further research and continuous data generation concerning the wetlands in the Lake Tana watershed. The establishment and sharing of a regional as well as national wetland data base is important.
- There are other polices such as water and environment which address the issue of sustainable wetlands management indirectly but there is no wetland policy that directly addresses the wetland issue. Preparation of policy document and more should be a prime action.
- There seems to be lack of awareness of the functions of wetlands and users of the wetlands want to maximize their benefits out of the wetlands. Therefore, creation of public awareness at all levels regarding wetland health and sustainable utilization and the provision of feedback to the policy makers on the management status of the wetlands need immediate attention.
- An integrated watershed management should be implemented for the lake and its surrounding environs, which should include wetland management plans.

References

Barbier EB, Acreman MC, Knowler D (1996) Economic valuation of wetlands: a guide for policy makers and planners. Ramsar Convention Bureau, Gland

Biru Y (2007) Land degradation and options for sustainable land management in the Lake Tana Basin (LTB), Amhara Region, Ethiopia. Dissertation, University of Bern

Cowardin LM, Carter V, Golet FG et al (1979) Classification of wetlands and deepwater habitats of the United States. FWS/OBS-79/31. Fish and Wildlife Service, USDI

Dugan PJ (ed) (1990) Wetland conservation: a review of current issues and required action. IUCN, Gland

Maltby E (1986) Waterlogged wealth: why waste the world's wet places?. International Institute for Environment and Development and Earthscan, London

Shimelis AZ (2013) Birds of Lake Tana area, Ethiopia. A photographic field guide. View Graphics and Printers, Addis Ababa

Shimelis A, Nowald G, Schroder W (2011) Observation on the biology and ecology of cranes: wattled cranes (grus carunculatus), black-crowned cranes (balearica pavonina), and eurasian cranes (grus grus) at Lake Tana, Ethiopia. INDWA. J Afr Crane Res Conserv 7:1–12
Stewart RE, Kantrud HA (1971) Classification of natural ponds and lakes in the glaciated Prairie Region. Bureau of Sport Fisheries and Wildlife, U.S. Fish and Wildlife Service, Washington, D.C., USA. Resource Publication 92
Tilahun S, Edwards S, Tewolde-Berhan GE (eds) (1996) Important bird areas of Ethiopia: A first inventory. Ethiopian Wildlife and Natural History Society, Semayata Press, Addis Ababa
Williams M (ed) (1990) Wetlands: a threatened landscape. UK, Institute of British Geographers, Oxford

Chapter 17
Exotic and Invasive Plants: Water Hyacinth

Alie Seid and Banchiamlak Getenet

Abstract This chapter is about biodiversity threats posed by water hyacinth in the pristine Lake Tana, discuss its potential problems and suggest possible solutions. Lake Tana watershed is within the East African Afro-montane Hotspot and the productive agro-ecosystems of Ethiopia. As a hotspot, the watershed is considered as global priority conservation area and Tana is proposed to be designated as a biosphere reserve. The catchment stretched from Lake Tana 1785 m above sea level to mount Guna 4150, and contains three distinct agro ecosystems. The national development strategy is changing the socioeconomic and biophysical landscapes. The expansion of a century old introduced *Eucalyptus* still has unsettled controversies. The lowland plane and the lake shore ecosystem are facing challenges of the worst invasive in 2004. Since then, the alien species is added on the two major environmental challenges namely anthropogenic activities and climate change.

Keywords Biosphere reserve · Invasive plant · Lake Tana · Waterhyacinth

17.1 Introduction

This chapter covers threats to biodiversity posed by exotic invasive plants in the Bahir Dar—Lake Tana region The Lake Tana watershed is one of the 34 Great Biodiversity Hotspots of the world (East African Afro-montane) and a productive agro-ecosystem of Ethiopia. The watershed is home to a number of endemic plants and considered as global priority conservation areas. It stretches from Lake Tana (1785 m above sea level) to mount Guna (4150 m above sea level), and contains

A. Seid (✉)
College of Natural Sciences, Bahir Dar University, Bahir Dar, Ethiopia
e-mail: alinabiot@yahoo.com

B. Getenet
College of Agriculture and Environmental Sciences, Bahir Dar University, Bahir Dar, Ethiopia
e-mail: banchugetneta2020@gmail.com

K. Stave et al. (eds.), *Social and Ecological System Dynamics*,
AESS Interdisciplinary Environmental Studies and Sciences Series,
DOI 10.1007/978-3-319-45755-0_17

three distinct ecosystems. However, the watershed's biodiversity is facing challenges of invasive alien species driven both by development activities and climate change.

The recent development and transformation strategy of Ethiopia considered the watershed as a major development corridor likely to change socioeconomic status and biophysical landscapes. Economic development and increasing human population mobility (including tourism) will inevitably lead into intentional and unintentional introduction of non native species. While expansion of *Eucalyptus* in the watershed already has caused controversies for decades, invasive alien species are increasing in number and geographic range. Invasive species threaten agricultural and natural ecosystems requiring concerted efforts and strategic actions to minimize damages. This risk of biological contamination may pose an even greater risk than well-recognized chemical pollution.

Currently, invasive alien weeds are of great concern, for both economic and ecological reasons, posing particular problems in agricultural lands, range lands, national parks, water ways, rivers, power dams, roadsides and urban green spaces. Foremost among these is the Parthenium weed (*Partenium hystrophorus*), although major problems are also being caused by water hyacinth (*Eichhornia crassipes*), mesquites (*Prosopis juliflora), Lantana camara*, and the parasitic weeds of *Striga*, *Orobanche* and *Cuscuta* species.

Water hyacinth, *E. crassipes* Martius (Solms-Laubach) (Pontederiaceae), is a perennial free-floating aquatic plant. In the absence of natural enemies, water hyacinth forms large mats on still and slow-moving water bodies, where it severely degrades aquatic ecosystems and limits their utilization. Water hyacinth has been identified by Holm et al. (1991) as the world's worst weed; the International Union for Conservation of Nature (IUCN) considers it one of the 100 most aggressive invasive species (Téllez et al. 2008) and possibly one of the top 10 worst weeds in the world (Shanab et al. 2010; Gichuki et al. 2012; Patel 2012). Thus, this chapter contains a comprehensive account of the plant's biological features and we identify important research gaps toward improved understanding.

Invasive species have been identified by the Environmental Policy of Ethiopia (EPA 1997) as posing a major threat to biodiversity and economic well being of the population. The water hyacinth has appeared (in the hotter parts of Ethiopia) where it was first reported in 1965 at the Koka Reservoir and in the Awash River. The Ethiopian Electric Light and Power Authority have managed to bring it under moderate control at the location, with a considerable cost in human labor. Other infestations in Ethiopia include many water bodies in the Gambela Region, the Blue Nile just below Lake Tana into Sudan, and Lake Ellen near Alem-Tena (Fessehaie 2005). Even though it is not clear when and how the weed entered Lake Tana, it appeared during the last 3 or 4 years in some pocket grazing and wet farm areas near the mouth of the Megech River and proliferating and covering shoreline habitats in the Rib and Dirma River mouths. According to Amhara National Regional State Bureau of Environmental Protection, Land Administration and Use (BEPLAUA 2015 water hyacinth distribution report, unpublished), the weed has occupied 20,000 hectare (ha) of the lake area and extended up to 1 kilometer (km) into the lake. Similarly, Wondie et al. (2012 water hyacinth infestation at Lake Tana area, unpublished) reported that the infested area have expanded to more than 10% of the shoreline area in the north-eastern part of the lake.

17.2 Nature of Invasive Alien Plants

Human migration and movement of goods around the world have also provided pathways for introducing alien species into regions beyond their place of origin. Most of the world's population depends on food from crops and livestock from other regions of the world. Thus, non-native species are fundamental to most human cultures, and strong economic incentives encourage many sectors of society to continue importing potentially useful new organisms. When the introduction of alien species becomes an invasive, either through intentional or unintentional release and establishment, the impact can be overwhelming for native species, ecosystems and ultimately people. Invasive alien species are a major global challenge requiring urgent action (Xu et al. 2012), for they are considered one of the key pressures on world's biodiversity: altering ecosystem services and processes, reducing native species abundance and richness, and decreasing genetic diversity of ecosystems (Rands et al. 2010; Vila et al. 2011).

In general invasive alien species (IAS) are those plants, animals and microbes which are introduced to and spread in new regions, mainly through human activities; the subsequent negative impact on biodiversity, agriculture, water resources, and human health represents a substantial challenge to economic growth and livelihoods.

However, only a few introduced species actually become problematic, with the vast majority of an estimated 480,000 introduced alien species never becoming established or spreading. In fact, beneficial species such as corn, wheat, rice, cattle, and others, provide more than 98% of the world's food supply with a value of more than US$5 trillion annually (USBC 2008). However, many alien species introduced intentionally for horticulture, agro-forestry, landscape restoration, and other purposes—as well as species which have been introduced unintentionally—pose a significant threat to economic development and ecological integrity. It is estimated that the impact and cost of managing IAS globally accounts for US$1.4 trillion annually or 5% of annual global GDP. Put in context, this is double the annual GDP for the whole of the African continent. The threat posed by IAS is increasing at an exponential rate as a result of increased international trade, transport or travel.

17.2.1 Pathways of Invasive Alien Introduction

The same few pathways and vectors are used by a large array of invasive alien species. So, exclusion efforts are best focused on interventions designed to block entire avenues of spread, rather than on intercepting individual species. Common pathways for the introduction of invasive alien species include:

Intentional introductions

- Plants introduced for agricultural purposes
- Exotic plants introduced for forestry use
- Non-native plants introduced for use as soil improvers
- 'Aid trade'
- Ornamental or hedgerow plants
- Germplasm
- Mammals or birds released for hunting purposes
- Animals released on islands as sources of food
- Biological control agents
- Fishery releases
- Pets released, or escaping, into the wild
- Aquarium trade
- Releases intended to 'enrich' the native flora and fauna.

Accidental introductions

- Contaminants of agricultural produce
- Seed or invertebrate contaminants of nursery plants
- Seed or invertebrate contaminants of the cut flower trade
- Organisms in or on timber imports
- Seed contaminants
- Soil inhabiting species
- Contaminated imports of machinery, equipment,
- Vehicles and military hardware
- 'Hitchhikers' in, or on, packaging materials
- 'Hitchhikers' in, or on, mail or cargo
- 'Hitchhikers' on aeroplanes
- Ballast water on ships
- Ballast soils
- Sediments in ballast water tanks
- Hull fouling on ships
- Debris
- Tourists and their clothing, footwear, luggage, or equipment
- Diseases in animals traded for agricultural or other purposes
- Parasites and pathogens of, or 'hitchhikers' on, aquaculture and mariculture.

Introductions via captivity

- Escapes from botanical gardens, for example, or zoos
- Feral domestic animals
- Escapes from aquaculture or mariculture
- Escapes from research institutions or facilities.

17.2.2 Invasive Alien Plants Distribution and Trends

Africa is home to hundreds of Invasive Alien Species (IAS) both plant and animal but the magnitude of the problem varies from country to country, and from ecosystem to ecosystem. In many parts, freshwater ecosystems are particularly at risk with IAS surpassing habitat loss as the number one cause of biodiversity loss. Invasive alien species are a problem in diverse ecosystems in northern, western, central, eastern and southern Africa and in the western Indian Ocean islands: they affect both savannahs and tropical forests and they are found on land, in freshwater systems, along the coast, and in the ocean (UNEP 2004).

Virtually all countries in the region are affected by IAS. In 2004, IUCN the World Conservation Union (IUCN) identified 81 IAS in South Africa, 49 in Mauritius, 44 in Swaziland, 37 in Algeria and Madagascar, 35 in Kenya, 28 in Egypt, 26 in Ghana and Zimbabwe, and 22 in Ethiopia (IUCN/SSC/ISSG 2004). In some countries there may be under reporting of the incidence of IAS. Many IAS found in Africa are included on a global list of the 100 worst IAS (IUCN/SSG/ISSG 2004). These include the infamous, *E. crassipes* (water hyacinth); economically important species including the Nile perch, *Oreochromis mossambicus* (Mozambique tilapia) and *Acacia mearnsi* (black wattle); species introduced for biological control, such as *Acridotheres tristis* (Indian myna) and *Bufo marinus* (cane toad); and ornamentals such as *Lantana camara*. There are many others IAS which present serious challenges to regional efforts to conserve the environment and to meet development objectives, the foundation of social, economic and environmental sustainability in Africa. In some countries, IAS has become a major ecological, social and economic problem despite the existence of legal measures and substantial funding to control them.

With increasing globalization, the threat posed by IAS is likely to increase through both intentional and accidental introductions. Human movement and the movement of goods are key drivers in the spread of IAS. With improvements in communications and infrastructure, this is likely to increase. Historically, IAS has been spread through colonization and exploration. Today, mobility through tourism, business travel and migration continues to be an important factor. Many IAS have been introduced to Africa in, for example, soil, plants, luggage, vehicles and aeroplanes (Kirby 2003).

Trade both legal and illegal particularly in, but not limited to, plants and animals, is particularly important. Many species have been introduced through trade in manufactured goods contaminated with seeds or insects. Trade has contributed not only to the introduction of species that colonize and fundamentally alter receiving ecosystems but that are also a factor in the growing incidence of disease. *Aedes albopictus* (Asian tiger mosquito), for example, is associated with the transmission of dengue fever and is believed to have been first introduced through a shipment of tyres from Japan to South Africa in 1989. By 1999 these mosquitoes were found to be present in Douala, Cameroon's main commercial harbor (Fontenille and Toto 2001). Invasive alien species have also been spread through the provision of humanitarian emergency food aid. For example, the weed *P. hysterophorus* is a recent introduction to Africa through grain shipments for famine relief to Ethiopia (McNeeley et al. 2001). The weed was first seen in 1988 near food-aid distribution centres in Ethiopia. Buried seeds of the weed can lie dormant for as long as 20 years before germinating (GISP 2004). Research activities and agricultural extension have also been a factor. Disturbed ecosystems are particularly vulnerable to invasion by alien species.

17.3 Water Hyacinth

The center of origin of Water hyacinth (*E. crassipes)* is from the Amazon basin in Brazil and Peru. It has been spread to most of South America and the Caribbean islands and was first recorded in the United States in New Orleans. According to Hill and Coetzee (2013) the plant has been spread around the world by humans since the late 1800s as an ornamental pond plant. It has established in most tropical and subtropical countries as well as in many warm-temperate regions between 40°N and 40°S (Fig. 17.1). It's beautiful, large purple and violet flowers make it a very popular ornamental plant for ponds. However water hyacinth has also been labeled as the world's worst water weed and has garnered increasing international attention as an invasive species (Zhang et al. 2010).

By the end of the nineteenth century, the plant was recorded in Egypt, India, Australia, and Java. The main mode of spread of water hyacinth throughout the world has been through anthropogenic means, via the horticultural and aquarium trades, due to the appeal of its attractive smooth, green foliage and beautiful purple flowers. It continues to be spread in this fashion.

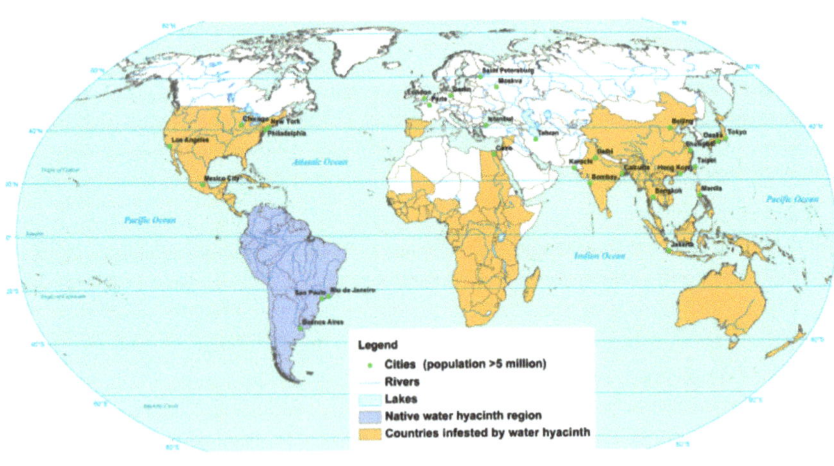

Fig. 17.1 Global distribution of water hyacinth (Map redrawn by Téllez et al. 2008 cited in UNEP 2006)

The success of this invasive alien species is largely due to its reproductive output. Water hyacinth can flower throughout the year and releases more than 3000 seeds per year (Gopal 1987; EEA 2012). The seeds are long-lived, up to 20 years (Gopal 1987). While seeds may not be viable at all sites, water hyacinth commonly colonizes new areas through vegetative reproduction and propagation of horizontally growing stolons. In the early stages of infestation, the weed takes foothold on the shoreline in the areas where native aquatic plants thrive (Gichuki et al. 2012). However, it is not restricted to shallow water, unlike many submersed and emergent macrophytes, because its roots are free-floating near the surface (Villamagna and Murphy 2010).

17.3.1 Distribution in Africa

Africa has particularly been affected by the introduction and spread of water hyacinth, facilitated in part due to a lack of naturally occurring enemies. In a review of water hyacinth infestation in eastern, southern and central Africa, Mujingni (2012) reports that the weed was first recorded in Zimbabwe in 1937. It colonized important water bodies, such as the in Comati River in Mozambique in 1946, the Zambezi River and some important rivers in Ethiopia in 1956. Historically, invasive species have been spread into Africa through colonization and exploration. As globalization increased Africa is becoming more exposed to invasive species (Fig. 17.2).

Water hyacinth has also spread to West Africa. It was first reported in Cameroon between 1997 and 2000 and since then the country's wetlands have become "home" for the weed (Forpah 2009). In Nigeria almost all river bodies have been dominated by water hyacinth (Borokoni and Babalola 2012). The water hyacinth problem is especially severe on the river Niger in Mali where human activities and livelihoods are closely linked to the water systems (Dagno et al. 2012).

It occurs throughout the Nile Delta in Egypt and is believed to be spreading southwards, due to the construction of the Aswan Dam which has slowed the river flow, enabling the weed to invade (Dagno et al. 2012). Infestation of water hyacinth in Ethiopia has also been manifested on a large scale in many water bodies of the Gambella area, Lake Ellen in the Rift Valley and Lake Tana. In Ethiopia, it has been problematic at Koka Dam along the Awash River, and in Gambela along Baro, Gilo, Pibor and Sobate rivers (Hedberg et al. 2006).

17.3.2 Water Hyacinth in Lake Tana

During the preliminary field monitoring end of October, 2012 the maximum cover of stationary mats of water hyacinth in Lake Tana covered about 10–15 ha and was distributed along 60–80% of the shoreline length where it is found. When it was

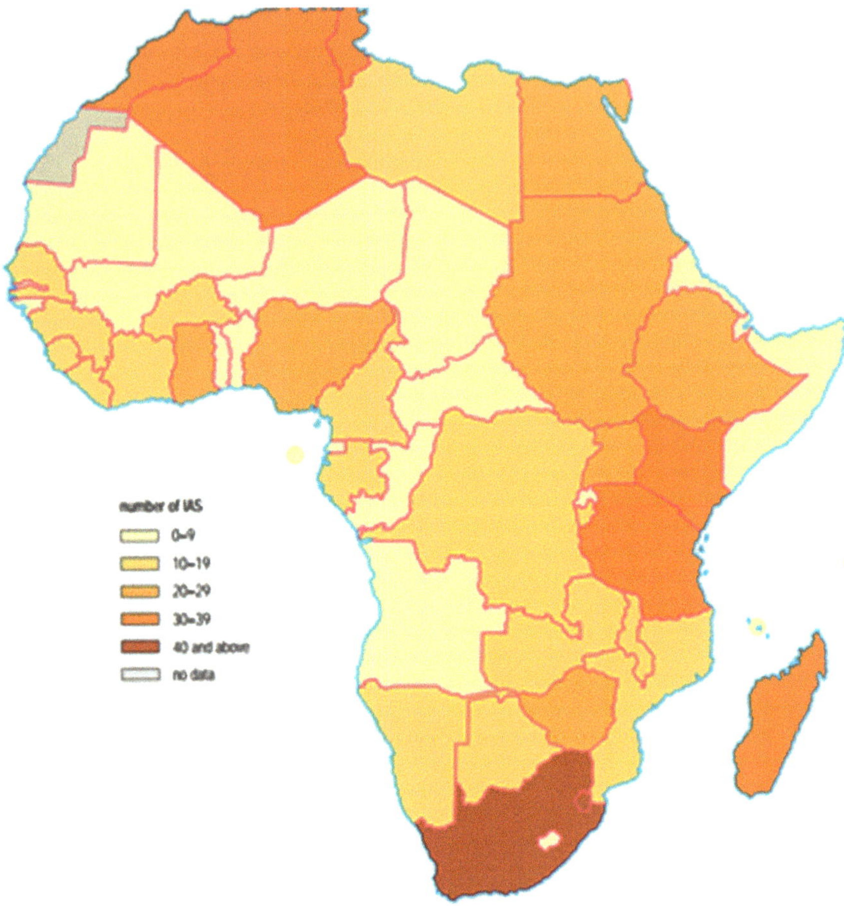

Fig. 17.2 The prevalence of invasive species Africa (IUCN/SSC/ISSG 2004)

first reported the weed remained confined to the northern extremities of the lake, probably because of the turbulence of the lake and the absence of extensive sheltered, shallow banks along other shores, where water hyacinth could anchor itself (Fig. 17.3). Maximum coverage of the mobile components of the weed in this lake is at least 3–4 times the area of stationary mats that cover about 20–30 ha. Surveys demonstrate that large mats of the water hyacinth were floating and moving into the lake especially near cattle grazed or disturbed areas.

However, no influx of the weed has been observed in the vicinity of river mouths, (possibly as a result of fragmentation by wave action). Thus, at present the weed is restricted in the lake shore and rice farm fringes; upstream rivers could be free from the invasive infestation.

Fig. 17.3 Water hyacinth (⊘) infested cites and potential distribution in Lake Tana (pin) free sites (Seid 2014, unpublished)

According to local informants, the stationary mats of the weed observed science 2010 were probably established around the Dirma river mouth, eliciting the local name "Afeshfasho" (Amharic) meaning delicate or weak, and later also named "Enboche" in 2014 with a similar meaning. In the newly invaded area, North-Eastern fringe, the weed has no name and local people perceive the weed as a gift of God "mena" (Amharic) because cattle can used to eat the leaves during times of grass (fodder) shortage. Other informants view the plant as a curse that destroys native grasses and invades rice farms.

17.3.3 Colonization Status in Lake Tana

Other plant species, including hippograss (*Vossia cuspidator*), papyrus (*Cyperus papyrus*) and morning glory (*Ipomea aquatica*), are now observed as the major associated plants promoted by the presence of hyacinths. However, the two popular tall sedges 'Dengle and Filla' (*C. papyrus* and *T. latifolia*) are often followed by hippo-grass or 'yeseytan ruze' (Amharic) (*Vossia cuspidator*) and water ferns (Lily and the like).

Currently in some of the infested places hippo-grass and water hyacinth are found side by side as the dominant weeds. As reported in other East African countries, the hippo-grass might be favored by the nutrients from dying water hyacinth and likely reduced with the removal of water hyacinth. The major associated animals were water fowl (*Alopochen aeguptiaca* L. (Egyptian Goose) and *Balearica Pavonina* L. (Black-crowned Crane) and abundant macroinvertebrates, such as leaches (Huridine species) and insects (Seid 2014) the biology of water hyacinth (*E. crassipes*: Invasive Gust of the Pristine Lake Tana, Ethiopia conference paper, unpublished).

Ecological succession can control stationary hyacinth mats along the shores and banks of rivers. In Lake Victoria, pure mats of water hyacinth are invaded initially by aquatic ferns/sedges (*Cyperus papyrus* and *Ipomea aquatica*) often followed by hippo-grass (*Vossia cuspidata*) which eventually dominates and shades out the remaining stressed and dying/rotting water hyacinth (Seid 2014).

17.4 Impacts of Invasive Alien Species

Invasive alien species may threaten native species as direct predators or competitors, as vectors of disease, or by modifying the habitat or altering native species dynamics (MA 2006). The threat posed to biodiversity by IAS is considered second only to that of habitat loss (CBD 2005). On small islands, it is now comparable with habitat loss as the lead cause of biodiversity loss (Baillie et al. 2004). Invasive species may out-compete native species, repressing or excluding them and therefore, fundamentally changing the ecosystem. They may indirectly transform the structure and species composition of the ecosystem by changing the way in which nutrients are cycled through the ecosystem (McNeeley et al. 2001). Entire ecosystems may be placed at risk through knock-on effects. Given the critical role biodiversity places in the maintenance of essential ecosystem functions, IAS may cause changes in environmental services, such as flood control and water supply, water assimilation, nutrient recycling, conservation and regeneration of soils (GISP 2004; Charles and Dukes 2007). Invasive may also affect native species by introducing pathogens or parasites that cause disease or kill native species. Among other things, both old and newly established IAS contribute to land degradation through soil erosion and the drawing down of water resources, reducing resources available to people and indigenous plants. Others produce leaf litter which poisons the soil, suppressing the growth of other plants, and in particular that of the under storey (UNEP 2004). They may alter the environment in directions that are more favourable for them but less favourable to native species. This could include altering geomorphic processes, biogeochemical cycling, hydrological cycles, or fire or light regimes (MA 2006; Levine et al. 2003).

Wattle trees and mesquite can sink their roots deeper into the soil than indigenous trees, sucking out massive volumes of water and out-competing indigenous plants for nourishment (Preston and Williams 2003). In some environments, invasive trees, like the black wattle, increase rainfall interception and transpiration, which cause a decrease in stream-flow (IUCN/SSC/ISSG 2004). The leaves and branches of the black wattle are believed to have allelopathic properties that are the chemical inhibition of growth and seed germination of other plants. Highly combustible, fire-tolerant alien plants may also alter the fire regime, and combined with competition for light, nutrients, water and space, this is believed to be an important factor in extinctions (Richardson and van Wilgen 2004). Marine IAS is a growing problem in Africa's coastal waters, estuaries and lagoons. Many of these introductions are related to sea vessels and aquaculture. *Hypnea musciformis* (hypnea) is red algae, originally from Trieste in Italy, and is now distributed throughout the world. It occurs in coastland, estuaries and marine habitats where it attaches to coral, stones or shells on sheltered tropical reef flats. Its success is related to its rapid growth rate, ability to epiphytize other algae and easy fragmentation. It is present in the coastal waters of several African countries including Ethiopia. Invasion pathways include aquaculture and dispersal by boats and other vessels (IUCN/SSC/ISSG 2004).

17.4.1 Positive Impacts (Utilization of Water Hyacinth)

Water hyacinth has some beneficial attributes and various bioremediation roles. It can be used in the production of paper, fiber boards, biogas, fertilizer, fish feed and in phytoremediation (the cleanup of polluted water bodies by aquatic plants like water hyacinth) (Khan and Sarwar 2002; Uka et al. 2007). Phytoremediation became popular when physical and chemical measures to remove aquatic pollutants were found to be more harmful than the pollutants themselves (Ndimele et al. 2010). Some of these pollutants are heavy metals and nutrients. Heavy metals are aquatic pollutant of major concern to ecologists because they are non-biodegradable that is, once they enter an aquatic ecosystem, they cannot be eliminated by ordinary biological processes.

There has been considerable research into the utilization of water hyacinth. Uses include in biogas production, animal fodder, as fertilizer, for mulch, the manufacture of paper and furniture, and water quality management (Mahamadi 2011). The main factor for arguing against successful utilization is that water hyacinth has up to 95% water content, which makes most utilization projects commercially not

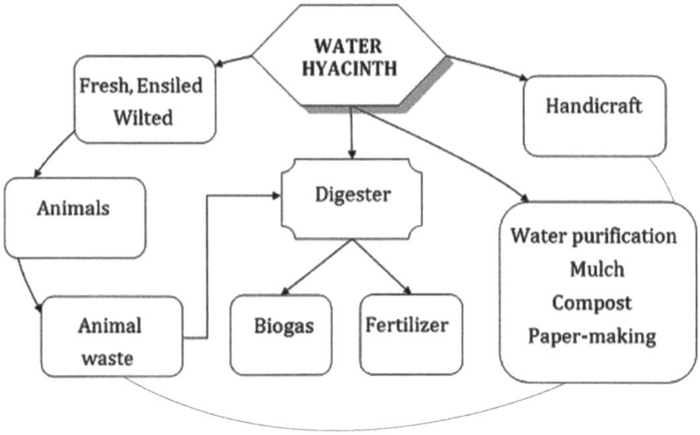

Fig. 17.4 Diagram showing water hyacinth utilization (*Source* Tham 2012)

viable on anything other than a cottage-industry level. Furthermore, no effective utilization program has been shown to control water hyacinth to acceptable levels except in Taiwan. Furthermore, at this moment creating reliance on using the weed as a resource could lead to its increased spread and to possible conflicts of interest between groups utilizing the plant and those wanting it controlled. However, an integrated scheme for water hyacinth control by utilization is also possible (Fig. 17.4).

17.4.2 Negative Impacts of Water Hyacinth

17.4.2.1 Impacts on Biodiversity

Coupled with this is its high rate of reproduction which has made it a serious threat to the continued use of the affected waters as a resource. This poses a great hindrance to the socio-economic potentials of these water bodies if appropriate and effective controls are not introduced. Dense mats of water hyacinth reduce light penetration into the water column, which negatively impacts submerged vegetation, reducing oxygen production in aquatic communities. The lack of phytoplankton production alters the composition of invertebrate communities.

The 'mat' of aquatic plants which covers aquatic ecosystem during severe infestation reduces dissolved oxygen by restricting the exchange of oxygen across the water interface. They also affect the chemistry of surface water which could render such water body unfit to support aquatic lives (Chukwuka and Uka 2007). Water hyacinth and other aquatic weeds also generate large amounts of organic matter because of their large biomass. When this macrophyte (water hyacinth) dies, sinks and decomposes, the water becomes more eutrophic due to the large release of nutrients. Water quality deteriorated, clean drinking water can be threatened and human health impacted. As the organic matter decomposes, biological oxygen demand increases due to the activities of decomposing bacteria. These bacteria use dissolved oxygen for their metabolic functions. The result is that the water quality deteriorates. This results in loss of aquatic biodiversity (Muli 1996).

Death of water hyacinth mats may influence changes in the composition, distribution and diversity of aquatic organisms as follows.

- Displacement of hydrophytes and depressed algal biomass (Twongo and Balirwa 1995).
- Increase in diversity and abundance of some macrofauna taxa, especially at the borders of the weed mats (Wanda 1997).
- Increase in the distribution and abundance of schistosome (bilharzias) snail vectors such as *Biomphalaria* spp. and *Bulinus* spp.
- Willoughby et al. (1993) reported that, based on studies on the Ugandan shoreline of Lake Victoria, mats significantly depressed the diversity of fish species and fish biomass. It was subsequently demonstrated that fish diversity, particularly small taxa, increased along the edge of water hyacinth mats (Twongo and Balirwa 1995).

17.4.2.2 Socioeconomic Impacts and Development Challenges

Many alien species, including some that are invasive, have had tremendous economic value for Africa. However, overall their impact on the sustainability of the resources, upon which livelihoods and development are often based, has been adverse, undercutting opportunities, human well-being and contributing to increased human vulnerability.

Invasive alien species are a serious impediment to the sustainable use of global, regional and local biodiversity (CBD 2005); this has implications for freshwater and marine resources, tourism, and forest and woodlands. Invasive alien species may affect livelihood and other economic opportunities in multiple ways. In addition to their impact on the supply of environmental goods, they also affect the integrity of ecosystems, undercutting essential environmental service. Thus IAS, through their impact on the environment, contributes indirectly to poverty, food insecurity, ill health and poor water quality (UNEP 2004; NEPAD 2003). They have multiple level and complex impacts on human wellbeing and the ability to achieve development targets, such as those set out in the Millennium Development Goals (MDGs).

The socio-economic effects of water hyacinth are dependent on the extent of invasion, the uses of the impacted water body, control methods and the response to control efforts (Villamagna and Murphy 2010). As much literature indicates, extensive mats of water hyacinth have a negative impact on aquatic ecosystems and cause problems for all aspects of water resource utilization. It has been found to drastically increase evapo-transpirational losses, as well as cause fish losses (Ndimele et al. 2011). These problems can be particularly severe in water-limited areas and small water bodies when water loss through evapotranspiration from the water hyacinth is 3.7 times that from open water (Timmer and Weldon 1967). The economic impacts of the weed in seven African countries have been estimated at between US$20–50 million every year. Across Africa costs may be as much as US $100 million annually (UNEP 2006).

In Ethiopia the weed poses serious problem on reservoirs, drainage structures, and water suppliers. For example since the 1996 flood, Wonji Sugar Factory incurs an additional cost for removing the weed from irrigation and drainage water structures. According to the information obtained from agricultural operation of the factory, the cost incurred to manage this weed increases from year to year. The factory spent more than 130,197 Ethiopia Birr (14,897 USD) over six years due to the water hyacinth (Yirefu et al. 2007).

Just after a year of invasion, the water hyacinth rapidly invades water ways and causes problems to activities dependent of the Lake Tana shore and flood plain complex, which includes the Dembia rice fields. It impedes water transportation preventing people from accessing their sources of livelihood. It also hinders access to fishing grounds especially during the rainy season which, unfortunately, is the period when fishing activities are highest and more profitable.

The socioeconomic impacts of water hyacinth include reduction in the quality and quantity of drinking water caused by bad odors, color, taste, and turbidity; increased incidence of waterborne, water-based, and water-related disease (e.g., malaria, encephalitis, and filariasis); increase in siltation and sedimentation of rivers, lakes, and impoundments; reduction of useful water surface area for fishing, recreation, and water transport; clogging of irrigation canals and pumps; drowning of livestock; interruption of hydroelectric power generation; and enhanced flood damage to road and rail bridges and impoundment walls. Finally, the most direct impact of dense mats of water hyacinth is on boating access, navigability, water supply systems, and drainage canals and on recreation (Villamagna and Murphy 2010).

17.5 Control of Water Hyacinth Invasion

Traditionally the control of water hyacinth has fallen into one of three broad categories: physical control (manual and mechanical removal), herbicidal control, and biological control. More recently, the emphasis has moved to an integration of these three methods.

17.5.1 Physical/Mechanical Control

Manual removal through hand-pulling or through the use of other handheld tools such as pitchforks is employed in a number of developing countries such as Guyana, South Africa, and China. This method is very labor-intensive (Mara 1976), is effective only for small infestations, and essentially is used as employment creation exercises.

Mechanical control using custom-designed machinery has been implemented, for example, on Lake Victoria, with limited success (Villamagna and Murphy 2010). However, the amount of biomass that has to be removed and the growth rate of water hyacinth render this method ineffective, except for very small infestations or to keep specific areas such as the waters around hydro intakes free of the weed. Furthermore, the remoteness of many infestations makes mechanical control impractical. Booms and cables have also been used to prevent water hyacinth from entering water abstraction pumps and hydropower coolant intakes (Wittenberg and Cock 2001). Cables have also been used to concentrate weed infestations behind them, making physical removal and herbicide applications more efficient.

17.5.2 Herbicide/Chemical Application

Herbicide control has been successfully used against infestations of water hyacinth since the early invasion history. Although this control method is relatively expensive, it has the advantage of being quick and temporarily effective. Glyphosate and 2,4-D [(2,4-dichlorophenoxy) acetic acid] have been the most widely used herbicides and considered as effective and relatively safe herbicides (Chen et al. 1989). *E. crassipes* is also susceptible to 2,4-D, Diquat, Paraquat, and Glyphosate herbicides, which have resulted in successful control in small (<1 ha), single-purpose water systems such as irrigation canals and impoundments (Center et al. 2002).

Despite such effectiveness of herbicides, the major disadvantages are that they are non-selective and could cause major environmental problems if incorrectly applied (Wittenberg and Cock 2001). Chemical control needs to be carried out repeatedly as re-infestation of water hyacinth occurs from seeds or clonal multiplication of surviving plants (Chen et al. 1989; Charudattan 1986). Herbicide control provides only short-term results and requires regular follow-up applications. Moreover, it is dangerous for pristine ecosystems like Lake Tana.

17.5.3 Biological Control

Mechanical and chemical controls are viewed as short-term or immediate control options; biological control is perceived as the long-term or sustainable control option for this weed. When chemical control is economically unfeasible or harmful to the environment, biological control is recognized as a cost effective, permanent and environmentally friendly control method (Charudattan 1986). In some areas, they have provided considerable control, but this is not consistent in all areas. The principal drawback with biological control of water hyacinth is the time required to achieve control. In tropical environments, this is usually 2–4 years and is influenced by the extent of the infestation, climate, water nutrient status, and other control options (Wittenberg and Cock 2001).

The first agent for water hyacinth was the weevil, *Neochetina eichhorniae* Warner (Coleoptera: Curculionidae), which was released in Florida in 1972. The most widely used agents are the weevils, *N. eichhorniae* and *N. bruchi* Hustache, and the moth, *Sameodes albiguttalis* (Warren) (Lepidoptera: Pyralidae) (Wittenberg and Cock 2001). Other agents, released as classical biological control agents, are the moth, *Xubida infusella* (Walker) (Lepidoptera: Pyralidae), the water hyacinth bug, *Eccritotarsus catarinensis* (Carvalho) (Hemiptera: Miridae), the galumnid mite, *Orthogalumna terebrantis* Wallwork, and the fungal pathogen, *Cercospora piaropi* Tharp. Most recently, the leafhopper, *Megamelus scutellaris* Berg (Hemiptera: Delphacidae), has been released in the United States. Several other insects could be screened for release in regions of the world, where the other agents have not achieved the acceptable level of control or where the time taken to achieve control (1.5–3 years in the tropics) is perceived to be too long, ultimately affecting fish and other higher-order vertebrate populations (Mara 1976).

17.5.4 Popular Awareness

Public awareness and understanding may also be lacking. Members of the public, made aware of the issue and engaged in the preventive effort, can make an enormous difference. Well-informed travellers are the front-line in any campaign to prevent the dispersal and spread of invasive alien species. A well-informed public is more likely, moreover, to appreciate the need for preventive checks and other regulations, which otherwise might come across as just an inconvenient nuisance. Access to information then, in the form of posters or notices displayed at entry points, or published alerts in travel magazines and other media, is another important aspect of effective exclusion strategy. Inculcation of awareness at all levels of society social, economic and political is of course an essential first step towards mounting a successful campaign to prevent the influx and spread of invasive alien species.

Yet, alien invasions are often the result of entrenched human values, habits and patterns of behaviour. So awareness alone may not be enough to bring about the behavioural changes that are required for such campaigns to be effective. We know, from humanity's response to the impacts of global climate change, how difficult it can be, even against a backdrop of almost universal awareness (of the need to reduce anthropogenic greenhouse gas emissions, say), to usher in behavioural adaptations that might help to mitigate these impacts. Awareness regarding the invasive species' threat is, in the continuing absence of a global treaty (such as that of the Intergovernmental Panel on Climate Change), still far from universal. Yet, even in nations where awareness campaigns have succeeded in raising the profile of the invasive threat through the dissemination of information and educational materials, follow-up actions have not necessarily resulted. The aim of a social marketing campaign is to build motivations and partnerships through which awareness of a pressing social need might be translated into concrete actions which address that need.

17.5.5 Risk Assessment

Another important aspect of control is determining the level of invasive risk associated with the introduction of any species which may be new to a country. For this, a sound regulatory framework is needed, representing and ruling on the wider interests environmental and social, national and regional of alien species' introductions that are proposed by agri-business or by commerce. Under such a framework, the introduction of species deemed to pose an unacceptable invasive threat to ecosystems and societies as a whole can be prohibited under international trade law, irrespective of how useful, or profitable, the species in question might be to a minority of would be importers or investors. A risk assessment is the standard procedure for determining whether or not the proposed introduction of an alien species can be authorised. Invasiveness cannot always be reliably predicted, however. Species that show no invasive tendencies in one region may prove invasive in another, and vice versa. It may also take many years for the invasiveness of an introduced species to become apparent.

The most reliable indicator for invasiveness is to know whether or not a species has become invasive elsewhere in its introduced range, particularly in ecosystems that are comparable and which boast similar climatic and geographic conditions. Certain attributes among plants may create grounds for suspicion. Such attributes typically summarized by Rejmánek and Pitcairn (2002) as:

- Fitness homeostasis or the ability of an individual or population to maintain relatively constant fitness over a range of environments. This is equivalent to Baker (1974, 1995) "general-purpose genotype."
- Small genome size-usually associated with short minimum generation time, short juvenile period, small seed size, high leaf area ratio, and high relative growth rate. Dispersed easily by humans and animals.
- Ability to vegetative propagate. This is an especially important characteristic in aquatic environments (Auld el al. 1983) and at high latitude (Pyšek 1997).
- Alien plants belonging to exotic genera are more invasive than are alien species with native congeners. This may be partly because of an absence or limited

number of resident natural enemies for that species (Darwin 1859; Rejmánek 1999).
- Plant species without dependence on specific mutualisms (root symbiosis, pollinators, seed dispersers, etc.) (Baker 1974; Richardson et al. 2000).
- Tall plants tend to invade mesic plant communities. Persistent seed banks-seeds with different inherent dormancies that provide a random appearance through time and guarantee their survival and persistence.

Thus, plants with production of abundant seeds, capable of utilising highly effective agencies of dispersal Species whose foliage, outside their home environments, is poisonous to animals, or which are allelopathic (meaning they release chemical toxins into the soil that inhibit growth among plants of other species, or prevent the seeds of other plants from germinating), are usually also a high risk.

Plants belonging to certain taxonomic families, such as the Fabaceae (legumes) and the Asteraceae (Composites), which are disproportionately well represented globally among invasive species, are best treated with particular caution. A decision authorising introduction of a useful alien species that may also become invasive should not be taken carelessly, for as such decision, once taken may not reversible. If an invasive species introduced, then future generations as a whole will bear the impacts. Such costs will be very substantial as globally known well from the heavy price we are already paying for invasions control that resulted from ill-informed deliberate and accidental alien species introductions.

17.6 Research Gaps

Observed regional differences in water hyacinth biology exist and more research needs to be done to investigate the reasons for those differences. However, regardless of the slight regional differences the explosive growth rate of water hyacinth makes it a viable candidate for further research and development of water hyacinth benefit and control researches. Thus, the plant biology that relate to the growth, reproduction, ecological interactions, impacts on biodiversity, fishery, management approaches and other issues of water hyacinth are basic research priorities. For future research it would be beneficial to study the placement of water hyacinth in dryer areas for land preservation and waste water abatement.

It is speculated that the biomass can be used for crafting, waste water treatment, heavy metal and dye remediation, as substrate for bioethanol and biogas production, electricity generation, industrial uses, medicines, animal feed, agriculture and sustainable development. However, seldom does utilization provide a sustained solution to the spread and impact of water hyacinth, and in fact could provide a perverse incentive to maintain the invasive plant to the detriment of the environment and production systems at high economic and social costs. There is no example from anywhere in the world where utilization alone has contributed to the management of any invasive plant (EEA 2012). While researchers continue to investigate the perceived potential uses of water hyacinth, the current negative impacts of the weed far outweigh its benefits. The use of water hyacinth as raw material in cottage industry should not encourage propagation of the weed, but rather help control its growth.

Bioenergy production through the growth and managed harvesting of water hyacinth can benefit the region along with other coastal regions and give added ecological benefits which offset the higher economic costs of biofuel production while also saving invasive plant species abatement program. For future work it would be beneficial to analyze the wave energy absorption ability of water hyacinth. In future research it would be beneficial to explore several engineering methods of producing bioenergy from water hyacinth.

In order for policy makers to make informed decisions, much more economic information is required on the costs and benefits of environmental programs. For example, it is frequently stated that there are insufficient resources to control hyacinth. However, if the costs of improved water treatment are compared with the costs of decreased fish catches and the costs of increased water-borne diseases, it is likely that resources needed for hyacinth control are modest in comparison to potential losses from its proliferation.

17.7 Conclusion

Water hyacinth is the most invasive and damaging aquatic plant despite the fact that there are a number of effective ways to control it. Even though good progress has been made in controlling around the world, it still poses a threat to aquatic ecosystems and to human activities. The long-term control of this plant will require an integrated management approach utilizing all appropriate control methods, but with special emphasis on the need to reduce the inflow of nitrate and phosphate pollutants into aquatic environments.

17.8 Recommendations

Water hyacinth infestation in Lake Tana can be a symptom of broader watershed management and pollution problems. It calls for a concise national and trans-boundary invasive species policy designating noxious weed to aquatic and terrestrial systems. Given the complexity of control options and the potential for climate change to assist the spread of water hyacinth, it is critical to develop comprehensive management strategies and action plans needed.

A multidisciplinary approach should be designed, which includes the highest levels political and administrative to recognize the potential seriousness of the weed. Plans should also state clearly the role of each government department, stakeholders, municipal councils and local community involved in the fight against water hyacinth.

Awareness needs to be raised amongst local communities and all stakeholders on the inherent dangers of water hyacinth infestation and to mobilize communities towards control measures. One practical approach is to involve communities in manual and biological control activities, for example, in rearing weevils. There are excellent examples of community involvement in the first popular mechanical control champagne to control the hyacinth around Lake Tana.

Methods for water hyacinth control should include reduction of nutrient load in the lake through leaching of fertilizer from highlands, waste waters flowing from sewage works, urban wastes and factories. Changing land use practices in the catchment communities through watershed management will help reduce agricultural runoff as a mechanism for controlling the proliferation of water hyacinth. This is considered by many as one of the most sustainable long-term management actions.

References

Auld BA, Hosking J, McFadyen RE (1983) Analysis of the spread of tiger pear and parthenium weed in Australia. Aust Weeds 2:56–60

Baillie JEM, Hilton-Taylor C, Stuart SN (2004) IUCN red list of threatened species. A Global Species Assessment. IUCN—the World Conservation Union, Gland

Baker HG (1974) The evolution of weeds. Annu Rev Ecol Syst 5:1–24

Baker HG (1995) Aspects of the genecology of weeds. In: Kruckeberg AR, Walker RB, Leviton AE (eds) Genecology and ecogeographic races. Pacific Division, American Association for the Advancement of Science, San Francisco, pp 189–224

Borokoni T, Babalola F (2012) Management of invasive plant species in Nigeria through economic exploitation: lessons from other countries. Manag Biol Invasions 3(1):45–55. doi:10.3391/mbi.2012.3.1.05

CBD (2005). Invasive Alien Species. Convention on biological diversity. http://www.biodiv.org/programmes/cross-cutting/alien/. Accessed on 6 Jan 2014

Center TD, Hill MP et al (2002) Water hyacinth. In: van Driesche R, Blossey B et al (eds) Biological control of invasive plants in the Eastern United States. Forest Health and Technology Enterprises Team, West Virginia. http://www.fs.fed.us/foresthealth/technology/pdfs/BiocontrolsOfInvasivePlants02_04.pdf. Accessed 4 June 2015

Charles H, Dukes JS (2007) Impacts of invasive species on ecosystem services. In: Nentwig W (ed) Ecological studies. Biological invasions, vol 193. Springer, Berlin. http://globalecology.stanford.edu/DGE/Dukes/Charles_Dukes_inpress.pdf. Accessed on 6 Feb 2015

Charudattan R (1986) Integrated control of water hyacinth (*Eichhornia crassipes*) with a pathogen, insects, and herbicides. Weed Sci 34:26–30

Chen YL, Chiang HC, Wu LQ et al (1989) Residues of glyphosate in an aquatic environment after control of water hyacinth (*Eichhornia crassipes*). J Weed Sci Technol 34(2):117–122

Chukwuka KS, Uka UN (2007) Effect of Water Hyacinth (*Eichhornia crassipes*) Infestation on Zooplankton Populations in Awba Reservoir, Ibadan South-West Nigeria. J Biol Sci. doi:10.3923/jbs.2007.865.869

Dagno K, Lahlali R, Diourte M et al (2012) Fungi occurring on water hyacinth (*Eichhornia crassipes* (Martius) Solms-Laubach) in Niger River in Mali and their evaluation as Mycoherbicides. J Aquat Plant Manag 50:25–32

Darwin C (1859) The origin of species by means of natural selection. Murray, London
EEA (2012) The impacts of invasive alien species in Europe. EEA technical reports no 16/2012. Publications Office of the European Union, Brussels, Luxembourg. http://www.eea.europa.eu/publications/impacts-of-invasive-alien-species. Accessed 12 Sep 2014
EPA (Environmental Protection Authority) (1997) Environmental Policy of the Federal Democratic Republic of Ethiopia. EPA, Addis Ababa
Fessehaie R (2005) Water hyacinth (Eichhornia crassipes): a review of its weed status in Ethiopia. In: Rezene Fessehaie (ed) Arem, vol 6, pp 105–111
Fontenille D, Tato JC (2001) 'Aedes (Stegomyia) albopictus (Skuse), a Potential new dengue vector in southern Cameroon. Emerg Infect Dis 6(7):1066–1067
Forpah N (2009) Cameroon prepares a national strategy for the control of water hyacinth (exotic species). In: Proceedings on the elaboration of a national strategy for the control of water hyacinth in Cameroon, 15–18 Sept 2009, Douala. http://www.unep.org/pdf/UNEP_GEAS_APRIL_2013.pdf. Accessed 12 June 2014
Gichuki J, Omondi R, Boera P et al (2012) Water Hyacinth (Eichhornia crassipes) (Mart.) Solms-Laubach Dynamics and Succession in the Nyanza Gulf of Lake Victoria (East Africa): implications for water quality and biodiversity conservation. The Scientific World J Vol 2012. doi:10.1100/2012/106429
GISP (2004) Africa invaded: the growing danger of invasive alien species. Global invasive Species Programme, Cape Town. http://www.gisp.org/downloadpubs/gisp%20africa%202.pdf. Accessed on 6 Feb 2015
Gopal B (1987) Water hyacinth. Elsevier, Amsterdam
Hedberg I, Kelbessa E, Edwards S et al (eds) (2006) Flora of Ethiopia and Eritrea, vol 5. Addis Ababa, The National Herbarium, Addis Ababa University
Hill M, Coetzee J (2013) Water hyacinth. In: Borgemeister C, Langewald J (eds) Biological control in IPM systems in Africa. Wallingford
Holm LG, Plucknett DL, Pancho JV et al (1991) The world's worst weeds, distribution and biology. Krieger Publishing Co., Malabar, Florida
IUCN/SSC/ISSG (2004) Global invasive species database. IUCN—the World Conservation Union Species Survival Commission, Invasive Species Specialist Group. http://www.issg.org. Accessed on 6 Feb 2014
Khan S, Sarwar KS (2002) Effect of water-hyacinth compost on physical, Physicochemical properties of soil and on rice yield. J Agron 1:64–65
Kirby A (2003) Alien species cost Africa billions. BBC News Science. http://news.bbc.co.uk/2/hi/science/nature/2730693.stm
Levine JM, Vila M, D'Antonio CM et al (2003) Mechanisms underlying the impacts of exotic plant invasions. Proc R Soc Lond B Biol 270:775–781
MA (2006) Ecosystems and human well-being: current state and trends, vol 1. Millennium Ecosystem Assessment. Island Press, Washington. http://www.millenniumassessment.org//en/products. Accessed on 6 Feb 2015
Mahamadi C (2011) Water hyacinth as a biosorbent. Afr J Environ Sci Technol 5(5):1137–1145
Mara MJ (1976) Estimated costs of mechanical control of water hyacinths. J Environ Econ Manag 2(4):273–294
McNeeley JA, Mooney HA, Neville LE et al (2001) Global strategy on invasive Alien Species. IUCN—the World Conservation Union, Gland
Mujingni C (2012) Quantification of the impacts of Water Hyacinth on riparian communities in Cameroon and assessment of an appropriate method of control: the case of the River Wouri Basin. Thesis, World Maritime University
Muli JR (1996) Environmental problems in Lake Victoria (East Africa): What the international community can do. Lakes Reservoirs: Res Manag 2:47–53
Ndimele P, Kumolu-Johnson C, Anetekhai M (2011) The invasive aquatic macrophyte, water hyacinth (Eichhornia crassipes (Mart.) Solm-Laubach: Pontedericeae): problems and prospects. Res J Environ Sci 5:509–520

Ndimele PE, Jenyo-Oni A, Ayodele AI et al (2010) The phytoremedation of crude oil-polluted aquatic environment by water hyacinth (*Eichhornia crassipes* (Mart.) Solms) Afr J Livest Extension 8:48–52

NEPAD (2003) Action plan for the environment initiative. New Partnership for Africa's Development, Midrand. http://nepad.org/2005/files/reports/action_plan/action_plan_english2.pdf. Accessed 6 June 2015

Patel S (2012) Threats, management and envisaged utilizations of aquatic weed *Eichhornia crassipes*: an overview. Rev Environ Sci Biotechnol 11:249–259. doi:10.1007/s11157-012-9289-4

Preston G, Williams L (2003) Case study: the working for water programme: threats and successes. Serv Deliv Rev 2(2):66–69. http://www.dpsa.gov.za/documents/service_delivery_review/vol2no2. Accessed on 6 Feb 2015

Pyšek P (1997) Clonality and plant invasions: can a trait make a difference? In: de Kroon H, van Groenendael J (eds) The ecology and evolution of clonal plants. Backhuys, Leiden, pp 405–427

Rands M, Adams W, Bennun L et al (2010) Biodiversity conservation: challenges beyond 2010. Science 329:1298–1303

Rejmánek M (1999) Invasive plant species and invasible ecosystems. In: Sandlund OT, Schei PJ, Vilken A (eds) Invasive species and biodiversity management. Kluwer, Dordrecht, pp 79–102

Rejmánek M, Pitcairn MI (2002) When is eradication of exotic pest plants a realistic goal? In: Veitch CR, Clout MN (eds) Turning the tide: the eradication of invasive species. IUCN/SSC Invasive Species Specialist Group. IUCN, Gland, Switzerland, and Cambridge, pp 249–253

Richardson DM, Allsopp N, D'Antonio C et al (2000) Plant invasions—the role of mutualisms. Biol Rev 75:65–93

Richardson DM, van Wilgen BW (2004). Invasive alien plants in South Africa: how well do we understand the ecological impacts? S Afr J Sci 100:45–52. http://www.dwaf.gov.za/wfw/Docs/Papers/SAJSFeb2004richardson.pdf. Accessed on 6 Feb 2015

Seid A (2014) A review on the biology and control of water hyacinth (Eichhornia crassipes): invasive Gust of the Pristine Lake Tana, Ethiopia. Proceedings of the Second Annual Science Conferance, ASC 2014, pp 161–180

Shanab S, Shalaby E, Lightfoot D et al (2010) Allelopathic effects of water hyacinth (*Eichhornia crassipes*). PLoS ONE 5(10):e13200. doi:10.1371/journal.pone.0013200

Téllez T, López E, Granado G et al (2008) The water hyacinth, *Eichhornia crassipes*: an invasive plant in the Guadiana River Basin (Spain). Aquat Invasions 3:42–53

Tham HT (2012) Water hyacinth (*Eichhornia crassipes*)—biomass production, ensilability and feeding value to growing cattle. Doctoral thesis, Faculty of Veterinary Medicine and Animal Science, Department of Animal Nutrition and Management, Uppsala, Sweden

Timmer CE, Weldon LW (1967) Evapotranspiration and pollution of water by water hyacinth. Hyacinth Control J 16:34–37

Twongo T, Balirwa J (1995) The water hyacinth problem and the biological control option in the highland region of the Upper Nile Basin: Uganda's experience. Paper presented at the 2002 Nile conference, "Comprehensive Water Resources Development of the Nile Basin," Arusha, Tanzania, 13–17 Feb 1995

Uka UN, Chukwuka KS, Daddy F (2007) Water hyacinth infestation and management in Nigeria inland waters: a review. J Plant Sci 2:480–488

UNEP (2004) Invasive aliens threaten biodiversity and increase vulnerability in Africa. Call to Action 1(1). United Nations Environment Programme, Nairobi

UNEP (2006) Africa Environment Outlook 2. Division of Early Warning and Assessment, United Nations Environment Programme, Nairobi

USBC (2008) Statistical Abstract of the United States 2008. U.S. Census Bureau, U.S. Government Printing Office, Washington, DC

Vila M, Espinar J, Hejda M et al (2011) Ecological impacts of invasive alien plants: a meta-analysis of their effects on species, communities and ecosystems. Ecol Lett 14:702–708

Villamagna A, Murphy B (2010) Ecological and socio-economic impacts of invasive water hyacinth (*Eichhornia crassipes*): a review. Freshw Biol 55:282–298. doi:10.1111/j.1365-2427.2009.02294.x
Wanda FM (1997) The impact of water hyacinth *Eichhornia crassipes* (Mart.) Solms (Pontederiaceae) on the abundance and diversity of aquatic macroinvertebrates in northern Lake Victoria, Uganda. Thesis, International Institute of Infrastructural, Hydraulic and Environmental Engineering
Willoughby NG, Watson IG, Lauer S et al (1993) An investigation into the effects of water hyacinth on the biodiversity and abundance of fish and invertebrates in Lake Victoria, Uganda. NRI Project Number 10066 A0328. Accessed 23 Aug 2013
Wittenberg R, Cock MJW (2001) Invasive Alien species: a toolkit of best prevention and management practices. CAB International, Wallingford, Oxon, UK
Xu H, Qiang S, Genovesi P et al (2012) An inventory of invasive alien species in China. NeoBiota 15:1–26. doi:10.3897/neobiota.15.3575
Yirefu F, Tafesse A, Gebeyehu T et al (2007) Distribution, impact and management of water hyacinth at Wonji-Shewa Sugar Factory. Eth J Weed Manag 1(1):41–52
Zhang Y, Zhang D, Barrett S (2010) Genetic uniformity characterizes the invasive spread of water hyacinth (Eichhornia crassipes), a clonal aquatic plant. Molec Ecol 19:1774–1786

Part III
Socioeconomic System Characteristics and Land Use

Chapter 18
Demographic Characteristics of the Lake Tana Basin

Mesfin Anteneh

Abstract This chapter examines the demographic characteristics of the Lake Tana basin through document analysis techniques. It uses demographic indicators such as age-sex ratio, dependency ratio, population distribution, population density and population growth rate to provide a demographic profile at the watershed level. The study also looks at the changes in the indicators over time to establish patterns and trends. The analysis shows that the age group ranging from 0 to 14 years old is the largest population category. As the result the dependency ratio of the population in the watershed is very high. In addition the number of women in the reproductive age group is also very high which predicts even higher the potential growth of population in the watershed. If uncontrolled this future population growth will lead to greatly unbalanced relationships between resources and population in the watershed.

Keywords Lake Tana basin · Demography · Population density · Reproductive age

18.1 Introduction

Demographic variables such as population growth, age-sex ratio and the spatial distribution and fertility level of a population, determines the development level of any geographical area. This is because all the social service and development agendas designed by a government are based on the data of these demographic variables.

In Ethiopia the three national censuses conducted in 1984, 1994, and 2007 indicated a very rapid population growth such that the country's population has almost doubled in 23 years. The growth rates were extremely high, 3.4% per annum

M. Anteneh (✉)
Department of Geography, Faculty of Social Sciences, Bahir Dar University, Bahir Dar, Ethiopia
e-mail: mesfin74@yahoo.com

K. Stave et al. (eds.), *Social and Ecological System Dynamics*,
AESS Interdisciplinary Environmental Studies and Sciences Series,
DOI 10.1007/978-3-319-45755-0_18

from 1984–1994 and 2.9% from 1994–2007. Although, the rate is declining it is still one of the highest in the world.

The Lake Tana basin, which is located in the North western part of Ethiopia is among the most densely populated parts of the country. In general food supplies and agricultural production must be greatly increased to meet the needs of a rapidly growing population; this limits the allocation of resources to other economic and social sectors. Moreover the rapid increase in population means that there will be an increase in the dependency ratio. The chapter explores the effect of the high growth rate on the demographic characteristics in the Lake Tana basin.

18.2 Historical Settings of Human Settlement in the Lake Tana Basin

The region of northeast Africa is considered as the home of the ancient Afro-Asiatic language family that gave birth to Omotic, Cushitic and Semitic language groups (Bahiru 1991; Marcus 1994; Ehret 2002; Hudson 2000). Omotic and Cushitic language speakers are considered the earliest and the most dominant people in southern and northern Ethiopia respectively (Taddesse 1972, 1988; Levine 1974; Ehret 1979, 2002; Phillipson 1993). The Semitic speakers, on the other hand, emigrated from South Arabia and settled in northern Ethiopian plateau and the Red sea coast of Africa since the early first millennium BC (Simoons 1958; Bahiru 1991; Erlich 1994). Some scholars attribute special place to the role of South Arabian settlers on ancient civilization of Ethiopia including agricultural technologies and developments (Messing 1957; Ullendorff 1973; Galperits 1981).

However, as mentioned above Cushitic speaking people are the oldest inhabitants of northern Ethiopia. Probably from 3500 BC to 1000 BC, much of north and north central Ethiopia were thus occupied by their descendants such as Agew and Beja upon which the South Arabian settlers later exerted significant and lasting cultural and political impact (Kaplan 1992; Qurin 1998). Prior to the emergence of Aksum, the Agew came to dominate many parts of Northern and Central highlands. They also moved south to the Blue Nile River and the upper Awash River valley and integrated some of Omotic people of the region (Levin 1974; Ehret 2002).

With regards to Gojjam, where the major portion of Lake Tana is found, Dombrowski (1971) argues that the Cushitic speaking Agew were the ancient inhabitants of the province perhaps since around 5000 or 4000 BC. When the Semitic people later occupied much of northern Ethiopia, the Nilotes and Cushitic speaking people were either assimilated into the Semitic speakers or were forced to survive only in isolated pockets of settlement in some parts of north and northwest Ethiopia. Inscription of the Aksumite period also indicates that the Cushitic speaking Agew were settled in Gojjam in the sixth century AD (Taddesse 1988). Taddesse (1972, 1994) further suggests that the Nilotic speaking Gumuz people of the western hot lowlands might be the ancient inhabitants of the highlands of Gojjam.

The Semitic speaking Amhara subsequently put pressure on the Cushitic speaking Agew which in turn pushed the Gumuz people far to the western margin of the province.

The Gafat people had probably settled in much part of central, northern and eastern portion of Gojjam long before the spread of Christianity and Semitic speaking Amhara into the province (Taddesse 1994). Their language was spoken in Southern Gojjam until the nineteenth century AD (Henze 2000). In addition, the Shinasha also crossed the Blue Nile River from the south and settled in the western part of Gojjam probably after the sixteenth century movement of the Oromo population (Taddesse 1994). Huntigford (1989) suggests that in the seventeenth century AD, the Omotic speaking Gonga or the Kaffa people (the Shinasha are the member family) were located on both sides of the Blue Nile River (the river originated from Lake Tana). Other scholars argue that the Omotic speaking Shinasha might have extended as far as Lake Tana region since ancient times. They therefore constitute one of the surviving entities of the northern most branches of these people (Taddesse 1972; Levin 1974; Ashenafi and Kidus 1994; Ehret 2002).

From such brief review of the history of the people in Lake Tana and its surounding, it can safely be concluded that the Semitic speaking Amhara seems to be relatively late arrival into the region. Henze's description supports this assumption. He argues that during the Aksumite times and after its collapse, the Lake Tana areas as well as the upper Blue Nile region was inhabited by the ancestors of the Wayto, the Agew, and the Gumuz. Even the pagan kings of Gojjam who fought with the Christian emperors of Ethiopia from the late thirteenth century to early fifteenth century AD were Agew (Henze 2000). Taddesse (1972, 1988, 1994) shows that early attempt to conquer and evangelize Christianity in Gojjam could have been made after the middle of the twelfth century AD. But, effective expansion of the Christian state in the area took place since after the reign of King Amde Siyon (1313–1344). Eastern Gojjam was brought into the Christian state empire then on, and it gradually became Christian and Semitic speaking Amhara with significant number of Amhara settlers from the Christian empire. It also became a key part of the central state and completely transformed "into a distinctively Christian and Amhara Country." Due to fierce resistance from the powerful kingdom of Gojjam, the Agew around the Lake Tana and in Agewmidir in the southwest began to feel the direct impact of the central state and its institutions only after the sixteenth century AD. Therefore, much of the Cushitic speaking Agew of the region were relatively free from the intensive and direct influence from the North until the sixteenth and seventeenth centuries AD. Nevertheless, after their incorporation into the socio-cultural values of the Christian Amhara dynasty, the Agew became bilingual, devoted Christians and have developed profound Christian identity. Before this process, however, the general features of pagan worship centring on a sky god, with numerous good and bad spirits inhabiting the mountainous, trees, rivers and lakes was prevalent in the region. The spirits manifested in various forms of natural phenomena, such as fire, storm, famine, life and death. The smallest details of the daily life of the people were presented to be under the control of these spirits. It is believed that an angry spirit could strike hard at minor divergence from traditional forms of ritual worship. It was largely on this background that Christianity was superimposed in the highlands in general (Taddesse 1972). Yet, Awngi, the southernmost language group of the Cushitic speaking people still survives among the Agew of Gojjam area of the Lake Tana basin (Tosco 2000). Therefore the present day Lake Tana basin is inhabited by dominantly by the Amhara and Agew population. The Amhara population holds the majority portion of the watershed both in Gojjam and Gondor, while the Agew is found in the south western parts of Lake Tana in some parts of Gojjam. Although their number is quite insignificant the woytos are also the other group of people living in the shores of Lake Tana.

18.3 Change in Population Size and Growth Rate

At the beginning of the twentieth century, the world population was about 1.7 billion, and Ethiopia's population estimated at 11.75 million. One hundred years later, at the beginning of the new millennium, 2000, the world's population was more than 6 billion and Ethiopia's population was 63.5 million. As Table 18.1 shows, Ethiopia passed through about 23.6 million in 1960 (CSO 1964), 42.8 million in 1984 (CSA 1987) and 53.4 million by 1994 (CSA 1998). By 2007 the population of Ethiopia attained 79 million (CSA 2008), and is now estimated to be 94.1 million (World Bank 2015).

As noted in Table 18.1 prior to 1900 the annual average rate of population growth of the world and Africa were about were about 0.5 and 0.4% respectively. The annual average rate of population growth of the world has rapidly increased from about 0.5% in 1900 to its peak of 2.17% in 1970 (see Table 18.1). Since then it has gradually declined and reached slightly in 1999 (UN 1999). In the case of Ethiopia, however, the average growth rate was about 0.2% by the beginning of the twentieth century. The average growth rate of Ethiopian population has continued to rise and reached to 2.9 and 3.0% in 1984 and 1994 census (CSA 1985, 1995), after which it was declined and dropped to 2.6% in the 2007 census (CSA 2008).

Table 18.1 Estimates of various population growth rates and reconstructed populations of Ethiopia, Africa and the world

Period	Population (million)			Reconstructed growth rate (%)		
	World	Africa	Ethiopia	World	Africa	Ethiopia
1900	1550–1762	–	11.75	0.5	0.4	0.2
1950	2520	224	19.19	1.0	2.0	2.0
1960	3021	282	23.55	1.9	2.2	2.47
1984	4846	549	42.83	2.17	–	2.9
1994	5716	728	53.48	1.5	2.7	3.0
2007	>7000	810	79.22	1.3	2.5	2.6

Source UN (1973, 1999), Abdulahi (1989), CSA (1985, 2008)

As there was no population census in Ethiopia before 1984 and no dependable estimates made of the number of Lake Tana basin it is difficult to reconstruct the size of the Lake Tana basin population in the past. According to the 1984 Housing census report the total population of the Lake Tana watershed was about 1.8 million, this number increased to 2.5 million in 1994 (CSA 1985, 1995). The 2007 population and housing census reported that the total population in the basin increased to 3, 434,623 (CSA 2008). Furthermore, based on medium variant population projection adopted for the Amhara National Regional State (ANRS), the population of the Lake Tana basin was expected to reach 4.1 million in 2015 (CSA 2008). It appears that the population of the watershed increased by more than 40% in the last two decades. Thus because of such unprecedented increase in the absolute size of the population, the pressure on the available resources such as land forest and water of the watershed increased significantly.

18.4 Change in Population Composition

In demography the terms, population structure and population composition are often used interchangeably to describe the distribution of population characteristics such as age, sex, the size and composition of families and household and religion.

18.4.1 Age Composition

Age data are useful for demographic analysis and for different types of socio-economic development planning. It they are determined by the effects of past fertility, mortality and migration as well as potential fertility that is the capacity of women to give birth. Changes in the age structure the proportion of the population which may be grouped as youth (0–14 years old), adults (15–64 years) and elderly (65 years and over), and changes in the proportion of the working age population in particular are generally related to levels of fertility and mortality (Clarke 1965). Except where migration causes distortion, the proportion of the adult population does not vary from region to region within a country. The main regional differences are the proportions of children and old people. Typically in developing countries where fertility is high, over 40% of the population is under the age of 15. This is true in Ethiopia. The distribution of the national population by age group shows that, the proportion of young population under age 15 has declined from 45.4% to in 1994 to 43.8% in 2007. At the same time, the proportion of the working age population 15–64 years increased from 49.3% in 1994 to 51.6% in 2007. The proportion of population aged 64 and over was 5.3% in 1994 and declined 4.6% in 2007 census. The age composition of the population in the Lake Tana basin is a direct reflection of the national population distribution. Accordingly, the 1994 population and housing census report indicated that the population of the Lake Tana basin was dominated by the working age population at 50.3% followed by the age group between 0 and 14 years old which accounts for about 44.9%. The elderly section of the population accounts for less than 4.8%. The proportion is essentially the same in the 2007 report with the young age group (0–14 years) decreasing very slightly to 44.6% to the proportion of the working age group remained constant at 50.3% and the proportion of the elderly people rose slightly to 5.1%. The highest proportion of the very young age group between 0–14 points to the huge potential down the road, or even an explosive growth if unchecked, due to the built in momentum. The impulse behind it is evidenced by the very high percentage of women in the various reproductive age groups, or soon to be in reproductive age groups. The lowest, 5.1% of the population in the 64+ age category in the watershed is reflecting the lowest life expectancy rate, the very young age structure as well as high fertility and mortality in the past, which know is keeping the age pyramid very wide at the bottom and tapering quickly toward advanced age.

18.4.2 Age and Potential Fertility

The reproductive potential and hence growth of population significantly depends on the proportion of women in their reproductive age, between 15 and 49. Based on the 2007 population and Housing Census of the Lake Tana watershed the reproductive of women in their reproductive age range accounted for 49% of the female population. This is higher than the national average,which was 46% in 1994 (CSA 2008). More over the proportion of women in the age group 20–29 years old, which is presumed to be the most fertile, accounts for about 30% of women in their reproductive ages (CSA 2008).

18.4.3 Age and Dependency Ratio

The patterns observed in the age structures above are also reflected in the dependency ratio both at the national level and in the watershed. At the national level the total dependency burden, defined as the ratio of the total number of children and aged people divided by the working population, decreased from 112.3% in 1984 to 94.6% in 1994 and 88% in 2007. The youth dependency ratio declined from about 98.9% in 1984 to 82.2 in 2007, the old age dependency ratio dropped from 13.4% in 1984 to 5.6% in 2007. However for the Lake Tana basin the total dependency and youth ratio, were about 107 and 95.8% in 1984, and declined to 84.2 and 80% in 2007 respectively. Therefore the dependency burden in general or categorically appears lower in the watershed than in the national level. The youth age dependency of the Lake Tana basin is much lower than the national level by about 10%.

This may be partially attributed to the role of rural out migration, which drains away a large number of both the youth population and also working age population from the basin.

18.5 Sex Composition

The sex or masculinity ratio (number of males per female) changes over space and time depending up on population dynamics. In most cases the sex ratio is below 100, sometimes it may deviate from 100 because of the preponderance of male births, higher mortality of males and migration Plane and Rogerson (1994). In Ethiopia the sex ratio at the national level is higher than in regions. According to CSA reports, based on the 1984 and 1994 population and Housing census it was 99.4% in 1984 and increased to 101.3% in 1994. This figure declined to 100.1 in the 2007 census. In the Amhara region and Lake Tana basin taken together, the sex ratio was 94.1% in 1984 and 93% in 1994. The 2007 census also reported 93% which is almost similar to the 1994 data (Table 18.2).

18.5.1 Household Size

Household size also shows both temporal and spatial variations. The average household size of rural Ethiopia was 4.5 people in 1984 (CSA 1985). By 1994 it had increased to 4.8 people to the rural population (CSA 1995). In the Lake Tana basin a figure of 4.3 was reported in 1984. However, based on the results of the 1994 census it increased to 4.6 people and 5 people in the 2007 census. This increment of house hold size has its own implication in demand for extra food and other necessities.

Table 18.2 Population age-sex ratio structures of Ethiopia and Lake Tana basin

Age group	Ethiopia						Tana basin					
	1994			2007			1994			2007		
	M	F	T	M	F	T	M	F	T	M	F	T
0–14	23.3	22.1	45.4	22.2	21.6	43.8	22.6	22.3	44.9	22.5	22.1	44.6
15–64	24.1	25.2	49.3	26.2	25.4	51.6	23.1	27.2	50.3	26.1	24.2	50.3
64+	2.9	2.4	5.3	2.9	1.7	4.6	2.8	2.0	4.8	3.0	2.1	5.1
Total	50.3	49.7	100	51.3	48.7	100	48.5	51.5	100	51.6	48.4	100

Source CSA (1995, 2008)

18.6 Spatial Distribution of Population

The population is most commonly shown using choropleth maps and dot maps which can give an absolute population size and visual impression of density. Different measures of population density are used to generalize and assist the analysis of the diversity of the distribution of population over space. Population density relates the number of people to the space inhabited by them (Clarke 1965). Moreover, various population density measures may be used to indicate population pressure on land resources at various levels. Hence, both the arithmetic population density and the agronomic density measures have been used to show the variation in population distribution and thus are indices of population pressure.

The population pressure in Ethiopia and the Lake Tana basin, defined in terms of population density, has shown in significant increase. The density of population in Ethiopia was 45.5 people per square kilometer in 1994 and 58 people per square kilometer in 2007. According to the 1994 and 2007 data the population density in the Lake Tana basin is 166 people/km^2 and 228 people/km^2, respectively. Consequently, the population density and hence the burden on the land resources of the region has increased by 4.7 people per annum between 1994 and 2007. Therefore it appears that population pressure in the watershed has increased more than 3 times that of the national average.

18.7 Conclusion

In this chapter a modest attempt has made to explore the demographic profile of the Lake Tana basin. As a result it is possible to understand that the demographic profile of the basin is characterized by high population growth, the concentration of younger age group with almost equal number of male and female group as well as high proportion of women with active reproductive age ranging from 15 to 49 years old. All the above mentioned demographic characteristics clearly indicate that the watershed will experience rapid population growth that can produce huge number of people in the future.

18.8 Recommendation

As indicated in the analysis the demographic profile of the Lake Tana basin clearly indicated that the number of population will put tremendous strain on the resource of the watershed in the future. In order to make a positive balance between the available resource and the number of people in the watershed, the following points should be take into consideration by the concerned organizations.

- Effective strategies to provide and promote effective family planning service should be adopted decreasing the population growth rate.
- The government should adopt natural resources and population policies that take into consideration population growth, demographic pattern and access to and availability of resources.
- The local administration bodies should commit to increase the decision making role of local groups and communities in the design and implementation of population assistance project. This may be aided by increasing representation of women in the highest administration levels of the government.
- Advances in demographic modeling are needed to develop a new population/household model with moderate data requirements, manageable complexity, and explicit representation of demographic events, and output that includes sufficient information for population-environment studies.

References

Abdulahi H (1989) Fertility level differentials in Ethiopia: with reference to Metu, Alemaya. University of Dar Es Salaam, Addis Ababa

Ashenafi T, Kidus K (1994) Aspects of Omotic tonogenesis: Shinasha. J Ethiop Stud 27(2):1–19

Bahiru Z (1991) History of modern Ethiopia 1855–1974. Addis Ababa University Press, Addis Ababa

Clarke JI (1965) Population geography. Pigmon Press, Oxford

CSA (1985) The 1984 population and housing census of Ethiopia. Census Supplement, Addis Ababa

CSA (1987) Population and housing census, analytical report on Addis Ababa. Central Statistical Authority, Addis Ababa

CSA (1995) The 1994 population and housing census of Ethiopia for Amhara Region. Central Statistics Authority, Addis Ababa

CSA (1998) The 1994 population and housing census of Ethiopia results for SNNP Region. Central Statistical Authority, Addis Ababa

CSA (2008) The 2007 population and housing census of Ethiopia for Amhara Region. Central Statistics Authority, Addis Ababa

CSO (1964) Statistical abstract of Ethiopia. Central statistical Office, Addis Ababa

Dombrowski JC (1971) Excavation in Ethiopia: Lalibela and Natchabet caves, Begemidir provience. Unpublished PhD desertation, Boston University, Boston; Dombrowski JC (1981) Ethiopia: population, resource and economy. Progress Publisher, Moscow

Ehret C (1979) On the antiquity of agriculture in Ethiopia. J Afr Hist 20:161–177

Ehret C (2002) African civilization to 1800. Virginia University Press, Charlottesville

Erlich H (1994) Ethiopia and the Middle East. Lynne Rienner Publisher, London

Galperits G (1981) Ethiopia: population, resources and economy. Progress Publisher, Moscow

Henze PB (2000) Layers of time a history of Ethiopia. Hurstand Company, London

Hudson G (2000) Ethiopian semitic overview. J Ethiop Stud 33(2):75–86

Huntingford GW (1989) The historical geography of Ethiopia from the first century AD to 1704. Oxford University Press, Oxford

Kaplan S (1992) The Beta Israel (Felasha) in Ethiopia: from the earliest times to the twentieth century. New York University Press, New York

Levine ND (1974) Greater Ethiopia the evolution of multiethnic society. Chicago University Press, Chicago
Marcus H (1994) A history of Ethiopia. University of California Press, Berkley
Messing SD (1957) The highland plateau Amhara of Ethiopia. Ph.D. dissertation, University of Pennsylvania, Michigan
Phillipson WD (1993) The antiquity of cultivation and herding in Ethiopia. In: Shaw T et al (eds) The archaeology of Africa food, metals and towns. Rutledge, London, pp 344–357
Plane AD, Rogerson R (1994) The geographical analysis of population. Wiley, New York
Quirin J (1998) Caste and class in historical north-west Ethiopia: the Beta Israel (Falasha) and Kemant 1300–1900. J Afr Hist 39(22):195–220
Simoons JF (1958) The agricultural implements and cutting tools of Begemider and Seymen, Ethiopia. Southwestern J Anthropol 14(4):385–406
Taddesse T (1972) Church and state. Clarendon Press, Oxford
Taddess T (1988) Process of ethnic interaction and integration in Ethiopian history: the case of the Agew. J Afr Stud 129:5–18
Taddesse T (1994) Ethiopia in miniature: the peopling of Gojjam. In: Marcus GH (ed) The 12th international conference of Ethiopian studies, vol 1, pp 951–62
Tosco M (2000) Cushitic overview. J Ethiop Stud 33(2):87–121
Ullendroff E (1973) The Ethiopians an introduction to country and people. Oxford University Press, London
UN (1973) The determinants and consequences of population trends. United Nation, New York
UN (1999) World population data sheet. UN Population Reference Bureau, Washington, DC
World Bank (2015) World development report 2015. Washington, D.C: The World Bank and Oxford University Press

Chapter 19
Gender and Rural Livelihood in the Lake Tana Basin

Sewmehon Demissie and Azanaw Abebe

Abstract This chapter deals with gender and rural livelihood in the Lake Tana Basin, which lies within the Lake Tana watershed of Amhara region in the North West part of Ethiopia. It includes areas around the lake with strong ecological, socioeconomic and cultural linkages to the lake itself. In the basin, mainly integrated crop production and livestock rising and fishery activities are undertaken. As it is one of the major food producing areas of Ethiopia, it has critical regional and national significance. In the basin, women and men have different access to, and control over, productive assets. Economic capacities and incentives are also strongly gender-differentiated in ways, which affect supply response, intra household resource allocation, labour productivity, and welfare. Women are still working in subsistence agriculture where they are in-charge of food production, processing and marketing. Moreover, all domestic and reproductive roles like taking care of the health and education of children, fetching water, collecting fuel wood, and cooking, are entirely done by women and girls. These conditions have become constraints on women's self-enhancement and productivity. Women have weak access to credit, cash, appropriate technology, and information that are essential to increase their productivity and income. Girls are victims of various harmful traditional practices such as early marriage, female genital mutilation, etc. On the other hand, most young farmers have also limited access to resources like farm land and finance for their livelihoods. Hence development interventions need to consider all these gender gaps.

Keywords Rural Livelihoods · Gender · Lake Tana

S. Demissie (✉)
Amhara Regional Agricultural Research Institute, Bahir Dar, Ethiopia
e-mail: sewmehon2010@yahoo.com

A. Abebe
College of Agriculture and Environmental Sciences, Bahir Dar University, Bahir Dar, Ethiopia
e-mail: azanawabe_2006@yahoo.com

K. Stave et al. (eds.), *Social and Ecological System Dynamics*,
AESS Interdisciplinary Environmental Studies and Sciences Series,
DOI 10.1007/978-3-319-45755-0_19

19.1 Introduction

There is no a clearly defined spatial unit portraying exact boundaries of Lake Tana Basin (ANRSBIUD 2012). The Lake Tana Basin boundary comprises portions of four zones of the Amhara region (Bahir Dar, South Gondar, North Gondar and West Gojjam), portion of 11 Woredas (North Achefer, Bahir Dar Zuria, Bahir Dar Town, Libo Kemkem, Fogera, Dera, Takusa, Gonder Zuria, Dembia, Alefa, and Chilga) and 131 kebeles inside these Woredas.

Lake Tana, located at an elevation of 1875 m.a.s.l, is the source of the Abbay River (the Blue Nile). It is the largest freshwater lake in Ethiopia with an area of around 3500 m^2 and volume of 28,000 million cubic meters with approximately 85 km wide and 65 km long (Ashinie 1998). The Lake Tana Basin is cut off from the lower Blue Nile by the 40 m drop at Tis Issat Falls that drains the Blue Nile along its journey into Sudan, Egypt, and eventually, the Mediterranean Sea. The region lies within the Lake Tana watershed (LTW) and includes those areas around the lake with strong ecological, socioeconomic but also cultural linkages to the lake itself (Friedrich 2012).

The Lake Tana Basin covers an area of 15,123 km^2 and is fed by four perennial rivers (Gilgil Abbay, Megech, Ribb, and Gumera) and about 60 seasonal streams. Lake Tana has a mean depth of about 9 m and maximum depth of about 14 m, and stores nearly 28 billion cubic meter (BCM) of water. It is a well mixed lake with a detention time o f 1.5 years. Its basin has 1600 km^2 of fluvial wetlands that provide natural filter against sediments and pollutants and habitats for fish spawning, birds, wildlife and plants (World Bank 2008).

Lake Tana sub-basin's water, land, livestock, forests, fishery, cultural, and other environmental assets offer considerable social and economic benefits to the livelihoods of its 3 million people scattered in numerous towns and settlements (Kebeles) (World Bank 2008). Socioeconomic data have not been compiled specifically for Lake Tana Basin. Although one data set will not capture the entire basin, by looking at a variety of data from various spatial subsets of the area, a general picture of the gender and livelihood situation of the Lake Tana basin can be cautiously inferred.

19.1.1 Agriculture, Livestock and Fishery

The major crops grown on about half a million hectares of cultivated land (on 450,000 ha rain fed, and 6000 ha floodplain irrigation, and 500 ha small-scale irrigation, with an additional 7000 ha irrigated area being developed at Koga) are cereals (77%), pulses (17%), oilseeds (6%), and vegetables, root crops and fruits (<1%). The sub-basin has nearly 5.9 million cattle for local use and trade. 4–5 large multi-purpose dams (flood control, irrigation, etc.) are planned to be constructed on the Gilgil Abbay, Megech, Ribb, and Gumera rivers upstream of Lake Tana

(World Bank 2008), from which dam construction on Rib and Megech rivers are already undergoing currently.

The livestock population in Lake Tana Basin is composed of cattle, sheep, goat, mule, horse and donkeys) which totaled 1,116,666 heads in (590,326.72TLU), which includes 748,907 cattle, 134,486 sheep, 162,882 goats, 1622 horses, 64,819 donkeys, 1,014, 157 poultry, and 109,146 bee colonies. Dedicated grazing land or this population of livestock is small and estimated at 43,534 ha. Cattle account for 67.1% of the total livestock in the sub-basin, sheep 12%, goat 14.6% and equine species 6.3%. Cattle serve for the production of milk, meat, and draft power for crop production.

The potential fish production of Lake Tana as estimated by the Ministry of Agriculture is about 13,000–16,000 tons per annum. However, the current production is only about 1000–1400 tonnes per year (Tadesse 2008). The Lake Tana fishery created job opportunities for 3514 fishers in Amhara region (ANRSLRDPA 2011).

19.1.2 Main Livelihood/Demographic Issues

Lake Tana basin is known to have high population density (158 per Km^2) (Friedrich 2012) because of the cultural, historical, and socio-economic resources. Lake Tana basin human population is generally homogeneous linguistically and consists of the main ethnic families of Amhara that accounts 91.5% of the total regional population). According to Central Statics Agency 2007 census and Woredas, the area has a total population of 3,103,231 with male 1,563,276 and female 1,539,955. The very big proportion or 75.8% of the population of the area is living in rural areas where as the remaining 24.2% are concentrated in urban and semi urban centers (Mequanent and Sisay 2015).

The average size of land holding in the lake Tana Watershed is 1.25 ha (BoARD 2003). The size of landholding of male-headed and female-headed HHs is 1.30 and 1.13 ha, with the mean range of 0.81–2.05 and 0.63–2.06 ha respectively.

Rain-fed farming is the mainstay of the economy for the mainly rural population in the basin. The Lake Tana basin is one of the major food producing areas of Ethiopia, and hence has critical regional and national significance. Crop production and livestock rising are closely integrated. The main crops are cereals (Eragrostis tef, finger millet, maize, rice, etc.), pulses (faba bean, field pea, chick pea and grass pea), oil crops (nouge—Guizotia abyssinica), Ethiopian mustard (Brassica carinata), lentil, flax and others) and some vegetables. Cereal dominates the crop production system (cultivated on over 70% of the cropland) and there is a notable absence of perennials. The fine grained cereal teff in particular requires repeated cultivation for a fine seedbed preparation and provides little ground cover during the most erosive storms of June, July and early part of August that exacerbates soil erosion and soil structure collapse (ANRSBIUD 2012).

19.2 Rural Livelihoods and Gender Linkages

Person's livelihood is comprised of 'the capabilities, assets (material and social) and activities for a means of living'. A livelihood is sustainable when it can cope with and recover from stresses and shocks, and maintain or enhance its capabilities and assets both now and in the future, while not undermining the natural resource base (Chambers and Gordon 1992, p. 5).

Livelihood is largely affected by gender. Men and women often have different access to assets, resources, and markets. They often have different skills and participation in decision-making. In agriculture, particularly, women's participation in policy-making or governance is low. They have unequal access to power and low opportunities to voice their concerns. Gender asymmetries play out at many levels, including the household, market, community and the state level.

In Lake Tana basin, women and men have different access to, and control over, productive assets. Economic capacities and incentives are also strongly gender-differentiated in ways, which affect supply response, resource allocation within the household, labor productivity, and welfare. In the basin, women are still working in subsistence agriculture where they are in-charge of food production, processing and marketing. They are also responsible for the health and education of children together with all domestic roles. These conditions have become constraints on women's self-enhancement and productivity, which is similar to other regions of the country (IFAD-EPLAUA 2007).

Further (IFAD EPLAUA 2007) report elucidated that there exists gender imbalance. The reproductive activities like fetching water; collecting fuel wood, cooking, and childcare are entirely done by women and girls. Women have weak access to credit, cash, appropriate technology, and information that are essential to increase their productivity and income. Girls are victims of various harmful traditional practices such as early marriage, and female genital mutilation.

19.3 Gender Characteristics of Lake Tana Basin

19.3.1 Female-Headed Households

According to IPMS (2006), households headed by women are common in Ethiopia, accounting for over 20% of all households, and is about 15–20% in Amhara Region. In addition to the constraints facing women in general, in terms of accessing inputs, services and information for example, FHHs face additional constraints which prevent them from reaping the full benefit from the land they cultivate. FHHs are often neglected by development initiatives and are in a weak position economically (for example, if they lose their land following the death of their spouse). They tend to be unable to access agricultural inputs, training and information from the Office of Agriculture and Rural Development (OoARD) and credit, when their access was formerly through their husbands. They are highly dependent on others, particularly male relatives or share cropping partners for labour, skills and inputs, since they are, by convention, unable to plough, sow, harvest and store on their own. Under share cropping arrangements, they tend to end up with less than half of the produce. They usually make decisions to adopt new technologies or practices after consulting male relatives.

However, with the appropriate support, leading FHHs can break the mould, demonstrating their competencies in testing new ways of doing agriculture, supporting and encouraging similar FHHs to do likewise, and challenging the existing extension service delivery mechanisms positively to serve their needs and development purposes. For instance:

> In Fogera, a woman learnt indigenous beekeeping skills from her father as a child and has continued to adapt them through the challenges of time and new technical innovations. At present she uses both the modern and traditional skills of apiculture side by side. She accessed credit from Amhara Credit and Saving Institution (ACSI), which she used as seed capital to engage in farm and nonfarm income generating activities, including trading. This in turn has created the opportunity for her to learn financial management and encouraged her to access modern financial institutions including banks. In Metema, a woman farmer helps FHHs (who are unable to access land after the death of their husbands) and poor households to access plots, agricultural inputs and labour and thereby enables them to develop a productive socio-economic life.

19.3.2 Married Women

It is also important to be clear about the distinction between women in male-headed households and women in female-heading households. While the latter may be poorer, they may enjoy better access to resources than married women. However, married women usually are food secure and may benefit indirectly from development initiatives that reach men. It is often assumed there is a trickle across of ideas, skills, knowledge and at least some share of the benefits arising from productive activities from husbands to wives (IPMS 2006).

19.3.3 Workloads of Women and Men

In most rural communities in Ethiopia, women work from dawn to dusk and, in contrast with men, have little time for leisure or socializing. Women are not only the major source of labour in the agricultural sector, they are also responsible for the vital tasks of caring for children, the sick and the elderly as part of their household responsibilities. Despite their immense contribution to society, women's productive, domestic and community related activities seem to be undervalued, are often misunderstood and are rendered invisible from official discourse and national statistics.

The overall length of the working day for women does not vary much between the wet and dry seasons. They work for between 10 and 12 h/day, half of which is spent on household tasks such as fetching water and firewood, preparing and cooking food, and caring for children. In rain fed farming systems, men's workload is lightest during the dry season because they participate to a very limited extent, usually, in household tasks. In contrast, members of households with access to both rain fed and irrigated lands are busy throughout the year. The busiest time for men with access to irrigated land is usually towards the end of the rain fed season, when they are harvesting, threshing and winnowing their rain fed crops and are simultaneously starting to prepare the land for cultivating irrigated crops (IPMS 2006).

19.3.4 Rural Livelihoods

In addition to working in the home and on the farm, rural women engage in a diverse range of off-farm livelihood activities. These partly reflect the local farming systems and are also influenced by resource endowments and wealth. Women from rich and middle wealth households often trade in agricultural products, whereas poorer women work as casual labourers on farms and in the homes of richer households; they also harvest natural resources for resale (fuel wood, grass, reed plants around the Lake shore especially southern and eastern part of the lake) or engage in low input activities such as cotton spinning or brewing and selling local alcohol.

Men also undertake a wide range of off-farm activities, the nature of which is closely related to wealth. Rich men are often involved with activities requiring capital such as trading in agricultural products, investing in processing equipment or property, or money lending. Poor men typically engage in casual labouring, harvesting and selling natural resources, or migrating temporarily for work locally and far distance to other regions.

19.3.5 Implications of Gendered Rural Livelihoods

As male and female livelihood problems vary, development initiatives should be designed with a gender perspective to ensure they are relevant to their context. For example, women generally are likely to be more responsive to activities that: can take place on a small area of land; can be undertaken close to the home (especially if they are caring for other household members, such as children, the elderly or the sick); do not require many resources, including labour; and do not expose them to too much risk if they fail to do the activity.

19.4 Gender and Agriculture in Lake Tana Basin

19.4.1 Gender and Land Issues

All land in Ethiopia belongs to the State, which does not recognize private ownership of land. Citizens may be granted user rights to land in the form of certificates, and may privately own assets on land such as houses (MoWE 2010). Accordingly, women in Ethiopia are essentially protected by the country's law, not only in political, civil and economic terms, but also with regard to access to natural resources, including land (TSLI-ESA 2013). The Amhara National Regional State (ANRS) government has enacted Proclamation No. 133/2006 on Rural Land Administration and Use, which gives priority in land allocation to women, disabled and orphans. It also provides for the establishment local level land administration and use committee(s) whose membership should be balanced between men and women. Accordingly land registration and certification was used to secure rural men and women with land access and use rights that brought an increase in women self-confidence.

However, women's direct access to, and ownership of, land in the Lake Tana Basin is affected by gender-biased inheritance laws and customary practices. Poor women mainly access and use agricultural land as sharecroppers, laborers, or de facto household heads due to male migration. Lack of land is known to be a major problem for young people, and is perceived to be a major factor in the recent encroachment of cultivation into wetlands, including the lakeshore zone. 13% of farmers rent their land to others, for reasons including lack of oxen for plowing (45%), lack of labor (44%) and seeds (9%). Typically, rent is paid in kind in the form of 1/3 or 1/4 of the crop (MoWE 2010).

Ownership of land or formal land rights is critical for the rural poor since ownership or formal rights provides access to key markets and nonmarket institutions such as community power structures. Land rights are a gender issue: direct access to land and formal rights are critical for female-headed households (headed by widows, abandoned and divorced women, or females becoming heads of household due to migration of the male head) to establish claims over household assets and labor. They often have to rent out their land due to lack of household labor resources to cultivate it. Sometimes this eventually results in permanent loss of the land as dominant males exert control. Lack of land then results in further cultural and economic marginalization.

Cases from the study by RIDP-ESIA, 2012 "A women anonymously named *from Sifatra (Diba Sifatra kebele of Libo Kemkim Woreda in the Lake Tana region) became landless the day when her husband discovered that she was avoiding pregnancy by using contraception (given by the local health post). He divorced her and she became landless. When she tried to join the association of sand extractors "Ye ashoa mahaber", she was refused because she was landless.*" This depicted that marriage and land inheritance practices determine the degree of female dependence, control over production and income, women's managerial authority over land and labor.

19.4.2 Access to Farm and Grazing Land

Although basin-wide data on access to agricultural land is limited, a study by Demissie (2013) regarding one district in the LTB gives a representative picture. This study examined access to farmland and grazing land in Fogera district from year 2008 to 2010 by gender and age group. The study found significant variation in access to farmland (mainly in size and quality) among groups. Young farmers and women headed households are the disadvantaged groups in land size distribution. Mostly Young male farmers would access land largely through getting a portion from their parents. Women headed households, despite joint titling of land in the region, still had access to smaller pieces of land compared to male headed households due to land scarcity nowadays and rural population pressure who badly demand land for their livelihoods. Most of the women farmers have smaller sized farmland (0.25–1 ha) while the better-off and medium farmers own between 1.5 to 3 ha. In addition, the better off farmers possess up to 2 ha of extra rent-in land. Irrigation land users in and those who have rice fields around Fogera area in the basin for example are the advantaged groups with regards to quality. They can produce crop two to three times a year using irrigation and residual moisture, thus they are relatively better to secure both food and animal feed requirements than non-users. Regarding gender variation, few women farmers have also access to irrigated land.

According to Demissie (2013), Regarding the grazing land access, open communal grazing land is accessible to all farmers from their entitlements. But few farmers have private grazing area (28.3% of 60 respondent farmers in Fogera district for example). However, other sources of grazing and browsing like portion of non-arable open and enclosed hills, farm boundaries and crop aftermath (stubble and weeds), gullies, natural waterways, small grazing areas near homesteads, and live fences are also used.

19.4.3 Gender and Agricultural Labor Markets

Gender inequities in the agricultural labor market are pervasive across all regions. Women are engaged in agricultural work in different ways, including as unpaid family workers, those who are paid in kind (exchange labor), self employed workers and wage laborers. Typically rural women in the Lake Tana Basin tend to have less control over their own labor than men, and their role in agriculture is assumed to be contribution along with their main reproductive role at home.

They also have less control over the labor of their household members/relatives, because they are unable to provide reciprocal labor or favors.

Assessing agricultural wage labor, especially by gender is difficult; because, agricultural labor demand is seasonal and for some specific activities, the market is also mostly informal. However, women's time constraint due to triple role remains a key issue in rural livelihoods improvement and development interventions. In the Lake Tana Basin where households are organized along corporate lines as a communal group, women are still responsible for the reproductive work such as taking care of children, collecting water and fuel wood and other domestic activities. This places constraints on how women engage in the labor market and perhaps explains why women in this region have been slow to move out of this sector compared to men. According to the World Bank (2008), this engagement of women in the reproductive activities has reduced women's bargaining power in the market.

Moreover, the actualization of agricultural production has led to a concentration of females in low-paid, insecure, labor-intensive work. The gender wage gap exists in all the sub-regions. For example, the average wage for casual labor for women is 30% lower than men (UNRISD 2005). Seasonal migration for agricultural work had contradictory effects on women. On the one hand, the experience of work itself is not empowering as women are in insecure, low-paid work with no opportunities for alternative employment. The reproductive burden on women, those who migrate with families, does not change (UNRISD 2005).

Labor is an important livelihood asset that determines the capability and interest to engaged in productive activities such as crop cultivation and livestock keeping and hence to improve productivity. It is also a major input in agriculture to benefit from land. When we see gender division of labor, herding cattle is the duty of children (boys and girls from 7 to 12 years old). Children, especially, boys usually do herding at grazing area and watering from rivers. While girls, in addition to their supportive role in feeding and watering livestock around homestead and cleaning the shed, they usually do help mothers in water collection for home uses and other domestic activities. However, availability and accessibility of labor varies seasonally and among households. In the sub basin for instance, mid November to end of December is peak season when children's labor is highly in need. During this time, most of them stop going to school temporarily which affects their educational activity (Demissie 2013).

Women headed households are disadvantaged group in labor accessibility than others as they lack mainly men's labor to do farming unless they have own son or other relatives. Religion of the communities also contributes to labor use differences. For example, in the basin Christian community, farmers have spare labor on religious holidays (about 10 days every month) as they do not work major farm activities (like plowing, weeding, and harvesting). Though this condition may affect the crop production activities, it can be considered as an opportunity to livestock in getting especial attention by men for well feeding, watering and health care. Children can also get free time to study and do their own activity. Women will also have opportunity to get support from their free children labor at home. Both men and women can do social activities during these holidays. However, in the Muslim community, since every day is a working day, this situation is different (Demissie 2013).

19.4.4 Gender and Rural Finance

There is a range of financial service providers focused on Lake Tana areas such as banks, microfinance institutions, cooperatives, etc. They offer a range of services to the poor such as credit, savings, etc. Studies have analyzed the benefits of microfinance programs on women. They have researched the impact on individual and household level well-being and women's empowerment. The studies have also looked at women's groups acting as a collective. The overall impact of credit on well-being has been positive, improving nutritional status, children's schooling, consumption, etc.

However, some studies also demonstrate impoverishment of the households as a result of debt recycling. Time analysis shows that the overall work load of women involved in group-based programs has increased. Women make time for income generation activities and also complete reproductive work.

Debates exist on whether credit is empowering for women. Some studies show that credit empowers women at the individual level increasing their decision making power. However, other studies illustrate that group based credit fails to promote broader empowerment at the political and social levels (Sharma et al. 2007; Kalpana 2008).

Micro finance institutions prefer women as clients as they are reliable in terms of repayment. Though poverty alleviation remains the key focus of these programs they are also seen as a pathway to empower women. How gender equitable the impact of microfinance is depends on the following aspects: female mobility; women's responsibility for subsistence work (which means they may invest in safe products); opportunity for investing in non sex-stereotyped activities; female literacy level and costs related to accessing information (Mayoux 2005).

Financial problem is a major limiting factor, especially for women and young male farmers to acquire farm assets like livestock. Institutions like Farmers' Cooperatives and Amhara Credit and Saving Institution (ACSI) in Fogera woreda (district) for instance, work in credit service to solve financial problems. However, ACSI credit system does not invite the very poor farmers to use its services. A research report from a study conducted by Tamir and Yitayew (2014) in West Amhara Region depicted that in mixed crop-livestock production system, the role of rural credit on livestock production was misunderstood and overstated. Because, mostly borrowers purchased cattle as an input for crop production (draft power) not to add value or produce it. The rural credit market especially for livestock production in Western Amhara region is largely support crop production than livestock, which can help the land less poor (Young and women farmers). Another study conducted in a district in Lake Tana basin also showed that the institution's credit system is group lending that systematically requires collateral (livestock or enough land) from farmers, which they do not have at hand, especially women and young farmers (Demissie 2013). In addition, the loan size and repayment duration do not fully consider the requirements of the livestock production.

19.4.5 Gender, Infrastructure and Water

Land and water rights are closely related. As women are responsible for domestic water collection, access to clean water may reduce their workload, and increase overall well-being. Water management influences the livelihood strategies of women, thus they need user and ownership rights over water as they are producers in the agricultural sector (IFAD 2007). Women spend 1–4 h even more than 6 h depending on the distance of water sources (communal pump, river, communal pond).

Many women are involved in farming as laborers and as *de jure* family heads (due to male migration). According to the World Bank et al. (2008), agricultural water management includes the following: irrigation, drainage, recycled water use, water conservation and watershed management. There is an overwhelming emphasis on the engineering and technological aspects of this sector, as well as on the environmental impacts. Policy making processes related to water bodies and management is a male dominated area, and in many cases has not taken into account women's knowledge, needs and unequal ownership/user rights. The reasons for this oversight are many: lack of staff capacity, absence of gender disaggregated data, and gender biased social and cultural norms (IFAD 2007). Women are rarely seen as farm owners. This perception even affects group-based strategies for managing water resources.

The access and use of production technology by poor women are constrained because they lack literacy skills and money to buy inputs from the markets and due to class (and gender) bias that operates in agricultural extension services. Women do not have the money to buy new seeds, fertilizer or other technological equipments.

19.4.6 Industry

The industrial sector is an important part of the economic infrastructure of both the Amhara region and Lake Tana Basin next to agriculture. This sector is supposed to bring economic growth to the region and beyond by offering investors a globally competitive combination of geographic position, infrastructure, services and labor. The industrial sector is seen not only to offer investment opportunities, but also to play a pivotal role in the support of the economic empowerment, development of small, medium and micro enterprises (SMME's) and job creation. Major industrial composition in Lake Tana Basin include the manufacture of food products and beverages, manufacture of textiles and clothing, manufacture of wood and products of wood, manufacture of, manufacture of rubber and plastic products, tanning and dressing of leather, manufacture of footwear, luggage, handbags, and electronics manufacturing (ANRSBIUD 2012). Among these lists, women are mostly participated in manufacture of food products and beverages, decorative furniture and manufacture of footwear, luggage and handbags.

19.4.7 Education

The two main educational systems that exist in the LTBs are the traditional and modern institutions, having their own ramifications and subsystems. The traditional system consists of various flexible approaches of orthodox Christian teaching and Islam. The major element of this kind of education is learning religious obligations of Christian and Islam religion. This is conducted at religious orientations located in churches and Mosques. The modern education consists of formal approaches of learning. The formal education system in LTBs runs the classical stratification of standardized schooling systems: that is primary, secondary and tertiary. In the urban settings of LTBs a good deal of education infrastructure, particularly schools are found unlike the rural areas (ANRSBIUD 2012).

According to (IFAD-EPLAUA 2007) report daughters above 12 years age most of whom do not attend schools, neighboring females or unrelated women on annual payment basis, normally assist them in this activity.

19.4.8 Health

Health servicing systems and infrastructure The health services in the LTBs are insecure, are dispensed by modern and traditional service provisions. Modern health service and extension is provided by hospitals, health centers', and junior health centers and at health posts. Categories of health facilities and services are operated by health personnel of specialist medical doctors, General Practitioner doctors, nurses and health extension staffs. On the other hand, treatment through traditional methods varies from faith healing and traditional medicine (ANRSBIUD 2012).

Traditional healing systems fall into two categories, namely, traditional medicine, and spiritual healing. The first type of treatments consists of package of knowledge transmitted from parents to children over the generations through oral tradition and by observation of practices. Therapists of this kind include herbalists, bone setters and surgeons. On the other hand, in the modern health systems, the health seeking population in the LTBs area resorts to health facilities, most of which are concentrated in Bahir Dar Town and Woreda town centers, and Kebele health facilities.

Health extension and counseling service This service is a strategy which has been started few years ago by the regional government. The philosophy is preventing the diseases is more effective than curing. For this reason, health extension personnel are trained and assigned to rural areas. These health workers are responsible to provide awareness to the community on basic health, hygiene and sanitation, reproductive health, and counseling services on HIV AIDS and other related issues (ANRSBIUD 2012).

19.4.9 Organizational Settings

The Kebele Organizational framework of the LTBs The capacity and efficiency of the administrative and political structures to undertake development programs, conservation of resources, good fiscal management and service delivery systems at kebele level is based on the kebele administration as a head and other sectoral nominees as a development facilitator (ANRSBIUD 2012).

Hence at kebele level, the main organizational entity include; the Kebele administration which is the superior structure and responsible for all political, economic, social and other development responsibilities. It is a decentralized structure of the government's administrative system which is next to the Woreda/district administration.

19.4.10 Gender and Agricultural Governance

Decentralization is generally considered to be a positive step towards ensuring participation of poor and marginalized groups in decision-making processes at the local level. However, research shows that projects that are administered through local level pubic administration are less likely to address women's needs when compared to projects that are community driven (Baden 2000). This is because women may have greater access to local decision making bodies and the issues they raise would have greater relevance. However, the local level patriarchal structure may be stronger and when combined with customary laws, these may make it harder for women to participate effectively and influence decision-making. Moreover, the structure, availability of resources and the politics of decentralization also affect what women can do at the local level. Women themselves may lack capacity, literacy skills and time. Community driven projects may create scope for inclusion of and participation by women. However, these projects are affected by local customs and class/caste/ethnicity based divisions. They may lack institutional capacity and may be implemented in a top-down manner.

Good governance also includes collective organization by people. Rural women in the basin participate in different types of groups related to agriculture; including self help groups, women associations, voluntary and business groups. There are also natural resource management groups, NGO credit groups, etc. There is variation in how women benefit from participating in these groups and whether they are able to exercise their voice to influence service provider's policies or hold organizations to account in matters that affect their livelihood.

According to Abe (2013), women's participation in decision making in Amhara region from region to district level is only about 16.3%. In the region, men clearly dominated managerial and leadership positions of both private and public institutions, while women occupied lower level positions regardless of capability, willingness, or birthright. A woman tends to be remained subordinate under the care of man all her life, beginning with her father and automatically transferring to her husband once she got married due to influences from custom and tradition.

19.5 Gender Roles and Share of Benefits in Crop and Livestock Enterprises

19.5.1 Gender Division of Labor in Crop Production

The division of tasks between women and men varies according to the crop grown, the farming system, the technology used and the wealth of the household. Though women participate from land preparation to harvesting of crops, they actively do weeding and harvesting. As most of women from female headed household rent out their land due to different cases, women from male headed households participate more in crop production activities than FHHs.

19.5.2 Gender Roles in Livestock Production

Gender is one of the important issues in resource sharing/allocation for performing different livelihood activities in the Lake Tana Basin. Labor and time are the two important human assets that need to be considered while defining the role of gender in meeting targeted household objectives. In livestock keeping, gender refers to men, women and children, whose labor contribution is significant. In a household, all members of the family are involved in different livestock-related activities, and have different role and responsibilities. The activities mainly include collection of water and feed (green grass/weeds, fodder/forages, farm residuals, crop residues and purchased feed), grazing/feeding and watering, shed cleaning and dung management, milk processing, selling and buying live animals/animal products, taking care of the sick, pregnant or other small animals, etc.

The intra-household roles and responsibilities change depending on the household structures. Since households by themselves are variable in their structure, resources (goods, time and responsibilities) allocation, and power distribution in decision making among the members, it is important to consider the intra-household structure and dynamism in order to target for an intervention (Rogers, undated). In the case of livestock productivity improvement interventions, it is also important to see both inter and intra-household characteristics. Time allocation decisions (for each intra-household activity) are affected by the characteristics of the households, the resources available and the constraints they have to face to satisfy household needs, which can be achieved through effective utilization of time (Demissie 2012).

19.5.3 Gender Roles in the Household, Marketing and Sharing the Benefits of Production

Gender roles and responsibilities are important determining factors of vulnerability and coping capacity of the male and female farmers in a household. This is because gender roles are specific for the different livelihood activities though some joint activities are still practiced by most households. According to Rogers (undated), the women's role is vital in farmers' households. In Sub Saharan Africa, women are responsible for 90% of food crop processing, providing household water and fuel wood, 80% of the food storage work and transport from farm to village, for 90% of hoeing and weeding, and 60% of the harvesting and marketing activities. Regarding responsibilities in livestock keeping, women are usually responsible for feeding animals, cleaning barns, milking, processing milk and marketing of livestock products. They also play a substantial role in managing confined animals throughout the year, and are involved in feed and water provision and shed sanitation and management. Children especially girls between the age of 7 and 15 years, are mostly responsible for managing calves, chicken and small ruminants while older boys are responsible for treating sick animals, constructing shelters, cutting grass and herding of cattle and small ruminants (Demissie 2012).

In the highlands of Ethiopia including the Tana Basin, smallholders rear cattle primarily for draft power. Milk production, cash source, manure and fuel are considered as secondary outputs that are mostly controlled and used by women. Cattle and equine are used in smallholder farms for crop cultivation and transportation. Equines are also jointly used by both women and men (Ahmed et al. 2003 in Belete et al. 2010: 5). Poultry keeping in most developing countries is the responsibility of women, which is also true of the central highlands of Ethiopia, where chickens are owned and managed mainly by women. The men also participate in some processes like selling of live chickens. Here it is important to mention that there are cultural and religious factors in some parts of Ethiopia that restrict contacts of women with extension workers. Consequently, women obtain information through their husbands, which impacts the flow of information (Dessie and Ogle 2001).

In the crop-livestock system of the Ethiopian highlands, women are more involved in cattle production than in arable farming. They clean cow sheds, milk cows, look after calves and sick animals, cut the grass and supervise feeding and grazing of cows, make dung cakes, butter and cheese and sell these products once or twice a week. Women decide on the allocation of milk for different uses. Men feed the oxen and take the animals for veterinary treatment when the need arises. Joint decisions by husband and wife are made regarding the purchase and sale of livestock, though men are responsible for taking the animals to the market. Boys and sometimes girls, generally graze the ruminant livestock. The same applies true in the Tana Basin, where children (boys and girls) who have age of 10–12 years are engaged in Whalen (1984 in van Hoeve and van Koppen 2005: 11; Tangka et al. 2000: 21).

In Fogera district in the Lake Tana Basin for instance, livestock management decisions, roles and responsibilities are different for men and women farmers among and within households. Men make important decisions like acquiring animals, herd disposal and control over income (from sale of live animals including sheep and goat) and expenditure. Based on the market condition, they fix selling and/or buying; but women exceptionally manage income from sale of butter, egg and chicken. In both the study sites, women manage shed sanitation and manure disposal, feeding and watering, taking care of newborn, sick animals and other activities at homestead. They also control milk processing and decide to allocate milk and its byproducts for consumption and/or sale. During farming season when there is a greater need for labor, women assist their husbands by keeping animals away from growing and harvested crops (Whalen 1984 in van Hoeve and van Koppen 2005: 11; Tangka et al. 2000: 21). Moreover, they take care of children's tasks during school time. Women also handle weeding and help with harvesting.

19.6 Technology Adoption and Preferences

19.6.1 Gender-Based Access to Technology

Women generally have extremely limited access to technologies and services associated with farming. There are very few items which they use to a greater extent than men (such as the use of local cows, donkeys and kitchen utensils). In contrast, men enjoy the use of a relatively wide range of resources and they control nearly all household resources. In particular, women's use of technologies which would reduce the drudgery of their workloads and possibly release their time for relaxation or more productive tasks has been extremely limited. 'Walking long distances to fetch scarce firewood and water, and cooking with inefficient stoves and crude utensils, are time consuming and strenuous'. This is partly due to the failure of research, policy and extension to acknowledge differing gender needs. Even where technologies have been specifically targeted at women, they tend to run into problems of limited acceptance (such as improved cooking stoves) or appropriation by men (such as water pumps or earnings from livestock fattening) (Seyoum 2000).

19.6.2 Gender-Based Access to Inputs and Services

Access to different inputs and services for different livelihood activities varies among farmers because of their gender or wealth status differences. Financial problems with very poor credit services for women headed and poor young farmer households and shortage of labor for women headed households in the LTB can be mentioned as among the major challenges. Limited awareness or experience and extension service in the basin is also another factor to motivate farmers to participate in productive activities such as livestock keeping and other off farm activities in addition to the crop production. Other social factors like theft and resource computation are also other limitation to women farmers.

19.7 Gender Livelihood Opportunities and Challenges

Government and policy supports regarding gender issues can be taken as important opportunities. After the 1995 constitution, a legal and political right that ensures gender equality is established. The National policy of Ethiopian Women published in 1993 also establish important footsteps to ensure women's Participation in policy and project planning, implementation, and decision making that directly or indirectly benefit them. Gender issues are also considered in every development interventions at different level. There are also implementing structures and experts in different sectors up to the district level in addition to the Ministry of Women, Youth, and Children in the government structures. However, when come to the implementation at the ground, things cannot work as planned and expected due to different technological, technical, institutional, socio-cultural, and economical constraints and challenges.

Hence, as men and women livelihood problems and needs vary; development initiatives should be designed with a gender perspective to ensure they are relevant to their context. For example, women generally are likely to be more responsive to activities that: can take place on a small area of land; can be undertaken close to the home (especially if they are caring for other household members, such as children, the elderly or the sick); do not require many resources, including labor; and do not expose them to too much risk if they fail to do the activity.

In addition, as gender is more dependent and the roles, responsibilities, and needs are contextually subjected to change, it is important to conduct gender analysis every time and revise the gender gap for targeting gender sensitive interventions that can improve rural livelihoods.

References

Abe A (2013) THE lived experience of women's holding leadership positions: the case of women in the Amhara National Regional State sector Bureaus. MSc thesis, Bahir Dar University, Bahir Dar, Ethiopia

Agarwal B (1994) A field of one's own: gender and land rights in South Asia. Cambridge University Press, Cambridge

Ahmed MM, Bezabih E, Jabbar MA, et al (2003) Economic and nutritional impacts of market-oriented dairy production in the Ethiopian Highlands. Socio-economic and policy research working paper 51. International Livestock Research Institute (ILRI), Nairobi, Kenya, 27 pp. In: Anteneh B, Azage T, Fekadu B, Berhanu G (2010) ILRI working paper

ANRSBIUD (2012) Amhara National Regional State Bureau of industry and urban development. Potential assessment, identification of opportunities and designing strategic plan for sustainable development of tourism and transport in Lake Tana and its environs, thematic report (vol IV): Socio-economics of Lake Tana and its Environs, Bahir Dar, Ethiopia

ANRSLRDA (2011) Amhara National Regional State Livestock Resources Development and Promotion Agency annual report of the year 2010–2011, Bahir Dar, Ethiopia

Ashine S (1998) Fisheries of Lake Tana of Ethiopia. A socio-economic study, MSc thesis, University of Hull

Baden S (2000) Gender Governance and the Feminaization of Poverty. In: Women and Political Participation: 21st century challenges, Newyork, UNDP

Belete A, Azage T, Fekadu B, and Berhanu G (2010) Cattle, Meat, and Milk Production and Marketing system and opportunities for Market-orientation in Fogera woreda, Amhara Region, Ethiopia: In IPMS (Improving Productivity and Market System) of Ethiopian Farmers Project Working Paper No. 19 ILRI (International Livestock Research institute), Nirobi, Kenya, 65 pp

Bureau of Agriculture and Rural Development (BoARD) (2003) Rural household socioeconomic baseline survey of 56 Woredas in Amhara region (Phase 1) Vol III main document, Federal Democratic Republic of Ethiopia, Amhara National Regional State, Bahir Dar

Chambers R, Gordon C (1992) Sustainable rural livelihoods: practical concepts for the 21st century. IDS discussion paper 296, IDS, Brighton, UK

Demissie S (2012) Water and rural livelihoods in the crop-livestock system of Amhara Region, Ethiopia: Multiple use system (MUS) approach for water productivity improvement. In: Proceedings of the 2nd national workshop on challenges and opportunities of water resources management in Tana Basin, Upper Blue Nile Basin, Ethiopia, vol II, 26–27 Mar 2012. Blue Nile Water Institute, Bahir Dar University (BNWI-BDU), Bahir Dar, Ethiopia

Demissie S (2013) Gendered livelihood implications of resource access for livestock productivity improvement in the mixed crop-livestock system of central highlands of Ethiopia. Ethiop J Soc Sci Humanit (EJOSSAH) IX(2)

Dessie T, Ogle B (2001) Village poultry production system in the central highlands of Ethiopia. Trop Anim Health Prod 33(6):521–537

Esther v. Hoeve, Barbara v. Koppen (2005) Beyond Fetching water for Livestock. A gendered livestock framework to assess livestock water productivity. ILRI, IWMI, Addis Ababa, Ethiopia

Friedrich ZH (2012) Feasibility study for a Lake Tana biosphere reserve, Ethiopia

Hobley M, Jay A, Mussa M (2004) Understanding vulnerability: opening political, social and economic space towards greater equality, report for DFID Ethiopia, Addis Ababa

IFAD (2007) Gender and water: securing water for improving rural livelihood. IFAD, Rome

IFAD-EPLAUA (2007) Amhara National Regional State community-based integrated natural resources management in Lake Tana Watershed, baseline information on water resource, watershed, water harvesting and land use, Bahir Dar, Ethiopia

IPMS (2006) Gender analysis: an overview of gender issues in the agricultural sector of Ethiopia. Part I

Jackson C (1997) Actor orientation and gender relations in a participatory project interface. In: Goetz AM (ed) Getting institutions right for women in development. Zed Books, New York

Kalpana K (2008) The vulnerability of self help: women and microfinance in South India. IDS working paper 303, IDS, Brighton

Mayoux L (2005) Women's empowerment through sustainable microcredit. Discussion draft. http://genfinance.net

Mequanent D, Sisay A (2015) Wetlands potential, current situation and its threats in Tana Sub-Basin, Ethiopia. World J Environ Agric Sci 1(1):1–14

MoWE (2010) Environmental and social impact assessment of about 20,000 ha irrigation and drainage schemes at Megech pump (SERABA), Ribb and Anger dam. Environmental and social impact assessment of the Ribb irrigation and drainage project

Rogers BL (Undated) The internal dynamics of households: a critical factor in development policy. Tufts University School of Nutrition, Medford, Massachusetts, USA

Seyoum S (2000) Gender issues in food security in Ethiopia. In: Panos Ethiopia, No 4, pp 37–53

Sharma J, Parthasharathy SK, Dewevedi A (2007) Examining empowerment, poverty alleviation, education within self help groups. Niranatar, Delhi

Tadesse A (2008) Application of the WEAP model to assess the water resource implications of planned development in the Lake Tana Sub basin, Ethiopia. Master's thesis Presented to Addis Ababa University

Tamir S, Yitayew A (2014) Credit supply and demands for livestock production and fattening practices in western Amhara region, Ethiopia (unpublished)

Tangka FK, Jabbar MA, Shapiro BI (2000) Gender roles and child nutrition in livestock production systems in developing countries: a critical review. Socioeconomics and policy research working paper 27. International Livestock Research Institute (ILRI), Nairobi, Kenya. 64 pp

TSLI-ESA (2013) Land and natural resources learning initiative for Eastern and Southern Africa (TSLI-ESA). Case study report: strengthening women's access to land in Ethiopia. Addis Ababa, Ethiopia

UNDP (2010) Rural livelihoods and gender. Asia-Pacific Human Development Report, background papers series

UNRISD (2005) Gender equality: striving for justice in an unequal world. A policy report on gender and development 10 year after Beijing, UNRISD, Geneva

Whalen IT (1984) in van Koppen (2005) ILCA's Ethiopian highlands programme: problems and perspectives in expanding the participation of women. Paper prepared for IITA/ILCA/Ford Foundation workshop on women in Agriculture in West Africa, Ibadan, Nigeria, 7–9 May 1984, 24 pp

World Bank (2008) Project appraisal document for a Tana & Beles integrated water resources development project, Ethiopia

World Bank, FAO, IFAD (2008) Gender in agriculture source book. World Bank, Washington DC

Chapter 20
Examining the Characteristics of Stakeholders in Lake Tana Sub-basin Resource Use, Management and Governance

Dessalegn M. Ketema, Nicholas Chisholm and Patrick Enright

Abstract The concept of broader participation of stakeholders and more inclusive decision making has been increasingly recognized in many countries both at national and international level. Effective stakeholder engagement in a complex resource system is fundamental to ensure the successful implementation of policies at the ground level. Particularly when the multi-purpose values of the whole Lake Tana ecosystem are considered, there are multiple stakeholders with diverse interests in and power over natural resource use, management and governance. Therefore, a clear understanding of the potential roles and contributions of the many different stakeholders of the Lake Tana social-ecological system is a fundamental prerequisite for successful participatory natural resource management and governance. Stakeholder analysis is a basic tool for achieving this understanding. The overall objective of this chapter is therefore, to provide a preliminary description of stakeholders' interest in and influence over natural resource use, management, governance and policy process in the Lake Tana sub-basin. It depicts the stakeholders' landscape based on focus group discussions, key informant interviews and stakeholder identification and analysis workshop which was carried out by a multidisciplinary team of experts from different stakeholder groups (governmental, research, academic and NGOs). The results can help inform and assist policy makers in the development of sustainable natural resource management policy, taking into account the interest and influence of a wide range of stakeholders. The chapter begins with a brief introduction of the Lake Tana context. It then presents the concept of stakeholder and stakeholder engagement followed by the rationale for stakeholder identification and analysis.

D.M. Ketema (✉)
Department of Rural Development and Agricultural Extension, College of Agriculture and Environmental Sciences, Bahir Dar University, Bahir Dar, Ethiopia
e-mail: dessmoll@gmail.com

N. Chisholm · P. Enright
Department of Food Business and Development, University College Cork, Cork, Ireland

K. Stave et al. (eds.), *Social and Ecological System Dynamics*,
AESS Interdisciplinary Environmental Studies and Sciences Series,
DOI 10.1007/978-3-319-45755-0_20

Keywords Stakeholder · Engagement · Participation · Common pool resource · Lake Tana · Ethiopia

20.1 Introduction

20.1.1 Background

The Lake Tana social-ecological system, located in the northwest highlands of Ethiopia, is the source of livelihood, economic and socio-cultural resources on which a wide variety of groups depend. It is also one of 250 lakes identified by Lake Net[1] as having globally significant biodiversity (Barker 2004). However, due to natural and human-induced calamities, the Lake ecosystem is under severe threat. A recent decadal trend analysis of Lake Tana over four decades (from 1968 to 2007) confirms that the water level was increasing for the first three decades and declining for the last decade. The lowest depth was recorded in 2002/2003 and after that time onwards the level of Lake Tana was not able to return to its original level. In the last 35 years, more than 6.2% of the lake's area was converted to other land covers like crop land (Minale and Rao 2011).

According to Vijverberg et al. (2009) despite the limited direct human influences on Lake Tana, common pool resources including fish, grazing lands, wetlands and surrounding catchment land have already been seriously damaged by human activities. Most of the original forest in the upper stream of the watershed has disappeared. A number of recent studies also suggest that future development will exacerbate pressure on the lake ecosystem. There will be a significant decrease in water levels and massive wetland degradation in the Lake Tana sub-basin in response to the planned water withdrawal and other development interventions (Dargahi and Setegn 2011; McCartney et al. 2010; Setegn 2010).

The main characteristic of lake ecosystem landscapes is that they contain resource systems with multiple-use value which are used by multiple actors. The relationship between these dynamic and multiple use landscapes and the multiple users can be analyzed and understood through how different stakeholders satisfy their needs and local people derive their livelihoods by having legitimate control over resources (Leach et al. 1999). Rural households in particular are highly dependent on common-pool resources (CPRs) for their livelihood. These resources are usually characterized by (a) multiple use values, such as consumptive, recreational, environmental and spiritual (Baland and Platteau 1999) and (b) multiple users with different powers and interests. When resource units are highly valued and many stakeholders benefit from harvesting them for consumption, exchange,

[1]LakeNet is U.S.-based nonprofit organization dedicated to bringing together people and solutions to protect and restore the health of the world's lakes.

or as a factor in a production process, the harvests made by one individual or actor are likely to create negative externalities for others (Ostrom 2008).

Swallow et al. (1997) note that most analyses of the efficiency of natural resource management and governance have failed to recognize that resources often have multiple uses and that there tend to be sub-groups of users who are characterized by their use patterns. For example the same water source can be used for irrigating, hydropower, fishing, navigating, washing, watering animals, or other enterprises by multiple stakeholders. They explain that some resource uses are complementary, others are competitive, most are somewhere between. Some groups of resource users are mutually exclusive, others are overlapping. However issues of accommodating multiple uses and multiple-users are especially critical in the case of the commons.

Steins and Edwards (1999) define complex, multiple-use CPRs as resources that are used for different types of extractive and non-extractive purposes by different stakeholder groups (multiple users) often with divergent interests and are managed under a mixture of property rights regimes. In such a scenario, collective action becomes increasingly complicated. Different resource uses will be regulated and governed through different decision-making arrangements by different user groups. Therefore when commons evolve into multiple-use resources, the institutional framework within which collective resource use takes place has to be re-negotiated among resource users and other stakeholders to avoid externalities associated with increased access of new users to the resource system, such as overexploitation, alienation of traditional users, and inter/intra-user group conflicts. Therefore, the analysis in this chapter looks at stakeholders in terms of their stakes and power levels, participation, legitimacy, and organizational capacity, pattern of conflicting interaction in the process of resource use, management and governance of the Lake Tana social-ecological system.

20.1.2 The Concept of Stakeholder and Stakeholder Engagement

There is a difference of opinion over who or what exactly stakeholders are. This is reflected in the term "stakeholder" itself, apparently first used in 1708, in the context of a bet or a deposit implying "a person who holds the stake or stakes in a bet" (Ramirez 1999). The word now refers to anyone significantly affecting or affected by someone else's decision-making activity. Many recent definitions of stakeholders build on Freeman's (1984) seminal work on stakeholder theory that distinguished between those who affect or are affected by a decision or action [sometimes referred to as active and passive stakeholders in the natural resources stakeholder literature (Grimble and Wellard 1997)].

20.1.2.1 Defining Stakeholder and Stakeholder Analysis

What is a Stakeholder?

- Freeman (1984) defines a 'stakeholder' as "any group or individual who can affect, or is affected by, the achievement of corporation's purpose."
- '*Stakeholders*' are people, groups, or institutions which are likely to be affected by a proposed intervention (either negatively or positively), or those which can affect the outcome of the intervention".
- Other terms sometimes used in a similar way to stakeholders are "actors" and "interest groups". The word "actors" stresses that stakeholders are active and interact with each other. The use of the words "interest groups" indicates that people can be grouped according to a common interest.

When talking about stakeholders, it is important to realize that:

- *Stakeholders are not only local people*. They include governments and their agencies, as well as people, organizations, institutions and markets, which are not necessarily located close to the natural resource that is being managed.
- *Stakeholders are not only organizations and formal groups*. They include individuals, communities and informal networks.
- *Stakeholders are not only the users of natural resources*. They include people and institutions that impact directly but also indirectly on the resources even without using them, and include people who may not even be aware that they have a stake in the management of these resources.
- *Stakeholders change over time*. New stakeholders can enter a resource management system, while others may lose their role or interest. For example, changes in the local economy and society, or in the status of resources, will inevitably provoke changes in the ways people use, and relate to, natural resources. Stakeholder identification is therefore an ongoing process, and it is one that must incorporate a historical dimension.

20.1.2.2 At What Institutional Level Do Stakeholders Exist?

Stakeholders can be at any level or position in society, from the international to the national, regional, household or intra-household level. Stakeholders include all those who affect and are affected by policies, decisions or actions within a particular system. For example Table 20.1 shows the stakeholders and their interests in a typical forest by institutional level.

Table 20.1 Institutional level at which stakeholders exists

Institutional level	Actors	Interests
Global/international	International agencies, donors, environmental lobbies	Biodiversity, conservation, climate regulation
National	National governments, NGOs	Timber extraction, tourism development
Regional	Forest departments, regional authorities	Forest productivity, soil conservation
Local off-site	Local communities, local authorities, logging companies	Access to timber supply, conflict avoidance
Local on-site	Forest dwellers, livestock keepers, agriculturalists, women firewood collectors	Land for cultivation and grazing, cultural sites

Source Grimble and Wellard (1997)

Table 20.2 Summary of stakeholder Classification and categorization

Eden and Ackermann (1998), De Lopez (2001)	Arthur and Claudia (2007)	Hughes et al. (2008)
Key players	Veto players	Core
Context setters	Key stakeholders	Involved
Subjects	Primary stakeholders	Supportive
Crowd	Secondary stakeholders	Peripheral

20.1.2.3 Types of Stakeholders

Depending on their objectives, different scholars and organizations uses a variety of ways to classify stakeholders and stakeholder analysis. Table 20.2 shows four frameworks for characterizing stakeholders.

The literature on stakeholder analysis frequently makes the distinction between primary and secondary stakeholders. This difference comes from the field of project planning and management, where it can indeed be helpful to distinguish between those who will be directly affected by a project, and those who are only peripheral to that intervention. In the practice of natural resource management and governance, however, this distinction is not particularly useful. Relationships between and among people and natural resources tend to be changing and complex, so fitting stakeholders in one of these two categories runs the risk of marginalizing some of the stakeholders and could exclude less obvious, powerless and voiceless groups.

20.1.2.4 What is Stakeholder Analysis?

Stakeholder analysis means many things to different people. Various methods and approaches have been developed in different fields for different purposes, leading to confusion over the concept and practice of stakeholder analysis (Reed et al. 2009). The origins of stakeholder analysis, however, belong to the history of business and managerial science development and policy literature in the 1930s and stakeholder analysis has since developed into a systematic tool with clearly defined steps and purposes for examining the organizational environment. The term stakeholder analysis was first used in management science for identifying and addressing the interest of different stakeholders in business. Stakeholder analysis is a way of understanding a system through its stakeholders. It looks at their interest, objectives, power and relationships. Nowadays, stakeholder analysis is frequently used for: policy formulation, project formulation, implementation and evaluation, understanding and analyzing complex situations in natural resource management.

- "Stakeholder analysis" can be defined as a methodology for gaining an understanding of a system, and for assessing the impact of changes to that system, by means of identifying the key stakeholders and assessing their respective interests (Grimble 1998).
- "Stakeholder analysis" refers to a range of tools for the identification and description of stakeholders on the bases of their attributes, interrelationships, and interests related to a given issue or resource (Ramirez 1999).
- According to Reed et al. (2009), "stakeholder analysis" is defined as a process that: (i) defines aspects of a social and natural phenomenon affected by a decision or action; (ii) identifies individuals, groups and organizations who are affected by or can affect those parts of the phenomenon (this may include nonhuman and non-living entities and future generations); and (iii) prioritizes these individuals and groups for involvement in the decision-making process.

20.1.2.5 Normative Versus Instrumental Approaches/Theories to Stakeholder Analysis

There have been numerous attempts to classify the different approaches to stakeholder analysis (e.g. Donaldson and Preston 1995; Friedman and Miles 2006 all cited in Reed et al. 2009) Perhaps the most significant difference is between normative and instrumental approaches. A third approach, descriptive stakeholder analysis, is rarely conducted for its own sake, since it has no purpose beyond describing the relationship between a particular phenomenon and its stakeholders (Donaldson and Preston 1995 cited in Reed et al. 2009). However, since normative and instrumental analyses require an understanding of the current state of affairs, descriptive analyses are in effect a necessary precursor to normative and instrumental analyses.

Normative approaches have been advocated increasingly as stakeholder analysis has been adopted in policy, development and natural resource management circles, emphasizing the legitimacy of stakeholder involvement and empowerment in decision-making processes. In this context, stakeholder analysis has been used to legitimize the decisions that are made, through the involvement of key and/or representative figures. Others have suggested that normative stakeholder theory needs to identify who decision-makers are morally responsible to in their legal and institutional context. Instrumental stakeholder research is more pragmatic, and largely devoted to understanding how organizations, projects and policy-makers can identify, explain, and manage the behavior of stakeholders to achieve desired outcomes.

Finally, it should be noted that normative justifications for stakeholder analysis may lead to instrumental outcomes. The normative basis suggests that stakeholders should be involved in decision-making processes and thus feel some level of ownership of these processes. By doing this, stakeholder analysis may serve instrumental ends if it leads to the transformation of relationships and the development of trust and understanding between participants. Although this may not necessarily lead to changes in attitudes and behavior, it may enable diverse groups of potentially conflicting stakeholders to appreciate the legitimacy of each other's views and see new ways of working together (Mathews 1994; Forester 1999 cited in Reed et al. 2009).

20.1.3 The Rationale for Stakeholder Identification and Analysis

Natural resource management and governance is typically highly complex and characterized by (e.g. Bressers and Kuks 2003) multiple levels of policy implementation; multi-actor character of policy implementation; multiple perceptions of the problem and the objectives of policy implementation; multiple strategies and policy instruments for policy implementation; and a complex multi-resourced and multi-organizational basis for implementation of policy. In particular, common pool resource (CPR) systems are rival and non-excludable. In other words, these resources are systems where it is difficult to exclude users through physical or institutional barriers and where the use of the resource by one person or group leaves less for another (Ostrom et al. 1994). To effectively manage and govern these situations, there have to be innovative ways of dealing with those multiple actors, multiple levels and multiple objectives in a sustainable manner.

In complex scenarios, such as the Lake Tana social-ecological system, people-centered development approaches are a prerequisite which often relies on participatory inquiry and decision-making processes. The broad aim of participatory development is "to increase the involvement of socially and economically marginalized peoples in decision-making over their own lives" (Guijt and Shah 1998).

However, many projects and policy interventions have not met their stated objectives because of non-participatory, non-co-operation or even opposition from key stakeholders, who believed they would be adversely affected by change. Moreover many interventions that have been perceived to be successful by their designers, have in fact achieved their success only at the expense of certain stakeholders- often local resource-poor people (Grimble 1998). The development in the 1990s of stakeholder analysis in natural resource management has largely stemmed from those concerns.

Development interventions are joint ventures which are negotiated, planned and implemented by many different stakeholders. Therefore, the importance of stakeholder analysis is widely recognized as a necessary means for gaining insight into the complex systemic interactions between natural processes, management policies, and local people depending on the resource. The growing popularity of stakeholder analysis in natural resource management partly reflects an increasing recognition of the extent to which stakeholders can and/or should influence environmental decision-making processes (Burroughs 1999; Brugha and Varvasovszky 2000; Duram and Brown 1999; Selin et al. 2000). Stakeholder analysis can be used to understand environmental systems by defining the aspects of the system under study; identifying who has a stake in those aspects of the system; and prioritizing stakeholders for involvement in decisions about those aspects of the system (Grimble and Wellard 1997).

The general purpose for stakeholder identification and analysis may be seen as providing a methodology for better understanding environmental and development problems and interactions of a given system through comparative analysis of the different perspectives and sets of interest of stakeholders at various levels (Grimble and Wellard 1997). The basic purpose behind stakeholder identification and analysis was primarily to generate knowledge about the relevant stakeholders so as to understand their behavior, intentions, interrelations, agendas, interests and the influence or resources they have brought or could bring to bear on CPR use and rule making processes, to identify different categories of stakeholder and anticipate the kinds of influence they could exert on the management and governance of CPRs, potential areas of synergy, collaboration, potential conflicts of interest among stakeholders and between stakeholder groups.

As noted by different scholars (Grimble and Wellard 1997; Engel 1997; Röling and Wagemakers 1998 cited in Ramirez 1999) the reasons for carrying out stakeholder analysis are; (i) empirically to discover existing patterns of interaction; (ii) analytically to improve interventions; (iii) as a management tool in policy making; and (iv) as a tool to predict conflict. For this study the primary focus is on the first and forth reasons.

A two-day stakeholder identification and analysis workshop was conducted from February 26–27, 2011 at *Wereta* town (*Fogera* district) in the Lake Tana basin. The purpose of the workshop was to identify stakeholders in the Lake Tana Basin social-ecological system and characterize their interests. Since resources, time, and finance for this research were limited, participants from different stakeholder groups were prioritized. Therefore, participants (who may have important knowledge about or perspective on the issues) were selected purposively from different organizations representing different stakeholder groups (governmental, non-governmental, academic and research) and various disciplines (for instance ecologists, environmentalists, hydrologists, fishery experts, animal scientists and economists). In addition, representatives from NGOs who are working in relation to natural resource management (for example, Ethiopian Wetlands and Natural Resource Association—EWNRA) and experts from districts were involved. However, it was too difficult to bring local people (the majority of whom are illiterate) together with the panel of experts in the workshop. Therefore, in order to capture the views of resource users and their interaction with other relevant stakeholders at grassroots level focus group discussions with resource users at different districts were held.

20.2 Stakeholders of the Lake Tana Social-Ecological System

20.2.1 Introduction

Although stakeholder-related considerations are receiving increasing attention within development scenarios and in particular, policy making processes, the actual management of stakeholders' identification and phase-specific involvement has not been covered sufficiently (Gerald 2008). Therefore, it is imperative to understand how different stakeholders interact to solve CPR problems and to reduce or eliminate externalities, who are the key players, what are their interests, power, agendas, characteristics, circumstances, the roles, existing pattern of interactions, involvement and responsibilities of stakeholders in the process of resource use, management, governance and policy making.

Theoretically the interactive perspective on governance proposes that societies are made up of large numbers of governance *stakeholders*, who are constrained or enabled in their actions by *structures*. *Stakeholders*, in this perspective, are any social unit possessing a stake/agency or power of action. These include individuals, associations, leaders, firms, departments and international bodies. *Structure* refers to the frameworks within which these stakeholders operate; these limit or widen their action potentials and which they therefore must take into account. These frameworks include culture, law, agreements, material and technical possibilities (Kooiman et al. 2008).

In analyzing an institutional arrangement in CPR management and governance, one must investigate who is involved, what their stakes and resources are, and how they are linked to one another and to outcomes. Specifically, the types of actions that stakeholders can take, the type of information available to them, how actions lead to outcomes, and how rewards and punishments are allocated in light of the outcomes achieved and the actions taken all require identification (Ostrom et al. 1993).

The understanding of stakeholders' involvement in decision making processes is important in bringing them into governance, using their competencies and capacities as necessary, and ensuring they are heard and have influence (Bavink et al. 2005). Moreover, natural resource management typically deals with conflicting interests of various stakeholders since they use the same resources for different purposes. It is therefore important to understand the different perspectives of the stakeholders involved (Reed et al. 2009).

The overall objective of this section is to provide a preliminary description of actors' interest in and influence over CPR use, management and governance, and interactions of actors/stakeholders in the process of resource use, management, governance and policy process. It depicts the stakeholders' landscape based on focus group discussions, key informant interviews and the stakeholder identification and analysis workshop which was carried out by a multidisciplinary team of experts from different stakeholder groups (governmental, research, academic and NGOs).

20.2.2 Identification of Stakeholders and Stakeholder Groups

Before the actual stakeholder identification and analysis, the stakeholder concept was pre-presented and discussed with workshop participants to ensure that participants had an appropriate level of understanding of the term as deployed in this study and to make sure all individuals have common understanding of stakeholder concepts and related issues. After a brief explanation of the overall concepts and definitions of CPRs, stakeholders, stakeholder analysis, tools for stakeholder identification and analysis, a flexible set of steps for conducting stakeholder analysis as suggested by Grimble et al. (1995), Grimble (1998) and Schmeer (1999) were followed for this research.

A list of possible stakeholders with input from participants was developed. All stakeholders, who could have an interest in and control over the selected CPR situations (water, fishery and wetland) including stakeholders outside this situation that could affect or be affected by those actions were identified. Potential stakeholders in different sectors, geographic or administrative areas within Lake Tana sub-basin were also considered; experts who know the sector (water, fishery and wetland), policy, CPR situations and players in and around Lake Tana helped a lot in this process.

Then 5 broad stakeholder groups (see Table 20.3) consisting of 23 specific stakeholders and sub-stakeholder groups who have legitimate interest in and power over common-pool resources in the sub-basin were identified. However, this identification cannot guarantee the exhaustive lists of stakeholders in Lake Tana's jurisdiction, because, stakeholders change over time, new stakeholders can enter a resource management and governance system, while others may lose their role or interest. Accordingly, in the wake of resource scarcity, undetermined demand for resources, dynamic institutional and structural adjustments; the nature of

Table 20.3 Stakeholder groups in Lake Tana Sub-basin by type, participation mode, and source of influence

Stakeholder groups	Mode of participation, rights and responsibilities	Relative power/influence determined by
Governmental organizations		
• Law makers and executives • Policy makers and planners • Line ministries and authorities	• Enacting legislations (proclamations, regulations and directives) • Planning and implementation of sectoral policies • Managing development plans	• Legal hierarchy (chain of command and resource flow) • Authority of leadership (formal and informal, political affiliations) • Control of strategic resources (like finance, or land) • Trusted by the central government
• Academic and research institutes • Regional Bureaus • Municipalities • Corporations • District offices and administrations	• Enforcement of Federal and Regional laws • Oversight of natural resource use, management and governance • Monitoring and evaluation • Administration of natural resources	
Local communities		
• Local farmers • Local fishermen • Commercial fishermen • Youths and Women • *Negede woyto*	• Livelihood diversification and security • Economic development • Management of natural resources	• Socio-economic and political status of stakeholders (i.e. Well-off groups are more influential than the poor and voiceless) • Social capital and informal influences within their group
Private (investors and enterprises)		
• Transport enterprises • Market enterprises and Investors • Boat owners	• Profit maximization • Enclosure of commons	• Their financial capital • Political affiliations
Non-governmental organizations		
• EWNRA • EOTC	• Environmental protection and conservation • Public awareness creation and education • Capacity building (both for community members and experts) • Promote and encourage stakeholders participation and collaboration	• Programme compatibility with the Government plan and agenda • Their financial capital • Knowledge and expertise

(continued)

Table 20.3 (continued)

Stakeholder groups	Mode of participation, rights and responsibilities	Relative power/influence determined by
International communities		
• FAO, GEF, RAMSAR, UNESCO	• Technical and financial support • Compliance with international agreement and responses • Promoting policy improvement at national and international level	• Their financial capital • Knowledge and expertise • Political capital (their recognition worldwide)

Source Wereta stakeholder identification and analysis workshop, February 26–27, 2011

stakeholder's mandate or interest, the bases of their stake, the stake value and the power level to take action can be also more dynamic.

Once stakeholders were identified, it was necessary to do further analysis to better understand their relevance and the perspective they offer, to understand their relationship to the issue(s) and each other, and to prioritize based on their relative usefulness for CPR use, management and governance. A list of criteria to analyze each identified stakeholder was selected by the participants. These are; modes of participation, rights and responsibilities, their interest (stake value) and influence (power level), engagement in the process of CPR use, management and governance, and their legitimacy to claim for engagement and influence decision making.

As can be seen from Table 20.3 the role played by stakeholder groups and the source of their power to influence the CPRs management and governance system is diverse among stakeholders and stakeholder groups. For instance governmental organizations (spanning from Federal to *Kebele* level) are usually responsible to oversight the natural resource use, management and governance, to enact legislations, developing policies and ensuring the enforcement and implementation of Federal and Regional laws. The relative power of these stakeholders to influence the natural resource management and governance system mostly esteems from the formal law.

20.2.3 Stakeholders' Participation in CPR Management and Governance

Bottom up and inclusive stakeholder approaches for CPR use, management and governance are likely to enhance the credibility of the decision making process at different levels. Higher levels of stakeholder involvement usually imply that participants will have the opportunity to communicate their views, agendas and judgments in detail (OECD 2004) so that there will be a consensus and common vision on CPR use, management and the governance decision making process. In reality however, most decisions regarding resource use, management and governance in Ethiopia emanate from above in a top down manner. In most cases natural resource policy makers follow a centralized approach in policy planning and development process. Relevant stakeholders are merely informed about any new policies, strategies but never engaged in its development and implementations. More often such policies and strategies fail to address real problems at grass root level and were at times in conflict with stakeholders' interest.

The stakeholder identification and analysis is shown in Table 20.4. The table clearly shows that there is low level of stakeholder involvement in key aspects of the CPR use, management and governance processes. Of 23 stakeholders identified in the analysis, only 8 of them were participating in all forms of involvement (information, consultation, decision making and as a co-partner) in one way or another. Moreover, only few (less than 4) government-affiliated key stakeholders are active in all forms of involvements, particularly in decision making which needs power to persuade others. This clearly shows that government affiliated stakeholders spanning from the Federal to *kebele* level are more dominant in any key decision regarding resource use, management and governance. Whereas the majority of stakeholders identified are not fully involved in CPR use, management and governance processes. Particularly the participation of local users (local community, fishers, youth's groups and women's groups) who are supposed to be active participants in any decision regarding their commons are not well engaged in the process.

According to workshop participants, stakeholders' involvement is fundamental to effectively use, manage and govern CPRs in the region. They also assert that it is important to recognize the roles that can be played by potential stakeholders and appreciate the extent to which different perspectives are valued within the CPR use, management and governance process. However, in the absence of strong political will and inefficiency of executive organs of the government at different level (particularly at regional and district levels), the key challenge for the CPR governance system is how to integrate the different perspectives, and at which stage different types of stakeholders can play roles and the ways that each might be involved.

20.2.4 The Power and Interests of Stakeholders

Policy analysts have long been aware of the importance of interest groups in the policy process, and the need to characterize and categorize levels of interest and power which influence, and therefore impact on, particular policies (Brugha and Varvasvsky 2000). *Power* to influence policies or institutions stems from, the control of decisions with positive or negative effects. Stakeholder power can be understood as the extent to which stakeholders are able to persuade or coerce others into making decisions, and following certain courses of action. Power may derive from the nature of a stakeholder's organization, the political power of the leaders of the organization or their position in relation to other stakeholders (for example, line ministries which control budgets and other departments). Other forms of power may

Table 20.4 Forms of stakeholder involvement in CPR use, management and governance in the Lake Tana Basin Social-Ecological system

Stakeholders	Forms of involvement					
	Information	Consultation	Decision making	As a coordination partner[a]	As cooperation partner[b]	As co-production partner[c]
Bureau of Environmental Protection	XX	XX	XX	XX	XX	XX
Bureau of Agriculture	XX	XX	XX	XX	XX	XX
Bureau of Water and Energy	XX	XX	XX	XX	XX	XX
Administration (Local-Federal)	XX	X	XX	XX	XX	XX
Local Community	X	–	–	–	X	X
Academic and Research Institute	X	XX	–	XX	XX	X
Institute of Biodiversity	XX	XX	–	XX	XX	XX
Fishers	X	–	–	–	–	–
Abbay Basin Authority	X	XX	X	XX	XX	XX
International Organization	XX	XX	–	XX	X	X
Ethiopian Orthodox. Tewahido Church	X	–	–	X	X	–
Civic Societies	X	XX	–	X	X	X
Ethio. Wild Life Conservation Authority	X	X	–	XX	X	X
Ethiopian Electric Power Corporation	X	X	X	X	X	X
Investment Bureau	XX	X	X	XX	X	X
Youth's group	X	–	–	–	–	–
Women's group	X	–	–	–	–	–

(continued)

Table 20.4 (continued)

Stakeholders	Forms of involvement					
	Information	Consultation	Decision making	As a coordination partner[a]	As cooperation partner[b]	As co-production partner[c]
Fish Prod. and Market. Enterprise	X	–	–	–	–	
Bureau of Tourism and Culture	X	X	–	–	X	X
Lake Tana Transport Enterprise	X	–	–	–	–	–
Municipality	XX	XX	X	X	X	X
Boat owners	X	–	–	–	–	–
Private Investors	X	–	–	–	–	–

Source Wereta Stakeholder identification and analysis workshop, February 26–27, 2011; X—passive, XX—active, –, no or very minimal participation
[a]Symmetrical exchange of information about intentions and plans
[b]Continual exchange of information and use of complementary resources
[c]Continual exchange of information and pooling of resources to achieve an agreed objective

be more informal (for example, personal connections to ruling politicians) (IIED 2005).

According to Majchrzak (1984), the power of the stakeholders is explained by decision making, resources at hand, their ability to mobilize resources, and their accessibility to policy decision makers in relation to CPR use, management and governance. It is the ability of the stakeholders to possess enough resources to make the outcomes they desire happen. Stakeholders exercise power based on physical, financial, and symbolic or social capital resources. Physical resources such as physical sanctions, forces and violence are ascribed as coercive power. Financial resources including material means, money, goods or services constitute useful power. Symbolic or social capital resources are normative symbols such as prestige, esteem and social symbols of acceptance (Mitchell et al. 1997). In CPR situations, interest can take many forms such as legal or moral rights, legal title, or ownership. The source of stakeholders' interests may derive from a variety of reasons; livelihood dependence, cultural and historical association, economic interest, institutional mandate, social obligations, value commitment and political interest are among the most important.

As suggested by Grimble and Han (1995); Grimble (1998) and Schmeer (1999), for each stakeholder, participants were asked to estimate how much value they place on each stakeholder's stake or interest in the CPR use, management and governance system. The value can be positive or negative or both (i.e. their stake may be to affect the system positively (+) or negatively (−), or both). The values were 1–5, 1 denotes non-essential and 5 denotes critical. At the same time, participants were assigned a power level to each stakeholder. The power levels were from 1–6, where 1 denotes lowest and 6 denotes complete control. This rates the ability of the stakeholders to take effective action and ensure their stake. Then the value of the stake to the stakeholder was multiplied by the power to take action. The result was an indication of the stakeholder's likely impact on CPR management and governance in Lake Tana sub-basin.

Table 20.5 identifies the nature of the stake, relative strength of influence and direction of influence for each stakeholder. Although the powers and responsibilities to administer and govern all land based natural resources are vested to the Federal and Regional governments, the lion's share goes to the Federal government and its line ministries and subsidiary administrative bodies at different level. And yet, other stakeholders that have an interest in and power over CPR management and governance were identified. As indicated in Table 20.5, a high score on both variables (stake value and power level) will make specific stakeholders a clear candidate for becoming involved in the natural resource governance decision-making process. Such stakeholders are likely to be the first to be consulted or represented. Variance in score among stakeholder groups may determine their relative influence and their formal status within the governing system. A low score on one attribute may be compensated by a high score on others. Thus, stakeholders may have less stake value and/or legitimate concerns, yet still enjoy a powerful governing position. Such a situation might easily challenge the participatory process and question the design of the governing system (Jentoft 2007).

Table 20.5 Relative levels of stakeholder power in CPR management and governance in the Lake Tana Sub-basin

No	Stakeholder	Nature of the stake	Stake value	Power level	Total	Impact on CPRs
1	Bureau of Agriculture (BoA)	Regulation/administrative	5(+, −)	6	30	Higher
2	Bureau of Water and Energy (BoWE)	Regulation/administrative	5(+, −)	5	25	Higher
3	Administration[a] (Local to Federal)	Regulation/administrative	4(+, −)	6	24	Higher
4	Bureau of Env. Prot. Land Use and Administration (Bo-EPLUA)	Regulation/administrative	5(+, −)	4	20	Higher
5	Local Community	Livelihood/cultural value	5(+, −)	3	15	Medium
6	Fishers	Livelihood/cultural value	5(+, −)	3	15	Medium
7	Academic and Research Institute	Knowledge generation and transfer	3(+)	5	15	Medium
8	Ethiopian Wild Life Dev. and Conservation Authority (EWDCA)	Regulation/administrative	3(+)	5	15	Medium
9	Institute of Biodiversity Conservation and Research (IBCR)	Conservation and protection	3(+)	4	12	Medium
10	Abbay Basin Authority	Regulation/administrative	3(+, −)	4	12	Medium
11	International Organization (FAO, UNESCO, RAMSAR etc.)	Cultural values and conservation	3(+)	4	12	Medium
12	Investment Bureau	Regulation, administrative	3(−)	4	12	Medium
13	Ethiopian Electric Power Corporation (EEPCo)	Public service/profit maximization	4(−, +)	3	12	Medium
14	Youth's group	Livelihood/cultural value	5(+, −)	2	10	Medium
15	Ethiopian Orthodox Tewahido Church (EOTC)	Livelihood/religious value	3(+)	3	9	Low
16	Fish Production and Marketing Enterprise (FPME)	Economic/profit maximization	4(−, +)	2	8	Low
17	Private Investors	Economic/profit maximization	4(−)	2	8	Low
18	Boat Owners	Livelihood/profit maximization	4(+, −)	2	8	Low
19	Lake Tana Transport Enterprise	Economic/profit maximization	2(+, −)	3	6	Low
20	Municipality	Administrative	3(−)	2	6	Low
21	Civil Societies and Local NGOs	Non-profit making	3(+)	2	6	Low

(continued)

Table 20.5 (continued)

No	Stakeholder	Nature of the stake	Stake value	Power level	Total	Impact on CPRs
22	Women's group	Livelihood/cultural value	5(+, −)	1	5	Low
23	Bureau of Tourism and Culture	Administrative/cultural value	2(+, −)	2	4	Low

Source Wereta stakeholder identification and analysis workshop, February 26–27, 2011
Note Stake value ranges from 1 up to 5 and power level from 1 up to 6 were assigned for each stakeholder and the value of the stake to the stakeholder was multiplied by the power to take action, the result is an indication of the stakeholder's likely impact on the common-pool resource management and governance. The + and − sign indicates whether the stakeholder affects the CPR system positively, negatively or both
Stake value—Critical = 5, Essential = 4, Necessary = 3, Desirable = 2, Non-essential = 1
Power level—Control-complete = 6, Very significant = 5, Influence-significant = 4, Moderate = 3, Low = 2, Appreciation-lowest = 1
[a]Administration includes Federal and State representatives (such as Amhara Regional State Council, President Office, Zone, district, *kebele* administrations, police, judiciary) excluding sector organizations

It is perhaps obvious that stakeholders with high levels of stake and high power to secure their stake (e.g. the Bureau of Agriculture) have the potential to have an extreme impact on the management and governance of the CPR system. However, this simple analysis helps explain relative power among stakeholder groups who are not so clearly influential. It shows that even when a stakeholder places a high value on their stake in a CPR management and governance system (whether positive or negative) but has a low power level to secure the stake in the resource the system may not serve their interests. (e.g. Youth's group). Conversely, a more influential stakeholder with perhaps only a moderate interest in a particular outcome may still have greater power to secure this outcome (e.g. Institute of Biodiversity Conservation and Research).

Even though they possess low power for CPR governance, local communities and fishers residing near to Lake Tana, who have the *de jure* and de facto rights or claims over using and managing the CPRs are the most important stakeholders in the resource system. Their high interests in CPRs are mostly driven by economic incentives or livelihood security. Under the power vested on the 1995 Constitution of Federal Democratic Republic of Ethiopia (FDRE), Article 40(3) all 'Ethiopian people and the State have exclusive right to ownership of rural and urban land, as well as of all natural resources, presumably including CPRs'. Their customary and ownership rights should help them to own and/or access these resources. According to key informants and group discussants however, they often have limited rights to use CPRs. Where they have gained ownership or secure access, they may be hampered by corrupted *kebele* and district administrators who are using CPRs for their self-interest, inadequate levels of public investment, inappropriate policies, or competition from greedy private investors unrestrained by rules and regulation. In some cases, government agencies and other powerful interests like private investors often claim exclusive access to these resources and their benefits, and these other claims are supported by government officials.

Some stakeholders have administrative and social obligations to manage and govern the natural resource system. Power to take action and secure the outcomes could be vested politically or as a result of their social/financial capital. For example, the Bureau of Agriculture (BoA) has been vested a power by Federal government under Proclamation No. 110/2007 and by Council of Amhara National Regional State to conserve fish biodiversity and its environment, cultivate fisheries resource with appropriate fishing equipment as well as prevent and control over exploitation of the fisheries resource.

The Amhara National Regional State, Fisheries Resource Development, Protection and Utilization Proclamation enforcement, Regulation No, 50/2007 Article 3 sub article 4 states that:

> The Bureau may particularly or fully cause the activity of fish production be terminated, as deemed necessary, at any water body whenever it believes that the potential of the fishery's resource may be disturbed and thereby be extinct, the area is ascertained to be that of delivery and reproduction, especially during breeding seasons, or where there happens to be a dwindling of the species of fish thereof.

Generally at Federal level, the Ministry of Agriculture (MoA) has both development and natural resource management responsibilities (vested under Proclamation No. 380/2004). Relating to natural resources management, MoA has the responsibility to prepare policy on land use and draft legislation on forestry and wildlife. Bureau of Agriculture is the key natural resource management institution at regional level, responsible for the management of land, forest, wildlife and biodiversity resources (Proclamation No. 110/2007). However in reality BoA is mainly focusing on crop production and promoting cultivation of more lands including wetlands which are important components for the fish reproduction and the kidney for the health of water bodies like Lake *Tana*.

Conversely, international stakeholders such as FAO and UNESCO can affect CPR management and governance through their interests in issues like ecosystem and biodiversity conservation. Basically their power over CPRs is driven by their financial capital, international agreements and conventions that individual countries have ratified.

20.2.5 Classification of Stakeholders

Following Eden and Ackermann (1998); De Lopez (2001) cited in Reed et al. (2009) based on their power and interests, stakeholders who have the stake on the CPRs in and around Lake Tana ecosystem were classified into '*Key players*', '*Context setters*', '*Subjects*' and '*Crowd*' (Fig. 20.1). '*Key players*' are stakeholders who should be actively organized and engaged in key decisions, because they have high interest in and power over CPR management and governance. '*Context setters*' are highly influential (powerful), but have little interest. Because of this, they may be a significant risk, and should be monitored and managed. '*Subjects*' have high

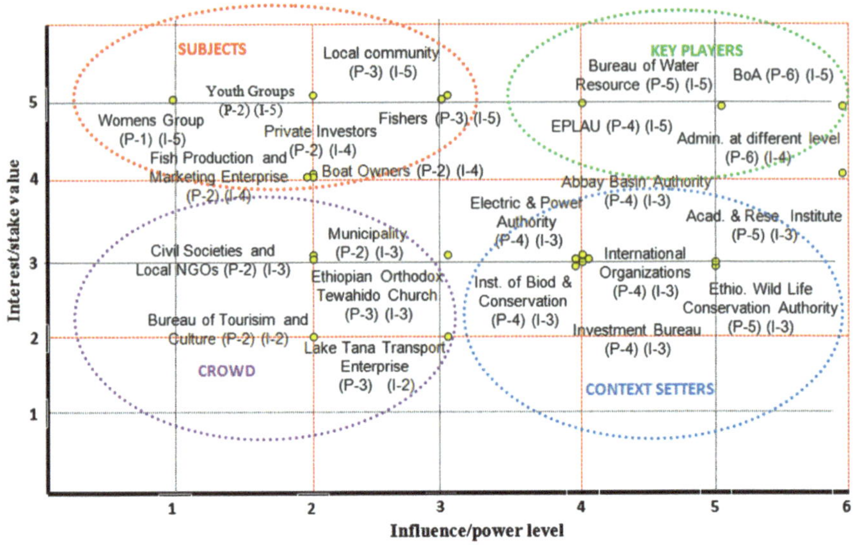

Fig. 20.1 Interest-influence matrix of stakeholders. *P stands for power level, I stands for interest level. Source* Wereta stakeholder identification and analysis workshop, February 26–27, 2011

interest in resource use but low influence on management and governance decision making. By definition they are supportive; they lack the capacity to affect the resource governance system significantly, although they may become influential by forming alliances with other stakeholders. These are often the marginal stakeholders that development initiatives seek to empower. The '*Crowd*' includes stakeholders who have little interest in or influence over desired outcomes and there is little need to consider them in much detail or to engage with them in key decision making. However, precaution should be taken not to ignore their interests.

Stakeholders can enter into and out of the system at different points in time, therefore interest and influence typically change over time and the impact of such change can be considered. The purpose of this classification is to reveal the interests and influence of the stakeholders and to understand synergies and conflicts between the stakeholders around their demand for Lake Tana ecosystem functions and services. However, interest and power are not static, and as stakeholders change position, tensions arise when key players have conflicting interests (Reed et al. 2009).

Stakeholders appearing in the *key player's* category in Fig. 20.1 have a high degree of influence/power on the management and governance of CPR systems and environmental decision-making processes, which are also of high importance/interest for the success or failure of technological and institutional innovations induced and/or created by government agencies, user groups and community members. This implies that natural resource management program and policy implementing organizations will need to construct good working relationships with these stakeholders, to ensure an effective coalition of support for the resource management and governance. These are the stakeholders the resource governance system must fully engage and make the greatest efforts to understand and address their concerns.

Stakeholders in the *subject's* category are of higher stake value/importance to the success of the management and governance of the CPR system and environmental decision-making processes, but with low influence/power to take action. This entails that they will require particular initiatives if their interests are to be restricted. Despite their low power, such stakeholders could be valuable allies in important decisions regarding CPRs. Therefore, it is advisable to keep them informed about the issues they are interested in. The governance system should keep these people adequately informed, and talk to them to ensure that no major issues arise (De Colle 2005). An example may be traditionally marginalized groups (e.g. Indigenous people like the *Negede woyto* community, youth's group, and women's groups), who might be solely dependent upon the resource system for their livelihood and beneficiaries of a new service, but who have little 'voice' in its development.

Context setters are stakeholders with high influence/power, who can therefore affect the management and governance of CPR systems and environmental decision-making processes and outcomes, but whose interest/stake value is not necessarily aligned with the overall goals of the environmental decision-making processes. In a multiple use-multi-stakeholder scenarios of resource management and governance, the relationships with these stakeholders can be very complex to manage. Most of the time, they behave passively and show a low interest in key management and governance decisions. It is therefore necessary to analyse potential intentions and reactions of these groups in all major developments, and to involve them according to their interests (De Colle 2005). They might be international organizations such as FAO, UNESCO, who can exercise considerable impact over funding disbursements and governmental organizations such as research and academic institutions who can exert valuable impact on the CPRs through their intellectual excellence. This conclusion implies that these stakeholders may be a source of significant risk in terms of financing environmental governance programmes and backing technical expertise, and they will need careful monitoring and management.

The stakeholders in the *Crowd*, with low power/influence on, or importance/stake value to the management and governance of the CPR system, may require limited monitoring or evaluation, but are of low priority. However it is advisable to monitor these stakeholder groups as they ability to influence and their interest could increase overtime (De Colle 2005).

As Bryson (2004) points out, power versus interest grids like Fig. 20.1 help determine which players' interests and power bases must be taken into account in order to address the problem or issue at hand (in this case CPR use, management and governance issues in and around Lake Tana sub-basin). This interest-influence grid clearly shows that governmental organizations such as the Bureau of Agriculture, the Bureau of Water resource, the Bureau of Environmental Protection

Land Use and Administration, and Administrations at different levels are key players with high power to secure their stake and have a significant impact on CPR use, management and governance issues in the sub-basin.

20.2.6 The Legitimacy of Stakeholders

Legitimacy in this research refers to the perceived validity of the stakeholders' claim to a stake in CPR management and governance in the Lake Tana sub-basin. Based on the mandate and responsibilities given by the Law of the Land (The Ethiopian Constitution), international treaties and customary laws of specific communities in the jurisdiction of Lake Tana sub-basin, stakeholders' legitimacy with regard to CPR management and governance were assessed and evaluated against criteria. During the stakeholder identification and analysis workshop, parameters that denote the legitimacy of a stakeholder or stakeholder groups were derived. However, this assessment is subject to the qualification that under complex and very dynamic systems, stakeholders' legitimacy may changed at any point in time, so that the current assessment is only representing the existing scenario.

Participants in the Wereta stakeholder identification and analysis workshop were agreed to evaluate each stakeholders legitimacy based on the following criteria;

- The formal and informal property rights vested in the stakeholder(s),
- The extent to which they have a historical or cultural relationship with the resource,
- The knowledge and skills that they are able to contribute to the management process,
- The level of commitment to the management process,
- The compatibility of their resource use patterns with sustainability and conservation requirements,
- The extent to which there is equity in access to, and in distribution of the benefits from, the use of the resource,
- The compatibility between local practices made by the respective stakeholder and national development and conservation policies and priorities of the country,
- The present and potential impact of the activities of the stakeholder on the resource base.

Each of the participants (representing different stakeholder group) of the workshop were requested to rank each stakeholders based on their perceived legitimacy. Table 20.6 summarizes the average perceived legitimacy determined by the participants for each other based on the eight different criteria of legitimacy. Examining the table reveals that the Bureau of Environmental Protection Land Use and Administration, Bureau of Agriculture, Bureau of Water and Energy, Administration (from Federal to local level) and the local community had a

Table 20.6 Stakeholder perceptions of the legitimacy of other stakeholder groups to manage and govern CPRs

	Criteria								Score/rank	
Stakeholders/stakeholder groups	The formal and informal property rights vested in the stakeholder (s)	The extent to which they have a historical or cultural relationship with the resource	The knowledge and skills that they are able to contribute to the management process	The level of commitment to the management process	The compatibility of their resource use patterns with sustainability and conservation requirements;	The extent to which there is equity in access to, and in distribution of the benefits from, the use of the resource;	The compatibility between local practices and national development and conservation policies and priorities	The present and potential impact of the activities of the stakeholder on the resource base	Total score (40)	Rank
Bureau of Environmental Protection	4	3	5	5	5	3	4	4(+)	33	1st
Bureau of Agriculture	5	4	4	3	3	3	3	5(+,−)	30	2nd
Bureau of Water and Energy	4	3	4	4	3	3	3	5(+,−)	29	3rd
Administration (Local-Federal)	5	3	3	2	4	3	4	4(+,−)	28	4th
Local community	5	5	4	3	2	2	2	5(+,−)	28	4th
Academic and Research Institute	2	3	4	4	4	2	4	4(+)	27	6th
Institute of Biodiversity	3	3	4	3	4	4	3	3(+)	27	6th
Fishers	5	5	3	2	1	2	3	5(+,−)	26	8th
Abbay Basin Authority	5	3	4	3	2	2	2	4(+,−)	25	9th
International Organization	2	1	5	4	4	3	3	3(+)	25	9th
Ethiopian Orthodox. Tewahedo Church	2	4	3	3	3	3	3	3(+)	24	11th
Civic Societies	1	1	4	3	4	3	4	3(+)	23	12th

(continued)

Table 20.6 (continued)

	Criteria								Score/rank	
Stakeholders/stakeholder groups	The formal and informal property rights vested in the stakeholder (s)	The extent to which they have a historical or cultural relationship with the resource	The knowledge and skills that they are able to contribute to the management process	The level of commitment to the management process	The compatibility of their resource use patterns with sustainability and conservation requirements;	The extent to which there is equity in access to, and in distribution of the benefits from, the use of the resource;	The compatibility between local practices and national development and conservation policies and priorities	The present and potential impact of the activities of the stakeholder on the resource base	Total score (40)	Rank
Ethio. Wild Life Conservation Authority	2	1	4	4	3	2	3	3(+)	22	13th
Ethiopian Electric Power Corporation	3	3	2	2	2	3	3	4(−,+)	21	14th
Investment Bureau	3	3	4	2	1	1	2	5(−)	21	14th
Youth's group	4	3	2	2	2	1	2	4(+,−)	20	16th
Women's group	4	3	2	2	2	1	2	3(+,−)	19	17th
Fish Prod. and Market. Enterprise	3	3	2	1	2	2	2	3(+,−)	18	18th
Bureau of Tourism and Culture	3	2	2	1	3	1	3	2(+)	17	19th
Lake Tana Transport Enterprise	2	2	2	2	1	1	2	4(+,−)	16	20th
Municipality	2	2	2	1	1	2	2	3(−)	15	21st
Boat owners	1	2	1	2	2	1	2	3(+,−)	14	22nd
Private Investors	1	1	1	2	1	1	1	5(−)	13	23rd

Source Wereta stakeholder identification and analysis workshop, February 26–27, 2011
Note Their right and responsibilities are evaluated as; 5 = very high, 4 = high, 3 = medium, 2 = low, 1 = very low

particularly strong and legitimate claim to use, manage and govern the CPRs in Lake Tana sub-basin. These stakeholder groups, ranked high on all the criteria, and had the highest total scores. This analysis can be useful in the CPR management and governance planning process as a way to convince other stakeholders that the legitimacy of these stakeholders has to be respected and integrated into the management and governance system. At the same time, stakeholders with high perceived legitimacy are also perceived to be responsible for and accountable to the overall success and/or failure of the management and governance system of the study area.

20.2.7 Organizational Capacity of Actors in the Action Arena

Organizational capacity is defined here as 'a set of attributes that help or enable an organization to fulfill its missions' (Eisinger 2002). Attributes commonly associated with high organizational capacity are governance and leadership, mission, vision, and strategy; common values, program and strategic relationships; resource development; internal operations and management. In addition, for effective organizational capacity, networking, political power, skill of leaders, trust and unity among stakeholders within the organization are essential. These interdependent factors all contribute to high performance of the organization to fulfill its mission. On the other hand, institutional instability is best understood as a pattern in which, in response to a given disturbance (i.e., exogenous shocks like changes in power), a particular institutional arrangement changes with greater frequency than other similarly designed ones (Levitsky and Murillo 2009).

A Strength, Weakness, Opportunity, and Threat (SWOT) analysis was conducted as part of the workshop. The results show that most key stakeholders (particularly formal organizations) in the study area have low organizational capacity and high institutional instability. According to workshop participants, institutional instability caused by frequent changes in organizational structures is one of the crucial factors that negatively affect the CPR management and governance system. Due to these changes, personnel find themselves working for agencies or jobs for which they are not properly qualified. Instability of the institutional structure leads to power shifts in decision making on CPR management and governance from one stakeholder to another. A lack of congruence between formal rules and informal norms also exacerbates institutional instability.

Documentary analysis of proclamations and regulations shows that frequent and often unplanned organizational re-structuring and institutional changes are quite common in Ethiopia particularly at ministry level. For instance, initially the FDRE Proclamation No. 4/1995 that defined the powers and duties of the executive organs established 15 ministries. After 10 years, Proclamation No. 471/2005 that repealed the previous proclamation re-structured the executive organs and 20 ministries were approved. Again after 5 years proclamation No. 471/2005 was repealed by proclamation No. 691/2010 and almost all (except 5 ministries that remain untouched) executive organs and ministries were merged, re-structured and re-organized with different and/or additional mandates. Thus, in the past 15 years, most ministries were separated, merged and/or dissolved, sometimes multiple times, to meet redefined mandates and responsibilities.

As reported by Vaughan (2004), regular separation, reemerging, and re-separation of ministries and sector offices like those dealing with agriculture, natural resource management, and environmental protection has occurred. As a result, institutional instability impinges on institutional capacity and development. This is characterized by overlapping mandates and responsibilities of agencies, as well as frequent changes in mandates. A typical example here is the Ministry of Agriculture which has been changing its structure for the last 3 or 4 decades. Even very recently, it has been changed from Ministry of Agriculture and Rural Development (MoARD) and retains its former name Ministry of Agriculture (MoA).

Frequent changes and often unplanned organizational re-structuring become detrimental to the attainment of the mission and goals of individual organizations and disruptive to the sustainability of linkages, collaborations and coordination of institutions among different stakeholders. With such dynamic organizational and institutional changes, it becomes hard to monitor the roles, responsibilities, and accountability of stakeholders in the process of CPR management and governance.

20.3 Understanding Patterns of Multi-stakeholder Interaction

Patterns of interaction emerge from the behaviour of stakeholders in the action arena (Polski and Ostrom 1999). Actors/stakeholders may differ in relation to types of actions that they are required, permitted, or forbidden to take in a certain action situation (Schlager and Blomquist 1998). The differences designate the pattern of interaction between the various actors/stakeholders which finally determine the outcome, in this research context, the way how stakeholder use, manage and govern the CPRs (Stellmacher 2007).

If a well managed and enabling environment is put in place, multi-stakeholder interactions facilitate stakeholders to work together, to solve societal and environmental problems. Multi-stakeholder interactions may lead to understanding of different interests and perspectives of stakeholders in the process of CPR use, management and governance. In a very complex scenario like Lake Tana's CPRs, however, the likelihood of positive outcomes from the multi-stakeholder interaction is rarely possible. Stakeholders make choices based on their own preferences, objectives or mandates (in the case of government agencies). These individual choices lead to aggregate patterns of interaction that affect user based CPR management and governance.

Lake Tana sub-basin is an example of a constantly changing and very complex CPR system whose water resource, commercial fishery, surrounding wetlands and other associated resources are managed and governed by different property rights regimes, involving a wide range of issues and multiple stakeholders with diverse power over and interests in CPRs. Hence to better understand this scenario the patterns of multi-stakeholder interaction and their outcomes were evaluated based on stakeholder participation in legislation, planning, policy making and implementation processes, linkage and collaboration among relevant stakeholders, organizational capacity and commitment, managing CPR related externalities (such as conflict of interest).

Multi-stakeholder interaction SWOT analysis reveals that Lake Tana sub-basin's CPR management and governance interaction is characterized by weak organizational capacity of key stakeholders, low organizational culture and commitment, overlapping mandates of stakeholders, weak linkage between informal institutions and formal governance structures, lengthy law making and implementation processes. As a result, these multi-stakeholder interactions lead to competition, overexploitation and degradation of CPRs that ultimately lead to conflict of interests among different stakeholders at different levels.

20.4 Patterns of Conflicting Interaction Among Stakeholders

There is no internationally accepted definition of what constitutes a natural resource conflict (Upreti 2004). Yet, sharp differences in power and in stake values across interested stakeholders make conflict inherent in CPR management and governance. So far, long-term maintenance of ecosystem health is in conflict with the short-term interests of many stakeholders and policymakers (Dietz et al. 2003; Bavinck et al. 2005). Although conflict is a feature of many resource management regimes, it is often assumed to reflect differences in material interests between stakeholders (Adams et al. 2003). As Upreti (2004) defined, conflict in this research however, could refer to observable differences in opinion, misunderstandings, clashes of interest, disagreements, complaints in public, and protests by arguments and physical assault, antipathy filing cases with the local administration, police and courts related to CPRs (with a particular emphasis on water, fish and wetlands).

As these commons evolve and are used for different purposes by different stakeholders and are governed under different management regimes, externalities such as competition, resource degradation and conflict amongst user groups are inevitable (Edwards and Steins 1999). In handling stakeholder interest over CPRs, it is exceptional that 'win-win' governance solutions can be found in key decisions that affect resource management and governance. Usually the decisions may not meet the needs and interests of every stakeholder and most decisions will end-up with 'win-loss' situations and in most cases, 'the win-loss' situation is against the interests of the poor and voiceless. The different stakeholders involved in the joint uses may act in the pursuit of various goals and private interests (e.g. commercial use, the strategic stance of a national and local authority) and choose their actions in order to satisfy those interests efficiently or the general public interest (e.g. Government and NGO with an objective of environmental concern). Furthermore, stakeholders can pursue local (within the boundary of the resource), regional, national or even global use interests (Knight 1992; Gerber et al. 2009).

In most cases, the combination of environmental and demographic change, developmental and market pressures, and multiple and often divergent interests of different stakeholders lead to natural resource competition and force local people to alter their livelihood strategies. Usually these factors are considered as the underlying causes for resource competition, both among community members, and between community groups and outside public and private organizations which ultimately leads to conflict (Warner 2000; Engel and Korf 2005). Increased competition and demand for natural resources of all kinds is in turn leading to increased conflicts generated by overlapping resource claims between local users, large-scale resource users (private investors) and government officials (Darby 2010). In some cases natural resource conflict could arise due to failure in governance (institutions, policies, laws) to resolve these tensions equitably that leads to specific groups being disadvantaged, and ultimately to conflict (Matthew et al. 2009).

Common-pool resource conflict may arise at international, regional, national and local level. The causes and nature of conflict also depend on knowledge, perspective and context of stakeholders involved in CPR management and governance at these levels. However for this research, CPR conflict analysis focused on national/regional (higher level) and local level resource users conflicts with a special emphasis on resource users. The conflict situations in the sub-basin further illustrate how multiple actors influence pattern of interactions and outcomes. The following sub-sections answer the question of why do conflicts/disputes arise among stakeholders, and what are the underlying causes of conflict and resolution mechanisms. The common patterns of interactions in CPR conflict and conflict resolution mechanisms across communities and at higher level of governance are examined.

20.5 Summary

This chapter examined a range of characteristics for 23 key stakeholders in Lake Tana basin common property resources. The stakeholders fall into five main groups: governmental organizations, local communities, ... [add rest from Table 20.7]. These stakeholders hold a diverse set of interests in the use, management and governance of the basin's resources, and they wield different levels of influence. They have different levels of perceived legitimacy. The basin resources

Table 20.7 Conflicts of interest among stakeholders in Lake Tana sub-basin

Level of conflict	Stakeholders involved	Causes of conflict	Dynamics of conflict	Conflict management options
Micro-user level (within)	Fishers, local farmers, local experts, administrators, women's group, Youth groups, Investors, Boat owners, Lake *Tana* Transport Enterprise, Ethiopian Electric Power Corporation etc.	Competition for scarce resource (fish and wetlands) to secure livelihood and/or make a profit, theft of properties (e.g. net for fishing), unclear boundary of resource system, absence of or unclear rules and regulations to govern the resources etc.	Increasing trends (both in frequency and type)	Mainly negotiation between disputing parties, Mediation by neutral, independent person (traditional dispute resolution methods), courts as a last resort (formal method)
Micro-meso interface (between)	User community and/or local administration versus Regional and/or Federal level governmental agencies	Incompatibility of livelihood and property rights security, not compensated for loss of land or other properties versus Development and/or conservation goals of each agency	As development pressure on commons and leasing of communal lands for investors increases, the likelihood of conflict between the two parties will increase	Economic compensation, negotiation to convince the local community, in case of disagreement appealing to the higher level
Meso-regional (within)	Regional Bureaus of (Agriculture, Environmental protection, Investment, Water resource), municipalities etc.	Unclear mandate and responsibilities and duplication of efforts, multiple interest, lack of linkage and information flows, bypassing the mandates of others, uncoordinated planning and implementation	Unless there is revision of mandates and responsibilities, there will be an increasing trend of conflict among stakeholders	Policy review, stakeholder workshops

(continued)

Table 20.7 (continued)

Level of conflict	Stakeholders involved	Causes of conflict	Dynamics of conflict	Conflict management options
Meso-macro interface (between)	Regional level governmental agencies versus Higher/Federal level Authorities and ministries	Information and knowledge gap, overlapping mandate, higher level authorities sometimes bypass the mandates of regional states/authorities	Rarely happened	Policy review, stakeholder workshops
Macro-national (within)	Policymakers, Legislators, Ministry of (Agriculture, Water and Energy…), Institute of Biodiversity, International organizations (UNESCO, FAO)	The paradox of conservation verses development goals, policies imposed without local participation, uncoordinated policy planning and decision making	On the wake of climate change, there is an increasing trend environmental degradation and international pressure	National consensus, International treaties, agreements, conventions

Source Own survey, Wereta stakeholder identification and analysis workshop, February 26–27, 2011

on which communities around Lake Tana sub-basin depend are under severe threat. Access and availability of basic CPRs has been declining, leaving the wider community vulnerable to resource scarcity and seasonal shocks. Development activities around Lake Tana, population pressure, agricultural intensification (crop land expansion at the expense of grazing/wetland), livelihood diversification, unclear property rights and ineffective often irresponsible resource governance system at the user level are considered as the major factors that affect the CPR management and governance. The stakeholder analysis presented here is a first step in understanding how to decrease threats and improve the management of CPRs.

When the multiple uses of the entire Lake Tana sub-basin are considered, we see multiple stakeholders with diverse interests in and power over CPR use, management and governance. Conflicts arise from different stakeholder uses. Generally, the interaction of stakeholders in the process of resource use, management and governance is characterized by low or insufficient stakeholder participation in key decisions regarding CPRs. There are insufficient policy forums that facilitate collaboration, linkage and policy dialogue among relevant stakeholders at different levels. In most cases lack of joint policy planning and implementation lead to conflict of interest among stakeholders.

For any development intervention in around Lake Tana sub-basin, the processes of stakeholder identification and analysis can and should be supported by an enabling policy environment. Identification and analysis of stakeholders by itself is not an end for sustainable resource use, management and governance. National policy statements, including natural resource management and environmental policies, National Sustainable Development Strategies, National Biodiversity Strategies and Action Plans, and plans for systems of natural resources, must make explicit commitments to the inclusion of stakeholders in planning and management processes. Legal instruments should also lay out the procedures that must be followed by public agencies to identify and allocate rights and responsibilities within specific management instruments and agreements. In particular, policy and legal instruments should provide formal safeguards against marginalization and exclusion.

20.6 Key Research Questions Remaining

While this analysis set the stage for understanding stakeholder characteristics, more research is needed about the performance and effectiveness of the resource governance system, particularly at the middle and lower levels structure of the system. The results and discussions of this research should be supplemented by research to identify and define the major CPR management and governance problems, challenges, gaps and alternative solutions in Lake Tana sub-basin. It is important to consider that respondents (both resource users and other key informants) may not disclose the realities with regard to sensitive issues like land policy, administration and good governance problems. The stakeholders analyzed and presented, and CPR cases selected in this research may not be fully comprehensive; rather they are intended to provide an initial springboard for further research and participatory law and policy making processes. Therefore in some cases the research results presented here are considered as indicative (needing further research and evidence) rather than definitive. More importantly, the results should 'sound an alarm bell' for those who have a 'stake' in Lake Tana so that they will act together to save the Lake ecosystem from the serious challenges that it faces.

This research addressed observational types of questions (what, who or which) rather than analytical (why or how much) questions. It is based on an integrated exploratory and descriptive research framework. The output of this research is expected to facilitate future research (providing more detailed explanations of variables) focusing on explaining why different stakeholders pursue goals only based on their interests in the process of CPR management and governance, and how and why local or traditional conventions are losing ground in governing and managing CPRs at different field settings.

Acknowledgement The authors would like to acknowledge Ministry of Agriculture of Ethiopia, Rural Capacity Building Project (RCBP) and Irish Aid, Department of Foreign Affairs and Trada, Republic of Ireland for financing the study.

References

Adams W et al (2003) Managing tragedies: understanding conflict over common pool resources. Science 302:1915

Arthur Z, Claudia M (2007) Multi-stakeholder management: tools for stakeholder analysis: 10 building blocks for designing participatory systems of cooperation. Federal ministry for economic cooperation and development

Baland JM, Platteau JP (1999) The ambiguous impact of inequality on local resource management. World Dev 27(5):773–788

Barker DR (2004) Address by the president of LakeNet. Lake Tana symposium, Bahir Dar University, Bahir Dar, Ethiopia

Bavinck MR et al (2005) Interactive fisheries governance. Eburon Publishers, Delft, 72 pp

Bressers, Kuks MM (2003) What does 'Governance' mean? From conception to elaboration. In: Bressers HTA, Rosenbaum WA (eds) Achieving sustainable development: the challenge of governance across social scales. Preager, Westport, Connecticut, London, pp 65–88

Brugha R, Varvasovsky Z (2000) Stakeholder analysis: a review. Health Policy Plann 15:239–246

Bryson JM (2004) What to do when stakeholders matter: stakeholder identification and analysis techniques. Public Manag Rev 6:21–53

Burroughs R (1999) When stakeholders choose: process, knowledge, and motivation in water quality decisions. Soc Nat Resour 12:797–809

Darby S (2010) Natural resource governance: new frontiers in transparency and accountability, transparency and accountability initiative, London. Available at: http://www.transparency-initiative.org/wp-content/uploads/2011/05/natural_resources_final1.pdf. Accessed Aug 2012

Dargahi B, Setegn SG (2011) Combined 3d hydrodynamic and watershed modeling of Lake Tana, Ethiopia. J Hydrol 398:44–64

De Colle S (2005) A stakeholder management model for ethical decision making. Int J Manag Decis Making 6:299–314

De Lopez TT (2001) Stakeholder management for conservation projects: a case study of Ream National Park, Cambodia. Environ Manage 28:47–60

Dietz T et al (2003) The struggle to govern the commons. Science 302:1907–1912

Duram LA, Brown KG (1999) Assessing public participation in US watershed planning initiatives. Soc Nat Resour 12:455–467

Eden C, Ackermann F (1998) Making strategy: the journey of strategic management. Sage Publications, London

Edwards V, Steins N (1999) A framework for analyzing contextual factors in common pool resource research. J Environ Policy Plan 1:205–221

Eisinger P (2002) Organizational capacity and organizational effectiveness among street level food assistance programs. Nonprofit Voluntary Sect Q 31(1):115–130

Engel P (1997) The social organization of innovation: a focus on stakeholder interaction. Royal Tropical Institute, Amsterdam

Engel A, Korf B (2005) Negotiation and mediation techniques for natural resource management. FAO, Rome, Italy

Forester J (1999) The deliberative practitioner. M.I.T. Press, Cambridge, Massachussetts

Freeman RE (1984) Strategic management: a stakeholder approach. Pitman, Boston

Gerald S (2008) Supporting sustainable innovation through stakeholder management: a systems view. Int J Innovation Learn 5(6). Available at: http://inderscience.metapress.com/app/home/contribution.asp?referrer=parent&backto=issue,1,7;journal,16,41;linkingpublicationresults,1:110863,1. Accessed 25 Jan 2011

Gerber JD et al (2009) Institutional resource regimes: towards sustainability through the combination of property rights theory and policy analysis. Ecol Econ 68:798–809

Grimble R (1998) Stakeholder methodologies in natural resource management. Socioeconomic methodologies. Best practice guidelines. Natural Resources Institute, Chatham, UK

Grimble R, Han MK (1995) Stakeholder analysis for natural resource management in developing countries. Nat Resource Forum 19(2):113–124

Grimble R, Chan MK, Aglionby J, Quan J (1995) Trees and trade-offs: a stakeholder approach to natural resource management. Gatekeeper Series no. 52. International Institute for Environment and Development

Grimble R, Wellard K (1997) Stakeholder methodologies in natural resource management: a review of principles, contexts, experiences and opportunities. Agric Syst 55:173–193

Guijt I, Shah M (1998) The myth of community: gender issues in participatory development. IT Publications, London

Hughes R, Black C, Kennedy NP (2008) Public Health Nutrition Intervention Management: Stakeholder analysis and engagement. JobNut Project, Trinity College Dublin

IIED (2005) Stakeholders power analysis: power tools. International institute for Environment and Development (IIED). http://www.policy-powertools.org/tools/understanding/docs/stakeholder_power_tool_english.pdf. Accessed 12 July 2012

Jentoft S (2007) limits of governability: institutional implications for fisheries and coastal governance. Mar Policy 31:360–370

Knight J (1992) Institutions and social conflict. Cambridge University Press, Cambridge, p 1992

Kooiman J et al (2008) Interactive governance and governability: an introduction. J Trans-disc Environ Stud 7:2–11

Leach M et al (1999) Environmental entitlements: dynamics and institutions in community-based natural resource management. World Dev 27:225–247

Levitsky S, Murillo MV (2009) Variation in institutional strength. Annu Rev Polit Sci 12:115–133

Mathews D (1994) Politics for people: finding a responsible public voice. University of Illinois Press, Urbana, Illinois

McCartney M et al (2010) Evaluation of current and future water resources development in the lake tana basin, Ethiopia. International Water Management Institute, Colombo, Srilanka, 39 p. (IWMI research report 134). doi:10.3910/2010.204

Majchrzak A (1984) Methods for policy research. Sage Publications, London

Matthew RA et al (2009) From conflict to peace building: the role of natural resources and the environment. United Nations Environmental Programme (UNEP), ISBN 978–92-807-2957-3. Earth print. http://postconflict.unep.ch/publications/pcdmb_policy_01.pdf. Accessed 29 June 2013

Minale A, Rao K (2011) Hydrological dynamics and human impact on ecosystems of Lake Tana, North western Ethiopia. Ethiop J Environ Stud Manag 4

Mitchell R et al (1997) Toward a theory of stakeholder identification and salience: defining the principle of who and what really counts. Acad Manag Rev 22(4):853–886

OECD (2004) Stakeholders involvement techniques. Short guide and annoted bibliography. ISBN-92-64-02087-x. Nea. No. 5418. Nuclear Energy Agency. Organization for Economic Co-operation and Development (OCED)

Ostrom E (2008) Sustainable development of common-pool resources. Environ Sci Policy Sustain Dev (July/Aug, 2008): http://www.indiana.edu/~workshop/colloquia/materials/papers/ostrom_paper1.pdf. Accessed 23 July 2012

Ostrom E et al (1993) Analyzing the performance of alternative institutional arrangements for sustaining rural infrastructure in developing countries. J Publ Adm Res Theory J 3(1):11–45

Ostrom E et al (1994) Rules, games, and common-pool resources. The University of Michigan Press, Ann arbor

Polski M, Ostrom E (1999) An institutional framework for policy analysis and design. Workshop in political theory and policy analysis working paper w98-27. Indiana University, Bloomington
Ramirez R (1999) Stakeholder analysis and conflict management. Cultivating peace: conflict and collaboration in natural resource management, pp 101–126
Reed M et al (2009) Who's in and why? A typology of stakeholder analysis methods for natural resource management. J Environ Manag 90:1933–1949
Röling N, Wagemakers M (1998) Facilitating sustainable agriculture: participatory learning and adaptative management in times of environmental uncertainty. Cambridge University Press, Cambridge
Setegn SG (2010) Modeling hydrological and hydrodynamic processes in the Lake Tana Basin, Ethiopia, KTH. TRITA-LWR unpublished, PhD thesis 1057
Schlager E, Blomquist W (1998) Resolving common pool resource dilemmas and heterogeneities among resource users. Crossing boundaries. In Seventh annual conference of the international association for the study of common property. Vancouver
Schmeer K (1999) Guidelines for conducting a stakeholder analysis, Nov 1999. Partnerships for Health Reform, ABT Associates Inc., Bethesda, MD
Selin SW et al (2000) Modeling stakeholder perceptions of collaborative initiative effectiveness. Soc Nat Resour 13:735–745
Steins NA, Edwards VM (1999) Platforms for collective action in multiple-use common-pool resources. Agric Hum Values 16:241–255
Stellmacher T (2007) Governing the Ethiopian coffee forests: a local level institutional analysis in Kaffa and Bale mountains. Shaker
Swallow BM et al (1997) Multiple functions of common property regimes. EPTD workshop summary paper no. 5
Upreti BR (2004) Resource conflicts and conflict resolution in Nepal. Mt Res Dev 24:60–66
Vaughan S (2004) Ethiopia: public service delivery capacity building program (PSCAP). Social Appraisal, May 2004
Vijverberg J et al (2009) Lake Tana: source of the Blue Nile. The Nile, pp 163–192
Warner M (2000) Conflict management in community-based natural resource projects: experiences from Fiji and Papua New Guinea. Working paper 135. Overseas Development Institute (ODI), London

Chapter 21
Lake Tana Subbasin's Economy and the Role of Natural Resources

Daregot Berihun

Abstract This chapter gives a brief description of the Lake Tana Subbasin's (LTB) economy analogized with economic development requirements and standards. It also highlights the LTB's natural resources role in local economy. Although there has not been any attempt to compute the Gross Domestic (Geographic) Product for the LTB, available regional and zonal tax revenue data shows the contribution of prominent sectoral activities in the basin such as agriculture and fisheries, tourism and navigation, trade and industry, and transport to the regional revenue is substantial. A significant proportion of the Lake Tana Basin's economy is attributed to the agricultural sector and the Basin harbors considerable untapped potential for irrigation development. This calls for a better investigation of the development of a resources efficient modern agriculture which will contribute for the economic development of the Basin. The rest of the sectors such as industry, are also in a very early stage of the economic activity, and need more attention to be developed with optimized use of the resources of the basin.

Keywords Gross domestic product · Local economic development · Total economic activity

21.1 Introduction

This chapter gives a brief description of the Lake Tana Subbasin's (LTB) economy analogized with economic development requirements and standards. It also highlights the LTB's natural resources role in local economy.

The description of the economy of any geographic region includes explanation of the economic activity and economic development. By economic activity, we mean actions that involve the production, distribution and consumption of goods and services at all levels within a society. An economic activity takes place when

D. Berihun (✉)
Department of Economics, Bahir Dar University, Bahir Dar, Ethiopia
e-mail: daregot21@gmail.com

K. Stave et al. (eds.), *Social and Ecological System Dynamics*,
AESS Interdisciplinary Environmental Studies and Sciences Series,
DOI 10.1007/978-3-319-45755-0_21

resources such as capital goods, labour, manufacturing techniques or intermediary products are combined to produce specific goods or services. Thus, an economic activity is characterized by an input of resources, a production process and an output of products i.e. goods or services (EU 2008).

For a watershed (basin), total economic activity is defined as the sum of direct/indirect use, option, and nonuse values coming from activities including market use and nonuse value of water supply, fishing, hunting, recreation, boating, ecotourism, agriculture, and navigation/port benefits in the basin (Ingraham and Foster 2008). Gross domestic product or GDP is one way of assessing the overall performance of economic activities of any nation. Alternatively, the overall economic status of any specific location including river basins can be described by its performance of the regional gross domestic (geographic) product indicator. The positive performance of economic activities measured by GDP, is termed as economic growth, which is one constituent of economic development.

Most standard text books such as Todaro and Smith (2012: pp. 22), define economic development as "both a physical reality and a state of mind in which society has, through some combination of social, economic, and institutional processes, secured the means for obtaining a better life". This definition suggests a distinction between economic growth and economic development in which economic growth represents an increase and expansion of economic activities while economic development can involve expansion both economic and non-economic dimensions of life in a region. This conceptual background frames the following description of regional economy and economic development standards in the Lake Tana Basin.

21.2 Description of Regional Economy

The Lake Tana Basin overlaps with parts of the North Gondar, South Gondar, Bahir Dar, West Gojjam, and Awi Zones of the Amhara region, and is considered to be the major economic corridor of the Amahara Region. It is home to an estimated population of 4.1 million in 2015, who are the prime owners and motivators of the economy (CSA 2003).

The regional gross domestic product for the Amhara Region has only been formally computed very recently, since 2012, and this regional estimation has not been declared as official by the respective authorities. There has not been any attempt to compute the Gross Domestic (Geographic) Product (RGDP) for the LTB. Instead, we use here the available regional and zonal tax revenue data as a proxy to describe the economic status of the basin. The tax and revenue collected in Bahirdar and Gondar City, the major population centers of the basin, amounted to 158 million Ethiopian Birr at 2012 prices, which comprised 15.24% of the Amhara region's revenue. We understand then the contribution of the basin to the regional revenue is substantial as compared to that of the entire total.

Lake Tana Basin contributes to the regional economic growth in the form of prominent sectoral activities such as agriculture and fisheries, tourism and navigation, trade and industry, and transport. These sectors are the source of livelihoods of the population in the basin.

21.2.1 Agriculture and Fisheries

In the Amhara region, where LTB is found, the percentage GDP contribution of agriculture and allied activities to the regional GDP in 2014 current price was 59.14% (BoFED 2014). Consequently, agriculture makes a relatively large contribution to the Lake Tana basin economy compared to other sectors. The Amhara Region in general and LTB in particular predominantly produces grain crops that accounted for about 96% of the area cultivated under all crops; showing that agriculture is the main component of the LTB economy. Mixed farming is widely practiced, in which animals are largely used for draught purposes. Out of the available land for farming, cultivated land in LTB in 2011 is estimated to be 1.36 million hectares (CSA 2011). Major crops grown include cereals (rice, wheat, barely, maize, teff, millet), fruits and vegetables. Endowed with good water resources, the basin has been identified as one of the best areas for irrigation site by the Ethiopian Growth and Transformation Plan (GTP) (MoWR, nd). Despite the huge potential for irrigation, irrigated agriculture development is still in its infancy. Large-scale irrigation is a very recent phenomenon. The first large-scale dam, the Koga dam, was built in 2010 to irrigate 7000 hectares of farmland. Small scale irrigated agriculture development is practiced very little in the Gilgel Abbay basin, and the Gumara and Ribb basins. In LTB, an estimated 163,430 farmers had access to irrigation, with a total irrigated area of approximately 21,000 hectares CSA (2011). The rest of the agricultural economy is built on rainfed farming.

In 2011, the total livestock number in the LTB is estimated as 10.6 million stock (2.68 Million total livestock unit-TLU), and in which cattle accounts for 32%, poultry 37%, sheep and goats 24% and horses, asses, mules, camels and beehives account for the remaining 7% (CSA 2009). The potential fish production of Lake Tana is estimated by the Ministry of Agriculture to be 13,000 tonnes per annum. However, the current production is only about 1000–1400 tonnes per year (Tadesse 2008).

21.2.2 Trade and Industry

21.2.2.1 Trade

In this overview, trade implies distributive trade, which is defined as economic activity including wholesale and retail trade; service trades such as repair of motor vehicles. Distributive trade is the exchange of goods and services from the point of production to the point of consumption to satisfy human wants (WTO 2010). It is the supply of material goods to consumers, through retailing and wholesaling or the act or process of buying, selling or exchanging goods and services at either wholesale or retail, within a country or between countries. This activity has been a constant part of human life and a vital element for the proper functioning of markets (CSA 2011).

Distributive trade in Amhara and LTB has a major impact on the everyday lives of the citizens. It engages a multitude of traders who provide the avenues for exchanging commodities and money between producers and consumers and transmit information from the final users to the producers on the trend in demand for and taste of the commodity. It also promotes new ideas, technologies, and investments through the prospect of making a profit (CSA 2011).

Despite the many households that are involved in it, this sector is also in an elementary stage. This is suggested based on the appearance of the sector the number and distribution of distributive trade enterprises including wholesale, retail trade, and motor vehicles trading. With such enterprises, the share of LTB on 2011 fiscal year was only around 5%. The gross value of income of trade enterprises of the basin constitutes only 3% of the Amhara Region's economy due to trade. These figures show that a very small proportion of the LTB economy is attributed to the trade sector. Thus, trade requires enough policy attention to have a relevant position in economic contribution in the basin.

21.2.2.2 Industry

The industrial sector is an important potential part of the economic infrastructure of both the Amhara Region and the LTB. This sector is to "bring economic growth to the region and beyond by offering investors a globally competitive combination of geographic position, infrastructure, services and labour" (CSA 2011). The industrial sector is seen not only to offer investment opportunities, but also to play a pivotal role in the support of economic empowerment, development of small, medium and micro enterprises (SMME's) and job creation. Major industrial composition in LTB include the: manufacture of food products and beverages, manufacture of textiles and clothing, manufacture of wood and products of wood, manufacture of furniture, manufacture of rubber and plastic products, tanning and dressing of leather; manufacture of footwear, luggage and handbags, and electronics manufacturing.

However, the industrial sector is not yet well-developed in this region. According to CSA (2011), there were a total of 233 large and medium scale manufacturing industries reported in 2009/10 in all of Amhara region, with 51 in the LTB. Thus, over 21.89% of the manufacturing industries were located in LTB. The 51 establishments provided employment for 2584 employees in the LTB, and had a capital expenditure of Birr 2.55 million. The total employees engaged in the manufacturing industry in the LTB account for 22.44% of all manufacturing industry employees in the Amhara region. The establishments have been investing their capital for acquisition of various fixed assets, of which, around birr 44.3 million of the total new capital expenditure was spent on new fixed assets. Almost all (96%) of the industries are private holdings.

The value of production is regarded as one of the important variables for measuring economic activity and development of industrial production. In 2013, manufacturing industries in LTB registered a total value of production amounting to 105.7 million birr, which is 7.76% of the Regional value of industrial production (CSA 2011).

Small and micro enterprises (SMEs) , as a constituent of small industries have a major role to play in the economy in terms of employment creation, income generation and output growth. It is estimated that more than 31,081 establishments were found until 2011 and more than 31.1 thousand people in LTB are actively involved in the SME sector (BoFED 2014).

21.2.3 Tourism and Navigation

Tourism and navigation are another economic activity in the Lake Tana area. Lake Tana is one of the country's inland freshwater resources that provide intensive transportation for public goods and tourists. The Lake Tana region is endowed with many cultural and natural assets including Fasil Castle in Gondar, ancient monasteries on islands in the Lake, Tis Issat waterfalls, and some wetland areas with abundant bird and fish species. Owing to this fact, the region is one of the best tourist destinations in the country. According to Tadesse (2008), an average of ETB 2,221,533.00 was collected yearly (1996–2005) from the Tis Issat Falls alone.

21.2.4 Economic Infrastructure

Both economic growth and development in the region depend on the noticeable economic infrastructures including transportation, electricity, telecommunications, road, and sewer infrastructure.

21.2.4.1 Electricity

Hydropower generation is one of the key components of the economic sector in the basin. The LTB, and particularly Bahir Dar, first acquired access to electricity during the period of the Italian invasion around 1941 from diesel generators. In 1953, the Tis Abay hydroelectric power plant was established with the help of the then-Yugoslavian Government. At present, there are two hydropower stations, which are supplying electricity energy. The first is located on the Abbay River 35 km downstream of Bahir Dar city, with a total electricity generating capacity of 84.4 MW, the second is the Tana-Beles hydropower station, having a total installed capacity of 460 MW (BcA 2012).

21.2.4.2 Transport and Telephone Service

The road infrastructure in the basin is asphalt, gravel, and cobblestone roads, which serve transport services via city bus, taxi, bicycle and other vehicles. The total amount of roads in each category has been changing rapidly in the last few years.

One of the systems of urban transport is city bus. Bahir Dar and Gondar in the basin both have a city-bus transport system, although they serve a limited number of passengers. Their carrying capacity is below 300 passengers in one round trip. Each of these buses makes a passenger transportation of not more than 1000 on average every day. Taxi transport service is increasing over time in the basin, which is dominated by Indian-made Bajaj taxis. Local bus services run between towns and cities in the Basin.

Two airports are found in the LTB, located in Bahir Dar and Gondar. Ethiopian Airlines operates scheduled flights between Bahir Dar and the capital, Addis Ababa, as well as with Gondar to the northwest. The two airports are located in the South and North of Lake Tana. As a result, these infrastructures provide support for the tourism sector.

Water transport service in the LTB, mainly in Bahir Dar began in the 1940's, mainly for trade purposes. The first route ran between Bahir Dar and Gorgora at the northern end of the lake. Water transport now runs from Bahir Dar to several locations around the lake. At present, it provides access for an estimated 2.3 million people in 10 Woredas near and around Lake Tana, many of which do not have any inland transportation connections. These areas are notable areas of natural tourist attraction.

The water transport service delivered by governmental boats in 2013, is transporting 76,000 people and 60,000–68,000 quintals of goods/commodities per year. The major transported goods are grains such as oilseeds and spices. Of the total 76,000 people transported, 10,000–15,000 are tourists. Currently, there are about four water transport routes from Bahir Dar for this purpose. Besides, there are also some seven additional trips on holidays transporting some 2000–3000 people in a single trip (LTE, nd).

Five tourist boats in Bahir Dar make multiple trips daily for short distances and one shuttle per two days for long distances. Generally, there are a total of 11 boats in the city (three of them are allocated for commodity transport and the other three are for tourist transport while the rest are mainly for local passengers. There are a total of seven ports (harbors) for these boats around the lake. Beside these operation ports, there are other two additional ports under construction expected to render service in the near future. Of all these 9 ports, the Bahir Dar port is the largest one which can harbor 9 big boats at a time. The Gorgora port is the second largest, which can serve about 4 boats at a time(LTE, nd).

Only the Bahir Dar and Gorgora ports have accompanying facilities such as rail facility, warehouses, loading and unloading services, and waiting facilities. The rest do not have full services.

21.2.4.3 Telephone Service

In the LTB, the telephone service is run with an automatic type of system with a 24-h capacity. The major telecommunication and related service types include telephone, telegram, fax and internet. The number of subscribers for each of these services is not known, although the majority of the urban residents are expected to have access to the service.

21.3 Indicators of Economic Development

Developing a clear picture about economic development requires intensive study and analysis of each of the components presented. Unfortunately, this kind of development data is very limited in the basin. The following section reviews a few of the indicators which are available to highlight the economic development status of the LTB.

The level of the livelihood contribution of the above economic activities are used as indicators of development. In this regard, agriculture accounts for approximately 80% of the employment in LTB (BoFED, 2014). In the basin, subsistence farming accounts for the majority of agricultural activities. Agriculture in the basin is essentially a smallholder activity. The area of each holding of nearly 75% of the holdings is less than 1 hectare. When one takes into consideration agriculture's contribution to the livelihoods of the rural population, where it is often the only source of income, this sector plays a very important role in human survival. Thus, it is possible to realize the economic development of the basin is at a subsistence level (CSA 2011).

21.3.1 Employment

Aggregated data on employment and unemployment level for the entire LTB is unavailable. Therefore, the CSA data of 2011 for Bahirdar and Gondar cities is used for initial consideration. In these two cities, 77% of the total economically active population is employed (CSA 2011). Bahirdar Town had a record 23% urban unemployment rate in 2011. Unemployment data for the Amhara Region as a whole indicate 19.3% of the urban labour force was unemployed. Thus unemployment in Bahirdar is higher than the regional value. In addition, the unemployment rate in Bahirdar is higher for female labour force (29.3%) than the male labour force (16%) (CSA 2011).

A closer look at the Bahirdar subdivision employment figures from the 2011 data of CSA indicated, 77% of the people who were urban active labour force was employed; employment in major industrial divisions showed that most were participating in the wholesale and retail trade (15.43%). The next largest group was involved in construction industries (14.64%) followed by manufacturing at 10.84%.

The hotel and restaurant (tourism industry) employed 9.25% in 2011. According to these results, this labour force participation highlights that more employment is in primary sector. The fact that majority of the employment is in the primary sector of the economy manifested that the basin is still in the under development stage.

21.3.2 Incidence of Poverty

Incidence of poverty is an overall indicator of the development of a region. In a study about the relationship between poverty and natural resource degradation, Daregot (2012) found that the incidence of poverty in the Lake Tana basin in terms of the cost of basic needs approach (CBN) was 29.4%. This incidence of poverty is moderate. It means that 29.4% of the basin's population were deprived in terms of basic food and nonfood necessities.

21.4 Natural Resources Role in Local Economy

No basin wide economic analysis on the role of the Basin's natural resources in local economy has been done at the scale of the whole Lake Tana Basin. However, Daregot et al. (2012) estimated the economic value of land and water resources for LTB and found that the basin's resources have significant economic value.

The economic value for water resources was determined based on the residual value methodology, whereas for land, one of contingent valuation techniques, mean willingness to pay, was applied. The results showed that the yearly value of the total economic land resource was about ETB 3.6 billion in 2010/11. Among these, the highest share is by agricultural land, which was Birr 2.70 billion. Likewise, the economic value of water resources in LTB was estimated to be ETB 4 billion in 2010/11 accounting period. Finally, the total economic value of land and water for LTB was estimated to be ETB 7.6 billion in 2010/11 accounting period.

This estimated yearly value of land and water accounts for 31.15% of the regional gross domestic product (RGDP) of the Amhara region, which was estimated 24.4 Billion in 2010 fiscal year (BoFED 2010). In this study, effort was made to include the community's willingness to pay and accept so that it makes the estimates more reliable.

21.5 Conclusion

This chapter presented an overview of the economic situation in the Lake Tana basin. It is necessarily a shallow description, since little rigorous economic analysis at the scale of the whole Tana Basin has not yet been done.

However, even from this broad-brush description, we can see that a significant proportion of the Lake Tana Basin's economy is attributed to the agricultural sector and that the Basin harbors considerable untapped potential for irrigation development. This calls for a better investigation of the development of a resources efficient modern agriculture which will contribute for the economic development of the Basin. The rest of the sectors such as industry, are also in a very early stage of the economic activity, and need more attention to be developed with optimized use of the resources of the basin.

We also conclude that detail natural resources investigations, valuation and economic analysis through economic models have to be developed to value the status, trend and forecasts of the Basin's resources. If this will be undertaken their significant contribution to the economy and livelihood of the population can be well documented, and applied for policy actions.

References

Bahir dar city Administration (BcA) (2012) Offical website from http://www.bahirdarcity.net. Accessed on March 2012

Bureau of finance and Economic Development (BoFED) (2010) Development indicators of Amhara Region 2007/08. Bahirdar, Ethiopia

Bureau of finance and Economic Development (BoFED) (2014) Development Indicators of Amhara Region 2013/14. Bahirdar, Ethiopia

Central Statistical Agency (CSA) (2003) Agricultural sample survey: reports for 2001/2002. Central Statistical Agency, Addis Ababa

Central Statistical Agency (CSA) (2009) Agricultural sample survey: reports for 2008/2009. Central Statistical Agency, Addis Ababa

Central Statistical Agency (CSA) (2011) Agricultural sample survey: reports for 2010/2011. Central Statistical Agency, Addis Ababa

Daregot B (2012) Natural resource degradation–poverty interactions and farmers' responses: implications for land and water resources conservation in Lake Tana Basin, Ethiopia. PhD dissertation presented to Haramaya University, May 2012

Daregot B, Ayalneh B, Belay K, Deginet A (2012) Poverty and natural resources degradation: Analysis of their interactions in Lake Tana Basin. J Int Dev. doi:10.1002/jid.2914

European Communities (EU) (2008) Statistical classification of economic activities in the European community. European Commission, Europe

Ingraham M, Foster S (2008) The value of ecosystem services. Ecol Econ 67:608–818

Lake Tana Transport Enterprise (LTE) (nd), offical website http://laketanatransport.com/en. Accessed on June 2015

Ministry of Water, Irrigation and Energy (MoWR) (nd) official website from http://www.mowr.gov.et/index.php?pagenum=3.3&pagehgt=1000px. Accessed on July, 2015

Tadesse A (2008) Application of the WEAP model to assess the water resource implications of planned development in the Lake Tana Subbasin, Ethiopia. Master's Thesis Presented to Addis Ababa University. 140p

Todaro M, Smith SC (2012) Economic development. Addison-Wesley, USA

World Trade Organization (WTO) (2010) Measuring trade in services https://www.wto.org/english/res_e/statis_e/services_training_module_e.pdf

Chapter 22
Land Use Distribution and Change in Lake Tana Sub Basin

Amare Sewnet Minale and Wubneh Belete

Abstract Lake Tana, the biggest lake in Ethiopia, is very important water resource for community living and depending on the lake's resources. It plays a role in balancing of the microclimate of local areas in the catchment of the lake. However, the recent development activities at the catchment areas have negatively affected the natural systems of the Lake Tana's catchment. The catchment system and its interface with atmosphere, lithosphere, hydrosphere, and biosphere assume greater importance to ecosystem management. The study of land use and cover changes will help to apply the appropriate land use and land use palanning. The changes in different land cover units such as forest, wood and bush lands, grass, wetlands and water bodies, and farm and settlements were analyzed from different sources. Population change, land tenure, poverty and lack of market and credit facilities in the watershed area were analyzed as causes of land cover changes. There is a need to work against the causative variables to solve the resource degradation in the catchment. Despite the large extent of land converted to farms and settlements, individual house holding of land and household incomes have declined. This is perhaps due to the absence of coordination between the land distribution policy and the family planning policy. There is degradation of vegetation such as forest and grasses and wetland vegetation (riparian vegetation). A detailed inspection of land use change (LUC) change is also crucial for those planners and decision makers to consider environmental and socioeconomic impacts of the land use cover changes (LUCC) in the LTB and design sustainable solutions.

Keywords Catchment Lake · Tana Land degradation · Land cover land use

A.S. Minale (✉)
Department of Geography and Environmental Studies, Bahir Dar University, Bahir Dar, Ethiopia
e-mail: amare1974@gmail.com

W. Belete
Amhara Design and Supervision Works Enterprise (ADSWE), Bahir Dar, Ethiopia
e-mail: wubnehb@yahoo.com

K. Stave et al. (eds.), *Social and Ecological System Dynamics*,
AESS Interdisciplinary Environmental Studies and Sciences Series,
DOI 10.1007/978-3-319-45755-0_22

22.1 Introduction

Land is one of the most valuable assets representing surface and space that provides food, filters and stores water, and is a basis for urban and industrial development, leisure, and a wide range of social activities. Land is one of our most precious assets and finite in its extent (Herold et al.,2008). The land surface represents the solid part of the earth's crust, in contrast to the areas covered by sea and oceans. Land is, moreover, constantly under threat of degradation, mainly as a result of intensive cropping, soil mining, and inadequate management, and of population pressure. It is this process which leads to land cover and use change. The growing demand for food cannot be met indefinitely through intensification of crop production and biotechnological progress; it also requires the extension of arable land. The overall result is an increasing competition for land and contributing to land cover change overtime.

In Ethiopia some micro level studies from aerial photo and satellite images have revealed agricultural land expansion at the expense of other land uses. According to these studies, it was confirmed that the significant increase in cultivated and settlement lands at the expense of forestland, wetlands, riparian vegetation, grass lands and open access areas (Amare and Kameswara 2011, 2012; Nyssen and Poesen 2004). Similarly, the studies by (Amare and Kameswara 2011, 2012) in Gilgel Abay catchment have shown continuous declining of forest, wetlands and expansion of farm and settlement area between 1973–2008. The simultaneous upward surge in both human and livestock populations would, therefore, bring about the depletion of the biological resources. Institutional issues like political, legal, economic, and traditional and their interaction with individual decision making are important in explaining land use/cover changes (Lambin and Geist 2003).

Land cover changes in watershed systems can cause local, measurable changes in watershed systems. Studies have indicated that the land cover/use change is also related to hydrological and climate changes at global, regional and local scales. It is also responsible for releasing greenhouse gases to the atmosphere and increases the release of carbon dioxide to the atmosphere caused by deforestation, especially when followed by agricultural expansion (Christy et al. 2006; Ezber et al. 2007; Nunez and Ciapessoni 2008).

Concerns about land-use change emerged on global environmental change several decades ago with the realization that land surface processes influence climate, modifies surface albedo and loss of biodiversity. This chapter reviews the land use, its change and the major cause of land use in Lake and Bahir Dar areas. Hence, main objective of this review is to assess the land use distribution and changes in the Lake Tana Basin.

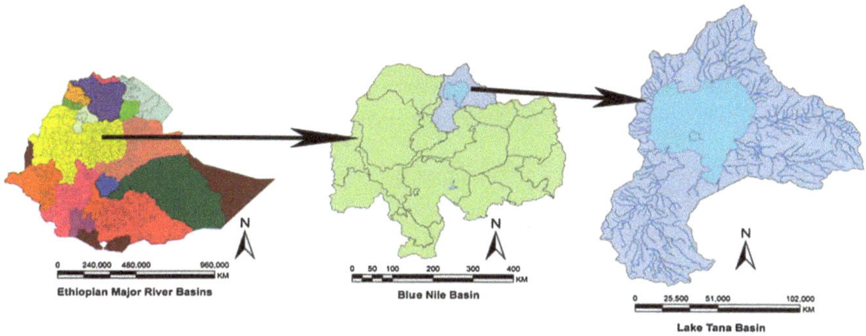

Fig. 22.1 The Lake Tana Basin

22.2 Study Area Description

The Lake Tana region is situated in the northwestern part of the Ethiopian Highlands in Amhara National Regional State. Geographically situated between latitude 10°58′–12°47′N and longitude 36°45′–38°14′E, the watershed consists of 347 Kebeles and 21 *Woredas* (districts) in four administrative zones (IFAD 2007) (Fig. 22.1). With a surface of 3156 km^2 stretching approximately 84 km north–south and 66 km east–west, Lake Tana is the largest lake of Ethiopia and one of the largest in Africa. The average annual rainfall in the watershed is about 1350 mm and is concentrated from June to October. The average annual temperature is about 28 °C. The topography of Lake Tana watershed consists of 48.4% plain, 36.1% rugged and 15.5% valley. The topography varies greatly on the peripherals of south and north sides of the watershed. It is generally flat on the east and west side of the Lake Tana. Undulating slopes are found in northern and southern part. In areas with flat topography, stream flow is relatively slow, and floodwaters tend to spread out into adjacent lands such the Fogera and Dembia floodplains around Lake Tana.

22.3 Terminology and Concepts on Land Use Change

The meaning given to land in different area is not uniform and straightforward. The simple term of land, for example, can be interpreted in a wide variety of ways according to outlook and perspective. In the strict sense it can stand for soil; in a broader sense as a consumer good or commodity, as location, property or a form of capital; or in an ecological view when it is associated to nature and ecosystems (Verheye 2004). These can generally be associated to three major concepts: land as a country or a homeland, land as a type of soil with specific biophysical and chemical properties, and land in terms of property and asset. This indicates that the meaning and the perception of society as the land is a gift from God that cannot be alienated to humans, but that can only be temporarily used by them. When dealing with land different people (farmers, real estate managers, scientists, politicians) look at land very differently, and in particular refer more often to what is happening at the surface of that land in terms of ecological, socio-economic or legal aspects, and what are the consequences of this in a regional or global context. This creates ambiguity and confusion in its understanding.

The other meaning is currently used in natural sciences areas, which associated the land to the concept of soil or ground. This part of concept on land embraces mainly its physical properties like depth, texture, wetness, etc. It is currently believed meaning of land and most commonly accepted by people from various disciplines. But after FAO (1976) publication of Framework for Land Evaluation, land has received a broader meaning than the narrow connotation of soil. The FAO definition of land is "an area of the earth's surface, the characteristics of which embrace all reasonably stable, or predictably cyclic attributes of the biosphere vertically above and below this area including those of the atmosphere, the soil and underlying geology, the hydrology, the plant and animal populations, and the results of past and present human activity, to the extent that these attributes exert a significant influence on present and future uses of the land by man" (FAO 1976).

Land in its economic context, is considered as one of the major factors of production, including all natural resources, but different from capital. Although the chief economic role of land is vital for the production or provision of food, water, fuel and shelter, it is also a consumer good, to the extent that it supplies space for residential or industrial sites, infrastructure or parks and recreation. Competition among users might create rent differentials, but these result only from differences in fertility or location; in rural areas rent is mainly determined by the yield potential of the land; in more industrialized countries rental differences are mainly due to the location of plots, the one compared to the other. Hence, the value or rent of land in city centers may reach astronomical levels, whereas plots in remote rural areas may have much less value. In general, land is one of our most precious assets and finite in its extent.

22.3.1 Land Use Change

Land use describes human uses of the land, or immediate actions modifying or converting land cover. It is the change of the original land to human settlements, protected areas and agriculture. It involves both the manner in which the biophysical attributes of the land are manipulated and the intent underlying that manipulation—the purpose for which the land is used (Turner et al. 1995). In a similar way, Meyer (1995) stated that land use is the way in which, and the purpose for which, human beings employ the land and its resources. Briefly, land use denotes the human employment of land (Turner and Meyer 1995). Skole and Cochrane (2004) stated that land use itself is the human employment of a land-cover type. FAO (1995) explained the land use concerns the function or purpose for which the land is used by the local human population and can be defined as the human activities which are directly related to land, making use of its resources or having an impact on them.

The land cover refers to the natural vegetative cover types that characterize a particular area. Land cover is the biophysical state of the earth's surface and immediate subsurface (Meyer 1995). Land cover embraces, for example, the quantity and type of surface vegetation, water, and earth materials. Moser (1996) noted that: the term originally referred to the type of vegetation that covered the land surface, but has broadened subsequently to include human structures, such as buildings or pavement, and other aspects of the physical environment, such as soils, biodiversity, and surfaces and groundwater. The conversion of natural lands to croplands, pastures, urban areas, reservoirs, and other anthropogenic landscapes represents the most visible and pervasive form of human impact on the environment (Turner et al. 1995). Today, roughly 40% of Earth's land surface is under agriculture, and 85% has some level of anthropogenic influence (Sanderson et al. 2002).

The removal or destruction of significant areas of forest cover has resulted in an altered environment with reduced biodiversity. In many countries, deforestation is ongoing and is shaping climate and geography. Deforestation results from removal of trees without sufficient reforestation, and results in declines in habitat and biodiversity, wood for fuel and industrial use, and quality of life (FAO 2005). A dramatic example is Ecuador whose Amazon region's forest canopy is facing rapid attrition owing to growing settlements of frontier farmers, although overall rural population is declining because of falling fertility and rapid urbanization (Pan and Walsh 2004).

By altering ecosystem services, changes in land use and cover affect the ability of biological systems to support human needs. It also determines the vulnerability of places and people to climatic, economic or socio-political perturbations. These are generally a reflection of the local climate and landforms, can be altered by human actions and natural factors. Take, for example, conversion of forested areas to crop lands, pasture or human settlements that can result in the loss of biodiversity, especially in the tropics. Inturn the biodiversity loss results in declines in ecosystem integrity, and also genetic losses that may impede future scientific advances in agriculture and pharmaceutics, impact hydrological processes, leading to localized declines in rainfall, and more rapid runoff of precipitation, causing flooding and soil erosion. Clearly, all of these changes impact society. This dual role of humanity in both contributing to the causes and experiencing the effects of global change processes emphasizes the need for better understanding of the interaction between humans and the terrestrial environment. This need becomes more imperative as changes in land use become more rapid. Understanding the driving forces behind land-use changes and developing models to simulate these changes are essential to predicting the effects of global environmental and local change (Veldkamp and Lambin 2001).

Land cover and use change plays a major role in climate change at global, regional and local scales. LCUC is responsible for releasing greenhouse gases to the atmosphere, thereby driving global warming and is influenced by growing concentrations of greenhouse gases, which absorb solar radiation and warm the atmosphere (Lambin and Geist 2006). Land-use and land-cover changes have impacts on local and regional hydrological balances. The impact of land-use and land-cover change on hydrological cycle is not yet adequately assessed. However, a massive removal of forest in the Amazon has led to a decrease in evaporation and precipitation in the region (Turner et al. 1995). Land-use and land-cover changes also, especially vegetation cover, affects water and energy balance (Houghton 1995). In the past 50 years, the construction of dams and reservoirs has become important part of human induced land cover changes. Impacts of land cover changes that occur due to artificial water bodies are beyond their proportion of aerial extent. The type of land cover, obviously, can affect both rate of infiltration and runoff amount by following the coming of precipitation (Houghton 1995). According to Turner et al. (1995), both surface and ground water flows are significantly affected by type of land cover.

22.3.2 Causes of Land Cover and Use Change

The degree as well as the extent of the problem of land covers and use change varies from the region to region, land cover and use change can be caused by both direct and indirect factors. The direct causes of land cover and use are clear and easily observable and these include production on steep slopes and fragile soils with inadequate investments in soil conservation, erratic and erosive rainfall patterns, declining use of fallow, and limited recycling of dung and crop residues to the soil, limited application of external sources of plant nutrients, deforestation, and overgrazing. Underlying these proximate causes include many other factors such as population pressure, poverty, high cost and limited access to agricultural inputs and credits, low profitability of agricultural production and many conservation practices, fragmented land holdings and insecure land tenures and farmers lack of appropriate techniques of technology. All these in turn are affected by the government policies related to infrastructural development, market development, input and credit supplies, land tenure, agricultural research and extension, conservation programs and local governance. They can be categorized as natural, socio-economic and institutional and policy factors (FAO 1995).

Natural Factors: The natural factors causing land degradation, rainfall intensity and steep slopes are the main ones. The impact of raindrops on bare unprotected soil with tremendous energy starts the process of erosion by water (Lambin et al. 2003; Sherbinin et al. 2002). The impact of rain drop causes splash erosion which seals off infiltration as the soil pores get plugged with the particles. Splash erosion grows into sheet erosion and then rill erosion. Finally, the rills end up with formation of big gullies, which accelerate the erosion processes. The impact of the rain drop depends on the level of plants cover, which is lacking due to socio-economic problems of deforestation, over cultivation and over grazing. The slope of the land that is its steepness and gentility affect the degradation of the land. Ethiopia is country more than 45% of the area is considered as highland. In combination with other factors the slope has influenced land degradation in Ethiopia.

Socio-Economic and Institutional Factors: Socio-economic and institutional factors affect land degradation through their impacts on farmer's decisions with respect to land use and land management practices such as plowing, fallow, use of manure and other sources of organic matter, fertilizer use and adoption of soil and water conservation measures. Some of these factors include population pressure, poverty, land tenure relations, the nature of markets, local institutions and organizations and farmer's perceptions and attitudes.

Population Pressure: The interrelationship between population growth and environmental degradation was controversial and debatable for several years. Similarly, in Ethiopia population pressure has been found to have negative effect on scrublands, riparian vegetation and forests in Kalu district (Kebrom 2000), riparian trees in Chemoga watershed (Woldeamlak 2003), and natural forest cover in Dembecha Wereda north-western Ethiopia (Gete and Hurni 2001). Unlike these studies, Muluneh (2001) reported for Sebat Bet Guraghe that population growth was found to have a positive impact on forest cover. Similarly, different micro studies from aerial photo and satellite images interpretation of forest cover changes, also have shown the remarkable difference between the past and the present forest cover in Ethiopia. For example, the comparison of forest coverage in Ankober between 1975 and 1986 (Woien 1995, cited in Nyssen and Poesen 2004), the land cover dynamics between 1957 and 1998 in Chemoga watershed (Woldeamlak 2003) and the impact of population on land cover change between 1957 and 1982 in west Guraghe land (Muluneh 2001) all showed a slow but steady expansion of planted forest due to increasing cover of eucalyptus trees.

Nexus between Poverty and Environmental Degradation: The World Bank defines poverty as "the inability to attain a minimal standard of living" (World Bank 1990). Poverty is generally defined in the following two ways: lack of "means" in relation to "needs" and lack of "means" in relation to "means" (Sen 1999). Sociologists also distinguish between relative and absolute poverty. Absolute poverty occurs when people fail to receive sufficient resources to support a minimum of physical health and efficiency, often expressed in terms of calories or nutritional levels. Relative poverty is defined by the general standards of living in different societies and what is culturally defined as being poor rather than some absolute level of deprivation. When poverty is defined relatively, by reference to the living standards enjoyed by the bulk of a population, poverty levels vary between societies and within societies over time. Poverty is a complex problem and is product, at least, in part, of political process and policy development. Poverty is not just about material deprivation but also about low levels of health and education, and vulnerability and exposure to risk.

Poverty is said to be both cause and effect of environmental degradation. The "poverty trap" or "downward spiral" theory suggests that poor people, as they live from hand to mouth, cannot afford to invest in resource conserving practices so instead they continue to plough up hillsides and overgraze land as they cannot wait for rangelands to recover. Thus, these practices will lead to increased land degradation and worsen poverty. However, this topic is frequently debated (Scherr 2000; Markandya 2001).

The circular link between poverty and environment is an extremely complex phenomenon. Inequality may foster unsustainability because the poor, who rely on natural resources more than the rich, deplete natural resources faster as they have no real prospects of gaining access to other types of resources. Moreover, degraded environment can accelerate the process of impoverishment, again because the poor depend directly on natural assets (Lindblade et al. 1998). Poorer people, who cannot meet their subsistence needs through purchase, are forced to use common property resources such as forests for food and fuel, pastures for fodder, and ponds and rivers for water. It also contributes to environmental degradation through over exploitation of natural resources like land, air and water.

However, many scholars argued on the interrelationship between the poverty and environmental degradation. Some scholars believed that poverty leads to environmental degradation as poor people are forced to use the surrounding natural resources for their survival. Bekele and Holden (2001) showed the intensified pressure on natural resources as a vicious circle in which resource degradation and drought affects household assets and household assets in turn affects degradation of the environment. They further argued that because of lower threshold opportunity cost, poor people may bear lower levels of productivity and environmental quality than those who are better off and have access to alternative livelihoods. This can lead to higher resource depletion than would be the case of if they had the choice to move on to alternative income generating methods. By supporting this view, the World Commission on Environment and Development Report (1987) said many parts of the world are caught in a vicious downward spiral: poor people are forced to overuse the environmental resources to survive from day to day.

On the other hand, some others argued that poor people also have to invest in the environment for their future survival, so cannot afford to degrade the environment beyond repair since they have no other means of survival. Harrison (1992) notes that "bigger farmers are more likely to use tractors, own more livestock and, if not properly managed, can do more environmental damage than the poor". Harrison also sees as for getting hold of large concessions of forest land, clearing and farming them with hired labor and tractors, the better-off are clearly the most likely to do that. There is much value in this analysis, as well as in the observation that "consumption and waste per person is also lowest among the poorest" and the conclusion that all in all, the poor probably tread lightest of all upon the earth, and do less damage to the environment than any other group. They are "victims, not perpetrators" (Harrison 1992; Alf Mortem 2002).

Poverty is common phenomena in the rural part of Ethiopia, because of lack of land to plow and technology to assist to produce more from small plot of land. The ultimate result is therefore, land degradation: deforestation, soil erosion, nutrient depletion and loss in productivity and income. Thus households are forced to find new land to farm and/or graze their animals and vicious circle of land degradation and conversion begins again. The study conducted by Alemu (2003) in Ethiopia showed that the poor people's perceptions of wel being are strongly related to the environment in terms of their livelihoods, health, vulnerability and empowerment to control their own lives. In Ethiopia GDP was very low but the incidence of poverty was high. This shows that dependency of the poor people on the environmental resource is high in the country.

Linkage of land tenure and land cover change: The total land area of the world is fixed, except for minor changes in the increase or decrease caused by coastal sedimentation, volcanic eruptions, earth-quakes, or sea level movements. The extension and production potential of cultivable land can nevertheless vary significantly: it can increase through drainage, land leveling, irrigation, and other human interventions, and it can decrease as a result of water and wind erosion, salinization, depletion of mineral nutrients, and reductions in biological activity. In analyzing statistical land data it is therefore important to define clearly what type of land area is meant.

Investment decision on land is affected by the tenure security (Besley 1995; Gavian and Fafchamps 1996). Since it determines access to land, it is critical variable in the management of natural and environmental resources, soil conservations, and water as well as wildlife management. With regard to factors which discourage land conservation efforts, it has been argued that only private ownership makes it worthwhile for peasants to care about sustainability of their farming methods (FAO 1983). Systems of land ownership as well as tenure and business arrangements that do not provide security to the farmers are held to be major obstacles to conservation.

Indeed a number of studies have illustrated the differences of behavior between individual owners and renters. Rented lands usually are the most degraded. But a closer look often reveals that renters with long-term use rights can be quite as inclined to improve the land as owners are. Security of tenure, not ownership, is decisive, because it enables farmers to reap the benefits from their investments. Conversely short-term land leases are "among the most pernicious" arrangements from this standpoint (FAO 1983).

The major problems of the pre-1975 land tenure in Ethiopia as indicated by different studies include exploitative tenancy, land concentration and underutilization, tenure insecurity, and diminution and fragmentation of holdings (Fesseha 1970). Tenure insecurity was cited as one of the limitations of the pre-revolutionary/reform land tenure system. It manifested in various forms ranging from endless litigation over land rights to complete eviction from holdings (Alemante 1970). The 1975 land reform by the Derge has been considered by many as a radical measure that has abolished tenant—landlord relationships in Ethiopia. The reform was designed to alter fundamentally the then agrarian relations and make those working the land owners; increase agricultural production; create employment; distribute land and increase rural income; and provide a basis for agricultural expansion (Yigrmew 2002). A survey of the literature on land tenure during the Derge regime (1975–1990) generally showed that diminution and fragmentation of holdings, tenure insecurity and all its consequences, land degradation, and inefficient allocation of land by way of restrictions on land transfer and to some extent lack of appropriate land use and administration are the most commonly cited problems.

Immediately after the downfall of the Derge, the land is announced as the property of the state. However, a study by Bruce and Shem Migot-Adholla (1993) identified as many people are considered as landless by their community; inheritance, sharecropping, cash rentals, disguised land sales and possessors mortgages have also been important means of gaining access to land. Other researchers have also concluded that there are problems with the current land tenure system. From recent studies in Amhara, Oromia and Tigray regions, Tekie (2000) considered that the government had only one imperative policy option: a movement away from the existing insecure tenure system towards a more stable and secured system. Thus, land tenure system existing so far in Ethiopia has not been secured and privatized. Side by side as many studies indicated mainly in highlands of Ethiopia land is seriously degraded owing to problems related to land tenure security and other factors of degradation.

Local market development and public policies: To a large extent, environmental degradation is the result of market failure, that is, the nonexistent or poorly functioning markets for environmental goods and services. In this context, environmental degradation is a particular case of consumption or production externalities reflected by divergence between private and social costs or benefits. Market distortions created by price controls and subsidies may aggravate the achievement of environmental objectives. The nature and development of markets for factors of production (land, labor, draft animals and credit), inputs and outputs can play a major role in determining patterns of land use and land management. Where markets are well developed and competitive, farmers can be expected to respond largely to the profitability of alternative land uses and management options and outcomes are likely to be relatively efficient, though not necessarily resource conserving (Singh 1996).

The pervasive "urban bias" in macro-economic management also played its role, for instance by promoting cheap food and fuel for urban consumers. To the extent that low producer prices discouraged more intensive production, and the unpredictable behavior of public marketing agencies increased farmers' risks, land-clearing for further extensive production and/or the shortening of fallow periods was promoted (Gorse and Steeds 1987). More generally, cheap food and low agricultural prices have kept land value low, making its conservation unattractive.

Improper agricultural activity: Direct impacts of agricultural development on the environment arise from farming activities, which contribute to soil erosion, land salination and loss of nutrients. The spread of green revolution has been accompanied by over exploitation of land and water resources and use of fertilizers and pesticides have increased many folds. Leaching from extensive use of pesticides and fertilizers is an important source of contamination of water bodies. Intensive agriculture and irrigation contribute to land degradation particularly salination, alkalization and water logging (Scherr 2000).

It has been suggested that agricultural intensification will lead to declining soil fertility, which may destroy the ecological basis of agriculture. Intensification of agriculture and intensive use of chemicals can be damaging to the natural flora and fauna (Ironmonger et al. 2002), notably birds and mammals, and can lead to the loss of specific habitats and species, and a rather monotonous landscape. Furthermore, these chemicals impact the quality of ground water and surface water and thus extend well beyond agricultural landscapes.

In Ethiopia a number of ways existed in which population increase impinges on land quality. Firstly, by increasing use of chemical inputs like fertilizers and pesticides, the productivity of once fertile land has been affected in many areas. Secondly, by increasingly converting marginal lands (e.g. deforested lands, converted wetlands, etc.) to crop cultivation they affect both the natural ecosystems as well as the productivity of such marginal lands. Thirdly, pressure is increased on non-crop ecosystems such as forests and wetlands to convert them to croplands. Thus, for example large parts of the floodplains have been converted from natural wetlands to relatively dry-lands for crop usually rice, in eastern part of Lake Tana (LTRC 2004). Some implication of land degradation and deforestation of lake catchments area are increased surface runoff, ultimately leading to slight rises in lake levels because of sedimentation (Geremew 2000). Most of the lakes that have undergone considerable changes are those located in terminal positions.

22.4 Lake Tana Basin and Its Land Cover/Use Changes

Ecological and economic evidence have shown that loss of biodiversity and decrease in land productivity are the major problems in Ethiopia. With continued population growth the problem is likely to be even more challenging in the future (FAO 1986; Sutcliffe 1993; Keyzer and Sonneveld 2001). In Ethiopia, few cases, for instance; the forest assessment made by EFAP (1993) has identified the causes for clearing of forest were related to population pressure, environmental policy, land tenure systems and poverty. As the studies in some parts of the country indicated that the general trend of land cover change was also an increase in cultivated land at the expense of woodland and grassland areas (Kebrom and Hedlund 2000; Gete and Hurni 2001). At the national scale, forest decreased from 16% in 1950 to 2.7% by the early 1990s (Million 2001). Similar findings with some exceptions (Woldeamlak 2003; Solomon 2005) were reported with respect to expansion of cultivation and reduction of forest and grazing lands elsewhere.

In other studies made in the central highlands, homesteads were reported to increase, during 1957 and 1986 (Woien 1995), which may indicate increasing population density. Girmay (2003) in his study, in Southern Wello, reported the decline of natural forests and grazing lands due to conversion to croplands. Similarly, Feoli (2002) also reported the expansion of bush land and evergreen vegetation with population increases. However, such expansions of cultivation, commonly into steeper slopes and marginal areas, may have been done without appropriate soil and water conservation measures. As a result, these lands become unproductive in short period of time, leading to soil erosion.

The undulating slopes and soils types facilitate the natural erosion and landslide problems that exist in the North and South highlands. The undulating slopes increase rain water runoff rates, which can increase erosion and sediment in the water bodies, flush stream banks, and deposit sediments in down streams. According to Nurelegn and Amare (2015) in the Ribb River watershed there was continued expansion of cultivated and settlement over years which has brought significant decrease in water bodies, forest and bush LULC classes. The rapid loss of habitat and biodiversity, removal of riparian vegetation, and deforestation are common threats in the watershed. Environmental degradation in Lake Tana Basin can be explained in terms of water, land and soil degradation caused by rapid population growth, improper agricultural activities, lack of proper environmental policies and urbanization especially at the southern gulf of lake. Land degradation in the basin has increased dramatically in recent years due to expansion and intensification of agriculture which was needed to feed rapid increase of population. Overgrazing and subsequent soil erosion is also another problem in the area. This threatens both the local highland users through a reduction in soil productivity, and lowlands through sedimentation.

High population densities within the basin of Lake Tana have been associated with a series of deleterious trends, in particular those arising from the clearance of vegetation for grazing and agriculture, resulting erosion and downstream nutrient and silt loading. In addition, most of the catchment escarpments and wetlands (flood plains) are mostly affected by deforestation and overgrazing as results of this a change in composition of vegetation and loss of water through evaporation and runoff is manifested (Girmay 2003). Ever increasing population, coupled with unsustainable land use and farming systems, catchment degradation and overgrazing has resulted in serious degradation of wetlands in the area. As a result, wetlands are converted into permanent cultivation and grazing fields. Due to deforestation and degradation of the surrounding catchment, water flow into the wetlands through infiltration has decreased and wetlands hold insufficient water. There is high siltation on wetlands due to erosion in the catchment. Another problem observed was the expansion of plantation of unsuitable plants in wetlands, including eucalyptus.

Different studies also indicated that there have been continues land cover changes that have been recognized in Lake Tana Watershed had shown continuous expansion of arable land in order to meet the increasing food demands of the growing population. Arable land expands at the expense of forest, grass and wetlands. In the watershed, there was declining of grass, wetlands and forest land covers from 1973–2008 (Amare 2013). The land covers of wood and bush lands and farm and settlement areas have been increasing for the whole periods (Amare 2013). Similar studies in 2002 has identified that about 54.5% in the area was cultivated land, water 21%, grass land 10.35%, shrub land was about 8.95%, 1.6% was wetland/swampy, plantation forest1 percent, rock only 0.53%, settlement 0.36%, bare soil 0.22 and afro-alpine only 0.15% (WBISPP 2002).

The study by Birru (2007) from the analysis of satellite images between 1985 and 2001 in Lake Tana Basin croplands were found to have increased to about 4.2% in 15 years (between 1985/86 and 2001/03), which largely occurred at the expense of grasslands and shrub lands. From the similar studies forest cover in the basin was found to have increased by about 0.23%. According to Alemayehu et al. (2009) in the years after 2009, there was declining in water level of Lake Tana influenced by the land use cover change of the watersheds. In Lake Tana basin, the steep slopes or mountains are cultivated. At present, about 27% of the rugged topographies with greater than 15% slopes were found to be threatened by severe soil erosion and 60 tons per hectare per year. Wondie (2010) indicated that land use/cover in the upstream of Ribb and adjoining watersheds are being degraded contributing for the flood hazard prevailing in the area. Since the Ribb watershed has been subjected to prolonged use for agriculture without conserving natural resources, forest degradation, loss of biodiversity, shortage of fuel wood and forage trees are vegetation related problems existing in the area (Getnet 2011).

Population and overgrazing also contributed for aggravating the loss of vegetation in Ribb Watershed (Getnet 2011). The natural forest was reported to have cleared significantly in the area due to the expansion of cropland, fuel wood consumption, timber production and construction for the increasing urban and rural population (Wood 2000; Teshome et al. 2004). These declining resources intern can cause decline in soil fertility and then agricultural productivity. In Ribb Watershed erratic rainfall, flooding, soil erosion and fertility decline, lack of potable and irrigation water, and water logging are the major problems related to soil and water. The introduction of rice in the lower parts of the watershed brought the grazing lands of Fogera plain changed to cultivated land (IWMI 2009).

Another study by Amare and Kameswara (2011) in Gilgel Abay Catchment area of Lake Tana Basin has found that poverty has become the major trigger to deforestation in the catchment. This has led to expansion of agriculture and settlement lands by clearing forest, grass and wetlands. In Gilgel Abay Catchment land cover and use changes occurred during the last 35 years. The farmers in the catchment, mainly at the upper part have not produced enough food per year, deforestation, declining of grass land and expanding of agricultural land at the expense of other land covers were the implications of the presence of land cover and land degradation in catchment. Demand for agricultural land, timber, and other forest products, as well as technological change in agriculture, significantly impacts the mode and rate of transformation of forested areas. Gilgel Abay catchment being situated in northwestern highlands of Ethiopia has faced rapid deforestation. The rate was high and about 72.3% of forest cover was lost from 1973 to 2008 by the loss rate of 2.1% per annum. At the beginning of the study year (1973) its proportion was 1.9% but at the end of the study year (2008) it lowered down to 0.5%. This shows deforestation rate was rapid (Amare 2013).

22.5 Conclusion and Recommendation

22.5.1 Conclusion

The catchment resources have been degraded as the result of mainly anthropogenic manipulation. The major irony being, a large extent of land has been converted to farms and settlements, but the individual house holding of the land has declined. This implies that despite more area under cultivation, the household incomes have reduced. This is perhaps due to the absence of coordination between the land distribution policy and the family planning policy. There is degradation of vegetation such as forest and grasses and wetland vegetation (riparian vegetation). A detailed inspection of LUCC change is also crucial for those planners and decision makers to consider environmental and socioeconomic impacts of the LUCC in the LTB and design sustainable solutions. But the extent, trends and nature as well as the causes and implications of the LUCC change over time and space was little understood in the study LTB.

22.5.2 Recommendation

There is an urgent need to limit the population growth rate in the LTB region with provision for incentives, for those who comply with the family planning policy, in the land distribution and other facilities like credit facilities, subsidized supply of fertilizers and seeds to liberate community from poverty. There should be land use planning by identifying the proper land for specific purpose so that the marginal lands will not put into use. Participatory catchment management should attempt at ensuring sustainability of the ecological, economic and social exchanges taking place in the catchment territory. There should be also detail land use study by using high resolution image.

References

Alemante GS (1970) Legal aspects of agricultural land disputes. MLRA, Hawaii, p 218–234

Alemayehu T, McCarthy M, Kebede S (2009) Simulation of water resource development and environmental flows in the Lake Tana Sub basin. In: Awulachew SB, Erkossa T, Smakhtin V, Fernando A (eds) Improved water and land management in the Ethiopian highlands: its impact on downstream stakeholders dependent on the Blue Nile, Intermediate Results Dissemination Workshop, IWMI, Addis Ababa

Alemu M (2003) The link between environmental change and poverty in gedio consultation papers on environment. In: Forum for Social Studies, vol 2. Addis Ababa, Ethiopia

Alf Mortem J (2002) Ownership and partnership: does the new rhetoric solve the incentive problems in aid? Forum Dev Stud 29(2):389–408

Amare S (2013) Retrospective analysis of land cover and use dynamics in Gilgel Abay watershed by using GIS and remote sensing techniques. Northwest Ethiop Int J Geosci 4(7):1003–1008
Amare S, Kameswara KR (2011) Hydrological dynamics and human impact on ecosystems of Lake Tana. Ethiop J Environ Stud Manage 4:56–74
Amare S, Kameswara KR (2012) Impacts of land cover/use dynamics of Gilgel Abay catchment of Lake Tana on climate variability. Northwest Ethiop J Appl Geomatics 4:155–163
Bekele S, Holden S (2001) Soil erosion and small holder's conservation decisions in the high lands of Ethiopia. In: Development, Vol 27 No. 4
Besley T (1995) Property rights and investment incentives: theory and evidence from Ghana. J Polit Econ 103:903–937
Birru Y (2007) Land degradation and options for sustainable land management in the Lake Tana Basin (LTB), Amhara Region. Ethiopia, Centre for Development and Environment
Bruce JW, Migot-Adholla SE (1993) Searching for land tenure security in Africa Dubuque, Kendall/Hunt, Iowa, USA
Christy JR, Norris WB, Redmond K et al (2006) Methodology and results of land-use change on climate. Nature 427:213–214
EFAP (1993) Ethiopian forestry action plan: The challenge for development. Ministry of Natural Resource Conservation and Development, Addis Ababa, Ethiopia
Ezber Y, Sen OL, Kindap T et al. (2007) Climatic effects of urbanization in Istanbul: A statistical and modeling analysis. Int J Clim 27:667–679, 148, MLRA
Feoli E, Gallizia LV, Zerihun W (2002) Evaluation of environmental degradation in Northern Ethiopia using GIS to integrate vegetation, geomorphological, erosion and socio-economic factors. Agric Ecosyst Environ 91:313–325
Fesseha, E (1970) The need for progressive taxation on unutilized lands in Ethiopia, pp. 129–148. MLRA
Food and Agriculture Organization (FAO) (1976) Land use Pal guide. Rome
Food and Agriculture Organization (FAO) (1983) Land, food and people., Rome
Food and Agriculture Organization (FAO) (1986) Highland reclamation study-Ethiopia. Final Report, vol. 1 and 2, Rome
Food and Agriculture Organization (FAO) (1995) Soil and water resource. Draft June 1995. Rome
Food and Agriculture Organization (FAO) (2005) Global forest resource report assessment 2005. Rome
Gavian S, Fafchamps M (1996) Land tenure and allocative efficiency in Niger. Am J Agric Econ 78(2):460–471
Geremew Z (2000) Engineering geological investigation and lake level changes in the Awassa Basin. MSc thesis, Department of Geology and Geophysics, Addis Ababa University, Addis Ababa, Ethiopia
Gete Z, Hurni H (2001) Implications of land use and land cover dynamics for mountain resource degradation in the Northwestern Ethiopian highlands. M. Res Dev 21:184–191
Getnet T (2011) Sedimentation modeling for Ribb dam school of graduate studies Addis Ababa. Institute of Technology
Girmay K (2003) GIS based analysis of land use/land cover, land degrdation and population changes: a study of Boru Metero area of South Wello, Amhara region. MA Thesis Geography Department, Addis Ababa University
Gorse JE, Steeds DR (1987) Desertification in the Sahelian and Sudanian zones of West Africa. World Bank, Technical Paper No 61. Washington
Harrison P (1992)The third revolution: environment, population and a sustainable world London and New York: I. B Tauris and Company Ltd
Herold,M.,Mayaux,P.,Woodcock,C.E.,Baccini,A.,Schmullius,C (2008) Some challenges in global land cover mapping: An assessment of agreement and accuracy in existing 1 km datasets. Earth Observations for Terrestrial Biodiversity and Ecosystems Special Issue:112, 2538–2556
Houghton JT (1995) Climatic change 1994: radiative forcing of climatic change and an evaluation of IPCC IS92 emission scenarios. Intergovernmental panel on climate Change, Cambridge University Press, UK

International Fund for Agricultural Development (IFAD) (2007) Community-based integrated natural resources management project in Lake Tana Watershed-Ethiopia. IFAD Project Document (Third Draft: 31. August 2007), Government of the Federal Republic of Ethiopia & International Fund for Agricultural Development

International Water Management Institute (IWMI) (2009) Importance of irrigated agriculture to the Ethiopian economy: capturing the direct net benefits of irrigation. Colombo, Sri Lanka: International Water Management Institute (IWMI). 40p. (IWMI Research Report 128)

Ironmonger DS, Aitken CK, Erbas B (2002) Economies of scale in energy use in adult-only households. Energy Econ 17:301–310

Kebrom T (2000) Land degradation problems and their implications for food shortage in South Wello, Ethiopia. Environ Manage 23:419–427

Kebrom T, Hedlund L (2000) Land cover changes between 1958 and 1986 in Kalu District, Southern Wello, Ethiopia. Mt Res Dev 20:42–51

Keyzer MA, Sonneveld GJ (2001) The effect of soil degradation on agricultural productivity in Ethiopia: a Non-paramatic regional analysis in economic policy reforms and sustainable land use in LDC's. Physica Verlag p 269–292

Lambin EF, Geist HJ (2006) Land-use and land-cover change: local processes and global impacts (Global change Â—The IGBP series). Springer, Berlin, Heidelberg, GE

Lambin EF, Geist HJ, Lepers E (2003) Dynamics of land use and land cover change in Tropical Regions. Annu Rev Environ Res 28:205–241

Lindblade A, Carswell G, Tumuhairwe K (1998) Mitigating the relationship between population growth and land degradation: land-use change and farm management in Southwestern Uganda. In: Ambio, vol. 27, No. 7, Royal Swedish Academy of Sciences Stable, p 138–214

LTRC (2004) Resource management in the Lake Tana Basin: research link project proposal. Bahir Dar University

Markandya A (2001) Poverty, environment and development. In: Folmer H, Gabel HL et al. (eds) Frontiers of environmental economics. Edward Elgar, Cheltenham

Meyer WB (1995) Past and present land use and land cover in the USA. In: Consequences, vol. 1, p. 25–33

Meyer W.B,Turner B.L (1995) Human population growth and global land use cover change. Annu Rev of Ecological Systems.23:39−61

Million B (2001) Country report: Ethiopia. In: Forestry outlook studies in Africa (FOSA)/WP/26. Corporate Document Repository, Rome, Italy

Moser M, Prentice C, Frazier S (1996) A global overview of wetland loss and degradation, p. 21–31

Muluneh W (2001) Impact of population pressure on land resources as reflected in land use\cover changes in Ethiopia. In: Assefa Taye (ed) Food security through sustainable land use population. NOVIB partners Forum, Addis Ababa

Nunez MN, Ciapessoni H (2008) Impact of land use and precipitation changes on surface temperature trends in Argentina. J Clim Sci 310:1674–1678

Nurelegn M, Amare S (2015) Land use/cover dynamics in Ribb watershed, North Western, Ethiopia, Ethiop J Dev Res. 35(1)

Nyssen J, Poesen J (2004) Human impact on the environment in the Ethiopian and Eritrean Highlands—a state of the art. Earth Science Reviews

Pan WK, Walsh SJ (2004) Farm level models of spatial patterns of land use and land cover dynamics in the Ecuadorian Amazon. Agric Ecosyst Environ 101:117–134

Sanderson EW, Redford KH, Vedder A et al (2002) A conceptual model for conservation planning based on landscape species requirements. Landscape Urban Plann 58:41–56

Scherr S (2000) A downward spiral? Research evidence on the relationship between poverty and natural resource degradation. Food Policy 25:479–498

Sen A (1999) Hunger in the World, the Suntory Centre. UK

Sherbinin De, Balk AD, Yager K (2002) Social science applications of remote sensing. A CIESIN thematic guide. Available at the web at http://sedac.ciesin.columbia.edu/tg/guide_main.jsp

Singh P(1996) Land degradation a global menace and its improvement through agroforestry. In: Signh P et al. (eds) Agroforestry systems for sustainable land use. Science publishers, Inc, USA. pp. 4–20

Skole DL, Cochrane MA (2004) Observations of Land use/ cover change in regional case studies. In: Gutman et al. (eds) Land change science: observing, monitoring and understanding trajectories of change on the earth's surface, Kluwer Academic Publishers, Dordrecht, pp. 461

Solomon A (2005) Land use and land cover changes in the head stream of Abay Watershed, Blue Nile Basin, Ethiopia. MSc. Thesis, Addis Ababa University

Sutcliffe JP (1993) Economic assessment of land degradation in the Ethiopian highlands: a case study. In: National Conservation Strategy Secretariat, Ministry of Planning and Economic Development, Transitional Government of Ethiopia, Addis Ababa

Tekie A (2000) Farmers' willingness to pay for tenure security. In: Mekonnen and D. Aredo, (eds) Institutions, resources and development in Ethiopia. Ethiopian Economic Association, Addis Ababa. pp. 87–112

Teshome S, Demel T, Sebsebe D (2004) Ecological study of the vegetation of GamoGofa, southern Ethiopia. Trop Ecol 45(2):209–221

Turner B.L, Davids S, Steven S, Gunther f, Rik L (1995) Land use and land cover change sciences /Research plan. International Geosphere and biosphere.Sweden Stockholem

Veldkamp A, Lambin EF (2001) Predicting land-use change agriculture. Ecosyst Environ 85:1–6

Verheye WH (2004) Land use, land cover and soil sciences, in land use, land cover and soil sciences, in Encyclopedia of life support systems (EOLSS), Eolss Publishers, Oxford, UK, http://www.eolss.net. Retrieved 20 Nov 2010

Woien H (1995) Woody plant cover and farming compound distribution on the Mafud Escarpment, Ethiopia. In: An aerial photo interpretation of changes 1957–1986. Working Paper on Ethiopian Development. No.9. University of Trondheim, Norway

Woldeamlak B (2003) Land cover dynamics since the 1950s in Chemoga Watershed, Blue Nile Basin Ethiopia. Mt Res Dev 22(3):263–269

Wondie, A. (2010) Current land use practices and possible management strategies in shore area wetland ecosystem of Lake Tana catchment. Ecohydrol Hydrobiol 10(2): 123–131

Wood AP (2000) Policy implications for wetlands Management. In Proceedings of National Workshop on Sustainable Wetland Management, Addis Ababa, Ethiopia

Woody Biomass Inventory and Strategic Planning Project (WBISPP) (2002) Report on natural grazing lands and livestock feed resources, Amhara National Regional State. Addis Ababa

World Bank (1990) African development indicators. World Bank, Washington

Yigrmew A (2002) Rural land holding, readjustment and rural organizations in West Gojiam, Amhara Region. In: A Summary Report. 4th Work Shop, Institute Of Development Research, Addis Ababa

Chapter 23
Agriculture in the Lake Tana Sub-basin of Ethiopia

Merkuz Abera

Abstract Agriculture is the mainstay of the Ethiopian economy in general and Lake Tana Sub-Basin in particular. The crop diversity in the Sub-Basin is high and the crop categories including cereals, legumes, root crops, oil crops, vegetables, fruit crops and other cash crops. The cropping pattern is also including the rain-fed, irrigation, residual moisture and minor recession cropping. More than 80% the cultivated land during the base-year is under rain-fed system and the remaining land is cultivated using irrigation and residual moisture. The farming system is characterized by crop-livestock mixed production system; crop production is the focus of farmers both for food and cash income. In recent years agriculture has shown a sustained increase in the use of improved inputs notably seed varieties and chemical fertilizer, pesticides and farm credit. In the Sub-Basin areas, horticultural crop production is a relatively new activity which is mainly triggered by the commencement of small scale irrigation scheme. Studies in area revealed that insect pests and diseases were, and still are, one of the major production constraints. Farmers also ranked pest problems are to be one of the major production constraints in the area.

Keywords Agricultural crops · Farming systems · Crop pests

23.1 Introduction

The Lake Tana Sub-Basin comprises three major watersheds namely, Gilgel Abay, Megech and Gumara-Rib watersheds. The crop diversity in the Lake Tana Sub-Basin is high and the crop categories including cereals, legumes, root crops, oil crops, vegetables, fruit crops and other cash crops such as sugarcane and chat. The cropping pattern also includes rain-fed, irrigation, residual moisture and minor

M. Abera (✉)
College of Agriculture and Environmental Sciences, Bahir Dar University, Bahir Dar, Ethiopia
e-mail: merkuzabera@yahoo.com

K. Stave et al. (eds.), *Social and Ecological System Dynamics*,
AESS Interdisciplinary Environmental Studies and Sciences Series,
DOI 10.1007/978-3-319-45755-0_23

recession cropping around Lake Tana. A sample survey by ANRS BoA (2013) on the cropping pattern of the three major watersheds of Lake Tana Sub-Basin—Megech, Gumara-Rib and Gilgel Abay indicated that 84.11% of the cultivated land during the base-year was under rain-fed system. The remaining 7.3 and 8.58% of the cropland was cultivated using irrigation and residual moisture, respectively. Also the rain-fed season is a cereal dominated cropping pattern so that cereals cover more than 74% of the cultivated land during the base-year. A study also showed that 7.3% of the cultivated land in the Lake Tana Sub-Basin was under irrigation during the base-year in which the source of water for irrigation are river diversion, dams, hand dug wells and springs (ANRS BoA 2013; ARARI 2009 Agricultural potentials, constraints and opportunities in the Megech and Rib rivers irrigation project, unpublished).

Unlike the rain-fed cropping pattern, the crop mix during the irrigation season is less diverse and dominated by root crops which covers more than three-fourth of the irrigation land of the Sub-Basin. On the other hand, the cropping mix of the residual moisture cropping pattern is dominated by legumes that cover more than fifty percent of crop land. The agro-biodiversity situation of cultivated crop is generally high both in terms of the type of crops and varieties grown in the farming system. In case of Input (fertilizer) utilization, the farmers often apply inputs during the rain-fed season than under irrigation and residual moisture production cycles (ARARI 2009 Agricultural potentials, constraints and opportunities in the Megech and Rib rivers irrigation project, unpublished; Getaneh 2011).

23.2 Crop Production

The farming system of the Lake Tana Sub-Basin sub watershed areas is characterized by crop-livestock mixed production system. Farm households depend mainly on crops and domestic farm animals. A relatively diverse set of crop types are grown in the area, which could be taken as a risk averting mechanism for the farmers to minimize potential risks of crop failure in time of natural calamities (ANRS BoA 2013; BNWI 2013 inception report for natural resources and socio-economic follow up survey, unpublished).

The crop production in the Sub-Basin is also dominated by small farmer production focusing largely on products for on-farm consumption. Over 98% of the cultivated area is rain fed or based on residual moisture after flooding. In the Tana Sub-Basin, population density is high, with an average of over 200 persons per square kilometer in some woredas. Household landholdings average is less than 2 ha, typically comprising several small plots. Much of the highland has been degraded through overuse and erosion. Crop production and the rearing of livestock are closely integrated on the small farms, with livestock utilizing crop residue and providing draft power for ploughing and transport. Purchase of improved seed and use of chemical fertilizer is common, but per hectare application is low. Most farms use traditional methods of cultivation, harvesting and threshing. Only one crop is produced per year in rain fed area. Crop losses are high both in the field and during harvesting and storage. The combination of a reliance on rain fed production, only one crop per year, small and degraded plots, a low use of purchased inputs, traditional cultivation methods and high post-harvest losses means that production is low and varies substantially from year-to-year. Most farm households in the Sub-Basin are accordingly exceptionally poor. They market only a small fraction of their total output, and often this is to meet urgent cash needs and requires the repurchase of staples later in the year. Food insecurity in the Sub-Basin is a matter of considerable concern (ANRS BoA 2013; BNWI 2013 inception report for natural resources and socio-economic follow up survey, unpublished).

23.2.1 Crop Produced in Lake Tana Sub-basin

In the Tana sub basin the areas are more suitable for cereals, pulses and horticultural crops and to a lesser extent to oil crops. Teff, maize, finger millet, barley, wheat, sorghum, and rice, among cereals; grass pea, chickpea, fababean and field pea, among pulses; niger seed (Noug) and safflower among oil crops; potato, garlic, pepper, onion, tomato, black and white cumin among annual horticultural crops are commonly grown in the area. Growing a diverse group of crops helps farmers to minimize potential risks of crop failure and to fulfill their household requirements. Major Field crops grown in the basin and their productivity and land coverage of crops in the Lake Tana Sub-Basin are listed (Table 23.1).

The major crops in the three major watersheds of the Lake Tana Sub-Basin are indicated for their crop productivity based on the three traditional agro-ecologies defined by altitude as Woina Dega, Dega and Wurich zones in which Woina Dega is 1500–2300, Dega is 2300–3200, Wurich zones above 3200 m above sea level (m a.s.l) respectively.

Productivity of cereals in watersheds

Cereal crops productivity is influenced by cropping patterns as Rain-fed, Irrigation and Residual moisture. Below is described the cereal productivity in the three major watersheds of the Lake Tana Sub-Bain (Table 23.2).

Productivity of Pulse crops in watersheds

In Tana Sub-Basin Pulse crops productivity is influenced by cropping patterns as Rain-fed, Irrigation and Residual moisture. Below is described the pulse crop productivity in which Lupine and castor bean indicated not grown in Gumara-Rib watersheds and otherwise in the three major watersheds (Table 23.3).

Table 23.1 Productivity and land coverage of crops in Lake Tana Sub-basin (ANRS BoA 2013)

Crop	Average of productivity	% of area covered by the crop
Tefff	7.6	21.38
Maize	17.8	12.91
Finger millet	12.0	10.41
Barely	12.1	8.93
Potato	59.7	5.56
Wheat	13.9	5.22
Grass pea	11.1	5.03
Chickpea	8.7	4.46
Faba Bean	10.0	4.03
Garlic	39.1	3.47
Sorghum	10.3	3.27
Pepper	20.1	2.96
Noug	4.9	1.92
Gesho	92.5	1.79
Rice	28.8	1.49
Onion	75.1	1.31
Field pea	5.8	0.96
Lupine	8.2	0.56
Linseed	5.7	0.51
Chat	61.5	0.40
Lentil	7.2	0.40
Papaya	71.2	0.34
Safflower	10.7	0.32
Fenugreek	8.3	0.28
B/cumin	8.0	0.27
Rapeseed	11.4	0.22
Oat	22.2	0.22
Har. bean	8.2	0.21
Tomato	105.0	0.17
Triticale	12.0	0.16
Guava	54.4	0.14
Shallot	32.7	0.14
H/cabbage	105.3	0.12
Mango	39.0	0.11
Coffee	8.6	0.11
Sugarcane	58.7	0.07
Cotton	20.5	0.03
Cas/bean	24.0	0.02
B/root	0.1	0.02
Besobila	104.5	0.02
Dinbilal	97.4	0.01
Apple	16.0	0.01

(continued)

Table 23.1 (continued)

Crop	Average of productivity	% of area covered by the crop
Carrot	24.8	0.01
Lettuce	4.0	0.01
S/potato	105.6	0.00
Orange	67.8	0.00
Lemon	71.1	0.00
Pumpkin	32.3	0.00
Banana	17.8	0.00
Peach	38.8	0.00
S/chard	156.3	0.00
Avocado	0.0	0.00
Ginger	30.0	0.00
Grand total	22.5	100

Table 23.2 Productivity of major cereal crops (qt/ha) in Lake Tana Sub-Basin in 2013 (ANRS BoA 2013)

Crop	Season	Agro-ecology	Major watershed			
			Megech	Gumara-rib	Gilgel abay	Grand total
Barley	Rain-fed	W/Dega	11.7	10.5	11.5	11.3
		Dega	13.0	14.7	9.3	12.4
		Wurich		14.9		14.9
	Rain-fed total		12.1	12.8	10.7	11.9
	Irrigation	W/Dega	10.0	12.2	19.5	15.1
		Dega		10.0	8.0	9.4
		Wurich		14.0		14.0
	Irrigation total		10.0	12.1	17.6	13.9
	Residual moisture	W/Dega	8.0	10.6	7.3	9.1
		Dega		18.5	12.7	15.6
		Wurich		16.1		16.1
	Residual moisture total		8.0	15.4	11.1	13.1
Barley total			12.0	13.0	11.0	12.1
Finger millet	Rain-fed	W/Dega	13.1	12.6	11.3	12.0
		Dega		7.0	14.4	11.9
		Wurich		8.0		8.0
	Rain-fed total		13.1	12.3	11.4	12.0
	Irrigation	W/Dega			16.0	16.0
	Irrigation total				16.0	16.0
Finger millet total			13.1	12.3	11.4	12.0
Maize	Rain-fed	W/Dega	17.1	15.1	20.8	18.3
		Dega	14.8	15.9	11.5	13.3

(continued)

Table 23.2 (continued)

Crop	Season	Agro-ecology	Major watershed			
			Megech	Gumara-rib	Gilgel abay	Grand total
	Rain-fed total		16.9	15.1	19.9	17.9
	Irrigation	W/Dega	14.0	20.1	20.0	19.4
		Dega			1.0	1.0
	Irrigation total		14.0	20.1	15.3	16.0
Maize total			16.9	15.2	19.7	17.8
Wheat	Rain-fed	W/Dega	12.4	10.2	15.8	12.4
		Dega	12.9	16.4	8.9	13.5
		Wurich		20.7		20.7
	Rain-fed total		12.7	16.3	10.4	13.8
	Irrigation	W/Dega			17.4	17.4
	Irrigation total				17.4	17.4
	Residual moisture	W/Dega			10.7	10.7
	Residual moisture total				10.7	10.7
Wheat total			12.7	16.3	11.6	13.9
Oat	Rain-fed	W/Dega		43.8		43.8
		Dega	24.0	16.0	14.7	16.8
		Wurich		16.0		16.0
	Rain-fed total		24.0	29.9	14.7	23.4
	Irrigation	W/Dega		18.8		18.8
	Irrigation total			18.8		18.8
Oat total			24.0	25.1	14.7	22.2
Rice	Rain-fed	W/Dega	30.0	28.6		28.8
	Rain-fed total		30.0	28.6		28.8
Rice total			30.0	28.6		28.8
Sorghum	Rain-fed	W/Dega	10.2	10.0	21.3	10.5
		Dega		3.8		3.8
	Rain-fed total		10.2	8.6	21.3	10.3
Sorghum total			10.2	8.6	21.3	10.3
Tefff	Rain-fed	W/Dega	7.3	7.7	6.9	7.3
		Dega	10.1	10.8	5.9	8.6
		Wurich		17.3		17.3
	Rain-fed total		7.5	8.6	6.7	7.6
Tefff total			7.5	8.6	6.7	7.6
Triticale	Rain-fed	Dega	10.4	20.0		12.0
	Rain-fed total		10.4	20.0		12.0
Triticale total			10.4	20.0		12.0
Grand total			12.2	13.4	12.7	12.8

Note qt/ha = quintal per hectare

Table 23.3 Productivity of major pulse crops (qt/ha) in Lake Tana Sub-Basin in 2013 (ANRS BoA 2013)

Crop	Season	Agro-ecology	Major watershed			Grand total
			Megech	Gumara-Rib	Gilgel Abay	
Chickpea	Rain-fed	W/Dega	12.0	14.4	8.9	11.0
	Irrigation		4.5	8.8	9.6	8.3
	Residual moisture	Dega		10.2	3.1	8.4
		W/Dega	8.7	8.4	8.5	8.5
	Residual Moisture total		8.7	8.6	8.4	8.5
Chickpea total			8.8	9.0	8.5	8.7
Fababean	Rain-fed	Dega	4.4	12.7	5.4	8.5
		W/Dega	9.2	13.3	10.7	10.9
		Wurich		9.0		9.0
	Rain-fed total		7.9	12.2	9.3	10.0
Fababean total			7.9	12.2	9.3	10.0
Field pea	Rain-fed	Dega	4.6	5.2	4.6	4.7
		W/Dega	5.0	6.8	6.7	5.6
		Wurich		10.9		10.9
	Rain-fed total		4.9	8.2	5.9	5.8
	Residual moisture	Dega		4.0		4.0
	Residual moisture total			4.0		4.0
Field pea total			4.9	7.9	5.9	5.8
Grass pea	Rain-fed	Dega		13.4	3.0	11.6
		W/Dega	11.3	8.0	11.7	9.9
	Rain-fed total		11.3	9.7	10.4	10.3
	Irrigation	W/Dega	8.0	12.0	14.3	12.6
	Irrigation total		8.0	12.0	14.3	12.6
	Residual moisture	Dega		9.2		9.2
		W/Dega	10.9	11.5	12.0	11.5
	Residual moisture total		10.9	10.9	12.0	11.1
Grass pea total			10.9	10.8	12.0	11.1
Haricot bean	Rain-fed	W/Dega	32.3	5.1	16.0	8.2
	Rain-fed total		32.3	5.1	16.0	8.2

(continued)

Table 23.3 (continued)

Crop	Season	Agro-ecology	Major watershed			Grand total
			Megech	Gumara-Rib	Gilgel Abay	
Haricot bean total			32.3	5.1	16.0	8.2
Lentil	Rain-fed	W/Dega	4.0	10.6	1.0	7.6
		Wurich		24.1		24.1
	Rain-fed total		4.0	12.1	1.0	8.8
	Irrigation	W/Dega		4.5		4.5
	Irrigation total			4.5		4.5
	Residual moisture	W/Dega	4.0	5.3	4.8	4.7
	Residual moisture Total		4.0	5.3	4.8	4.7
Lentil total			4.0	9.6	2.3	7.2
Lupine	Rain-fed	Dega	9.3	2.1		7.5
		W/Dega	6.0	16.0		6.5
	Rain-fed Total		6.8	6.7		6.8
	Residual moisture	W/Dega	17.6			17.6
	Residual moisture Total		17.6			17.6
Lupine Total			8.3	6.7		8.2
Caster bean	Rain-fed	Dega	8.0	40.0		24.0
	Rain-fed total		8.0	40.0		24.0
Castor bean Total			8.0	40.0		24.0
Grand Total			8.1	10.5	9.3	9.5

Productivity of Root crops in watersheds

In all three major watersheds of Lake Tana Sub-Basin root crops are found to be grown in all traditional agro-ecologies under rain-fed, irrigation and residual moisture conditions. Hence the productivity under different conditions of the watersheds is indicated below (Table 23.4).

Productivity of Oil crops in watersheds

In all three major watersheds of Lake Tana Sub-Basin oil crops are found to be grown under rain-fed, irrigation and residual moisture conditions. Among oil crops Niger seed is produced in Woynadega and Dega agro-ecologies only during the Meher season. Safflower, on the other hand, produced only in the Woynadega agro-ecology with rain-fed, irrigation and residual moisture. Hence the productivity under different conditions of the watersheds is indicated (Table 23.5).

Table 23.4 Productivity of major root crops in the Lake Tana Sub-Basin (ANRS BoA 2013)

Crop	Season	Agro-ecology	Major Watershed			
			Gilgel Abay	Gumara-Rib	Megech	Grand total
Potato	Rain-fed	Dega	32.0	68.3	46.1	48.6
		W/Dega	65.2	59.6	69.7	66.5
		Wurich		70.8		70.8
	Rain-fed total		49.2	67.1	62.2	59.0
	Irrigation	Dega	53.9	56.4	41.4	54.2
		W/Dega	62.8	76.3	43.5	64.7
		Wurich		57.8		57.8
	Irrigation total		58.4	65.4	43.0	59.8
	Residual moisture	Dega	40.0	69.2		63.4
		W/Dega	64.5			64.5
		Wurich		114.1		114.1
	Residual moisture total		52.3	88.4		80.4
Potato total			52.3	67.5	60.0	59.7
Onion	Rain-fed	Dega	0.0		16.1	8.1
		W/Dega	87.5	28.2	32.7	46.8
		Wurich		12.5		12.5
	Rain-fed total		65.6	23.0	30.4	38.9
Onion	Irrigation	Dega		142.5	16.0	110.8
		W/Dega	22.8	92.5	46.4	83.9
		Wurich		48.4		48.4
	Irrigation total		22.8	95.1	36.3	85.4
	Residual moisture	W/Dega		83.3		83.3
	Residual moisture total			83.3		83.3
Onion total			44.2	89.9	32.1	75.1
Garlic	Rain-fed	Dega	24.7	86.9	37.9	60.6
		W/Dega	28.6	51.6	54.5	42.6
		Wurich		50.0		50.0
	Rain-fed total		28.3	69.2	50.8	47.1
	Irrigation	Dega	24.6	26.4	20.1	24.0
		W/Dega	22.2	23.2	32.0	24.4
		Wurich		39.4		39.4
	Irrigation total		28.3	69.2	50.8	47.1
Garlic total			26.6	44.3	46.5	39.1
Carrot	Rain-fed	W/Dega		12.0		12.0
	Irrigation	W/Dega	37.5			37.5
Carrot total			37.5	12.0		24.8
Sweet Potato	Rain-fed	Dega	161.3			161.3
		W/Dega		50.0		50.0
	Rain-fed total		161.3	50.0		105.6

(continued)

Table 23.4 (continued)

Crop	Season	Agro-ecology	Major Watershed			
			Gilgel Abay	Gumara-Rib	Megech	Grand total
Sweet Potato total			161.3	50.0		105.6
Shallot	Rain-fed	Dega			60.2	60.2
		W/Dega		16.1	34.2	28.2
		Wurich		12.0		12.0
	Rain-fed total			14.1	42.9	31.4
	Irrigation	W/Dega	113.6	12.1	20.0	34.0
	Irrigation total		113.6	12.1	20.0	34.0
Shallot total			113.6	12.9	37.2	32.7
Grand total			47.4	65.7	54.2	56.4

Table 23.5 Productivity of oil crops in the Lake Tana Sub-Basin (ANRS BoA 2013)

Crop	Season	Agro-ecology	Major watershed			
			Gilgel Abay	Gumara-Rib	Megech	Grand total
Niger seed (Noug)	Rain-fed	Dega	2.0	4.0	26.8	7.0
		W/Dega	5.0	4.9	3.5	4.8
	Rain-fed total		4.9	4.8	5.2	4.9
Niger seed (Noug) total			4.9	4.8	5.2	4.9
Safflower	Rain-fed	W/Dega	14.7	11.1	10.5	11.3
	Irrigation			6.0		6.0
	Residual moisture			9.4	11.2	9.9
	Safflower total		14.7	9.4	10.6	10.7
Grand total			5.6	5.7	7.8	6.2

23.3 Farm Input Utilization in Lake Tana Sub-basin

It is common knowledge that high yielding varieties of seeds, the use of chemical fertilizers and pesticides, irrigation and improved planting and weeding practices provide higher yields than conventional technologies. In recent years in Tana Sub-Basin has shown a sustained increase in the use of improved inputs notably seed varieties and chemical fertilizer, pesticides and farm credit. Despite this, improved input use in the Sub-Basin is still lower than that of expected (Getaneh 2011; TBIWRDP 2009a growth corridor for Tana and Beles sub-basins report, unpublished).

Seeds

Seeds are one of the most important inputs to increase crop productivity. Farmers grow several varieties of crops with different traits that fit various needs and circumstances. In Tana Sub-Basin, farmers grow mainly their own varieties of most crops. A few other crops such as maize, wheat, rice and rarely teff are exceptions. The reason for not using improved varieties widely by farmers is that improved seeds are not available at planting time. In the Sub-Basin local crop varieties have different names and characters in different localities/woredas (Getaneh 2011).

Fertilizer

Input utilization that is DAP (Diammonium Phosphate) and UREA fertilizers during the rain-fed, irrigation and residual moisture cropping seasons of the base year is also shown in Tana Sub-Basin. DAP and UREA are said to be applied more for rain-fed crops and compost also applied for rain-fed crops. Generally it is indicated that input utilization both in terms of percent of respondents and quantity of utilization per respondent were lower during irrigation and residual moisture planting seasons than during the rain-fed. Input utilization also varies with wealth in terms of quantity and type of crops applied. Better off respondents apply relatively largest portion of DAP and urea on crops and large quantity of improved seed to wanted crops (TBIWRDP 2009a growth corridor for Tana and Beles sub-basins report, unpublished).

Pesticides

In Tana Sub-Basin a wide range of pesticides is used for pest management and vector control in agricultural areas. That is for insects, insecticide; for fungal disease, fungicides; for weed control, herbicides and so on. At present it is observed that the usage of pesticides is increasing mostly around the irrigable areas. However, besides their beneficial effects, pesticides are accepted as having potential environmental and public health impacts. But many farming communities in Sub-Basin are not adequately informed about the hazards associated with these pesticides. As a result, farmers observed using pesticides without full understanding of their impact on their health and the environment. Hence detail and comprehensive study on pesticide usage and application and their effect on Lake Tana particularly, and in general Tana Sub-Basin is important (Getaneh 2011; TBIWRDP 2009a growth corridor for Tana and Beles sub-basins report, unpublished).

23.4 Farming Practice in Lake Tana Sub-basin

Cropping pattern

Cropping pattern is the system by which farmers' grow crops in a particular sequence, as mixture or rotation. Among several cropping systems, farmers in the

Tana Sub-Basin mainly practice crop rotation, mixture, intercropping and some farmers around black soils practice double cropping by using residual moisture with some supplementary irrigation (Taddesse 2006).

Crop rotation

Almost all farmers in the Tana Sub-Basin areas practice crop rotation to restore the fertility status of the soil as well as to minimize the buildup of weeds, insect pests and plant diseases. The most common crop rotations are cereal-legume-cereal, cereal-cereal-cereal, oil crop-cereal-pulses, horticultural crops-cereal-pulse and cereal-cereal-pulse. In rice growing areas, most farmers grow rice for two to three years without rotation. This might be because cereals are the major food crops for the community. Absence of rotation can contribute to soil fertility depletion and buildup of diseases, weeds and insect pests (Merkuz and Getachew 2012b; TBIWDP 2009b Socio economic baseline survey key informant interview, unpublished).

Though it is not as widely as crop rotation, farmers in the Tana Sub-Basin practice some mixed and intercropping of different crops. As Intercropping of safflower with barley, grass pea and chickpea are common practices and farmers believed and justified that they practice intercropping to reduce environmental risk and to increase yield per unit area. In all cases the main and companion crops are planted simultaneously by broadcasting method. In rice growing area as Fogera and Libo Kemkem woredas, farmers rarely practice relay intercropping of rice and finger millet with grass pea. In both cases the rice and finger millet are planted in June, and at the time of weeding in August or September, grass pea is over sown. When the grass pea is at the vegetative stage, the main crops (rice or finger millet) mature and they are harvested. Sometimes, the grass pea continues to grow, which is harvested later (Merkuz et al. 2011).

Double cropping

In Lake Tana Sub-Basin some farmers also practice double cropping on black vertisols by using residual moisture. The most common double cropping system in these areas is: rice—chickpea/grass pea/maize, tefff—chick pea/grass pea/maize/fenugreek, and finger millet—chickpea/grass pea.

Land preparation and planting

Farmers allot a field to a given crop depending on soil fertility, soil type and the precursor crop. The local oxen-drawn implement, i.e., 'maresha', is used for ploughing at a shallow depth. Priority in land preparation is given to cereals (teff, sorghum, rice, finger millet, barley, wheat) and horticultural crops (pepper, garlic, shallot, potato). Less priority is given to pulses and oil crops. Land preparation starts right after harvesting crops of the previous season. Depending on the precursor crop and the degree of weed infestation of the fields, for most crops the number of ploughings significantly varies, which ranges from 2 to 5 times until the seedbed becomes fine and ready for planting. Farmers are not aware of recommended seed rates for most crops. Poor land preparation and low or high seed rate reduce productivity (Fentahun et al. 2014).

Cropping seasons

In Tana Sub-Basin, as mentioned above three types of cropping seasons are known, namely main season (rainfed agriculture), wet season with supplementary irrigation (residual moisture) and full irrigation in the dry season. Farmers near the lake are not used to growing crops using irrigation because the fields remain moist and boggy most of the year. Instead, they use the residual moisture that is always available and also use the land that is freed as the water recedes during the dry season (a system known as recession agriculture) (ANRS BoA 2013).

Main season cropping

The main cropping season is the most dominant system in the Tana Sub-Basin, when crops are entirely grown under rain fed condition. Each household who own irrigated land also has rain fed land inside or outside the irrigation scheme. Sorghum, teff, finger millet, maize, rice and barley, faba bean, field pea, niger seed and linseed are the most important crops grown in the main season. The cropping operation (from land preparation to harvesting/threshing) is accomplished from February of the first year to December of the next year, i.e., after the wet season and after the residual moisture crops are harvested (Merkuz and Getachew 2012a).

Residual moisture cropping

The wet season irrigated cropping is the crop which is planted with residual moisture in September. The residual moisture cropping system is mainly practiced more in black soil than in red soils. The most common crops grown are chickpea (also mixed with safflower), grass pea, fenugreek, wheat, emmer wheat, 'Fetto' (*Lepidium sativum*), teff, barley and maize. Supplementary irrigation may be given during the growing season as required (ARARI 2009 Agricultural potentials, constraints and opportunities in the Megech and Rib rivers irrigation project, unpublished).

Dry season cropping

Irrigation during the dry season starts with land preparation soon after harvesting the crops grown on main season or residual moisture in December and January. The fields then become ready for planting after two to three till around end of January. Planting dates can be staggered from January to March depending on crop type. Maize and vegetables (tomato, potato, onion, shallot, garlic, head cabbage and beet root) are planted with irrigation water during the season (ANRS Boa 2013).

Crop management

Crop management practices contribute a great deal to production and productivity. Farmers manage their crop fields starting from tillage up to harvest. Management practices include tillage frequency, planting system, sowing date, weeding time and frequency, herbicide application, fertilizer application, selection of crops in need of chemical fertilizer and compost, 'shilshallo' timely irrigating, harvesting and threshing. In case of fertilizer usage the recommended rates are not strictly followed. Farmers also use compost for maize, barley, wheat, teff and finger millet and herbicides against broad leaved weeds for different cereal crops (ANRS BoA 2013).

Weed management

Weed infestation of the cropping and grazing land varies from place to place depending on their abundance, crop type, farming system, rainfall and the intensity of flood. According to farmers, weed infestation and diversity is rapidly increasing. A huge build up and expansion of thorny weeds infesting grazing lands and farmlands in the in some places in the Sub-Basin is a common scene. Farmers practice different weed control techniques depending on the level of infestation and crop type. Most prominent are continuous tillage and one to four hand weeding. Herbicides can also be used according to farmers (ANRS BoA 2013).

Crop production trend

The production trend of the newly introduced crop triticale is increasing in the high altitude watershed areas, but pulse crops are decreasing. The reason given by farmers could be disease problem, which makes pulse crops out of production. Rice production is also increasing in the growing areas. On the other hand, teff and niger seed production is decreasing, water logging and declining soil fertility could be the cause. Besides to these, Chat (Catha edulis) production is increasing near and around Lake Tana using the Lake, diversion small ponds and hand dug wells. The other important aspect is expansion of irrigation with its positive impact on increasing the diversity and intensity of crops that are growing in the area. Such a diversified production has an important implication in the fight for food security in the area (TBSR 2011, Socio economic baseline survey Tana Beles report, unpublished).

Crop production constraints

Different constraints have been documented as pertinent factors in challenging crop production and productivity. Among these are Low soil fertility, lack of improved varieties (for most crops), inappropriate time of seed supply, insect pests, weeds, diseases, lack of improved agronomic practices such as appropriate seed rate and planting method, cropping system, frequency and time of irrigation are well known constraints. Even though in all the watersheds major constraints are similar but are observed with different degree of their influence (Merkuz et al. 2011).

23.5 Horticultural Crops Production and Management

Horticultural crops have comparative economic advantage over the production of field crops and can offer alternatives to livelihood to cereal-based subsistence economy. Horticultural crops are well adapted for small-scale production units and can provide relief for people at the individual household level while they also offer opportunities for trade and earnings of foreign currency (WHO 2005). In Tana Sub-Basin areas, horticultural crop production is a relatively new activity which is mainly triggered by the commencement of irrigation. Perennial horticultural crops are still rare and people rely on vegetables and other annual crops for their livelihood. Nevertheless, a few perennial crops like guava, mango, and orange are grown for subsistence in and cash income. In the Sub-Basin watersheds at present horticultural crops production is largely traditional. For their short-cycle and comparatively favorable growing environment, farmers focus much on vegetable crops than fruits. Tomato, allium, hot pepper, garlic, cabbage and potato are among the most widely cultivated vegetable crops in the areas that have high production prospects.

Fruit production, on the other hand, is poorly developed as it is the case in most other places. Generally, the production of horticultural crops appears to be feasible undertaking in the Sub-Basin. However, its development is constrained by a multitude of technological and management factors such as lack of appropriate varieties, poor management practices, soil related factors, pests and diseases, poor post-harvest management and marketing. Development and promotion of technological options that enhance productivity, quality and management of crops could help the horticulture business to be more widely adopted so that food security and income generation will be ensured and thereby improve the living conditions of growers. Given that a multitude of factors are interacting in restraining horticultural crop production, this calls for integrated intervention in order to deal with the problems effectively. Future development of fruits needs to draw on the identified constraints and opportunities to reinforce and build upon farmers' present endeavors for improved efficiency and productivity of the system. Technical backstopping and encouragement of growers, at its minimum, to access quality planting material, adequate water supply, markets and trainings would be vital (Monteiro 2003).

23.5.1 Fruit Bearing Vegetables

The irrigation target areas generally are warm (tropical) climate and are ideal for the growing of warm season crops like pepper and tomato. Hot pepper is one of the most widely grown crops in the Tana Sub-Basin. It is produced both for its green vegetable and dry uses mainly for seasoning of foods. The major hot pepper production comes from rain fed than irrigated fields. Production area coverage of hot pepper is higher in Sub-Basin. In some areas of the Sub-Basin, apart from the main season, pepper is planted from August to September and is grown with the residual moisture which is later supplemented with irrigation water to guarantee a continuous harvest of fruits till March and April. If the flooding and disease problems are successfully tackled, this crop seems to have a high prospect of expansion. Tomato is a major vegetable crop especially in Sub-Basin. It is grown staggered between about June and March. It is not grown in the rainy season because of disease problems. In the rainy season, the performance of tomato is generally very poor. In some places farmers claim that staggered planting is difficult even in the off-season because the fields become excessively wet and are occupied by other crops. Tomato production peaks towards the mid dry season causing a market glut and falling prices. In contrast, in the rainy season, supply suddenly falls and the prices soar up (WHO 2005).

23.5.2 *Leafy Vegetable, Bulbs, Roots and Tubers*

Head cabbage, 'Gomen' (the leaves of rape seed), lettuce and swisschard are the leafy vegetables grown in the Tana Sub-Basin. Head cabbage is the most important leafy vegetable, which enjoys high demand from the consumers (Monteiro 2003).

Bulbs, roots and tubers consist of four crops, i.e., potato, sweet potato, carrot and beet roots. In the genus Allium three types of crops, i.e., shallot, onion and garlic are grown. Both garlic and shallot are traditional crops. Shallot excels both in terms of popularity and in the extent of production. Garlic is a major food seasoning agent grown rain fed or under irrigation. To date, farmers prefer to grow garlic because it is not perishable and it enjoys good market access to other places. Onion, on the other hand, is a relatively new crop in the area, and its area coverage and relative significance varies from location to location. In some places in the Sub-Basin, onion is not very popular, mainly because of its low market demand, lack of access to seeds and low pungency (taste). On the other hand, in some places, onion production (for both seed and bulb) is steadily increasing (Monteiro 2003; WHO 2005).

Potatoes are becoming one of the most important crops in the Tana Sub-Basin. They are grown both in the main and off-seasons. Potato production is also rapidly expanding especially in the recession cultivation, under irrigation and in some places under residual moisture. Another crop introduced to the Gumara-Rib watershed at highlands of Libo Kemkem area for human consumption was 'enset' (known as false banana). Nonetheless, because people in the area are not used to eating it, currently the crop is being used to make ropes (ARARI 2009 Agricultural potentials, constraints and opportunities in the Megech and Rib rivers irrigation project, unpublished; Monteiro 2003).

23.5.3 Fruit Crops

Fruits contain almost all known vitamins and many essential minerals (Simitu 2005), give more yield per unit area and are more lucrative than ordinary farm crops. In the Tana Sub-Basin, fruit production is at its infancy especially in some areas. Otherwise tropical fruits as mango, papaya, guava, coffee, and etc. are produced and grown for consumption and income purposes. It is also observed that there is high demand for orange, banana and guava in the area. Beside to this attempts are underway to produce temperate fruits such as apple and plum in the highlands that have been introduced by non-governmental organizations such as Winrock International and World Vision (Monteiro 2003; WHO 2005).

23.5.4 The Production System

Horticultural crops flourish in moist and fertile soil. Four types of vegetable production systems could be distinguished across Lake Tana Sub-Basin:

a. Reined: hot pepper is a major vegetable crop grown under rain fed conditions. Closer to the lake, vegetables are commonly grown under rain fed conditions and using the land that becomes free as the lake water recedes during the dry season.
b. Residual moisture: crops grown under this includes shallot, garlic and garden cress. In some places, pepper is also grown under this system, but it needs some supplemental irrigation to extend the harvest as the dry season progresses.
c. Irrigation: two cycles of irrigation production are practiced; October to January and February to May. Typical irrigated vegetable and spice crops include Allium species, tomato, leafy vegetables and fenugreek. Tomato and Allium are the first in area and production under irrigation. Previously these two crops used to be grown from January to February. Later, their production period has dramatically extended through staggered planting. Nevertheless, since water becomes limiting as the dry season advances, and due to the upcoming main season rains, their production time could not be extended any further.
d. Following receding water: some vegetable crops, and not others, are particularly grown following the receding water of the lake. The principal crop under such system, grown after harvesting rice, is potato. To some extent, tomato and fenugreek are also produced under this system.

Others: apart from the above systems, it was reported that ponds and hand wells have been used to grow horticultural crops. Some of the vegetables are grown in one or more of the above systems. Potato is grown under all the systems discussed. Fruits are normally started in the rainy season and given supplemental irrigation water as needed in the dry season (ANRS BoA 2013; ARARI 2009).

23.6 Challenges to Horticulture Development

Horticulture is one of the areas of the greatest challenge and opportunity in the Tana Sub-Basin. Triggered by the commencement of irrigated agriculture, horticultural crops production raised a lot of hope that encouraged many farmers to enter into the business. Nevertheless, the venture did not fully succeed as yet due largely to various factors as bio-physical, socio-economic and cultural impediments. Some of the major constraints are discussed as below.

Climate: when it comes to climatic factors rainfall pattern and unpredictability appear to be the major hurdle for the production of horticultural crops. Production of some horticultural crops such as tomato is not entirely taken up in the main season for the heavy rainfall that predisposes crops to the ravage of pests and diseases. Crops like the bulbs of Allium become rotten right in the field because of the unnecessary and rather detrimental rainfall that commonly falls during crop maturity. Moreover, excessive irrigation, drainage impediment, flooding from rivers that swell and overflow during the main rainy season, and moisture stress at the critical stages of crop growth is quite common problems, which play major role in undermining horticultural crops production. Residual crops face moisture shortage towards the middle of the dry season. However, most important of all is in some places the flooding problem caused by river overflow that raises the water table and cause seasonal water logging.

Seasonality of production: currently available early maturing tomato and Allium varieties are not only limited but they also mature at the same time. They are, therefore, disposed off to the market within a narrow period of time that often results in market glut. On the other hand, in the rainy season, because of rainfall and ensuing pests and diseases, the supply of some crops especially tomato dramatically declines. As a result, the price of tomatoes shoot up to more than three times compared to that of normal seasons.

Low production and productivity: despite the high production potential of horticultural crops in the Sub-Basin, so far they are taken up as subsidiary commodities in diminutive scale. Besides, their productivity is far from ideal.

Lack of improved varieties: as mentioned earlier in this book, a steady supply of improved varieties does not exist, which forced most farmers to grow local varieties that are inferior in yield and quality.

Lack of knowledge of management practices: at present, horticultural crops production is undertaken based on farmers' local knowledge. This is particularly true for most fruit species, which are recently introduced. Their management is new to most people in the Sub-basin as the case is true elsewhere in other Sub-Basins. There are a multitude of technical problems that need to be tackled as early as possible.

Pests and diseases: aggravated by the widespread waterlogging condition and high atmospheric humidity, horticultural crops suffer from severe pest and disease damage than any other crop grown in the area (Azanaw and Merkuz 2014).

Socio-economic: competing priorities are a common problem. Irrigated horticulture faces stringent competition from field crops. This is because of concerns of food security of the family and the long gestation period and low prices of fruits. Therefore, farmers traditionally prefer cereals and pulses to horticultural crops. This suggests that keeping fruit production more remunerative is a necessary condition. When vegetable crops are produced in abundance, prices dramatically fall and fail to fetch the right price. Because horticultural crops are highly perishable commodities, they need to be marketed right after harvest. In most of Sub-Basin water sheds, feeder roads are not available. Therefore, farmers have no access to markets (to sell products and also purchase inputs) during the entire rainy season. Other socio-economic factors include: lack of attention to product quality and prevention of physical damage, lack of post harvest technology such as storage facilities and inadequate research and extension support (TBSR 2011, Socio economic baseline survey Tana Beles report, unpublished).

Cultural: people in Tana Sub-Basin, as elsewhere in other Sub-Basins, are accustomed to cereal and pulse based diets. As a result, they have developed certain stereotype against other commodities. They therefore give priority to the production of traditional low-value crops. Many horticultural crops and their utilization are all new to many inhabitants in the area. This has a huge negative impact on the expansion of these crops. People are not well aware of the benefits of fruits and vegetables.

23.7 Opportunities and Threats

Opportunities

- Farmer's great interest and aspirations to pursue on horticultural crops production and modest experience on irrigation culture.
- Favorable agro-ecology. The climatic and edaphic conditions of the study areas are within the range of the requirements of most horticultural crops growing environment. The elevation is 1800 m and above, which is conducive for growing most horticultural crops.
- The opportunity to plant horticultural crops at any time of the year as long as there is adequate moisture. This gives an opportunity for all-year-round production.
- Location advantage—the irrigation target areas have the advantage of asphalt road near-by so that they have good market prospects including access to foreign market outlets. Besides, they are near to ports of Sudan by road or by air from Bahir Dar and Gondar airports to European markets.
- As agricultural chemicals are hardly used there is a possibility for niche market exploitation under the brand name of organic produce.
- NGOs give leverage and synergy for horticultural crops promotion attempts by farmers and the Government institutions.

- Strong political will in the promotion of horticultural crops and irrigation culture. These opportunities indicate that it is possible to establish a vibrant, rather successful horticulture business in the Tana Sub-Basin more than anything else.

Threats

Looking into factors that might provide alternative future development of events concerning irrigated agriculture, it is assumed that there is a high prospect for horticultural crops production in the Sub-Basin. Nevertheless, in the face of escalating food prices, there is also a danger of shift towards cereal crops production so that the opportunity to promote fruit and vegetables would be lost. There is also a possibility that salinity will develop and restrict the growing of sensitive crops like tomato and can change the cropping system altogether.

23.8 Crop Pest Management

Cereals, pulses, oil crops, vegetables, fruits and spices are grown in the Tana Sub-Basin. In this Sub-Basin the dynamics of crop types and the varieties grown is changing with time largely due to the introduction of new crops (such as rice, triticale, chat) and small irrigation systems. Studies in Sub-Basin revealed that insect pests and diseases were, and still are, one of the major production constraints. Farmers in the area ranked pest problems to be one of the major production constraints in the area.

The species composition of pests was more or less the same among major watersheds in which all of them lie within the same ecological set up. Crop pest is defined to encompass insect pests, diseases, weeds, storage pests and vertebrate pests. In Ethiopia, nine of around 30 Insect Orders in the Class Insecta are known to cause damage to agricultural crops. Only six of the nine orders, i.e., Lepidoptera, Homoptera, Isoptera, Coleoptera Dipetra and Orthoptera, were reported to have caused damage in the Tana Sub-Basin. To be specific, a few of the major insect pests include stem borers, shoot fly, red teff worm, grasshoppers, sorghum chaffer and army worms on cereals; cutworm, African bollworm (ABW), and aphids on pulses; ABW on tomatoes, and other aphid species on cabbage and few of major diseases includes Rust, mildew, blotches and leaf spots on cereals; root rot/wilt, chocolate spot, mildew on pulses; root rot on hot pepper, bulb rot on shallot, rust on garlic and late blight on potatoes. Post-harvest pests also include those insect pests (weevils), fungus, bacteria and etc. under storage conditions.

Recently, cutworm and red teff worm have become the most damaging insect pests in most of the Sub-Basin. According to farmers, cutworm is more serious on crops grown on residual moisture bordering Lake Tana than the upper part of the catchment area. These crops include, among others, wheat, emmer wheat, maize, chickpea, lentil, grass pea and safflower. Attack starts immediately after crop emergence. According to local farmers' beliefs, cloudy weather condition exacerbates cutworm attack. Likewise, red teffworm is increasing its presence and damage every year.

Root rot of hot pepper is the most devastating disease that is experienced recently. Shallot root rot, potato bacterial wilt and so on are also reported the major devastating diseases in the Sub-Basin. Other vertebrate pests include porcupines (damaging potatoes), birds (attacking hot pepper pods) and rodents attacking cereals mainly.

The farmer in the Sub-Basin also notice change in the dynamics of pests as a result of the recently enhanced irrigation practices. Because of the expansion of irrigation farmers' start growing different vegetables crops in which some seedlings are from unknown source, hence disease occurrences and distribution become also increasing. Because of this at present in some of the irrigation sites farmers are started spraying fungicides intensively for the control of the diseases. Example for this problem is around Koga dam irrigation sites. Hence this condition implies that pest incidence will give attention in increase. In some places farmers do not give attention for pest dynamics. Ignorance about the possibility of pest resurgence as a result of changing farming systems is a grave concern for experts and researchers.

23.8.1 Existing Pest Management Practices of Farmers

While acknowledging the seriousness of pests, most farmers are unaware of the need for pest control. Many of them reported that they do nothing to control pests. For a few of them, such unwelcome pests were simply God's punishment. However, other farmers conduct some traditional control practices and they even use some modern chemical pesticides. Traditional practices include hand picking of larvae and beetles, uprooting and removing infested plants, applying livestock urine, smoking with hot pepper as a botanical insecticide against storage pests, etc.

Insecticides are widely used against aphids on grass pea and African bollworm on chickpea. However, their use had serious repercussions by killing honey bees. According to farmers, the bees did not return home (to the beehive) after their foraging trip. Farmers reported that long time ago they used to harvest honey in November because pesticides were not used as widely as to date and bees did not die of insecticide poisoning. It is common knowledge that pesticide drift not only harms bees and other biodiversity but also pollutes the larger ecosystem. Up until now, the use of pesticides is extremely controversial among the concerned professionals and at the higher policymakers' level alike. This indicates that insecticides should be used judiciously only as a last resort, and when other options fail or when infestations turn epidemic (outbreak) proportions. Besides, specificity of pesticides is not taken care of. For example, aphids are very important on grass pea in the Sub-Basin, which farmers routinely spray Malathion against. In fact, a very specific insecticide against aphids is available commercially such as dimethoate and pirimicarb. No efforts are being made to adopt more specific insecticides, which could contribute for better efficacy and lower pollution. More concerted effort should be paid to other pest management components including biological, cultural, physical, botanical, behavioral and other non-chemical based alternatives.

23.8.2 Physical Environment that Favored Pests

The crop types and their pests varied with soil type and the level of soil moisture. Farmers in Sub-Basin reported severe pest damage close to the Lake Tana where alluvial soils dominate and soil moisture is abundant. Moist soils and alluvial soils favor cutworms and red teff worm, which are the key pests in the Sub-Basin. Moist soils allow cutworms to move around easily within the soft soil, which they find it easy to dig. Moist soils also allow red teff worms to be able to easily dig and hide in the soil, diapauses during the dry season, eventually pupate and emerge as adults in the following season. Both cutworm and red teff worm are lepidopterous larvae, which are not designed to dig tough soil material, as a few other insects do. Therefore, they can only thrive well in moist environments like close to the lake. Abundant moisture also accelerates root rot of hot pepper, according to observation by farmers.

23.8.3 The Dynamics of Major Pests that Have Become Prominent Recently

A few of the major concerns of farmers are that cutworms attack a wide range of crops grown particularly on residual moisture close to the lake. Other important pests include red teff worm on teff, root rot on hot pepper and aphids on pulses.

References

ANRS BoA (Amhara National Regional State Bureau of Agriculture) (2013) Community-based integrated natural resources management project in Lake Tana sub-basin. In: Baseline Report on Crop Production, Vol 8, TCS, Bahir Dar

Azanaw A, Merkuz A (2014) Gender mainstreaming in agricultural extension: implementation and constraints in Amhara Region, Fogera District, Ethiopia. Pacesetter J Agri Sc R 2(3):25–35

Fentahun T, Dagninet A, Merkuz A (2014) Evaluation of single yoke implement, harnessed and drawn by horse for cultivation of farm lands in north-western Ethiopia. WebPub J Agr Res 2 (3):38–46

Getaneh K (2011) The impact of selected small-scale irrigation schemes on household income and the likelihood of poverty in the Lake Tana Basin of Ethiopia. Cornell University, Thesis

Merkuz A, Getachew A (2012a) Distribution and severity of sorghum covered kernel smut in North Western Ethiopia. Inter J Curr Res 4(04):041–045

Merkuz A, Getachew A (2012b) Evaluation of improved and local/landrace/sorghum varieties for covered kernel smut. Arch Phytopathol Plant Prot 45(6):717–723. doi:10.1080/03235408.2011.595904

Merkuz A, Seid A, Chemeda F et al. (2011) Effect of mustard green manure and dried plant residue on chickpea wilt (Fusarium oxysporum f.sp.ciceris). Arch Phytopathol Plant Prot (Taylor and Francis) 44(9):821–831. doi:10.1080/03235408.2010.490390

Monteiro CA (2003) WHO fruit and vegetable promotion initiative—report of the meeting. Paper presented Geneva, 25–27 August 2003

Simitu P (ed) (2005) Utilization and commercialization of dry land indigenous fruit tree species to improve livelihoods in East and Central Africa. In: Proceedings of a Regional workshop, KEFRI Kitui, Kenya 20–24 June 2005

Taddesse A (2006) Current land management in Lake Tana watersheds: Future challenges and opportunities for sustainable land management. In: Proceedings of the national consultative and promotional workshop on Lake Tana and its environs: Conservation, utilization, development and threats, EPLAUA, Bair Dar, 6–7 Nov 2006

WHO (World Health Organization) (2005). Fruit and vegetables for health. Report of a joint FAO/WHO Workshop, 1–3 September 2004, Kobe, Japan. http://www.fao.org/ag/magazine/fao-who-fv.pdf. Accessed 4 June 2015

Chapter 24
Challenges and Opportunities for Increased Farm Animal Productivity in the Lake Tana Sub-Basin

Kefyalew Alemayehu and Asaminew Tassew

Abstract Livestock contribute 19% of the national GDP, 45% of agricultural GDP and 20% of the export earnings in Ethiopia. They are also contributing to people's livelihoods as source of food, cash income, liquid asset, diversification of risk, inputs to crop production, cultural value and fuelwood. Livestock are also very important to nutritional and food security, incomes and livelihoods of people in Lake Tana Sub-basin. The livestock production systems in Lake Tana Sub-basin are mainly mixed crop-livestock. Despite the important contribution made by livestock to people's livelihood in Lake Tana Sub-basin, the productivity of livestock is low due to various challenges. The challenges include land use change, limited access and low use of feed technology, inadequate veterinary service provision, genetic limitation of the indigenous livestock breeds, limited access and high cost of improved breeds, market inefficiency/failure, inadequate research/technology generation, weak linkages among stakeholders, limited access to credit, and climate change are main once. On the other hand, livestock products demand is increasing. This could be due to human population growth and urbanization. This is a good opportunity for livestock keepers to increase livestock products. There have to be policies support, extension services and research to enhance the complementarities between crop and livestock production and integration with watershed management activities. For peri-urban and urban livestock production systems, access to land, access to credit, establishment of milk processing plant, and veterinary service have to be improved. Rural-peri-urban-urban linkages should also be enhanced.

Keywords Challenges · Lake Tana Sub-basin · Livestock production · Opportunity · Productivity

---

K. Alemayehu · A. Tassew (✉)
Department of Animal Production and Technology, College of Agriculture and Environmental Sciences, Bahir Dar University, P.O. Box 79, Bahir Dar, Ethiopia
e-mail: asaminew2@gmail.com

K. Alemayehu
e-mail: kefyale@gmail.com

K. Stave et al. (eds.), *Social and Ecological System Dynamics*,
AESS Interdisciplinary Environmental Studies and Sciences Series,
DOI 10.1007/978-3-319-45755-0_24

24.1 Introduction

The Ethiopian economy mainly depends on agriculture. The agricultural sector contributes more than 40% of the Gross Domestic Product (GDP) and 90% of the foreign exchange earnings. About 83% of the populations live in the rural areas (EEA 2012), where they practice subsistence rain-fed agriculture. Livestock contributes 19% of the national GDP and 45% of agricultural GDP, including the value of draught power (IGAD 2010) and 20% of the export earnings of the country (IBC 2012). In most parts of the country, different animals are kept for several reasons: draught power; consumption of animal products such as milk, egg, meat; social and cultural values; production of organic manure for fertilizers and fuel (Dessie et al. 2013). Cattle are mainly kept for draught power followed by milk, meat and hides production and threshing. Smaller animals such as sheep, goats and chicken are kept for cash income and consumption of their products. Equines (donkeys, mules and horses) used for transport and cropland cultivation and threshing, especially in the high altitude areas of the mixed farming system (Grünenfelder 2005; Effa et al. 2012). Camels also provide multiple functions such as milk, meat and draught power, especially in the pastoral and agro-pastoral systems (Dessie et al. 2013).

In Lake Tana Sub-basin, as in other parts of the country, livestock production comprises a large proportion of farming activities. This is due to its potential to provide food products for home consumption, generate cash income and serve as a capital asset against risk, draught power and manure production (Tassew 2007; Alemayehu et al. 2009b; Anteneh et al. 2010; Ali and Neka 2012). However, the productivity of livestock in the Lake Tana Sub-basin has been low due to various challenges. These include land use change, poor animal health services, limited access to credit and market, and low livestock technology adoption are among others. Thus, the livestock keepers haven't been benefited from the growing demand for livestock products in both domestic and export markets. Therefore, this chapter discusses the challenges and opportunities for increased farm animal productivity in Lake Tana Sub-basin and to suggest intervention strategies and show gaps for further research.

24.2 Role and Contribution of Livestock in Ethiopia

Livestock contributes to the National GDP and also to the country's export earnings to a significant level. For instance, Ethiopia's lowland cattle, sheep, goats and camels are highly demanded in neighboring countries as well as the strategic livestock markets of the Middle East (Fig. 24.1).

Indeed, the presence of huge number of livestock resources, proximity to Middle East markets (Negassa and Jabbarr 2008), conducive investment policies, the liberalization of the economy and the recent supports and attentions given by the government to export trade gives the country comparative advantages in livestock

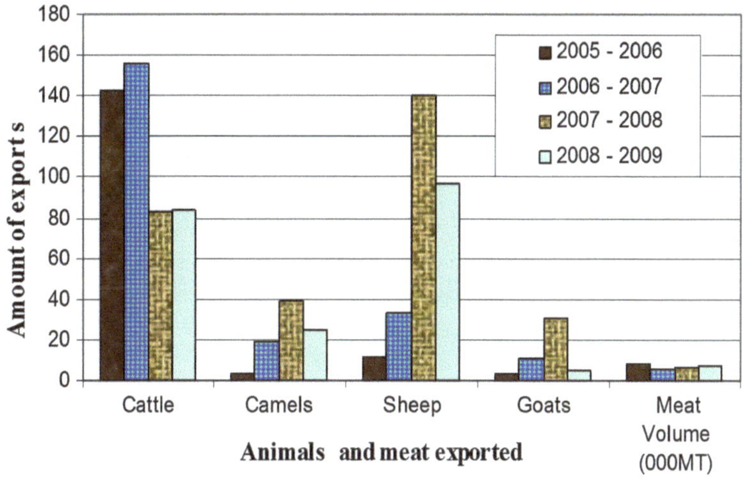

Fig. 24.1 Ethiopian live animals and meat exported through formal channels (in thousands from 2005–2009) (Sintayehu et al. 2010)

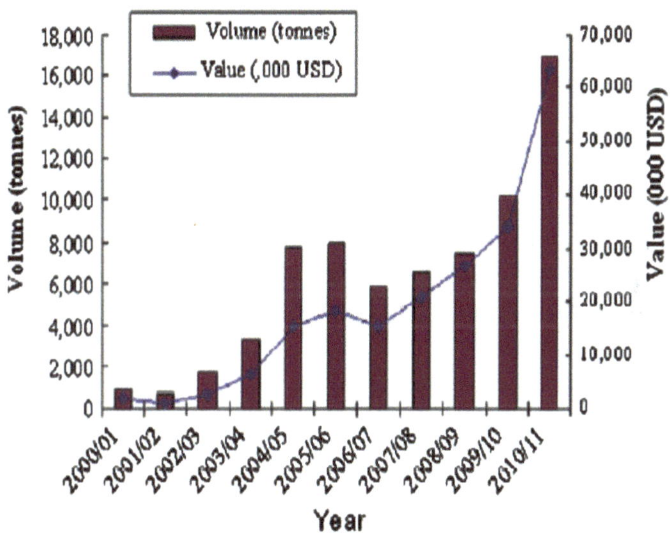

Fig. 24.2 Trends of meat export performance and value in USD from 2000–2011 (Trade Bulletin 5 2011)

trade (Figs. 24.2 and 24.3). However, inadequate market infrastructure, virtual absence of market information system, absence of market oriented livestock production systems, inadequate number of exporting firms with low level of capacities, inadequate knowledge of international trade, low level of quarantine facilities and procedures, prevalence of various diseases, and excessive cross-border illegal trade

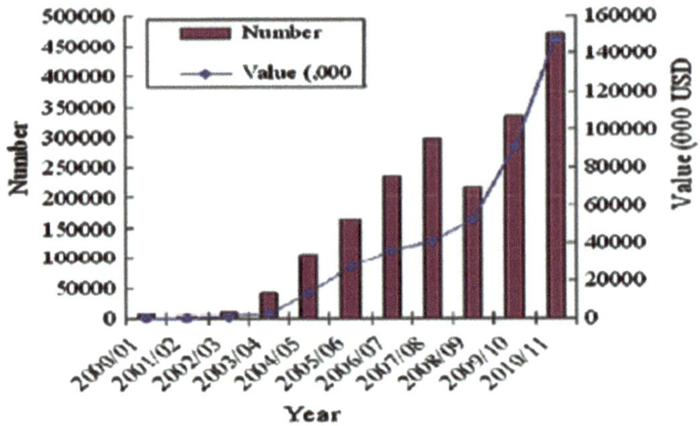

Fig. 24.3 Trends of live animal export performance and value in USD from 2000–2011 (Trade Bulletin 5 2011)

are the major challenges that hinder the smooth livestock trade of Ethiopia (Alemayehu and Ayalew 2013).

At household level, different animals are kept for different purposes (Grünenfelder 2005; Tassew 2007; Anteneh et al. 2010; Ali and Neka 2012; Dessie et al. 2013). Livestock are an important source of household food (notably milk, meat and egg). Income is generated through sell of animals, milk, meat, egg, hides, skins, and manure, and is a critical source of household income. They are also the major saving mechanism for smallholder farmers to buffer against crop failures and serve as a reserve that is readily convertible to cash. Livestock are also an important source of inputs to crop production. Draught power and manure are both very important inputs to crop production, especially in the mixed crop-livestock production system. Other contributions of livestock include transport services, cultural value and source of fuel (from animal dung).

24.3 Livestock Production in Lake Tana Sub-basin

24.3.1 Livestock Production Systems

The livestock production systems in Lake Tana Sub-basin are classified into three major production systems: smallholder mixed crop-livestock, peri-urban and urban (Tassew 2007; Alemayehu et al. 2009b). For these production systems, livestock species and breeds, land, animal feeds and market accessibility are the basis for the classification. The smallholder mixed crop-livestock production is found in rural areas. In this production system, the crop and livestock are interdependent. Cattle are the most important species kept by famers in the Sub-basin. Farmers derive several benefits from cattle keeping. This includes draught power, milk, meat, hides, cash, manure and social prestige. Small ruminants and chickens are also an important source of protein for farmers. They can also be a source of income as they are easy to sell when need arises compared to larger species like cattle. Other livestock species kept by farmers include donkeys, horses and mules, and these animals are used for transport purpose. In return crop residues contribute to animal nutrition and eaten in situ and/or conserved and fed to animal in times of feed scarcity, especially during the dry season (Tegegne and Assefa 2010).

The peri-urban livestock production system is found within smaller towns and outskirts of smaller and district towns and Bahir Dar city. In terms of feed resources, the peri-urban livestock production system has the features of the smallholder mixed crop-livestock production system. However, market accessibility and concentrate type of feed supplementation practices are better compared to the smallholder mixed crop-livestock production system. In Lake Tana Sub-basin, the urban livestock production system is found in Bahir Dar city. This system mostly depends on external inputs and keeps improved breeds.

24.3.2 Livestock Species

The livestock species kept in the rural areas include cattle, sheep, goats and equines. The cattle breeds are mostly indigenous ones such as Fogera and highland Zebu (Girma 2015) (Fig. 24.4). Some crossbred cows are also kept by few farmers. Similarly, the sheep, goats and equines breeds are indigenous. The dominant chicken ecotypes are local, however, exotic chicken such Rhode Island Red (RIR) and White Leg Horn (WLH) are also kept by farmers. In peri-urban livestock production system, the most important livestock species include both local and crossbred cows, local fattening animals and sheep, and exotic chicken. While in urban livestock production system, the crossbred cows are more to find them.

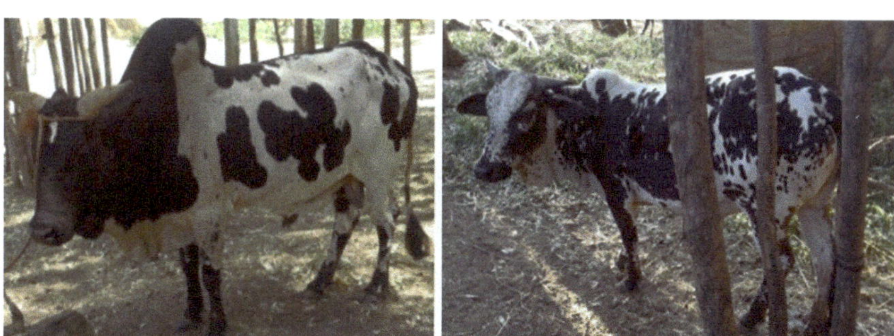

Fig. 24.4 Fogera breed bull (*left*) and heifer (*right*) (*Photo* Kefyalew A 2015)

24.3.3 Production Performance

In Ethiopia, the average milk yield per cow is 198.6 kg (kilogram) per lactation, while the average meat yield per cattle is 108.4 kg (Minstry of Agriculture and Rural Development, 2007 Unpublished). The average meat yield per sheep is 10, and 8 kg per goat. Egg production per indigenous chicken is between 40 and 60 (Moges et al. 2010; Hailu et al. 2013). Similarly, the production performance of all the livestock species is low in the Lake Tana Sub-basin (Tassew 2007; Alemayehu et al. 2009b; Anteneh et al. 2010). However, the production and productivity of dairy (Alemayehu et al. 2009b) and chicken are better in peri-urban and urban as compared to the mixed crop-livestock production system. This could be attributed due to improved breeds, better feeding and management practices and improvement in animal health services.

24.3.4 Feed Resources

In the mixed crop-livestock production system, the major sources of feed for livestock are natural pasture, crop residues, crop aftermath and non-conventional feed (*attela*, residue from local brewery) (Tassew 2007; Anteneh et al. 2010; Ali and Neka 2012). In the communal free grazing lands, all livestock species graze together and the grazing system is free and uncontrolled (Fig. 24.5). This practice does not allow smallholder farmers to improve the productivity of these grazing lands. However, nowadays awareness creation has been given to smallholder farmers to stop free grazing and to practice cut-and-carry system. The crop residues available in the Sub-basin include straw and stover. The straws include teff, finger

Fig. 24.5 Livestock grazing in communal grazing land (*Photo* Asaminew T 2012)

millet, barley, wheat and rice. The stover is from maize only. The use of improved forage species is limited due shortage of lands. The use of concentrate feeds is also limited due to low supplies and high cost. Chicken are raised on scavenging feeding system. They scavenge during the day time in and around homestead, but sometimes supplemented with maize and kitchen wastes.

In the peri-urban livestock production system, the major livestock feed resources are communal free grazing found in the outskirts of towns and Bahir Dar city. In addition purchased hay, crop residues and concentrate feeds are also used (Alemayehu et al. 2009b). The feed resources for cows and fattening animals in the urban livestock production system are purchased hay and concentrates. Whereas chicken are reared on purchased concentrates.

24.3.5 Diseases and Parasites

Foot and mouth disease (FMD), blackleg, anthrax, and bovine pasteurellosis are major diseases of cattle (Wuletaw 2004; Tassew 2007; Anteneh et al. 2010). Sheep and goat pox, ovine pasteurellosis, peste des petits ruminants (PPR), anthrax caprine pleuro pneumonia (CCPP) and FMD are major diseases of small ruminants. Newcastle disease (ND) and infectious bursal disease (IBD) are major diseases of chicken (Moges et al. 2010; Hailu et al. 2013). Predators are also affecting scavenging chicken, especially in rural areas. African horse sickness (AHS) and tetanus are major diseases of equines. Ticks and mites are major external parasites, while gastrointestinal nematodes and liver fluke are major internal parasites causing considerable morbidity and loss of livestock in Lake Tana Sub-basin.

24.3.6 Consumption of Livestock Products and Marketing of Live Animals and Livestock Products

In Lake Tana Sub-basin milk is produced from cows only. In rural areas milk mainly consumed at home. However, farmers those are near to towns and Bahir Dar city sell whole milk to individual consumers, restaurants, cafeterias and hotels (Tassew 2007; Alemayehu et al. 2009a). On the other hand, a substantial amount of milk is processed into butter and is used at home. It is also sold to individual consumers in rural areas, and to peri-urban and urban individual consumers as well. Restaurants, cafeterias and hotels also obtain milk from this source. Butter milk (a by-product of butter making) consumption is also common in rural areas. In peri-urban and urban areas, milk producers sell whole milk to individual consumers, restaurants, cafeterias and hotels. While the member of dairy cooperative sell whole milk to the dairy cooperatives. The dairy cooperatives in most cases separate the whole milk into skim milk and cream using cream separator. And these cooperatives sell skimmed milk and butter to individual consumers, restaurants, cafeterias and hotels.

In rural areas, people consume meat, from large and small ruminants mainly during holidays and festivities. But egg and chicken meat consumption is common in rural areas and also similar in peri-urban and urban areas. In peri-urban and urban areas, meat is consumed by slaughtering sheep and goats at household level and/or buying from butcher, especially cattle meat.

In the Sub-basin live animals are sold at local markets. The price of live animals varies significantly from time to time due several factors (personal communication).

24.4 Achievements, Challenges and Opportunities of Livestock Production in the Lake Tana Sub-basin

24.4.1 Achievements

Livestock research: a few studies done by Amhara Region Agricultural Research Institute (ARARI) and livestock researchers, MSc and PhD students in higher learning institutions and NGOs have described the livestock production systems in Lake Tana Sub-basin at different times (Wuletaw 2004; Tassew 2007; Alemayehu et al. 2009b; Anteneh et al. 2010; Ali and Neka 2012). However, the studies didn't address the entire Lake Tana Sub-basin. And also didn't address the trade-off/synergies of the level of livestock production ownership and its contribution to the community as well as the type of complementarities with crop production and natural resources management practices too.

Genetic improvement research on livestock has been concerned mainly in improving the milk production potential of local cattle through crossbreeding practice. The cross breeding mainly focuses with Holstein Frisian and Jersey, and multiplication and distribution of exotic chicken such as RIR and WLH. Public institutions are also involved in cattle crossbreeding, multiplication and distribution of exotic chicken. Different studies showed that crosses of 50–62.5% exotic blood level are appropriate for smallholder mixed crop-livestock production system; this is due to management implications. On the other hand, use of higher exotic blood level must be practiced through improved feeding and good level of management practice in peri-urban and urban area. This practice should be accepted when there is reliable market opportunity (Alemayehu et al. 2009b). So far, minimum efforts have been made to evaluate the production and reproduction potential of indigenous animals.

Research on feed resources development has been focused on improved forage species and accessions. Hence, different forage species and accessions have been tested in the different agro-ecologies and production systems (Table 24.1). The agronomic practices have also been developed for widely adapted and productive forage species (Assefa et al. 2011). However, research on characterization and development of the indigenous fodder plants and improvement of quality of roughage such as stover and straw are limited.

Livestock extension: in terms of organizational structure, the livestock extension systems were organized under Extension Department both at Ministry of Agriculture and Regional Bureau of Agriculture. However, in 2013 the Government of Ethiopia established a Livestock State Ministry within the Ministry of Agriculture and at agency level in Oromia and Amhara Region (Gizaw et al. 2010, LRDPA (Livestock Resources Development and Promotion Agency) 2010 Unpublished). And this agency have separate process owner both at zone and district levels. Similarly, training and deployment of livestock manpower in the public sector expanded considerably (Lemma et al. 2008). Thus, the change in the structural arrangement and deployment of human power at different levels are mainly to improve the provision of resources and to function of strategies.

24.4.2 Challenges for Improving Livestock Productivity

The major challenges for low productivity of livestock in the smallholder mixed crop-livestock production systems are discussed below:

Land use change: in different part of the country, the trend of feed availability has decreased from time to time due to shrinking of communal and private grazing lands (Benin et al. 2006; Ali 2009; Tegegne and Assefa 2010; Funte et al. 2010; Assefa et al. 2011; Mekasha et al. 2014). This phenomenon is also common in Lake Tana Sub-basin where grazing lands are changing into cultivated lands and settlement due to population pressure (Ali 2009; Fisseha et al. 2011).

Limited access and low use of feed technology: adoption of various forage species and accessions have been tried including Napier grass, vetch, oats, alfalfa, cow pea, pigeon pea, sesbania, and tree lucerne. However, only few farmers have practiced growing of improved forage species but constrained by land shortage and also unable to integrate with crop production and supported with watershed management practice (Mekoya et al. 2008). Moreover, utilization of commercial concentrate is limited due to low availability and its rising price. Urea treatment of poor quality feed is also another technology proven for smallholder farmers, but the technology transfer to farmers has not occurred yet.

Inadequate veterinary service provision: the prevalence of various livestock diseases including, infectious diseases, tick borne diseases, internal and external parasites are affecting the livestock development programs in varying scales, depending on ecological zones and management levels. The livestock health services provided is inadequate; the cost of drugs and acaricides is very high, while the diagnostic services are not readily available to the livestock keepers.

Genetic limitation of the indigenous livestock breeds: the main problem of milk and meat production in the country as well as in the Lake Tana Sub-basin is the low genetic potential of the indigenous livestock breeds, which gives rise to low milk, meat and egg output per animal. If improvement of the local livestock breeds for milk, meat and egg production is targeted, then it is important to have a well-designed selection and/or crossbreeding programs with the corresponding improvement in the environment.

Limited access and high cost of improved breeds: both crossbreed and pure exotic dairy cattle are usually in short supply and when available, the high cost is a major problem. Multiplication and distribution of crossbred cows/heifers have been carried out from Metekel ranch with a reasonable price, but have limited capacity to address the farmers demand. Prices of crossbreed cows and heifers from the local market are now unaffordable by the poor and the average smallholder farmers that would have liked to engage in the dairy business. Multiplication and distribution of exotic chicken (such as RIR, WLH) have been carried out from Andassa and Kombolcha Chicken Breeding and Multiplication Centers, but the availability of exotic chicken for farmers are a great challenge. So far, there is no effort to multiply and distribute improved sheep and goats breeds to farmers in the Sub-basin.

Market inefficiency/failure: the market constraints related to livestock development includes, infrastructure inadequacy, inadequate market services (poor market information system, lack of livestock and marketing standards), ineffective domestic demand and importation of livestock products. The support from the government and NGOs to organize livestock keepers' cooperatives and unions are also insufficient.

Inadequate research/technology generation: although research in livestock production is important, the efforts have been made so far not succeed in providing technological solution to improve the productivity of livestock. Still the support for livestock research is marginalized. Some of the major causes are limited budget, inadequate research infrastructure and facilities, shortage of skilled personnel, etc.

Inadequate extension and training services: effective and adequate extension services and advice on animal nutrition and feeding management, reproduction, animal health, farm management and livestock production efficiency are not always available to the livestock keepers.

Weak linkages among stakeholders: weak linkages among research, extension, and technology users are one of the critical factors that have hindered livestock development in Lake Tana Sub-basin.

Limited access to credit: livestock development enterprise needs higher level of investment and long-term loan. In Lake Tana Sub-basin, publicly owned Amhara Credit and Saving Institution (ACSI) and few NGOs affiliated micro-finance institution are providing financial services for livestock production investment. Most of the loan has been used for sheep production and fattening, and cattle fattening. However, the lending requirements as well as the loan size is small and hardly fit the needs and requirements of livestock development. This is a challenge to transform the subsistence livestock production into market oriented system.

Climate change: in Ethiopia, temperature has increased by 1.3 °C per decade. Daily average temperature records indicate the increasing trends in the number of hot days and hot nights. The spatial and temporal variability of rainfall is also high. Climate change also expected in Lake Tana Sub-basin. With regard to livestock production, climate change has direct and indirect impacts. Climate change induced drought and increased temperature could directly increase stress on the physical physiological activities of livestock. The indirect effects of climate change include increased temperatures and changing rainfall patterns, which could have far reaching consequences on livestock production through effects on availability of feed and water, spread of diseases and parasites.

The major challenges for low productivity of livestock in the peri-urban and urban livestock production systems include:

- Low feed availability and its high cost,
- Land shortage,
- Limited access to credit,
- High costs of animal health services,
- Lack of milk processing and marketing facilities, and
- Poor networking and organization of the producers.

24.4.3 Opportunities for Improving Livestock Productivity

The opportunities for improving livestock productivity in Lake Tana Sub-basin are described here under: *Emphasis given for livestock development*: the livestock sector has overlooked for many years and undermined the role it can play in contributing to the national economy (Benin et al. 2006; Tegegne et al. 2010). In recent times, the government showed recognition to develop and raise the contributions of livestock to both the domestic and export markets (EEA 2012). Similarly, training and deployment of livestock manpower in the public sector expanded considerably. In the Amhara Region where the Lake Tana Sub-basin is found, the livestock extension activities were structured together with crop in the regional extension program, but since 2011 the livestock extension were transferred into Livestock Resources Development and Promotion Agency, Amhara Region and have separate process owner both at zone and district levels and the organizational structure is under Amhara Region Bureau of Agriculture, Zone Office of Agriculture and District Office of Agriculture (LRDPA 2010 Unpublished). These practices are relevant for successful implementation and sustainability of livestock development through various support services.

Increased market demand for livestock products: the human population growth and urbanization in Lake Tana Sub-basin are increasing from time to time. Moreover, the Sub-basin areas have road access to Sudan and this create opportunities for selling live animals and livestock products. Two export meat abattoirs, one in Bahir Dar and the other one in Mekelle city are also functional. These offer a huge local and export market opportunities for live animals and livestock products.

Conducive climatic condition: in the Sub-basin, the altitude ranges from 1785 to 4135 m a.s.l (meter above sea level). The climate of the Sub-basin is 'tropical highland monsoon' with main rainy season between June and September with mean annual rainfall of 1326 mm. The average temperature is 19 °C is favourable condtion.

Large water resources: the Sub-basin areas have plenty of water resources (Lake Tana, rivers, and underground wells). The presence of such water resources directly or indirectly could benefit livestock production.

24.5 Suggested Interventions

The interventions suggested to improve the productivity of livestock in the mixed crop-livestock production system are discussed below:

- Improved feed production: addressing the livestock feed constraints faced by smallholder farmers in the Sub-basin is difficult. However, the most feasible solutions require use of what is already known feed technology. The feasible solutions are described below:
  - *Integration of crop production with improved forage species*: intercropping crop production with different legumes is a well known practice. Such practice enables to produce good quality feed for livestock. In addition, the legumes improve soil fertility and enhance crop yields with minimal application of inorganic fertilizers. The biomass could also be used as fuel. Improved forage species could also be sown within the crop farm boundaries.
  - *Improving utilization of crop residues*: straw and stover are the major feed resources in the Sub-basin. However, the nutritional quality of both straw and stover are poor. Thus, timely collection of the crop residues from field and improving the nutritional quality through urea treatment are relevant.
  - *Maintaining and improving the productivity of communal grazing lands*: first, policy support is crucial for maintaining communal grazing lands. Second, local institution is very important for better management of communal grazing lands. Third, the extension and research systems have to support the farmers especially in terms of improved forages supply.
  - *Improving productivity of protected areas*: as in other parts of the country, farmers in the Sub-basin have been participated in constructing soil and water conservation structures on hill areas. These areas protected from animal entrance. Hence, it is pertinent to integrate the physical soil and water conservation works with biological methods through planting and/or sowing improved and/or indigenous fodder species.

- Livestock diseases and parasites control: the possible ways to control livestock diseases and parasites include:
  - Through appropriate use of a wide range of anthelmintic drugs and vaccines. To do so the livestock extension department should plan for the procurement of anthelmintic drugs and vaccines and this important for timely provision of the animal health services.
  - Training the animal health extension workers and community-based animal health workers.
  - Training the smallholder farmers on livestock diseases and parasites control.
  - Promoting private veterinary services.
- Establishing livestock' cooperatives: promoting small-scale to medium-scale cooperatives is important to process the livestock produce.
- Financial services: organizational innovation is required by the financial institutions to increase the loan size and period and include additional services like livestock insurance as one options to improve the finance service.
- Strengthening coordination among stakeholders: the extension workers, researchers, NGOs, and smallholder farmers have to work together to improve the livestock production systems. Further, the livestock research and extension should be integrated with other research and extension activities. Because improving the livestock production systems means that smallholder farmers have access to reliable and affordable support services, offering them access to knowledge and inputs, including credit and marketing information.

For peri-urban and urban livestock production systems, it is suggested that access to land, access to credit, establishment of milk processing plant through encouraging the private sector, and system for monitoring and supervision of the veterinary service providers have to be improved in order to improve the productivity of livestock.

24.6 Research Gaps

- There is a good knowledge base on zero grazing livestock production systems; the challenge is how to mainstream the available information into decision making at local and regional levels.
- Technical interventions have had a limited impact on overall livestock production. Hence, there is need to focus on breeding, health, feeds and feeding as wells intensified and integrated production systems and translate these technical information into policy options.
- Demographic pressure and climate change are creating important changes in land use and land covers, access to resources and on livelihood strategies of

livestock keepers. Therefore, adaptation and mitigation strategies should be designed and implemented step by step.

- Improving feeding systems within the landscape and low-cost feed formulations.
- Forage species differ in adaptability and productivity. Hence, selection and testing of forage species appropriate for intercropping, over sowing and rehabilitation of degraded grazing lands and closures need to be recommended from research.
- Estimating and reducing the economic losses of livestock from animal diseases and parasites.
- Methods for the conservation and improvement of indigenous livestock need to be designed.
- The existing and future's livestock technologies have to be fine-tuned to suit local conditions.

24.7 Conclusions

Livestock are very important to nutritional and food security, incomes and livelihoods of people in Lake Tana Sub-basin. However, the livestock sector has been facing challenges such as lack of resources and limited support schemes. On the other hand, the increased demand for livestock products could represent sustained and perhaps increased revenues for livestock keepers. Thus, to improve livestock productivity, one is the government policies should be supportive for livestock keepers. The other is the livestock keepers have to increase the efficiency of their operations and the productivity of their animals. Hence, integration of livestock with crop production and watershed management is considered as feasible and environmental sound solution. This is to improve livestock production and productivity, especially in the mixed crop-livestock production system. The integration should be analyzed from holistic/integrated and participatory perspectives. The researchers also have to work effectively with farmers and other relevant stakeholders to develop technologies and systems for removing productivity challenges. Moreover, there have to be rural-peri-urban-urban linkages to be effective in marketing perishable livestock products such as milk and milk products.

Appendix 1

See Table 24.1.

Table 24.1 List of recommended improved forage species for the different agro-ecologies of Ethiopia (Ethiopian Agricutural Research Organization 2000 Unpublished)

Improved forage species	Scientific and common names	Adaptations
Grasses	Avena sativa (Oats)	Mid to high altitude areas
	Pennisetum purpureum (Elephant, Napier grass	Low to high altitude areas
	Chloris gayana (Rhodes grass)	Low to high altitude areas
	Panicum species	Low to high altitude areas
	Melinis minutiflora (Molasses grass)	Low to mid altitude areas
	Sorghum species (Sudan and Columbus grasses)	Low to mid altitude areas
	Cenchrus ciliaris (Buffel grass)	Low to mid altitude areas
Herbaceous legumes	Vicia spp. (Vetch)	Mid to high altitude areas
	Triflium spp. (annuals and perennials clovers)	Mid to high altitude areas
	Medicago sativa (Lucerne, Alfalfa)	Low to high altitude areas
	Lotus corniculatus (Birdsfoot trefoil)	Highlands
	Melilotus altisimus	Highlands
	Lablam purpures (Lablab)	Low to mid altitude areas
	Vigna unguiculata (Cowpea)	Low to mid altitude areas
	Desmodium species (Silver and green leaf desmodium)	Low to mid altitude areas
	Stylosanthes spp. (Stylo)	Low to mid altitude areas
Browse trees	Chamaecytisis palmensis (Tagasaste, Tree Lucerne)	Highlands
	Gliricidia sepium (Gliricidia)	Highlands
	Cajanus cajan (Pigeon pea)	Low to mid altitude areas
	Calliandra callothyrsus (Calliandra)	Low to mid altitude areas
	Acacia spp (Acacia)	Low to mid altitude areas
	Leucaena spp (Leucaena)	Low to mid altitude areas
	Sesbania sesban (Sesbania)	Low to mid altitude areas
Root crop	Beta Vulgaris (Fodder beet)	Highlands

References

Alemayehu K, Ayalew T (2013) Meat and live animal export in Ethiopia: status, challenges and opportunities. Global Advanced Res J Food Sci Techno 2(4):054–059

Alemayehu Y, Wurzinger M, Tegegne A et al (2009a) Handling, processing and marketing of milk in the North western Ethiopian highlands. Livestock Rese Rural Dev 21(7):2009

Alemayehu Y, Wurzinger M, Tegegne A et al (2009b) Performance and limitation of two dairy production systems in the North western Ethiopian highlands. Trop Anim Health Prod 41:1143–1150. doi:10.1007/s11250-008-9294-3

Ali H (2009) Land use and land cover change, drivers and its impact: a comparative study from Kuhar Michael and Lench Dima of Blue Nile and awash basins of Ethiopia. Bahir Dar University, Thesis

Ali M, Neka M (2012) Livestock husbandry and economic-sustainability of small farmers in peri-urban areas: a case study from West Gojjam region, Ethiopia. EJESM 2(5):2012

Anteneh B, Tegegne A, Beyene F et al (2010) Cattle milk and meat production and marketing systems and opportunities for market-orientation in Fogera woreda, Amhara region, Ethiopia. IPMS Working Paper 19. Nairobi (Kenya): ILRI. IPMS Working Paper, 19 ILRI (International Livestock Research Institute), Nairobi, Kenya

Assefa G, Kehaliew A, Alemayehu M (2011) Overview of major feed resources availability and utilization in the highlands of Ethiopia. In: Proceedings of a national stakeholder workshop on innovations, actors and linkages in the dairy value chain in Ethiopia held at Food and Agriculture Organization of the United Nations Sub Regional Office for Eastern Africa, Addis Ababa, 28 May 2010

Benin S, Ehui S, Pender J (2006) Policies for livestock development in the Ethiopian highlands. In: Pender J, Place F, Ehui S (eds) Strategies for sustainable land management in the East African highlands. International Food Policy Research Institute, Washington, DC, pp 141–164

Dessie T, Effa K, Mirkena T (2013) The resource base, strategies for improvement and use of the Ethiopian livestock. In: Proceedings of the 20th annual conference of the Ethiopian society of animal production (ESAP) held in Addis Ababa, 03–05 October 2012

EEA (Ethiopian Economics Association) (2012) Report on the Ethiopian economy: transport sector development in Ethiopia: performance, policies and its role in the economy. Ethiopian Economics Association, Addis Ababa

Effa K, Dessie T, Han JL et al (2012) Morphological diversities and ecozones of Ethiopian horse populations. Animal Genetic Resources, 50, 1–12. Food and Agriculture Organization of the United Nations

Fisseha G, Gebrekidan H, Kibret K et al (2011) Analysis of land use/land cover changes in the Debre-Mewi watershed at the upper catchment of the Blue Nile Basin, Northwest Ethiopia. JBES (1)6:184–198

Funte S, Negesse T, Legesse G (2010) Feed resources and their management systems in Ethiopian highlands: The case of Umbulo Wacho watershed in southern Ethiopia. Tro Subtro Agro Eco (12):47–56

Girma E (2015) Phenotypic characterization, population viability and breeding strategy for Fogera cattle in selected districts of Amhara region, Ethiopia. Thesis, Bahir Dar University

Gizaw S, Tegegne A, Gebremedhin B et al (2010) Sheep and goat production and marketing systems in Ethiopia: Characteristics and strategies for improvement. IPMS (Improving Productivity and Market Success) of Ethiopian farmers project working paper 23. ILRI (International Livestock Research Institute), Nairobi

Grünenfelder J (2005) Livestock in the Simen Mountains, Ethiopia: its role for the livelihoods and land use of local smallholders. University of Berne, Thesis

Hailu A, Mazengia H, Wuletaw Z (2013) Indigenous chicken production system and breeding practice in North Wollo, Amhara region, Ethiopia. Poult Fish Wildl Sci 1:108. doi:10.4172/pfw.1000108

IBC (Institute of Biodiversity Conservation) (2012) Ethiopia's strategy and plan of action for conservation, sustainable use and development of animal genetic resources. Addis Ababa, Ethiopia www.cbd.int/doc/world/et/et-nr-05-en.pdf. Accessed 20 Oct 2014

IGAD (Intergovernmental Authority on Development) (2010) The contribution of Livestock to the economies of IGAD member states: study findings, application of the methodology in Ethiopia and recommendations for further analysis. Working paper no. 02–10, Odessa centre www.fao.org/fileadmin/user%85/IGAD%20LPI%20WP%2002-10.pdf. Accessed 20 Oct 2014

Lemma T, Tegegne A, Puskur R et al (2008) Moving Ethiopian smallholder dairy along a sustainable commercialization path: Missing links in the innovations systems. IPMS (Improving Productivity and Market Success) of Ethiopian Farmers Project. ILRI (International Livestock Research Institute), Nairobi

Mekasha A, Gerard B, Tesfaye K et al (2014) Inter-connection between land use/land cover change and herders'/farmers' livestock feed resource management strategies: a case study from three Ethiopian eco-environments. Agri Eco Env 188:150–162. doi:10.1016/j.agee.2014.02.022

Mekoya A, Oosting SJ, Fernandez-Rivera S et al (2008) Farmers' perceptions about exotic multipurpose fodder trees and constraints to their adoption. Agroforest Syst 73:141–153. doi:10.1007/s10457-007-9102-5

Moges F, Mellese A, Dessie T (2010) Assessment of village chicken production system and evaluation of the productive and reproductive performance of local chicken ecotype in Bure district, North West Ethiopia, African. J Agri Resea 5:1739–1748

Negassa A, Jabbarr M (2008) Livestock ownership, commercial off-take rates and their determinants in Ethiopia. Research Report 9. ILRI (International Livestock Research Institute), Nairobi

Sintayehu G, Samuel A, Derek B et al (2010) Diagnostic study of live cattle and beef production and marketing: Constraints and opportunities for enhancing the system. ILRI and IFPRI, Adiss Ababa, Ethiopia www.ata.gov.et/%85/Ethiopia-livestock-Value-Chain-Diagnostic-July-2010%85 Accessed 15 October 2014

Tassew A (2007) Production, handling, traditional processing practices and quality of milk in Bahir Dar milk shed area, Ethiopia. Thesis, Alemaya University

Tegegne F, Assefa G (2010) Feed resources assessment in Amhara National Regional State. Ethiopia sanitary and phytosanitary standards and livestock and meat marketing program (SPS-LMM), Texas A and M University www.systemborlaug.tamu.edu/%85/Feed-Resource-Assessment-in-Amhara-National-Re%85 Accessed 20 Oct 2014

Tegegne A, Gebremedhin B, Hoekstra D (2010) Livestock input supply and service provision in Ethiopia: Challenges and opportunities for market-oriented development. IPMS (Improving Productivity and Market Success) of Ethiopian farmers project working paper 20. ILRI (International Livestock Research Institute), Nairobi

Trade Bulletin 5 (2011) Ethiopia's meat and live animal export, produced by Ethiopia Sanitary and Phytosanitary Standards and Livestock and Meat Marketing Program, quarterly bulletin, July issue, Addis Ababa, Ethiopia www.borlaug.tamu.edu/2011/%85/ethiopias-meat-and-live-animal-export-sps-lm%85Accessed 15 Oct 2014

Wuletaw Z (2004) Indigenous cattle genetic resources, their husbandry practices and breeding objectives in North-western Ethiopia. Alemaya University, Thesis

Chapter 25
Urban Areas and Planning in the Lake Tana Region

Genet Gebreegziabher

Abstract Ethiopia is currently one of the least urbanized countries with just under 20% of its population living in urban areas, but it is expected to urbanize rapidly (around 4.8% per annum), over the next 40 years. Projections for the Lake Tana region are similar. This chapter reviews the distribution of urban areas and towns in the Amhara Region, gives an overview of the type of infrastructure available, and explains the approach to planning for these urban areas. With a growing population and high rate of urbanization, careful coordination of plans and resources will be critical in the region for sustainable development.

Keywords Amhara Region · Lake Tana · Urban areas · Urban planning

25.1 Introduction

Ethiopia has a 2500 year old history and culture of urbanization and city development. Currently it is one of the least urbanized countries, with 19.5% of the population living in urban areas (World Bank 2017). Over the next 40 years, it is expected to urbanize rapidly (around 4.8% per annum) as its population grows to over 150 million by 2050. It is the second populous country in Africa next to Nigeria with an annual growth rate of 2.6% (CSA 2008). In 2015, the total population of the country was estimated to be around 99.4 million (World Bank 2017). Similarly, the demographic situation of the Amhara Region shows an annual growth of 2.17% with a total population size of 19.21 million as of 2011/2012. The population of the region accounts for over 20% of the total population of the country while in terms of area the region contributes around only 15% (Table 25.1).

Due to its huge rural population, the nation as a whole, and the Amhara Region will experience a dramatic and rapid rural-urban transformation in the coming years. Regarding the settlement pattern, the overwhelming majority, i.e. nearly 85%

G. Gebreegziabher (✉)
Institute of Urban Planning, Bahir Dar, Ethiopia
e-mail: ggeyesus@yahoo.com

K. Stave et al. (eds.), *Social and Ecological System Dynamics*,
AESS Interdisciplinary Environmental Studies and Sciences Series,
DOI 10.1007/978-3-319-45755-0_25

Table 25.1 Amhara Region urban and rural population

Type of residence	2008	2009	2010	2011	2012	2013
Urban	2,236,998	2,368,998	2,497,998	2,641,997	2,782,997	2,952,997
Rural	15,388,002	15,638,002	15,903,002	16,166,002	16,437,002	16,673,002
Total	17,624,999	18,006,999	18,400,999	18,807,999	19,219,999	19,625,999

Source Computed by BoFED Population Affaires Core Process Based on 2012 inter-censual survey

Table 25.2 Population estimate and density by zone, 2011/12

Zone	Total population	Area (km^2)	Density (person/km^2)
West Gojjam	2,381,671	13295.90	179.13
East Gojjam	2,358,051	14009.74	168.32
North Gonder	3,403,299	44742.90	76.06
South Gonder	2,268,821	14055.47	161.42
South Wollo	2,899,420	18727.24	154.82
North Wollo	1,526,024	11335.69	134.62
North Shewa	1,991,163	15954.50	124.80
Awi	1,130,123	8584.68	131.64
Wag Himira	495,954	8781.96	56.47
Oromiya	507,936	4192.10	121.17
Bahir Dar	257,536	368.66	698.57
Lake Tana	–	3078.01	–
Amhara total	19,219,999	157126.85	122.32

Source Computed by BoFED Population Affaires Core Process Based on 2012 inter-censual survey

of the region's population, resides in rural areas and is engaged mainly in agriculture. In addition, population distribution is uneven among zones and woredas. In Amhara, for example, North Gonder stands first in terms of population size while Wag Himira is last. In terms of population density, West Gojjam is relatively densely populated while Wag Himira is more sparsely populated than the other zones. Generally, the highlands are more densely populated than the lowlands (Table 25.2) and (Fig. 25.1).

Nevertheless, the country today is one of the least urbanized nations of Africa. Currently, the settlement pattern is characterized by scattered rural settlements with increasing populations which also demand more productive land for subsistence farming. The pressure from the growing rural population with its subsistence farming coupled with local and global environmental challenges exposes the limited land and water resources for further risk.

In comparison to other sub-Saharan Africa (SSA) countries, Ethiopia's 19.5% urban population is low-about half of the average for all SSA, which is approximately 38% (World Bank 2017). In keeping with the pattern of urban growth of the

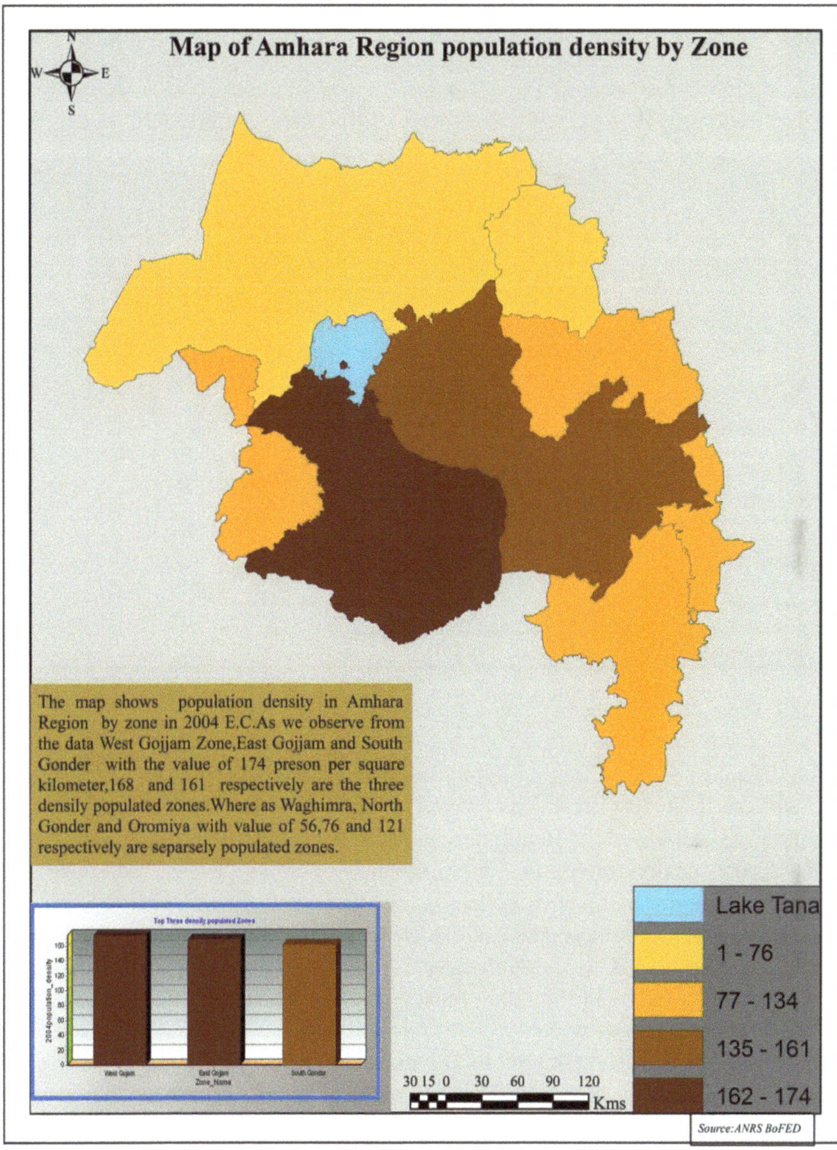

Fig. 25.1 Map of Amhara Region population density by zone

least urbanized countries, Ethiopia is currently witnessing one of the fastest rates of urban growth in the world. According to World Bank (2017) estimates, its 2015 rate of 4.8% annual urban growth was the 10th fastest in the world. It also shares some of the major characteristics of urbanization in SSA: rapid urban population growth with informal development and urban poverty, and urban primacy and rural-urban ambiguity.

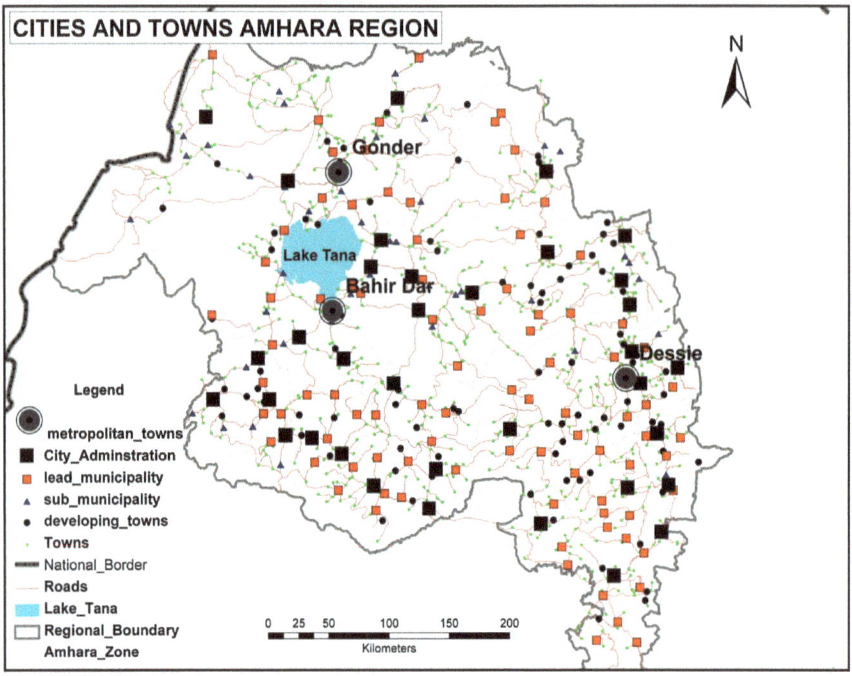

Fig. 25.2 Cities and towns in the Amhara Region

According to Amhara Region regulations, the designation "urban" applies to an area with at least 2000 residents, where at least 50% are engaged in off farm activities. As stipulated in Regulation number 65/2000 of the Zikre-Hig of the ANRS, urban centers are further categorized by population size, administrative importance, amount of revenue collected, occupation of the citizens and strategic importance. Based on these there are four levels: Metropolitan city (Bahir Dar, Dessie and Gonder), City administration, Leading municipality, and emerging towns. As shown below in Fig. 25.2, most of the towns are evenly distributed in all direction except in north west.

There are 422 towns in the region. Towns are encroaching on both sides of Lake Tana so it needs serious attention. They use the wetlands for different activities such as planting chat, for investment, and connecting sewer and drainage lines.

25.2 Urban Planning in the International Context

25.2.1 Definition

Urban planning and urbanization must go parallel so this paper will start by defining urban and urbanization. Urbanization is a phenomenon that happens as the result of three forces: natural rise, migration and reclassification of rural settlement as Urban. Urbanization is a decisive phenomenon of the 21st century over the entire planet. It presents great opportunities for economic growth and poverty reduction but also carries potential threats, particularly urban poverty and environmental crisis. Countries urbanize at different rates for various reasons however the rate and scale of urbanization expected.

Ethiopian urbanization is highly marked by primacy of the capital city, Addis Ababa, with almost 1/3 of the national urban population, and unbalanced urban system (Tarekegn 2014). Dire-Dawa, the second town is 12 times less populated than Addis Ababa. In this regard, regional capitals should play a significant role towards improving the unbalanced urban hierarchy both at regional and national level.

There are 900 urban centers and small towns: 340 towns have populations of less than 5000 inhabitants. The urban system is dominated by villages rather than true urban centers. The size limitation of urban centers is a constraint which minimizes the advantages of economy of scale. This advantage is a prerequisite for efficient services provision. Therefore, there is a need to encourage urban development of small towns with sustainability principles, sound urban development and efficient urban management systems.

Though the continent of Africa seems late in the process, the world has been transforming into an urban civilization especially since the beginning of the 19th century. By 2050, the United Nations projects that almost three-quarter of the world's population will call urban areas their home. Most of this growth will be centered in the developing countries (Habitat-UN 2006).

According to Hall (2002: 3), urban planning refers to "planning with a spatial, or geographical, component, in which the general objectives is to provide for a spatial structure of activities (or of land uses), which in some way is better than the pattern existing without planning. Such planning is also known as 'physical' planning is perhaps a more neutral and more precise term."

For Adams (1994: 2), urban planning is best defined as "a form of state intervention in a development process dominated by the private sector and market imperfections." When markets are likely to fail and competition is not perfect, urban planning can be used by the state to overcome market inefficiencies.

Adams (1994) further contends that urban planning in practice is less about finding the best technical solution that serves the public interest than about bargaining, negotiation and compromise over the distribution of scarce environmental resources. In attempting to mediate between conflicting claims on land, the planning authority may promote some interests over others.

25.2.2 *Overview of Urban Planning in the International Context*

The history of city planning, in the sense of their location and land-use being laid following a geometric pattern, is as old as the history of cities. Modern

Urban and Regional planning has arisen in response to specific social and economic problem triggered off by the industrial revolution at the end of the 18th century. In the evolution of urban planning theory in England since then, Hall (2002) identifies three different stages. The first, developed from the earliest times down to the mid 1960s and well exemplified in the early development plans coming after the 1974 town and country planning act, could be called the Master Plan or blueprint era.

25.2.3 Urban Planning and Development

In the most active way, urban planning interacts with the development process. Authorities seek to stimulate development and investment within their areas by promoting and marketing locations, making land available to developers and providing grants and subsidies. Such important activities, as Adams (1994) pointed out, are often neglected in academic accounts of urban planning, since they are not statutory duties under town and country planning legislation and are often undertaken in specialist units or departments that have no responsibility for development planning and control. However as a form of intervention in the development process, development promotion is conceptually no different from plans or control and in many parts of the country, may be far more effective (ibid).

25.2.4 Urban Planning in the Developing World

Under the influence of colonial powers, the Master Planning approach was exported in the 1950s and 60s to Africa and the rest of the developing world where it is still in use. In Tanzania for example, Master Plans are prepared even today under the Town Planning ordinance, which draws from the 1947 British Town and Country Planning ordinance (Nnkya 1998). These plans are based on Geds "Survey-analysis-Plan" method and server for a period of 20 years, but subject to revision on every five years (ibid). In Nigeria, the popularity of Master Plans among Policy markets and Planners has encouraged efforts to prepare these type of plans for most settlements of urban status in the country. Likewise, Ethiopia also has a history of Master Plans.

25.2.5 Urban Planning in Ethiopia

Ethiopia is also one of the few countries with a long history of urbanization dating back to more than 2000 years. Axum, the main city of the Axumite kingdom with its magnificent stone monoliths is a living testimony of this reality. Despite this early exposure to the world of urbanization, this country has remained predominantly rural.

Regarding the history of urban planning Ethiopia, there is little evidence on whether ancient Ethiopia cities such as Axum had any geometric planning at all. The haphazard distribution of the few remnants of this great civilization (the stone monoliths and other structures) seems to suggest in this direction.

The revival of further urbanization with the city of Lalibela in the 13th century and Gonder in the 17th century did not bring any formal planning even when the latter was it its splendor accommodating a population of 80,000 in the 18th century. So, the period until the foundation of Addis Ababa in 1886 (or until the Italian occupation) could be characterized by organic planning whereby city development was taking place through a spontaneous but conscious settlement formation system around a Royal palace, a market and a place of worship: what in some literature is known as 'the urban trilogy'.

As to the origin of modern urban planning Ethiopia, some experts in the field put the date to the foundation of Addis in 1886 when Empress Taitu, on deciding to come down from the cold Entoto mountain to the hot springs, started the first formal settlement in today's Filwouha area. Other historians associate it with the Italian occupation, the period when many of the cities occupied by the colonialists acquired their first ever official Master Plans. Whichever theory is valid, urban planning in its trajectory has taken two different directions: that of Addis Ababa and the rest of the country.

The planning history of other urban centers in the country goes back to the short period of Italian occupation (1936–1941) where few of the acquired their first ever Master Plan. The earliest planning initiative by the Ethiopian government started in the late 60s when an Italian consulting firm was commissioned by the imperial. So Bahir Dar established in the 14 century, it have three master plan which was prepared 1965, 1996 and 2006.

Ever since, different Ministries were given the responsibility of preparing Master Plans until the National Urban Planning Institute (NUPI) was established in 1987 as a sole planning body. With the federalization of the country and proclamation No 41/1993, urban centers were given the responsibility of managing and promoting urban development including the preparation of plan for their growth. Using this provision, some regional governments were involved in the making of urban plans. Established as a rural village around the 14th century Bahir Dar has developed into one of the current largest city of the country. Its fast development and transformation into a modern town ship was made during the Italian occupation period of 1928–1933 since it was used as a major military base for their expeditions in the region. In order to introduce the efficient use of land and facilitate the development of the city, the municipality initiated and financed the preparation of the 1965 and 1996 master plans and the 2006 integrated development plan (IDP) of the city.

25.2.6 Type of Urban Plans

Three types of plans now prepared in the country are Master Plans, Development Plans and Action (Detail) Plans (Belachew 2001).

Master Plans: contain the general land use of the urban center and the details of schemes necessary for the improvement of its physical features for a period of 20 years. These plans are prepared at a scale of 1:5000 and 1:2000.

Development Plans: consist of a general land use map at a scale of 1:5000 and a relatively detailed map at a scale of 1:2000 or 1:2500. Their main difference from Master Plans is the time span of their service, which is 10 years. The shorter timeframe is intended to reduce uncertainties in long term planning.

Detail Plans indicate whether a site is for school, worship or any other social function. This plans prepared in order to facilitate implementation. The scale at which they are prepared is 1:2000 or 1:2500 and their service time is for 20 year.

Due to the rigidity of Master Plans, they have no room for cities to make changes to accommodate unexpected activities. Thus, the current framework was developed with four types of plans (Structural, Strategic, Basic and Sketch plan) and three implementation plans (neighborhood, urban design and block plan) in the 2005 Federal Urban development urban plan preparing and implementation strategies. Between the old and the new types of plans, 968 towns (62% of the total towns in Ethiopia) have plans. In Amhara Region 380 out of 423 towns have plans. The remaining 43 will prepare plans in the upcoming budget year. These plans differ according to the city size with 5 standard levels as indicated in Table 25.3.

All these plans are supported by an explanatory report produced in English. This comprehensive report contains information ranging from situation analysis of the city in question (from a regional and local perspective) to a detailed explanation of the planning method, assumptions taken, conclusion reached, and recommended strategies.

Table 25.3 Types of plans required by size of the city

Standard	Small town	Medium town	Large town	City	Metropolitan
Size	2000–20,000	20,001–50,000	50,001–100,000	100001–1 million	>1 million
Plan	Basic plan sketch plan	Strategic plan and NDP	Strategic plan and NDP (neighborhood develop plan)	Strategic plan NDP and urban design	Structural plan NDP and urban design

Source Urban planning preparation and implementation strategy

25.2.7 Problems Related to the Planning System and New Trends

In a regional workshop on "integrated strategic planning by urban local authorities in Eastern and Southern Africa", Solomon (1999) presented an extensive evaluation of the problems of the plans produced by the institute. He indicated that urban plans in Ethiopia (including those prepared by NUPI) are characterized by strong physical emphasis. They determine how land should be used and define the structure of the road network. All these plans, he noted, have zoning control and development goals that disregard the integration required between physical and social economic programs with consequent lack of institutional linkages and public participation. Although this integration has improved somewhat, it is still a challenge.

Integration is time consuming during preparation, community participation is difficult and technical proposals are expensive and difficult to implement. The single most important limitation of the planning approach described by Solomon (1999) was its unresponsiveness to the multifaceted socio-economic and environmental problems faced by our urban centers. In order to improve the responsiveness of urban plans, the ministry of urban development housing and construction prepared strategy which consist principles, norms and standards to alleviate the problems stated above.

Tegegne (2011), reported that urban centers in the country are nowadays facing a serious socio-economic and environmental problem. High level of urban poverty (60.5% of the total urban population in 1994), unemployment (18% of all urban areas, poor urban service and infrastructure delivery, deplorable urban environment and other social evils) have become the images of our cities.

While the formal economic sector is shrinking the informal sector is blossoming. Local streets in major cities have become the platform for street vending and other related activities. The dismal aspect of this face is that the contributions of urban plans have been either very limited or non-existent. Of course from their very nature, the way Master and other plans are prepared in this country were not meant to solve urban problems of such dimensions, but to allocate land for different urban activities.

25.3 Infrastructure

25.3.1 Introduction

Infrastructure is a system of structures that are necessary for dynamic and integrated operation of towns/cities in particular and nation and the globe at large. Infrastructure could broadly be classified into two, physical and socio-economic infrastructure. Included in physical infrastructure are telecom, water supply, power supply, storm and waste water collection. Transport lines, schools, hospitals, banks, insurances, etc., belong to socio-economic infrastructure.

Due to their large population size and high density of settlement, urban areas require a network of different types of infrastructure that provide communication, supply and collection/removal services. Common services include transport, power, telecom, water supply, drainage and sewerage.

Transport routes, especially roads are unique in that in addition to the main purpose for which they are provided, they also serve as a space for installation of other physical infrastructure (both overhead, surface and underground).

Provision of dense infrastructure, sharing same line, naturally causes overlap and interference. Manifestation of the overlap and interference varies from unsightly layout to serious damages to one infrastructure from another, physical damage and overlap of incompatible uses (toilets and water supply lines for instance).

Infrastructure may be taken as one of the measures of the level of development of a nation. Infrastructure is also the leading requirement to attract investment. This being the case, the gap between the demand for and the supply of infrastructure is getting wider and wider in, particularly, the developing countries. The prevailing shortage is being aggravated by lack of optimized planning (and design), absence of integrated management holistic approach and improper utilization. In these regards, all infrastructure providing companies and other concerned institutions are increasing their efforts to induce holistic approaches and meet anticipated future demand through supply-oriented options such as the construction of new power stations loads and reservoirs.

25.3.2 Water Supply

Although the region is well endowed with a substantial amount of water resource potential, the performance of potable water supply and distribution is found to be low. Based on 2011/12 budget year data the regional water supply coverage was not exceeding 61.85%. This indicates that 38.15% of the people have no access to clean water.

This coverage is also less than the standard which is set by World Health Organization with the daily requirement of 45 L/person. Therefore, people who don't have access are forced to use unsafe drinking water from unprotected wells, rivers, and ponds. As a result, many people suffer from water borne diseases. This, in turn, needs additional budget to provide health services both at household and government level. At the same time it accelerates poverty since it reduces healthy and productive manpower. Besides, women bear the burden of traveling long distances to fetch water; which further reduces their economic contribution.

In order to improve the water supply status of the region, the regional government has made an effort to construct a number of hand dug wells, springs, shallow and deep wells, both in urban and rural areas. So far, in 2011/12 the water supply coverage of the region is estimated to be 61.85% (60.79% for rural and 70.65% for

Table 25.4 Water connections in Bahir Dar city

No.	Connection type	Number of water connections			
		2001	2002	2003	2004
1	Private yard	15,755	16,604	19,888	21,753
2	Commercial	1011	1042	1256	1108
3	Governmental + public	492	526	579	626
4	Industrial	71	75	94	81
5	Public tap	36	52	–	34
	Total	17,365	18,299	21,869	23,483

Source Bahir Dar water service office

urban areas). This coverage seem to be tremendous changes as compared to 2010/11 (53.43%) crude performances (51.95% for rural and 65.95% for Urban areas), still requires a great effort to change the deplorable situation of the rural household concerning potable water supply. So the government planned to increase to 90 for cities and 80 to rural areas.

Bahir Dar city gets its water supply from two springs and seven boreholes having a capacity of producing 19128 m^3/day. The water demand of the city is 25656 m^3/day and the city's water coverage is 74.6%. To achieve the 90% coverage in the year 2030 the city's water and sewerage authority has planned to dig 13 boreholes around Meshenti and is constructing a reservoir at Yibabe Iyessus. The following Table 25.4 shows summary of water connection in Bahir Dar City.

25.3.3 Roads

In general, the total length of all-weather rural roads under the jurisdiction of Ethiopian Road Authority (ERA) is about 5154 (60.08%) km and about 3424 km (39.92%) is under the jurisdiction of the Rural Road Authority (RRA). The total road network of the region is about 9425 km, which makes the regional road density to be 60 km per 1000 km^2. This implies that quite a number of people especially in rural areas are still traveling long distances to reach the main roads. The organization planned to upgrade to 10,055.58 and 64 km per 1000 km in 2015.

The road network in a city is considered a backbone for the city's development and makes neighborhoods more accessible to the residents. The city has been investing huge amount of money to provide access roads to the residents. There is also engineering equipment in the municipality which helped to construct the new roads and for maintenance purpose. The following Table 25.5 shows the total length of the road network that exists in the city.

Table 25.5 Roads in Bahir Dar by type

No	Surface type	2001 (km)	2002 (km)	2003 (km)
1	Asphalt	32.5	36.19	44.867
2	Gravel	75.7	121.81	186.68
3	Earth	174.7	379	390.28
4	Cobble stone	1.8	10.2	17.2
Total		284.7	547.2	639.027

25.3.4 Education

Education is a means to sustain and accelerate the overall development in a country and it has a direct effect on individuals' productivity and earnings as well. Hence, the challenge of development is the challenge of education. Currently, in the region, there are about 398 Kindergartens, 7327 primary schools (Grades 1–8), 278 General Secondary Schools (Grades 9–10), 128 Preparatory secondary Schools (Grades 11–12), 10 Teachers' Training Colleges, 65 Government 49 Private 10 Polytechnic and Vocational Education and Training colleges (TVET).

25.4 Health Service

Health is one of the fundamental social development indicators of a country. Getting health service is a human right and without it economic development of a country becomes inconceivable.

In the region, there are about 17 hospitals, 520 health centers and 2941 health posts which are rendering health services. Accordingly, on the average one health post for 5371, one health centers for 34,986 and one district hospital for 1,819,279 people are giving services.

Concerning the health professionals of the region, on the average one medical specialist serves 395,495, one medical doctor serves 187,554 and one nurse serves 3969 people.

25.5 Conclusion and Recommendation

25.5.1 Conclusion

Although preparing urban plans for towns was started over a decade ago, there are still challenges in preparing and implementation urban plans in this region. The core point is a lack of capacity especially in the way water bodies are integrated with development. In this chapter the main focus is Lake Tana. Lake Tana and the Blue Nile River are increasingly being encroached by buildings and urban agriculture without protecting the natural wetland and leaving a sufficent protective buffer. A critical issue is that many sewer and drain lines are connected to the water bodies. So protecting LakeTana and the Blue Nile River is directly linked to protecting Bahir Dar city and other rural and urban areas along the lake and river through sound urban planning.

25.5.2 *Recommendation*

- There must be special rules and regulations regarding the integration of water bodies with city development.
- Buffers will be compulsory for water bodies along their side (minimum 30 m).
- The water body must bounded by pedestrian and green areas.
- Future plans must integrate water bodies with the city.
- Continuous awareness training should be conducted about planning for the public, officials and professionals.

References

Adams D (1994) Urban planning and the development process. UCL press limited, University College London, London

Belachew KK (2001) Detail planning as a link to plan making and implementation. Paper presented at the 2nd national conference on urban development planning and implementation. Towards paving the way for partnership, organized by NUPI, 2001, Addis Ababa

Ethiopia CSA (CSA) (2008) Summary and statistical report of the 2007 population and housing census. Addis Ababa, Ethiopia: federal democratic republic of Ethiopia Population Census Commission, pp 1–10

Habitat-UN (2006) State of the World's Cities 2006/7. New York: United Nations

Hall P (2002) Urban and regional planning, 4th edn. Routledge, London

Nnkya TJ (1998) Land use planning practice under the public land ownership policy in Tanzania. Habitat Int 23(1):135–155

Solomon K (1999) Planning through decentralization: towards adopting an action oriented urban planning in Ethiopia. Paper presented at the Regional Workshop on "Integrated Strategic Planning by Urban Local Authorities in Eastern and Southern Africa". Municipal Development Program Eastern and Southern Africa Regional Office, National Urban Planning Institute (NUPI), May 31–June 4, 1999, Addis Ababa, Ethiopia

Tarekegn GY (2014) Addis Ababa, urbanization and urban planning practice in Ethiopia a historical perspective

Tegegne GE (2011) Livelihood and urban poverty reduction in Ethiopia: perspectives from small and big towns. African Books Collective

World Bank Open Data (World Bank) (2017) http://data.worldbank.org. Accessed 24 Feb 2017

Part IV
Management

Chapter 26
Environmental Protection in the Lake Tana Basin

Solomon Addisu Legesse

Abstract The Lake Tana Sub-basin has a splendid plenty of nature that should be conserved for future generations and not be spoiled. One effort toward its protection is the establishment of The Lake Tana Biosphere Reserve, Ethiopia's fourth natural heritage site established by UNESCO, after Simien National Park, Lower Valley of the Omo and Lower Valley of the Awash. However trend analysis shows that a significant change is taking place in the climate of the region at the same time that growing population increases pressure for agricultural expansion and development, and industrial activities to meet growing food and energy demands. Although policies exist that could help protect the catchment's resources, policy gaps, institutional problems and low policy implementation limit policy effectiveness. The major problems in the area can be summarized into three: Firstly, the policies and strategies are too general and do not give adequate direction/guideline to those who are implementing them. Secondly, even when there are laws that could be implemented, there are cases where there are no institutions or capable institutions that implement them. Thirdly, there are overlapping and, sometimes contradicting objectives of different institutions and hence there is little cooperation and coordination among different institutions. These gaps need to be addressed in order to manage and use natural resources in a sustainable manner. This chapter presents the key environmental trends in the region, and reviews vulnerabilities, threats, policies, and environmental protection gaps, in addition to potentially researchable thematic areas.

Keywords Climate change · Hazard management · Livelihoods · Environmental sustainability

S.A. Legesse (✉)
Department of Natural Resources Management, College of Agriculture and Environmental Sciences, Bahir Dar University, Bahir Dar, Ethiopia
e-mail: soladd2000@yahoo.com; solomonal@bdu.edu.et

K. Stave et al. (eds.), *Social and Ecological System Dynamics*,
AESS Interdisciplinary Environmental Studies and Sciences Series,
DOI 10.1007/978-3-319-45755-0_26

26.1 Overview of Lake Tana Basin Environment

Environmental protection is a practice of protecting the natural environment by individual, organizational or governmental levels, for the benefit of both the natural environment and human. Due to the pressures of population and technology, the biophysical environment is being degraded. This has been recognized, and governments have begun placing restraints on activities that cause environmental degradation. Progress in agriculture and industry is taken a general criterion of development of any country. This craze resulted in unlimited exploitation of every bit of natural resource. The splendid plentifulness of nature is a heritage that should be conserved for future generations and not to be spoiled. The main issues for developing countries are that the protected areas suffer from encroachment and poor management. African governments face several challenges in implementing environmental protection mechanisms and due to climate change hazards being day to day phenomenon, most of the environmental protection challenges focus on availability and management of resources.

The Lake Tana biosphere reserve includes Lake Tana, the source of the Blue Nile and the largest lake in Ethiopia, measuring approximately 84 km long and 66 km wide. The biosphere reserve covers a total area of 695,885 ha and is a hotspot of biodiversity. The biosphere reserve will seek to rekindle traditional communities' appreciation of their cultures, knowledge and skills, which reflect a sustainable lifestyle in harmony with the environment. The Lake Tana biosphere reserve is Ethiopia's fourth natural heritage inscribed by UNESCO, after Simien National Park, Lower Valley of the Omo and Lower Valley of the Awash. The Lake Tana biosphere reserve is among the 20 new sites added to UNESCO's World Network of Biosphere reserves world heritage site bringing their total number to 651 sites, including 15 trans-boundary sites in 120 countries. These additions were made by the International Coordinating Council of UNESCO during a meeting taking place in Paris by the year 2015.

This chapter consists of three broad sub-topics: general overview about the environment, baseline characterization and Environmental threats of Lake Tana Sub-basin. The introduction sub-session could address a briefs description about the environment, environmental protection, and lake Tana Sub-basin. The baseline characterization of Lake Tana Sub-basin included the climate trends, vulnerabilities, impacts and adaptations. Moreover, it included the legal and institutional frameworks of environmental protection, the environmental management plan and environmental impact assessment. The third sub topics, major environmental challenges and threats, included the current environmental protection gaps, potential researchable thematic areas, and possible recommendations were drawn from the past and ongoing researches in Lake Tana Sub-basin.

26.2 Baseline Characterization of Lake Tana Basin Environment

26.2.1 Climate Change Trends

Based on the meteorological station data, there was a general increasing annual maximum and minimum temperature change from 1961 to 2013 at Lake *Tana* Sub-basin (Solomon 2014). The trend line shows that the average annual maximum temperature was increased approximately by a factor of 0.0326. This value was indicated by the slope equation given by y = 0.0326x + 25.94. To the average, the annual maximum temperature was found to be 25.94 °C, however; this value was not kept constant because of the change in climate (Fig. 26.1). Similarly, there was a general increasing annual minimum temperature change as indicated by the trend line. To the average, the annual minimum temperature was found to be 10.118 °C, however; this value was not kept constant as a result of the change in climate by a factor of 0.059 °C. This value was also computed by using the slope equation, "y = 0.059x + 10.33". Accordingly, the change in annual minimum temperature change was found to be 2.16 °C. The annual minimum temperature shows a great difference than the annual maximum temperature change.

As shown by Fig. 26.2, there was a general decreasing trends in the average amount of annual rainfall. The trend line shows that the average amount of annual rainfall has been decreased by a factor of 3.947 in the lower sub-basin and by 1.561 in the upper one. As indicated in the trend lines equations, the slope has a negative value, which implies a decreasing of average amount of annual rainfall.

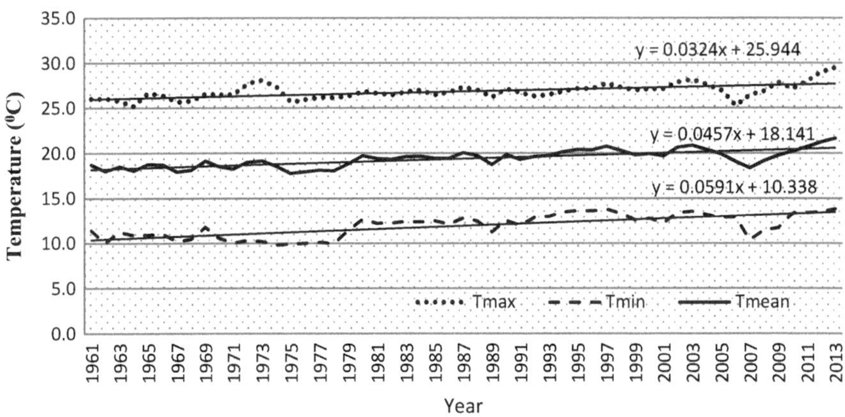

Fig. 26.1 Annual minimum, maximum and mean temperature trend in the lower sub-basin (Solomon 2014)

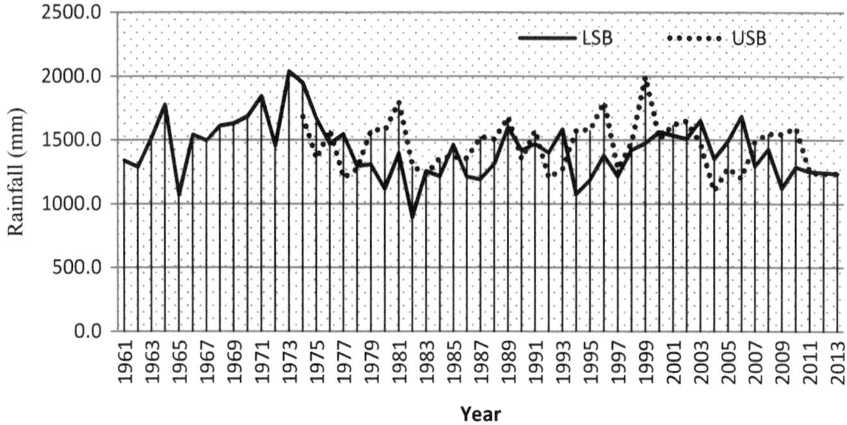

Fig. 26.2 Annual average rainfall amounts (Solomon 2014)

26.2.2 Hazard Magnitude, Frequency and Trend Analysis

Drought and rainfall variability are key climatic hazards to be reckoned in the Lake *Tana* Sub-basin. Generally accepted definition of drought is temporary moisture/water shortage in space and time (Gray and Mueller 2012). Drought can be categorized as meteorological, hydrological and agricultural. The meteorological drought is defined by below normal precipitation from long-term mean, hydrological drought is below normal water level (surface and underground) from long-term mean, and agricultural drought is below normal soil moisture for plant growth depending on plant water requirements (Gray and Mueller 2012). Rainfall variability is the level to which the amount of precipitation of a certain location observed in a specific time is above or below the mean. When the observed rainfall is below the mean for a considerable time scale, it can represent meteorological drought. Magnitudes of both drought and rainfall variability were measured by the standardized precipitation index (SPI). Figure 26.3 provides information on standardized precipitation index (SPI) as a measure of drought and rainfall variability of the mean average recorded annual rainfall amount for the past 30 years in the Lake *Tana* Sub-basin. The straight line from the point 0 represents the 30 years mean average annual rainfall (1404 mm) of the Lake *Tana* Sub-basin. SPI values above and below this line (standard deviation) represents the point of wetness or dryness of rainfall years.

Based on the SPI values in Fig. 26.3, Lake Tana Sub-basin experienced moderate to mild drought years in 1992, 1994, 1995, 2002, 2003, 2004, 2009 and 2012. The extent of drought was relatively higher in the years 1995, 2002, 2009 and 2012. This demonstrated that the occurrence of drought was frequent. According to national meteorological service agency classification, the magnitude of drought is mild if its SPI values are between 0 and 1, moderate if the SPI values fall between

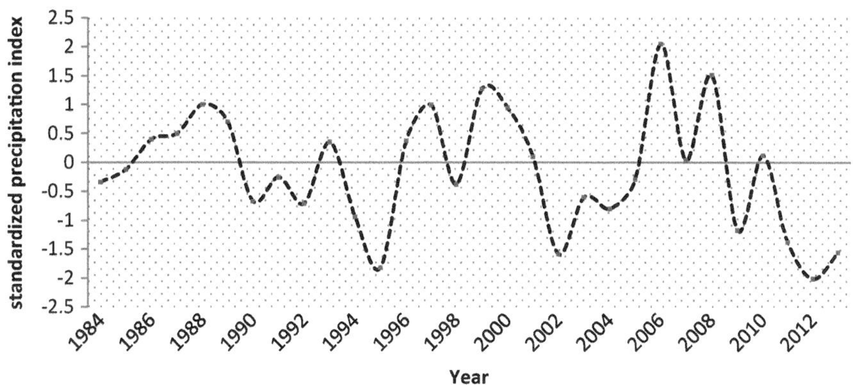

Fig. 26.3 Standardized precipitation index of Lake *Tana* Sub-basin, (Solomon 2014)

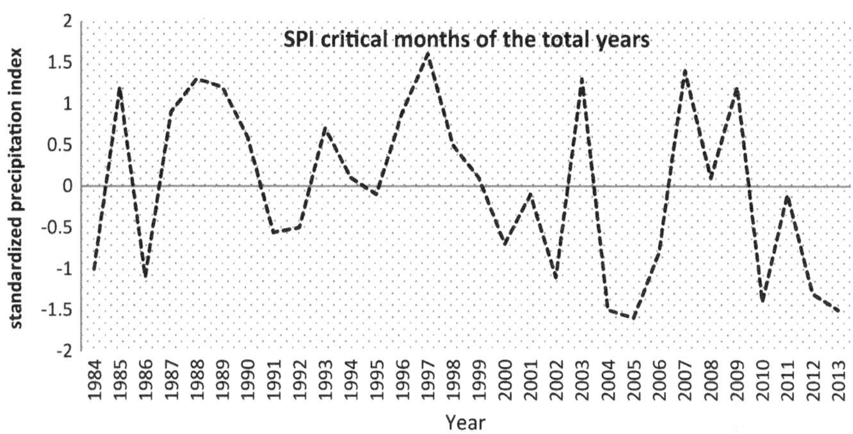

Fig. 26.4 Standardized precipitation index of critical rainfall months in Lake *Tana* Sub-basin

1 and 2, and sever if SPI value is greater than 2. Hence, moderate drought had occurred in 1995, 2002 and 2009. In 2012, a severe drought was happened.

Measuring drought and rainfall variability in inter-annual basis can be misleading as rainfall highs of certain months could offset rainfall lows in others, thus hiding inter-monthly variability. Moreover, aggregate rainfall values do not inform the time of onset and cessation of rainfall. Critical crop and pasture growing months of Lake Sub-Tana basin are therefore worth considering for rain-fed agriculture. As a result, five critical rainfall months (May, June, July, August and September) independently and in aggregate were taken into the analysis of rainfall variability and drought. Figure 26.4 has depicted the aggregate SPI values of critical rainfall months of lake Tana Sub-basin while Fig. 26.5 has shown SPI values in May and September.

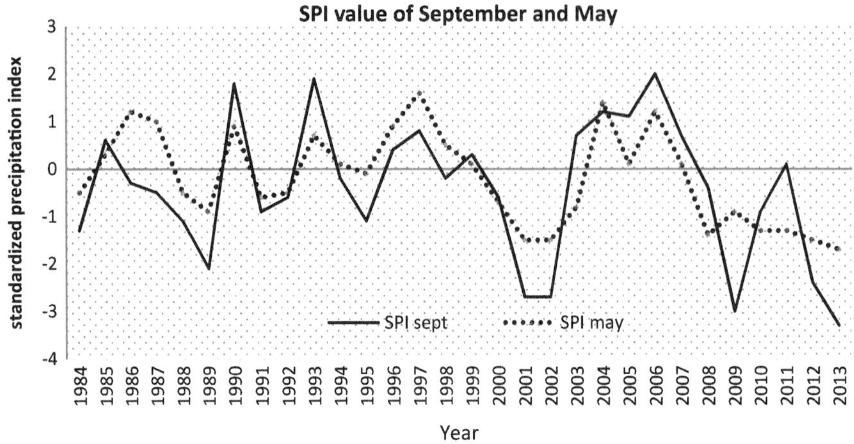

Fig. 26.5 Standardized precipitation index of May and September rainfall in Lake *Tana* Sub-basin

Standardized precipitation index values of critical rainfall months communicate differently from annual SPI values (Fig. 26.4). Only 1992 was the year with mild drought and above or normal rainfall distribution until 2000. After the year 2000, however, both the variability and drought severity increased. Based on critical months' analysis, it is possible to infer that 2000, 2002, 2004, 2005, 2010, 2011, 2012, and 2013 were the years when Lake Tana Sub-basin receives rainfall amount less than the long-term average. And moderate drought occurred in 2003, 2004, 2005, and 2009. Inter annual rainfall variability trend has also increased after 2000.

Experiences in the Lake *Tana* Sub-basin showed that May is the month which represents the most normal onset of rain. September represents the critical month of moisture requirement for flowering and maturing of crops. In light of this, Fig. 26.5 demonstrates the fact that May rainfall was below the long-term average between the periods of 2000 to 2005 and in 2009. In the period from 2002 to 2004 and the year 2009–2013, severity of drought occurrence in May was moderate. The September rainfall amount was in general more than the long-term average from 2002 onwards, with the exception of 2004 and 2009. Critical shortage of September rainfall with an SPI value less than −2 was observed in September 2009.

26.2.3 Lake Tana Sub-basin Livelihood Vulnerability to Climate Change

The LVI uses multiple indicators to assess exposure to natural disasters due to climate change or variability, social and economic characteristics of households that affect their adaptive capacity, and current health, food, and water resource characteristics that determine their sensitivity to climate change impacts.

Two approaches have been presented: the first expresses the LVI as a composite index comprised of seven major components while the second aggregates the seven into IPCC's three contributing factors to vulnerability as exposure, sensitivity, and adaptive capacity (Solomon 2014). This approach differs from previous methods in that it uses primary data from household surveys to construct the index. It also presents a framework for grouping and aggregating indicators on the sub-basin level, which can be critical for development and adaptation planning. By using primary household data, this approach helps to avoid the pitfalls associated with using secondary data. Another advantage is the reduction in dependence on climate models, which despite recent advances are still presented at too large a scale to provide accurate projections at levels useful for community development planning (Patz et al. 2005; Sullivan 2006).

Accordingly, Solomon (2014) stated that USB has a higher LVI than LSB (0.392 vs. 0.368, respectively). This suggests that relatively greater exposure to the adverse shocks of climate change in the USB than the LSB. The results of the major component analysis are presented collectively in a spider diagram (Fig. 26.6). The scale of the diagram ranges from 0 (less vulnerable) at the center of the outside edge, increasing to 0.5 (more vulnerable) at the outside edge in 0.1 unit increments. USB is more vulnerable in terms of livelihood strategies, social networks, health profile and water resource while LSB is more vulnerable in terms of socio-demographic profile, food and natural disaster related to climate variability.

The LVI–IPCC analysis yielded similar results. Almost all the three major components of the IPCC vulnerability approach value found to be higher in both agro-ecology of LTSB. Relatively, the USB households are vulnerable more than the lower one. The broken line in Fig. 26.7 indicates that the USB is on the outer edge of the triangle as compared to the solid line (LSB). It implies that from the three IPCC vulnerability components, USB appeared greater. In other words, USB has been highly exposed, sensitive, and has a higher adaptive capacity as well (Solomon 2014).

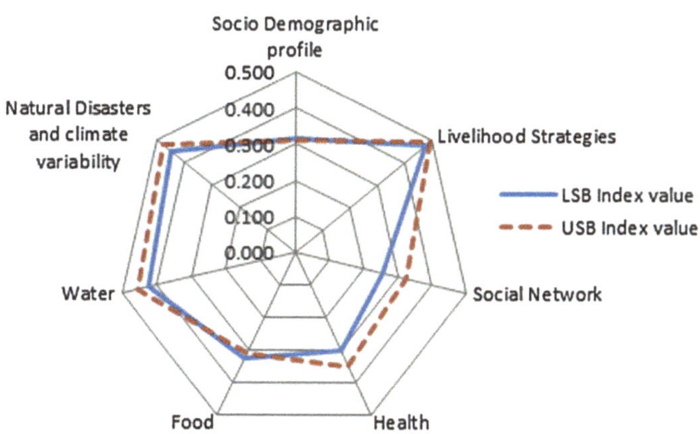

Fig. 26.6 Standardized vulnerability spider diagrams of livelihood major components

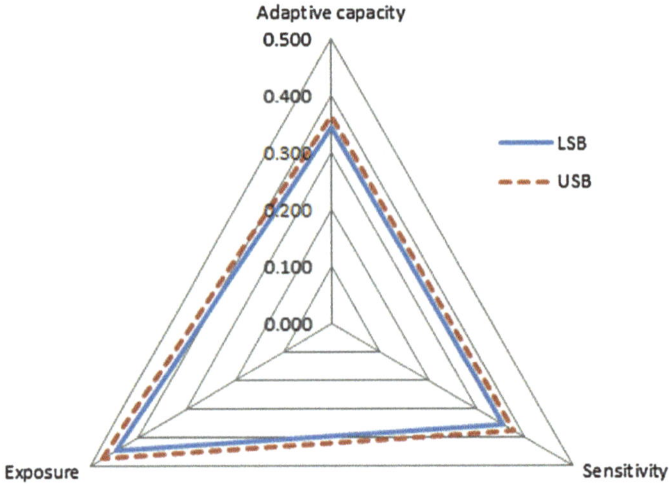

Fig. 26.7 Vulnerability triangle the contributing factors of the LVI–IPCC (Solomon 2014)

26.2.4 Climate Change Impacts on Rural Livelihoods

Solomon (2014), on his study reported that, crop livestock mixed production was the most common livelihood in the Lake *Tana* Sub-basin. Crop production and livestock raring were inseparable livelihoods due to the fact that one is depending on the other. In mixed farming system of the upper and lower basin of Lake Sub-*Tana*, where crop production is common; cattle are the most important livestock species for cultivation, threshing and manure. Most farmers in the study area own two oxen, a cow, few sheep and a donkey. Cattle are mainly kept for draught power and manure. Impact of climate change on field crop production pattern was significantly lower in the USB than in LSB (Table 26.1), where as its effect on yield reduction is significantly greater by 9.6% in the USB than LSB at 10% level of significance (p-value = 0.095). It implies that households were diversifying and adjusting the crop patterns to overcome climate change consequences. Regarding the fruits and vegetables, a significant number of households changed the patterns of fruits and vegetables in the USB (51.59%) than LSB (30.50%). On the other hand, yield reduction was significantly smaller in the USB (48.41%) than LSB (69.50%). More number of respondents in LSB reported that the type of climate change impact was manifested by yield reduction. However, Cattle production pattern change was significantly greater in the USB (55.93%) than the LSB (22.56%). It implies that large number of respondents changed the pattern of cattle production system in USB. Moreover, yield reduction from cattle is significantly smaller in the USB (44.07%) than LSB (77.44%). Small ruminants, Beekeeping production and fish production pattern change were significantly greater in USB than the LSB. Whereas, Small ruminants, beekeeping production and fish production yield reduction were greater for LSB agro-ecology. Mining and poultry production pattern change were significantly smaller in the USB than LSB agro-ecology. Whereas, mining and poultry production yield reductions were significantly greater in the USB than the lower (Table 26.1).

Animal husbandry practice in the sub-basin is extensive type, favoring disease outbreaks and parasite infestations. Disease and parasites affect the productivity of farm animals through direct mortality and morbidity issues. Disease and parasites could be a big burden on production and productivity of farm animals through their indirect effect on animals' fertility rate, growth rate, physical condition and traction power (Fig. 26.8).

Moreover, there were some local breeds guaranteed for better production in the sub-basin. However, most of the locally adopted indigenous breeds do have low genetic potential. During focus group discussion, most of the farmers appreciated

Table 26.1 Responses on climate change impact types on rural livelihoods

CCI on the Types of livelihood (%)	Climate change impact type, *n* = 300				Test statistics	
	Pattern change		Yield reductions		t-value	*p*-value
	W/dega	Dega	W/dega	Dega		
Field crop production	54.6	44.8*	45.4	55.2*	1.674	0.095
Fruits and vegetable production	30.50	51.59***	69.50	48.41***	3.575	0.000
Cattle production	22.56	55.93***	77.44	44.07***	6.088	0.000
Small ruminants	25.74	44.72***	74.26	55.28***	3.256	0.001
Poultry production	25.74	44.72***	74.26	55.28***	3.256	0.001
Beekeeping production	16.25	29.27*	83.75	70.73*	1.680	0.096
Fish production	23.73	97.73***	76.27	02.27***	11.213	0.000
Mining	29.07	09.52***	70.93	90.48***	4.214	0.000

Sources Solomon (2014)
*Significant at 10%, **significance at 5% and ***significant at 1% level of significance

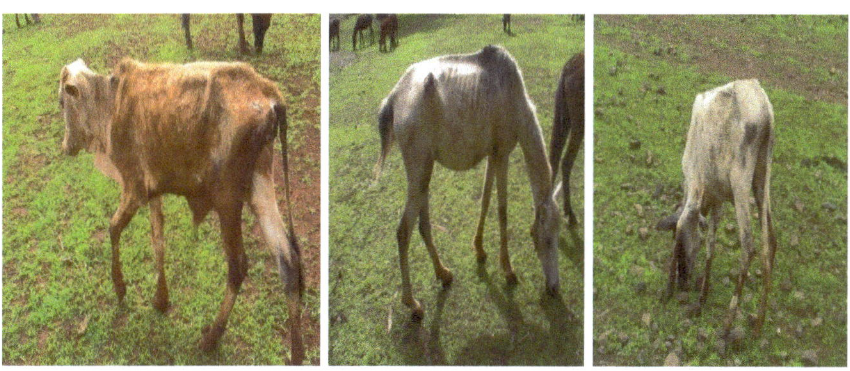

Fig. 26.8 Livestock at Gumara-Rib watershed of USB, (TCS 2013)

the better productive and reproductive performance of crossbreeds. Thus, appropriate breeding strategies with minimal gene pollution should be followed to benefit the sector.

26.2.5 Climate Change and Livestock Feed Balance

The land use land cover map analysis of the landsat image in the Gummara-Rib watershed showed that the area of grazing land (grassland, forestland, wetland and alpine) were 84,525 ha. Whereas; the current status of grazing land cover an area of 55,377 ha. This figure indicates that there were a change of 29,148 ha of grazing land to cultivated land including settlement and urban expansion. Feed balance analysis was carried out to estimate the proportion of feed that pasturelands, crop residues and crop aftermaths can provide from the total annual dry matter required for livestock maintenance. The amount of annual feed available was calculated based on satellite image analysis on the coverage of pasture (grazing/browsing) lands and other feed sources available in the watershed. Both degraded and un degraded grass lands, shrub lands and seasonal wetlands, etc., were considered for the analysis. The total land available as a natural feed/forage source in the watershed was estimated to be 361,655.5 ha. From these forage sources; it is possible to produce 100,706.17 tons of DM per year. On the other hand, crop resides and aftermaths are also important source of feed. Therefore, it is possible to produce 579,068.54 and 172,925.59 tons of DM from crop residues and aftermath, respectively.

Based on the report by FAO (1999) the conversion of grazing land to other type of land use type drastically decreases the livestock feed resources. The conversion factor for grassland is highest as compared to the rest feed sources. The total livestock population of the study area were obtained from the regional statistical agency. *Dera* and *Fogera* districts are completely found within the watershed, whereas, few administrative kebeles from libokemkem and Farta districts are not included. These were replaced by the kebeles which are found in the watershed from East and West *Estay* districts.

For the sake of convenience and uniformity, the different livestock populations were change into tropical livestock unit (TLU). Cattle and mule have the largest conversion factor and the TLU value of cattle was found to be the largest (445,079.60). This is because, cattle in Lake Tana Sub-basin is an important and integral component of the agricultural sector supplying draft power for cultivation, food and income to the households, including insurance against risks. Feed requirement was calculated considering the average number of livestock per household and total household population in the watershed.

The total number of livestock in the watershed was estimated to be 519,704.3 TLU, indicating a stocking density of 9.4 TLU/ha of pastureland. Considering DM requirement for maintenance to be 6.25 kg/day/TLU or 2.28 ton/year/TLU, the total DM requirement was estimated to be 1,184,926 ton/year. This shows that the available feed/forage sources addresses only 72% of the annual DM requirement and 28% were the gaps. This would be due to the fact that with the normal temperature and rainfall changes of the study area. To fill the gap, some farmers use concentrates like oil cake and residues of local alcoholic drinks. Others use locally available forage trees and homestead grown improved forage trees. Many farmers have also started feeding their livestock through cut-and-carry system by using grass from protected dense bushes/forests. In other words, the number of livestock population and its productivity are declining due to the shortage of feed which is indirectly controlled by the climate conditions of the study area.

26.2.6 Climate Change Impact on Biophysical Environment

The Lake *Tana* water is used for recession farming during the dry season while the volume of water declined and retreats. The gradual decline of rainfall, the extended nature of the dry season, sediment depositions by rivers and surface runoff caused the reduction of the lake level during the dry season. Following the water level reduction, the local residents are exercising agriculture in the outer periphery of Lake *Tana*, which leads to wetland degradations. Based on the analysis of raw data from the minister of water resources of Ethiopia, the water level of the last 28 years, (Fig. 26.9), the annual maximum (1787.34 m) and minimum (1785.79 m) water level were recorded in 2000 and 2003 respectively with mean average level 1786.64 m. From the year 1986 to 2000, the water level of Lake Tana was almost constant and maximum in 2000 due to *Chara Chara* weir construction (Yohannes 2007). However, starting from 2001, the lake level trend has got fluctuation and extremely declined in 2003 because of drought. After the construction of the *Tana Beles* hydroelectric power project, the lake level has shown improvements.

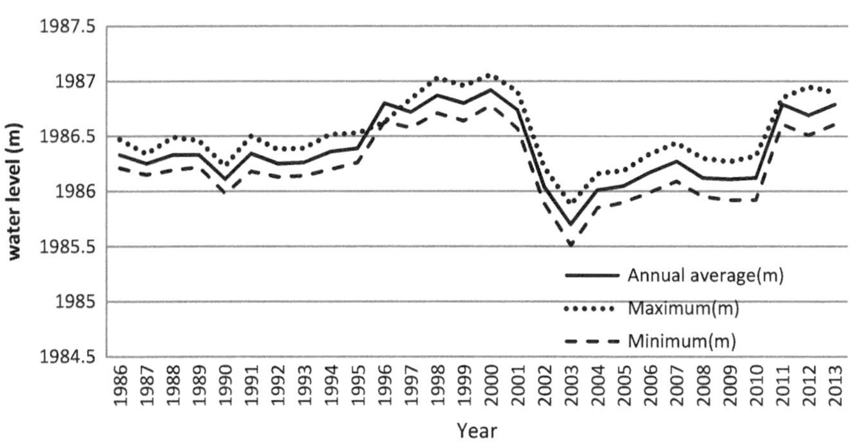

Fig. 26.9 Lake *Tana* water level trends (Ministry of water Ethiopia)

Fig. 26.10 Invasive water hyacinth (*Eichhornia crassipes*) problems of Lake *Tana* Invasive water hyacinth (*Left*) and fisherman (*right*) TCS (2013)

This could be due to the controlled release of the lake water as well as the high deposition of the sediment.

Sediment load and flooding are also the main consequences of climate change in the LSB. The erratic characteristics of the rainfall in the area leads to the sever occurrence of flood which damages the livelihoods of the households. The sediment depth measurement shows that the sediment load reduced from the lake center of the mouth of the four rivers: (*Gelda, Gumara*, old and new *Rib*). According to our field visits and key informants, flood inundation in 2011 had displaced and damaged crops for many people living in floodplain areas of the eastern periphery of Lake *Tana* (*Fogera*, *Dera* and *Libo Kemkem*). Invasive water hyacinth (*Eichhornia crassipes*) problems of Lake *Tana* are also currently the main problem in and around the Lake *Tana* socioeconomic activities (Fig. 26.10).

26.2.7 Adaptation Options to the Adverse Impacts of Climate Change in Lake Tana Sub-basin

Because of high climatic variability, communities residing in marginal environments of Ethiopia have developed strategies to cope with the adverse impacts of climate change. The high vulnerability of people in Africa to climate variability is attributed largely to their low adaptive capacity, which results from a deteriorating ecological base, widespread poverty, inequitable land distribution, a high dependence on the natural resource base (Hulme et al. 2003; IPCC 2014). Improving adaptive capacity is important in order to reduce vulnerability to climate change (Elasha et al. 2006). The most common climate variability and climate change adaptation strategies in Ethiopia are; Diversification of herds and incomes, growing of drought and heat resistant and early maturing crop varieties, use of small-scale irrigation, water harvesting and storage and improved water exploitation methods, labor migration and response farming (season-customized farm management practices). Moreover; increased agro-forestry practices, changes in farm location, reduction in herd and farm sizes, and food storage, crop and animal diversification, controlled grazing, Selling of assets, herd supplementation, communal holding of grazing lands which facilitate free mobility in pastoral areas, Culling of animals, and indigenous early warning and forecasting systems were proposed (Elasha et al. 2006; Admassie et al. 2008).

According to Solomon (2014) study at Lake *Tana* Sub-basin, the common adaptation methods taken for their livestock are pattern change (change from large ruminates to small ruminants), mobility, use of tolerant breeds, selling during hard seasons, use of vaccination and feedstock for hard season. however, crop diversification, intercropping, use of farm inputs, high cropping intensity, adjustments of planting dates, use of improved verity and irrigation agriculture are commonly practiced as an adaptation mechanism for crop production system. Moreover, households have practiced to diversify their income from non-farm and off-farm activities. The adaptation methods practiced by the households resulted in improvements on the biomass, water sources, the adaptive capacity of drought improvement, soil erosion reduction, soil fertility improvement, sediment reduction, landslide reduction, yield improvement, rehabilitation of degraded land, restoration of forest, wildlife return, micro-climate improvement and flood occurrence reductions.

In the lower sub-basin, rice is recently introduced crop by North Korean, in 1978 and production crop since the early 1984s, and is highly accepted as a means of adaptation strategies due to its high productivity and climate change resilient crop relatively to the indigenous crops. Any crop replaced by rice in the Lake *Tana* Sub-basin was perceived as positive change for people's livelihoods. Because of its resistance against flood inundation and less susceptibility to pests and the attractive

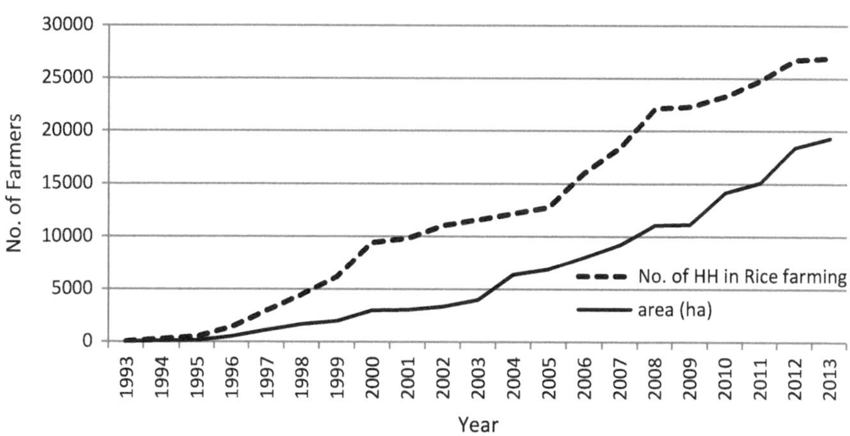

Fig. 26.11 Trends of farmers cultivating rice in the LSB, Solomon (2014)

market for rice cultivating farmers produce rice and benefit a lot to improve their livelihoods. Even conversion of communal lands (wetlands and grasslands) to rice cultivation was perceived as positive change to tackle the adverse impacts of climate change and to secure food (Fig. 26.11).

26.3 Legal Frameworks of Lake *Tana* Basin Environmental Protection

There are already a number of existing national and regional policy initiatives, sectoral policies, programs and strategies that may directly or indirectly address environmental issues. To mention some, Environmental policy of Ethiopia, Environmental Management Strategy, Environmental impact assessment, National Adaptation Program of Actions (NAPA), Energy policy, Forest Policy, Regional Food Security Strategy, Soil and water conservation policy (SWC), Plan for Accelerated and Sustainable Development to End Poverty (PASDEP), Growth and Transformation Plan (GTP), Agricultural growth plan (AGP) and so on. The direct responsible legal frameworks, polices and plans are stated in the following sections.

26.3.1 Environmental Policy and Conservation Strategy of Ethiopia

The Environmental Policy of Ethiopia was issued in 1997 to provide overall guidance in the conservation and sustainable utilization of the country's environmental resources in general. The Environment Protection Authority (EPA) which was founded in 1994 is the main actor in the design of this policy. The goal of the policy is 'to improve the health and quality of life of all Ethiopians and to promote sustainable social and economic development through the sound management and use of natural, human-made and cultural resources and the environment as a whole.

The conservation strategy of Ethiopia (CSE) was launched in 1989. The CSE Project is now under the Environmental Protection Authority of Ethiopia. This is one of the most important strategic documents as far as environmental or conservation of natural resources is concerned. Its aim was to study the natural resources, environmental imperatives and development demands in the country and to harmonize them. The harmonization process was to be activated through the formulation of an appropriate environmental policy. It provides a comprehensive and rational approach to environmental management in a very broad sense, covering national and regional strategies, sectoral and cross sectoral strategy, action plans and programs, as well as providing the basis for development of appropriate institutional and legal frameworks for implementation.

26.3.2 Biodiversity Conservation and Research Policy

After ratifying the conservation of biodiversity (CBD), Ethiopia has taken a number of fundamental measures. A National Policy on the Conservation and Research of biological resources was issued based on the rationale that the conservation of biodiversity is one of the conditions of the overall socio economic development and sustainable environmental management goals.

Some of the objectives of the Policy are to:

- ensure that the Ethiopian plant, animal and microbial genetic resources and essential ecosystems as a whole are conserved, developed, managed and sustainably utilized;
- assert national sovereignty over the country's genetic resources and develop a mechanism for a fair exchange, safe movement and proper management of these resources;
- integrate biodiversity conservation and development programs into Federal and Regional agricultural, health, industrial and overall national economic development strategies and plans;
- recognize, foster and augment the indigenous knowledge and methods relevant to the conservation, development and sustainable use of biodiversity, and promote and encourage the development and putting into practice of new and emerging technologies such as biotechnology;
- encourage the participation and support of local communities in biodiversity conservation, development and utilization;
- ensure equitable sharing of the benefit accrued from the use of genetic resources and/or associated indigenous knowledge;
- create a functional and efficient organizational structure that will ensure inter-institutional linkage and coordination in biodiversity conservation, development and utilization; and
- promote regional and international cooperation in biodiversity conservation, development and sustainable use.

Ethiopia is endowed with rich biodiversity. Ethiopia's protected areas are increasingly degraded. Land is being converted for subsistence and commercial agriculture, timber used for fuel wood and construction. The loss of forests and other protected land is caused by a growing population, unsustainable natural resource management, and poor enforcement of existing legislation. Recognizing the degradation of these important resources, Ethiopia has prepared its National Biodiversity Strategy and Action Plan (NBSAP) document in 2005, as demanded by Article 6 of the CBD. The NBSAP document "defines the current status of, pressures on, options for, and priority action to ensure the conservation, sustainable use, and equitable share of benefits accrued from the use of biological diversity of Ethiopia. The NBSAP serves as a roadmap for supporting the environmental component on Ethiopia's journey to sustainable development". The main actors, the Institute of Biodiversity and the Ethiopian Wildlife Conservation Authority, have also been strengthened with more power and mandate in conservation of biodiversity and sustainable use.

In order to govern the conservation, sustainable use and access and benefit sharing of the country's biodiversity and associated indigenous knowledge, the following policy and law instruments were issued:

- Institute of Biodiversity Conservation and Research Establishment Proclamation (No. 120/1998),
- Environmental Impact Assessment Proclamation (No. 299/2002),
- Institute of Biodiversity Conservation and Research Amendment Proclamation (No. 381/2004),
- Plant Breeder's Right Proclamation (No. 481/2006).
- Forest Development, Conservation and Utilization Proclamation (No. 542/2007),
- Access to Genetic Resources and Community Knowledge, and Community Rights Proclamation (No. 482/2006) and Regulation (No. 169/2009).

26.3.3 Environmental Impact Assessment (EIA)

A number of proclamations and supporting regulations were made that contain provisions for the protection and management of the environment that reflect the principles of the constitution and EPE. The Environmental Impact Assessment Proclamation No. 299 of 2002 was adopted by the House of Peoples' Representatives. Among other proclamations EIA (No. 299/2000) is the one that provides proactive and reactive provisions designed to ensure sustainable development. Its aim is to prevent, reduce or offset the significant adverse environmental effects of development proposals, and to enhance the positive ones. The environmental policy lays the foundation for environmental impact assessment in the country. The environment policy stipulates the country's policies regarding EIA. It provides for the enactment of a law which requires that an appropriate EIA and environmental audits should be undertaken on private and state development projects; and the development of detailed technical guidelines that direct the undertaking of EIA and environmental audits in the various sectors. It also provides for the establishment of an institutional arrangement responsible for undertaking, coordinating and approving EIA and the subsequent environmental audits. It states that EIA should consider not only physical and biological impacts, but also address social, socioeconomic, political and cultural conditions; and that environmental audits should be undertaken at specified intervals during project implementation to ensure compliance with terms of EIA authorization. It also state that environmental impact statements should always include mitigation plans for environmental management problems and contingency plans in case of accidents; and that the EIA procedure should provide for an independent review and public comment on environmental impact statements before they are considered by decision-makers.

26.3.4 Sustainable Land and Water Management Program

This program has its objectives to support scaling up of best land management practices and technologies in sustainable land management and the adoption of these management practices and technologies by smallholder farmers in the "high potential/food insecure" areas that are becoming increasingly vulnerable to land degradation and food insecurity. Land Administration and Use Proclamation No. 456/2005 stated that all land in Ethiopia is under government ownership and that the right to hold and use rural land may be acquired by individuals, groups (communal holdings), and private entities. The Proclamation shall apply to any rural land in Ethiopia, i.e. any land outside of a municipality holding or a town designated as such by the relevant law. The Proclamation provides rules relative to acquisition and use of rural land by peasant farmers or pastoralists, transfer of rural land use rights including rights of inter-generational tenure transfer, and rights of exchange for land and limited leasing rights, distribution of rural land, resolution of disputes, restrictions on the use of rural land; and defines responsibilities of the Ministry of Agriculture and Rural Development and Regions. The rural land administration and land use laws are being implemented by the regional states. But each region has its own land policy depending on the region's specific situation.

Ethiopia Water Resources Management proclamation is also established to ensure that the water resources of the country are duly conserved and protected from harmful effects and utilized for the highest social and economic benefits of the country. Accordingly, the proclamation describes the measures that must be taken for the conservation and protection of waterways and the conditions under which water resources may be exploited. The proclamation prohibits the release of any waste that endangers the lives of humans, animals or plants into water bodies. In addition, it prohibits the clearing of trees or vegetation and the construction of residential houses along the banks of water bodies so as to ensure their protection.

26.4 Major Environmental Challenges and Threats of Lake *Tana* Basin

26.4.1 Environmental Challenges

Population pressure calling for agricultural expansion and development, and industrial activities to meet the food and energy demands in the Lake catchment poses challenges for the Lake water level and quality. Lake *Tana's* water level fluctuation and quality deterioration is mainly caused by the development and expansion needs due to population pressure. The breeding and feeding habitats in the littoral region of the Lake are in severely degradation due to the water level fluctuation and quality deterioration. Lake *Tana* fisheries are sources of livelihoods for thousands local communities living around Lake *Tana*. It serves as a supplement to the daily diet, source of employment and income for the pro-poor. However, these fisheries are facing different challenges. These challenges include: Population pressure, water level fluctuation, water quality deterioration, breeding and feeding habitat destruction, fishing pressure during breeding season and poor implementation of fisheries' legislation.

Water quality deterioration is mainly caused by the wastewater discharges and sediment load entering into *Lake Tana* as a result of urban and rural activities. *Bahir Dar* town's current practice of discharging untreated industrial and domestic waste into the Lake is affecting adversely the quality of Lake-water. The Lake shores along *Bahir Dar* Gulf are polluted due to the waste water discharges, the unpleasant smell perceived on the Lakeshore, and intensive growth of weeds observed on the shores as well as in the Lake water. In addition to this, massive blooms of algae in the southern gulf have developed, and the phytoplankton became increasingly dominated by the potentially toxic cyanobacteria (Ayalew et al. 2009). The pollutant load (causing suffocation), toxic cyanobacteria blooms and weeds growth (causing shelter which means preventing sunlight entering to the underwater) push the fish far away from the Lakeshore habitats.

The other water quality deteriorating agent is the sediment load entering through the inflowing rivers. The total average annual sediment load of the four major rivers entering into the Lake shows an increasing trend despite the government's effort to reduce soil erosion. Annual flow of over 32 million tons of soil from the upper catchment enters to the Lake Tana. The sediment entering the Lake reduces penetration of sunlight deeper into the Lake and forces reduction in green algae (microphysics) productivity, resulting in the decline of the fish stock which feed on these. Water quality change is observed favoring the growth of highly invasive weed called water hyacinth (*Echhornia crassipes*). Recently water hyacinth is spreading along several parts of the Lake's shorelines. It also blocks sunlight from reaching native aquatic phytoplankton and hampers oxygen supply to the water body. In addition to these, the weed interrupts local subsistence fishing by blocking access to the shores.

26.4.2 Environmental Threats

- Despite the government's effort to reduce soil erosion, the ever-increasing sediment and nutrient load entering the Lake continues to threaten the Lake by affecting primary production, favoring aquatic weed growth.
- Recession agriculture, grazing and fuel collection during the dry season due to the decrease in the water level continues to degrade the littoral region and wetlands of the Lake.
- The rapid expansion of *Bahir Dar* town and the current practice of discharging untreated industrial and domestic waste into the Lake will have adverse effects on the quality of the Lake water leading to the lake ecosystem imbalance.

26.5 Environmental Policy Gaps

The federal and regional governments have issued various policies and strategies to deal with problems related to natural resources managements and environmental protection. It is also argued that, there is quite good framework for natural resource management in Ethiopia, and a number of established institutions monitor the environment. However, the extent of degradation of natural resources in most parts of the country shows that the country as a whole is unable to ensure the protection and sustainable use of its resources such as forests, biodiversity and wetlands as well as its water resources. This is mainly due to that limited on-the-ground implementation of policy as well as marked limitations in stakeholder participation. The major problems in the area can be summarized into three. First, the policies and strategies are too general and do not give adequate direction/guideline to those who are implementing them. Second, even when there are laws that could be implemented, there are cases where there are no institutions or capable institutions that implement them. Thirdly, there are overlapping and, sometimes contradicting objectives of different institutions and hence there is little cooperation and coordination among different institutions.

26.6 Potential Researchable Thematic Areas

1. What are the adverse impacts of climate change on rural livelihoods?
2. What adaptation options are taken?
3. what are the sources, causes and effects of global, regional and local Environmental pollutions?

These questions are selected because very little or no research has been conducted in Lake Tana Sub-basin. The environment is not given attention while the government, the investors and the local community are trying to develop the region. Researchers should have to conduct problem solving research by targeting the environment at all levels.

26.7 General Recommendations

Policy gaps, institutional problems and limitation in policy implementation are among the key factors that affect the environment and livelihoods of people in Lake Tana sub-basin. These gaps need to be addressed in order to manage and use natural resources in a sustainable manner. The regional government should be able to deal with the different challenges related to natural resource management and ensure conservation and sustainable utilization of biodiversity, wetlands, land and water resources in Lake Tana sub-basin. There is a need to closely examine policies, strategies and action plans, particularly those needing clarification or harmonization. Lack of integration and coordination to achieve common goal and running their development activities independently leads to further environmental deterioration and land degradation. Forums should be arranged to hear from stakeholders, to identify gaps and overlapping, and there should be a system of monitoring, evaluation and feedback to check level of implementation of policies. There is a need to create an enabling environment for successful implementation of integrated watershed management. Prevention of development induced land degradation, rehabilitation of gullies and degraded areas, Lake water conservation and utilization, biodiversity conservation, wetland conservation, Forestry and agro-forestry, focus on non-farm employment, raising awareness and participation of stakeholders should be actively practiced for sustainable use of environmental resources.

References

Admassie A, Adenew B, Abebe T (2008) Perceptions of stockholders on climate change and adaptation strategies in Ethiopia. International Food Policy Research Institute (IFPRI). Retrieved April 2013, from Series of IFPRI Research Briefs website http://www.ifpri.org

Ayalew W, Amy C, Getachew T (2009) Tana Beles watershed development. Appendix for Natural Resources Survey. Volume II Appendix. Bahir Dar, Ethiopia

Elasha BO, Medany M, Niang-Diop I et al (2006) Background paper on impacts, vulnerability and adaptation to climate change in Africa for the African Workshop on Adaptation Implementation of Decision 1/CP.10 of the UNFCCC Convention, Accra, Ghana, 21–23 Sept 2006. 54 pp

FAO (Food and Agriculture Organization) (1999) The state of food insecurity in the world. Rome, Food and Agriculture Organization, Italy. Retrieved Aug 2013, from FAO database

Hulme M, Conway D, Xianfu L (2003) Climate change: an overview and its impacts on the living lakes. A Report Prepared for the Eighth Living Lakes Conference. Tyndall Centre for Climate Change Research. 51 pp, Conway

IPCC (2014) Climate change synthesis report summary for policy makers. Retrieved 2014 from http://www.ipcc.ch/pdf/assessment-report/ar5/syr/SYR_AR5_SPMcorr1.pdf

Gray C, Mueller V (2012) Drought and population mobility in rural Ethiopia. Retrieved 22 June 2013, fromwww.elsevier.com/locate/worlddev.40(1), pp 134–145. doi:10.1016/j.worlddev.2011.05.023

Patz JA, Campbell L, Holloway D et al (2005) Impact of regional climate change on human health. Nature 438:310–317

Solomon A, Prasada R (2013) Farmers' perceptions and adaptations to climate change in Ethiopia, IEEE Potentials ISSN, Vol 35, no 5, pp 30–31, USA

Solomon A (2014) Climate change and sustainable livelihoods of rural people of lake Tana Sub-Basin, Ethiopia. PhD dissertation, Andhra University

Solomon et al (2013) Review of sustainable environmental resources management policies, plans, proclamations and strategies: Amhara National Region State, Ethiopia. Int J Sci Res Manag 1 (2):63–75. ISSN: 2321–3418

Sullivan C (2006) Calculating a water poverty index. World Dev 30:1195–1210

TCS (Temesgen Consultancy Service) (2013) Baseline survey of community-based integrated natural resources management project in Lake Tana Watershed. Bahir Dar Ethiopia

Yohannes D (2007) Remote sensing based assessment of water resource potential for Lake Tana basin. MSc. thesis, Addis Ababa University

Chapter 27
Institutional Analysis of Environmental Resource Management in Lake Tana Sub-basin

Belachew Getenet and Berihun Tefera

Abstract Natural resource degradation in the form of soil erosion, deforestation, and wetland resource depletion, is a major problem in Lake Tana Sub-Basin, Ethiopia. Various measures have been taken by the government and its development partners in response. This study investigates whether measures such as formulation of policies, strategies, laws and proclamations as well as establishment of organizations contribute to the protection and/or management of natural resource management of the Lake Tana Sub-Basin or not. Government reports, journals, articles and other relevant sources were used to examine whether the measures taken fulfil their objectives. The findings of the study indicate that the policy measures and the efforts of translating them into practice do support conservation of natural resources in Lake Tana Sub-Basin. Various sustainable land management (SLM) technologies introduced by the government and development partners helped to address soil degradation. The remnants of forests around the lake and in the upper sub basin have been preserved and protected. Yet, land degradation in various forms still outweighs the rehabilitation of degraded lands and conservation of the natural resources in the sub basin. Lack of genuine participation of the local people, technical and policy failure, lack of enforcement of some policies/laws, overlapping and conflicting responsibilities of different government institutions and conflict of interests accounted to failure of achieve the desired outcomes. The findings suggest ways policy makers and development actors can address the multifaceted barriers to improve policy effectiveness.

B. Getenet (✉)
Environment and Development Studies College, Addis Ababa University,
Addis Ababa, Ethiopia
e-mail: belachew_getnet@yahoo.com

B. Getenet
Faculty of Social Sciences and Blue Nile Water Institute, Bahir Dar University,
Bahir Dar, Ethiopia

B. Tefera
Agricultural Economics Department, College of Agriculture
and Environmental Sciences and Blue Nile Water Institute,
Bahir Dar University, Bahir Dar, Ethiopia
e-mail: berihunt2@gmail.com

K. Stave et al. (eds.), *Social and Ecological System Dynamics*,
AESS Interdisciplinary Environmental Studies and Sciences Series,
DOI 10.1007/978-3-319-45755-0_27

Keywords Environment resource management · Lake Tana · Institutions analysis

27.1 Introduction

As the report of the Millennium Ecosystem Assessment (World Resource Institute 2005) notes: "Humans have changed ecosystems over the past 50 years more rapidly and extensively than in any comparable period of time in human history, largely to meet rapidly growing demands for food, fresh water, timber, fiber, and fuel." Despite substantial net gains in human well-being and economic development, the changes that have been made to ecosystems have resulted in irreversible loss in the diversity of life on Earth (World Resource Institute 2005). The deterioration of world ecosystems due to economic development has become a critical issue for the civil societies, non-governmental organizations (NGOs) and academics since second half of 20th century. Environmentalists have campaigned against the deterioration of global environment and its negative effects on human life since 1960s, and a number of very influential studies about impacts of development on the physical environment, published since early 1970s, have influenced politicians and the international community. Environmentalists' campaign and academic publications have received substantial support from the media and people. The publications of studies and pressure from the environmental movement have also forced the governments and international organizations to recognize the deterioration of environmental resources as a critical development agenda. Consequently, more than 18 Multilateral Environmental Agreements have been signed (Sands 2003). The major aims of these Multilateral Environmental Agreements are to protect natural resources and biodiversity from over-exploitation and to safeguard the environment. Governments of most of developed and developing countries have ratified these multilateral agreements and included in their different macro level policies and laws. Ministers, commissions and departments tasked with the enforcement of policies on the environmental resource management have established at national level, and many regional and local governments have also developed programs and strategies.

Like many other countries, Ethiopia has recognized the inseparability of environmental resource management and economic development. The current government believes that poverty alleviation and economic growth would not be realized without the care use and stewardship of environmental resources (FDRE 1997a, b). Cognizant of this fact, various measures have been taken to conserve environmental resources as a key prerequisite for success of socio-economic development. A wide range of policies, strategies, laws or proclamations related to environmental protection and management have been developed in the past two decades. In order to put the policies, strategies, laws and proclamations into effect, certain implementing organizations have established that involve diverse stakeholders in the realization of the government policies. The aim of this paper is to investigate how the policies, strategies, laws, proclamations, and organizations hereafter called "institutions" contribute to the protection and/or management of natural resource of the Lake Tana Sub-Basin, in the upper Blue Nile watershed. Specifically, the paper assesses the institutional and organizational arrangements governing natural resource management in the Lake Tana Sub Basin, assesses the role of organizational structures, and investigates the challenges these institutions and organizations face.

In institutional analysis, differentiating institutions and organizations is difficult mainly due to the difficulty of separating rules and regulations from the organizations that affect them (Kayambazinthu et al. 2003). Some authors (e.g., Mukamuri et al. 2003; Bandaragoda 2000) view organizations as part of institutions, defining institutions as "the combination of policies, laws and regulations; organizations; operational plans and procedures and; norms, traditions, practices and customs." Others (e.g., World Bank 1999/2000) view institutions and organizations as distinct entities, defining institutions as written laws, rules and, informally established procedures, norms, practices and patterns of behavior while organizations are formal and informal agencies governed by institutions. In this paper, we adopt the World Bank's distinction in order to investigate which organizations are working in the realization of which policies, laws and regulations.

27.2 Functions of Institutions in Resource Management

Behaviors of society have been guided by moral precepts and customs than formal laws for thousands of years (Beyer 2006). However, the rise of industrialization and urbanization, and the resulting increase in the complexity of environmental and socio-economic problems has challenged the relevance of societal norms/customs in governing human behaviors (Beyer 2006). Modern laws and policies have evolved to address the complex societal and environmental problems, reflecting new scientific knowledge, application of new technologies and change in political consciousness (Sands 2003). Organizations involved in enforcement of policies and laws are established at different levels of the governments. The policies, strategies and laws cannot be put into practice unless organizations are strengthened. Swaminathan (1986) argues that policies and strategies can fail to achieve the intended outcomes unless organizations are well organized. He states (1986): "good policies and strategies will remain only on a paper and in conference halls unless the key agents are organized in order to convert policies or laws into accomplishment." Another scholar, Thrupp (1998) argues that unless the governments build the capacity of organizations and reforms organizational structures to support its policies, the goals and objectives of policies and strategies will not be achievable. He recommends that along with policy and strategy promulgation, governments should build organizational capacities at local, regional and national levels; and decentralize the management and institutional structures.

In addition to reforming and establishing public organizations to support policies, civil society and community-based organizations (CBOs) can contribute greatly to implementing policies and strategies. FAO (2005) and Swaminathan

(1986) argues that farmers' organizations, women's associations and other Community Based Organizations (CBOs) can promote the necessary partnership between farmers and public institutions as well as private organizations to make the implementation of the policies and strategies effective. Uphoff (1992) also states that CBOs have a paramount roles to promote effective public participation in natural resource conservation by regulating norms and making the members to reach consensus, by creating common expectations and a basis for cooperation that goes beyond individual interests, by mobilizing resources and regulating their use with a long-term base for productive activity and sustainable development, and by generating and using available resources efficiently and sustainably. For effective participation, rural people should be organized either in externally introduced organizations or the organizations established by farmers themselves (Oakley 1995). Oakley finds that the organizations that are established by community themselves are more successful in promoting participation of the rural people.

27.3 Institutional Arrangement of Environmental Resource Management in Ethiopia

Natural resources are the foundation of economic development in Ethiopia as they are the sources of goods and services needed for poverty reduction and economic growth. Managing environmental resources and economic development are, therefore, mutually reinforcing imperatives and have to be implemented together in Ethiopia's development initiatives (FDRE 1997a, b). Nevertheless, the worth of natural resources management for development of a country had been neglected for many decades (Shibru and Kifle 1998). The Imperial Regime (1930–1974) perceived that there were plenty of natural resources available so that there was no need to protection. This view also assumed that the natural resources of the country must be exploited as much as possible for the purpose of economic growth and creation of wealth (Shibru and Kifle 1998). Inadequate attention paid to conservation of natural resources together with the long-term cumulative impact of the actions of land usersduring this period, however, led to the degradation and depletion of these resources and resulted in reduction of agricultural productivity. After having been perceived as the bread basket of Africa in the 1950s and 1960s, Ethiopia did not produce a sufficient amount of food for its fast-growing population in the 1970s (Markos 1997). The concern for the proper management of natural resources became clearly visible in this time. Management of natural resources began to be incorporated as a distinct and separate part of government effort around 1970, following the recommendations of the Third Five Year Plan (Shibru and Kifle 1998). Capacity development of Ministry of Agriculture and its branch offices regional, provincial and district level was given priority to support participation of the local communities in conservation. Expansion of an effective extension service was also another strategy designed to mobilize the land users in soil and water conservation (SWC) practices.

The Derge Regime (1974–1991) also believed natural resource management was valuable for economic development (Kassahun et al. 2004). During this time, grassroots peasant organizations such as the Kebele Peasant Association, Peasant producers' cooperatives and service cooperatives were established, to implement natural resource conservation programs (Fasil 1993 and Kassahun et al. 2004). The grassroots peasant organizations were intended to mobilize the community in resource conservation programs (Kassahun et al. 2004). Unfortunately, they became the mechanisms of mobilisation to fulfil the interests of central government rather than serving the interests of their members (Fasil 1993; Mehret 2002 and Kassahun et al. 2004).

The role of natural resource management for achievement of economic growth and poverty alleviation in the country has also been acknowledged by the government led by Ethiopian People Revolutionary Democratic Front (EPRDF), which came to power in 1991. Building on the agrarian nature of the country's economy, the government has adopted the Agricultural Development Led Industrialization Strategy (ADLI) (FDRE 2003a, b). ADLI views agriculture as the driving force of the economy, and argues for investment in agriculture as both a motor for economic growth and a means of addressing widespread poverty (FDRE 2003a, b). The Government has realized that the country's sustainable economic development goals based on agriculture-based strategy cannot be achieved without sustainable management of natural resources (Mulugeta 2004). Thus, ADLI emphasizes the significance of natural resource management and incorporates natural resource management as one of its major components.

The development and implementation of a wide range of policies, laws and programs on environmental resources over the last two decades also shows the commitment of the government of Ethiopia for natural resource rehabilitation and management. These macro-level policies, strategies and laws have given significant emphasis for harmonization of the management of environmental resources and socio-economic development. Table 27.1 presents the aims and strategies of major policies, proclamations and regulations related to the management of environmental and natural resources.

The government has established organizations to put resource management related policies, strategies, laws and proclamations into effect. The government of Federal Democratic Republic of Ethiopia is a multi-layered organizational structure with territorial divisions at national, region; zone, district, township and kebele levels. It is composed of numerous government units (ministries, commissions, bureaus and departments) responsible for diverse issues. These functional units exist at the national level and typically replicate their functions in a vertical chain through successively lower territorial levels of government. The organizations of environmental resource protection/management reflect the basic features of the government. The government organizations solely responsible for environmental protection existed at the central level as the chief agency addressing the federal environmental issues. The organizations/ministries that currently exist at the federal

Table 27.1 Policies, proclamations and regulations related to the management of environmental resources of Ethiopia

Policies/proclamation/regulation	Description
1. Federal	
1.1. Federal policies	
• Environmental Policy of Ethiopia (EPE) (1997a)	– The overall goal to promoting sustainable social and economic development through the sound management and use of natural, human-made and cultural resources and the environment as a whole – The policy touches upon a wide range of resource sectors and cross sectional issues
• Ethiopian Water Resources Management Policy (2000)	– The overall goal is to enhance and promote efforts towards an efficient, equitable, and optimum utilization of the available water resources and contribute to the country socioeconomic development on the sustainable basis
• Agricultural and Rural Development policy (2003a)	– It recognizes that the objective of sustainable development of agricultural sector can only be achieved by addressing natural resources depletion (deforestation, soil erosion) through conservation and rehabilitation of degraded lands
• Biodiversity Conservation policy (2004)	– Aims to ensure the conservation and sustainable utilization of genetic resources and essential ecosystems of the country; to integrate biodiversity conservation with sectoral and cross-sectoral strategies and programs; and encourage public participation in biodiversity conservation, development and utilization
• Science and Technology Policy (2007b)	– Aims to ensure rational efficient utilization of natural resources and conservation of environment by searching, choosing, procuring, and exchanging appropriate and environmental sound technologies
• Wild Life Development and Conservation Authority Establishment Proclamation (2008)	– Its overall goal are the preservation, development, management and sustainable utilization of Ethiopia's wildlife resources for social and economic development and for the integrity of the biosphere (FDRE 2007a, c)
• Food Security Strategy (2002a)	– Declares that the objectives of achieving food security at household level and achieving food self-sufficiency of the country can be only achieved by exerting efforts on the investment of natural resource

(continued)

Table 27.1 (continued)

Policies/proclamation/regulation	Description
	conservation and rehabilitation (FDRE 2002a–c)
• Energy Policy	– Aims to achieve the sustainable use of natural resources as well as reduce GHG emissions by promoting renewable energy technologies
• Forest Development, Conservation and Utilization Proclamation (2007d)	– It gives due attention to conservation of natural forests, reafforestation and prevention of deforestation for better management and development of forest resources of the country (FDRE 2007a, c)
• Climate Resilient Green economy strategy (2010b)	– It aims to foster economic development and improve resilience to climate change by implementing climate change adaptation and mitigation strategies of all sectors
1.2 Federal proclamations	
• The Constitution of Federal Republic of Ethiopia (1995)	– Proclaims that all citizens shall have a right to live in a clean and healthy environment (article 44) and the Government and citizens shall have a duty to protect the environment (article 92)
• Environmental Impact Assessment Proclamation (FDRE 2002c)	– Proclaims environmental impact assessment at all tiers of government to ensure sound and sustainable development practices that guard against depletion of soils, loss of critical biodiversity, increased greenhouse gas emissions, or significant diversion of water from other human or environmental uses
• Environmental Pollution Control (FDRE 2002b)	– Addresses the management of hazardous waste, the establishment of environmental standards for various environmental media (air, water, and soil), and the monitoring of pollution
• Water Resource management Proclamation (FDRE 2000)	– Aims to manage water resources and prevent it from harmful effects for sustainable socio-economic development of the country
• Rural Land Administration Proclamation (FDRE, 1997b)	– Provides for environmental protection obligations on the land users; such as to plant trees around the land and properly protect them, to take care of water sources, to exercise care for wildlife and birds found around the user's holding and the obligation to plough the land far from river or gully

(continued)

Table 27.1 (continued)

Policies/proclamation/regulation	Description
• Fisheries Development and Utilization Proclamation (2003b)	– Aims to conserve fish biodiversity and its environment as well as prevent and control over-exploitation of fish resources. It prohibits fishing using illegal fishing materials (FDRE 2003a, b)
• Conservation and utilization of wild life	– Aims to conserve, manage, develop and properly utilize the wildlife resources of Ethiopia; and create conditions necessary for discharging government obligations assumed under treaties regarding the conservation, development, and utilization of wildlife
• Solid waste management proclamation	– Proclaims the enhancement of the capacities at all levels to prevent the possible adverse impacts while creating economically and socially beneficial assets out of solid waste (FDRE 2007a, c)
1.3 Regional proclamations	
• Administration and Use of Watershed Rehabilitated with Community Participation Proclamation (2013)	– Proclaims the promotion of integrated watershed management programs; and establish a suitable condition of the rehabilitated natural resource use, administration and protection on private and communal lands (FDRE Amhara Regional State 2013)
• Lake Tana and Its Environs Biosphere Reserve Delineation and Administration Proclamation (2014)	– Recognizes the sub-basin as a development zone for realization of the building of climate change resilient green economy strategy and proclaims the conservation of its environmental resources by delineating management zones (FDRE Amhara Regional State 2014)

level and are directly involved in the development and management of natural resources include: Ministry of Water Resources and Energy development; Ministry of Agriculture; Ministry of Forest and Environmental Protection, and other affiliated authorities and agencies. The most important organizations at the federal level in relation to natural resource management and descriptions of their roles are the following (Table 27.2):

The Regional organizations have been established with similar designations and responsibilities as the federal ministries described above. These organizations include Bureaus of Environmental Protection, Land Use and Administration (BoEPLAU), Water Resource Development, Agriculture; Culture, Tourism and Parks Development (BoCTPD); Agricultural Research Institute and Forest Enterprise of Amhara Region. BoEPLAU is an extension of the Federal Ministry of Forest and Environmental Protection. Hence, the major responsibilities for the

Table 27.2 Organizations of environmental resource management at federal level

Ministry of forest and environmental protection Major responsibilities for the protection of environment are vested on Ministry of Forest and Environmental Protection. The ministry is responsible for administration of environmental protection at the federal level and deals exclusively with environmental matters. Its tasks include the administration of main environmental issues such as environmental pollution control and environmental impact assessment. It also develops national environmental policy and regulation frameworks in consultation with different stakeholders. Furthermore, it is responsible for preparing and implementing proactive environmental management systems; enforcing environmental regulations; creating awareness and fostering the participation of the people in decision making by availing information and offering trainings; and identifying and availing environmentally sound technologies and best practices (FDRE 2010a)
Ministry of agriculture It is another organ of federal government with different responsibilities relevant to the protection and management of environmental and natural resources. It is responsible to prepares policies and strategies for land use, soil, landscape as well as conservation and utilization of forests, wildlife and other natural resources. The responsibility of conservation of fish biodiversity and its environment as well as prevention and controlling over exploitation of the fisheries resources is also vested on Ministry of Agriculture
Ministry of water resources, irrigation and energy It is the chief agency responsible for planning, management, utilization and protection of water resources and energy in the jurisdiction of federal government. Specifically, the ministry has the following powers and duties. These include inventory of the country's surface water and groundwater resources; basin level water management and benefit sharing; develop water infrastructure; issue permits and regulate the construction and operation of water works; and administer dams and hydraulic structures (FDRE 2010a)
Institute of biodiversity conservation and research (IBCR) Is responsible for management, conservation and research of flora, fauna and microbial organisms' genetic resources of Ethiopia. It particularly biodiversity policy and legislations and enforce their implementation; conserve the country's biological resources using ex-situ and in situ conservation methods; device a strategy to harmonize biodiversity conservation program with socioeconomic policies; encourage and support public participation in conservation, development and use of biological resources; and create awareness among the public on biodiversity conservation, development and utilization by closely working with mass media and educational establishments (FDRE 2004)

(continued)

Table 27.2 (continued)

The Ethiopian wildlife development and conservation authority It is an umbrella organization of Ministry of agriculture and responsible to prepare policies and laws relating to the development, conservation and utilization of wildlife resources; and follow up the implementation of the policies and laws; develop and administer wildlife conservation areas established and control illegal activities; provide support to regions with respect to the development and conservation of wildlife, and; prevent and control incidents of wildlife diseases within or outside of conservation areas (FDRE 2008)

protection of environment are vested on this bureau. Its main responsibilities include development of policies, strategies and laws related to the protection and conservation of natural resources in the region; and coordination of environmental protection and management in the region. Bureau of Water Resource Development is the regional water sector office mandated to develop region-wide policies, strategies, plans directives, standards concerning the water resource management in line with the federal policies and laws; and develop the water resource of the region. Implementation of sustainable land management and irrigation schemes; and conservation of fish biodiversity and its environment as well as prevention of and control over-exploitation of fisheries are under the jurisdiction of Bureau of Agriculture. BoCTPD is responsible for development and administration of wildlife areas established in the region and controlling illegal activities. The departments of the Regional Bureaus described above have established at zonal and *woreda* level. Zonal and *woreda* department of these sectors are engaged in the overall activities of environmental and natural resource management. Likewise, courts and justice offices at different tiers of regional government and office of police at district level are involved in natural resource protection especially land administration.

A variety of organizations directly or indirectly involved in resource management also existed at the grassroots level. These organizations are both formal and informal. The formal ones are organizations that have been established by the government rules and regulations and serve specific economic, social and political purposes. The main formal institution that established by the government is Kebele Administration. The kebele administration plays a decisive role in terms of local governance. This role includes the overall administration and management of community resources issues within the kebele jurisdiction. It has the ultimate decision making power in all political, social, economic and security issues within the kebele. Its responsibilities go beyond administrative purposes and include mobilizing and promoting participation, solving disputes and security issues, managing local institutions, planning and implementation of development activities. Kebele administration comprises kebele council, executive organ committee, social court and socio-economic and security bodies. In addition, *sub* committees are established that aimed to support the kebele administration in relation to land administration and natural resource management. Some of them are land use and administration committee, communal land administration committee, watershed committee and forest protection committee. The land use and administration committee is organized in each kebele administration of the sub basin and has the mandate to control land use pattern, resolve conflicts on land, register and certify land holding rights, and participate in relocation and redistribution of land as necessary. The watershed committee is established at both kebele and watershed level to ensure active involvement of the local people in planning, implementation and monitoring and evaluation of integrated watershed management activities. The forest conservation committee is established by the kebele administration representing kebele leaders, working managers, DAs, priests, youngsters, elderly people, women's associations, school principal and health centre toprotect and rehabilitate the remnants of forests, woodlands and bushes.

Civil society groups are also established in many Kebeles of the sub-basin such as women's associations, youth association, or water users association. These associations are organized by the office of the *Woreda* Cooperative Promotion based on the Regional State Cooperative Societies Establishment Proclamation. Although the main purpose of these associations is ensuring the right of their members to participate in socio-economic and political aspects of kebele life, their role in environmental protection could be significant, particularly in awareness creation and mobilization of resources.

Informal community based organizations (CBOs) and faith based organizations (FBOs) also exist in many kebele administrations of the sub-basin. They are governed by behavioural norms in society, family, or community and include sanctions, taboos, traditions, and codes of conduct (Leidreiter 2010). The main purpose of these organizations are: (1) a welfare function which is useful in times of social crises and economic hardship; (2) a social function which plays an instrumental role in the organization of social ceremonies and religious rites and (3) a developmental function which fosters a cooperative spirit among people (Leidreiter 2010). The main informal community institutions are religious institutions (church and mosques, *senbete*, and *mahber*, in Amharic), and social institutions (Webera and *Iddir*, *ekub*, group of elders). These institutions could play an indirect but essential role in natural resource management through resource mobilization, conflict resolution, gov-ernance, and risk reduction.

Natural resource conservation has also been a cross-cutting issue for many NGOs, bilateral and multilateral organizations. Some bilateral and multilateral organizations have provided considerable support to capacity building of government agencies. Some others like the Tana-Beles Integrated Water Resource Project have provided financial and technical support and material contributions to regional government as well as zonal and *woreda* administrations for rehabilitation and management of natural resource of the sub-basin. Some NGOs have been directly engaged in the rehabilitation and natural resource conservation activities. For example, SOS Sahel and FARM Africa have been working on Participatory Forest Management (PFM) in state forest of Tara Gedam and Alem Saga (Moreaux 2013). The Organization for Rehabilitation and Development in Amhara (ORDA) has also involved in forest rehabilitation, protection and sustainable use within the sub-basin. Moreover, the organization has implemented several projects around Lake Tana for soil and water conservation based on watershed management. Similarly, German Technical Cooperation (GIZ) and CARE Ethiopia have implemented sustainable land management practices such as protection of natural forests and woodlands, construction of physical soil conservation measures, afforestation, gully rehabilitation, establishing grass strips, planting on bunds, and area closure. Nature and Biodiversity Union (NABU) has been working in sustainable wetland management around Lake Tana. These NGOs have adopted a community-based natural resource management approach to promote the adoption of a variety of conservation practices. They recognize that community—based natural resource management approach can be put into practice through community based organizations. Thus, some NGOs like ORDA organized Irrigation Water Users Associations and SOS Sahel established Youth Bee Keepers Associations in protected forest areas of the sub-basin and collaborated with them as development partners.

27.4 Conservation Efforts and the Roles of Governmental and Nongovernmental Organizations

Investment in soil conservation and land rehabilitation measures has been undertaken since 1974 in some parts of sub-basin. The main conservation measures that were constructed in the 1970s and 80s were terraces and soil bunds, cut-off drains, check dams and afforestation. The coverage of implemented conservation structures was restricted in areas where land degradation was very serious. The large parts of the sub-basin which were assumed to be surplus producers of food crops were neglected (Gete et al. 2006 and Demeke 2003).

Although soil conservation and land rehabilitation measures have practiced in the sub-basin since late 1970s, the area covered by soil and water conservation practices in the recent four years has outnumbered the area affected by conservation efforts made before 2010. As part of the national Growth and Transformation Plan (GTP), massive conservation works were carried out in Lake Tana sub-basin from 2010/11 to 2014. Based on annual reports of Amhara Bureau of Agriculture (BoA 2014), between 2010/11 and 2013/2014, physical soil conservation structures (terraces and bunds) were constructed on hundreds thousands hectares of lands, close to 5000 ha of land was closed for natural regeneration, hundreds thousands hectares of trees were planted, 6193.5 km of check dams were constructed on gullied lands and 14,823 km of diversion ditches were constructed. Tens of millions of person days have been involved in these conservation works through free labour contribution in the past four years.

A number of government organizations at different tiers whose mandates are directly related resource management have played significant roles in the implementation of such large scale conservation practices. The most relevant institutions that have been involved in massive soil and water conservation programs are MoA,

Bureau of Agriculture and Bureau of Environmental Protection. The federal and regional institutions focus on building the capacity of local level administrations, providing financial, material and technical supports and undertaking monitoring and evaluation. However, the actual implementation of soil and water conservation projects has been shouldered by the *kebele* and *woreda* governments. *Woreda* agricultural and development experts provide consultation to the farmers about how to sustainably use natural resources, and conserve and rehabilitate the degraded lands. Work groups (developmental teams) are mobilized in the farming community at the kebele level to implement physical soil and water conservation works and afforestation.

Government organizations are also involved in efforts in the conservation of common pool resources. For example, BoEPLUA and its *woreda* branch have been involved directly in the campaign to eliminate the invasive water hyacinth in wetlands by providing materials support and offering awareness training for local government bodies and the community. *woreda* office of environmental protection, land use and administration (WOEPLUA) has made efforts to stop illegal land encroachment by demarcating the boundaries of farm lands and communal lands. The contribution of BoA and its *woreda* office in the protection of the remnant forests and woodlands is significant. Various organizations at the kebele level have been involved in efforts to safeguard communal resources. Preventing illegal encroachment by demarcating the boundaries of farm land and communal lands, mitigating encroachment by charging encroachers a fine, destroying crops which grows in illegally cultivated fields by allowing animals to graze, protecting the remnants of forests, woodlands and papyrus by developing community bylaws; coordinating the weeding of water hyacinth infestation; and creating awareness about the benefits of and threats to wetlands are some of the measures that have been taken by these institutions.

The involvement of NGOs and multilateral organizations in promotion of natural resource management practices in the sub basin has brought significant achievements. The most important achievement brought by their interventions is attitudinal change of farming communities towards conservation of natural resources and the enhancement of their participation in the overall process of conservation activities. The environmental education, technical training and other advocacy programs have contributed for the attitudinal change and enhancement of the participation of farming community. Conservation measures undertaken by them have also contributed to reverse land degradation. A range of appropriate technologies developed and introduced by ORDA in different *woredas* in the sub basin and GIZ in Farta *Woreda* helps to control floods, conserve the remnants of natural forests, stabilize gullies and restore the disrupted hydrological balances in catchment. As a result, gullies were transformed into productive land, where the concentration of fertile soil and sufficient moisture allow the production of high-value crops that generate income. The implementation of biological conservation measures ranging from soil fertility management such as composting and manuring, to vegetative measures such as planting trees, shrubs and herbaceous grasses and legumes enhance soil nutrients. In the fragile and degraded parts of the sub basin, the efforts of NGOs' tree planting and physical conservation measures have improved the vegetative cover of the hilly slopes that used to be barren in the past.

The contributions of informal institutions for conservation of natural resources are also significant. Ethiopian Orthodox church's role in conservation of the remnants indigenous trees of is one example. Alemayehu (2007) documented the role of Ethiopian Orthodox Church (EOTC)'s contribution in the conservation of forests in sub basin. His findings show that the church has saved indigenous old age trees by its commitment based on strong theological thoughts and a biblical basis. The Church's contribution for maintainingthe remaining forests of the Lake Tana Regionis also reported by Moreaux (2013). The findings of her study indicate that the church has preserved a high floral and faunal diversity with many indigenous and rare species in the monasteries of Lake Tana and the churches of Lake Tana region by guarding forests as remnants of the holy property of the church and conducting reforestation activities. Though iddirs are organized to serve funeral purpose and support to the bereaved families, it plays an important role for the protection of forest and regulated grazing in different kebele administrations adjacent to wetlands (BNWI 2014).

27.5 Weaknesses of Institutions in the Natural Resources Management of the Sub Basin

Despite the significant conservation activities already undertaken to reverse land degradation and maintain natural resources, land degradation in the sub-basin is still continuing. Land use/cover analysis done in one of the sub-basin watershed (Gilgel Abay) by Sewnet (2013) that farm and settlement lands of the watershed have highly expanded by 57.5% for the last 35 years (between 1973 and 2008). The expansion of farm and settlement lands caused for the shrinking of grassland, forest land, and wetlands. The natural forest experienced the largest decline, falling by about by 72.3% of its initial extent in 1973. Likewise, the grassland and wetlands have declined by 55.1 and 47.2% respectively. The study made in another watershed of lake Tana Sub-Basin (Angreb) by Haile and Assefa (2012) shows that the area coverage of the pasture land and forest have declined by 30.9 and 29.3% within 26 years (1985–2011) due to the expansion of built up area and farmland by 866.1 and 20.7%, respectively. The rapid expansion of cultivated land at the expense of the decrease in grasslands, forests and wetlands has implications for the increasing rate of soil erosion. Based on the study findings of Soil Conservation Research Project (SCRP) reported by Hurni (1988), soil loss rates vary considerably according to different land-use/cover types. Soil loss from grassland and woodland or bushlands is 3–10 times less than from cultivated land and; from forest land is lower by 40 or more times than from cultivated land. Thus, the total soil loss of the sub basin in 2010s would be much greater than the total loss of 1970s and 80s, due to land use change. Despite the massive conservation activities carried out in the past three decades, land degradation still outweighs rehabilitation of the degraded lands.

The land use/cover change studies also show that land use change has altered hydrological processes in the sub-basin. A study on the hydrologic impact of land use change in the upper sub basin of Lake Tana i.e. Gilgel Abay by Gumindoga (2010) showed that the expansion of agricultural land at the expense of natural vegetation caused peak flow to increase by 51% between 1973 and 1986 and by 44% between 1986 and 2001. A similar study done Angereb Watershed by Haile and Assefa (2012) linked the decrease of natural vegetation (forest, woodland and shrublands) due to expansion of agricultural land and built-up area with increasing surface run off and sediment carried to the lake, rivers and wetlands. The data collected though focus group discussions and key informant interview by BNWI confirmed the degradation of wetlands' ecosystem around Lake Tana. As to focus groups, the grazing pastures in the wetlands have highly shrunk over the past 20 years. The loss of other wetland resources (papyrus, fish, thatch grass, plants, etc.) was also extensive. The qualitative data collected by BNWI reveals the deteriorating rate of wetland resource. The key informant explains how the wetland resources around Lake Tana have been declining in the past three decades as:

> Hunting activities were carried out before early 1980s when there were abundant trees, grasses and papyrus in the wetland and river bank. Almost all of households had collected thatch grasses 15-20 years ago. Papyrus had been harvested 10 years ago as well by large number of people for making of ropes, traditional boats and construction. Most of the farmers had harvested fish for household consumption when the availability of papyrus, grasses and different plant species was abundant. During those days, everybody could collect large fish stock within a short time without netting. Large density of fish was found far from the lake since June to November. The availability of papyrus, grasses and plants was the fertile ground for breeding of fish. These important wetland resources, however, have been declined and lost in the past thirty years. The rate of the loss of wetland resources has been accelerated by uncontrolled encroachment of cultivation since the transition period between the Derge and EPRDF led government.

There are different challenges that contribute to the failures of institutions in achieving the desired objectives of natural resource conservation. The details are discussed below.

27.5.1 Top-Down Planning Approach

Though it is now widely agreed that genuine participation of the community in the overall process of natural resource management contributes to the dissemination of proper land management practices and sustainability of conservation measures, conservation programs in 1970s and 80s in Ethiopia tended to follow centralized top-down planning instead (Gete et al. 2006; Gebremedhin 2004). Little attention was given to community-level participation in conservation programmes. Community interests regarding land and water management were ignored.

Local residents were neither contacted nor included in design of the conservation programmes. The involvement of farmers in those conservation activities was limited to their required labour contribution.

The weaknesses of top-down planning approaches that are largely based on a numerical quota system for promoting the adoption of conservation technologies are now recognized by the current government, donors and Non-Governmental Organization (Gete et al. 2006). Since the 1990s, it is realized that the top down planning approach that focused on technical and physical works alone would not lead to the desired environmental objectives (Gete et al. 2006; Shimelis 2012). The government in its several macro-level policies and strategies planned to make the rural community genuine participant in the natural resource management and other rural and agricultural development programs. Several NGOs and bilateral organizations have also adopted participatory conservation planning approach in their respective areas of intervention and in close collaboration with government partners (Dessalegn et al. 2008; Samuel 2006).

Despite the recognition of the need to increase farmer participation in the planning as well as implementation of conservation activities, the planning approach is still largely top-down as it was in the 1970s and 80s. Conservation efforts still give priority to quick solutions rather than sustainability, quantity rather than quality, area coverage rather than impacts (Gete et al. 2006). The government has not learned from the failures of past soil and water conservation programs that lacked genuine people's participation. Although various policies, strategies and laws promise to involve the community in the whole process of natural resource management, in practice, they are generally not included on the ground. Conservation planning is done with very minimal consideration of farmers' interest and resources, or the capacity of agricultural offices in terms of manpower, technical knowledge, and available resources (Gete et al. 2006; Springsguth 2013). The total area of land which is covered by conservation structures and the types of conservation practices carried out in each district is decided by the Regional Governments. District officials and extension personnel act as controllers and enforcers of conservation plans prepared by regional governments rather than facilitators of conservation technology transfer. The target beneficiaries are largely passive recipients of externally introduced conservation technologies. They have limited opportunity to be actively involved in selection and planning of the conservation measures (Springsguth 2013). Farmers have participated in the construction of soil and water conservation projects more to meet the demands of the government's conservation plans, rather than for the sake of conserving their own soils and lands. Community-based institutions, which are believed to be playing a paramount role in promoting effective people's participation, have not provided the means where by people can effectively participate in the whole process of conservation works because the local government has never considered them as partners of government sponsored conservation programs.

27.5.2 Lack of Effectiveness

Ineffectiveness of the institutions responsible for implementation of conservation programs is reported as factor for the failure of natural resource conservation. A socioeconomic assessment conducted by BNWI (2014) reported that the efforts that have been made by *woreda* and *kebele* administrations are insufficient to safeguard communal resources including natural forests and wetlands from continued degradation. Instead, illegal encroachment of communal lands and unlawful use of wetland resource have increased. Inefficient organizational structures, understaffing, under-equipping and lack of organizational units are the sources of ineffectiveness at *woreda* level (Haileslassie et al. 2008). Leadership weakness is also reported as an impediment to institutional effectiveness. Lack of commitment of the members of *kebele* administrative councils and a desire to avoid conflict is one of the main leadership weaknesses. Some of the kebele leaders and security bodies are not daring enough and would like to be seen as 'good people' (Springsguth 2013). Thus, most illegal land occupants are not challenged. They are left free leading to more conversion of communal lands. There are also local government leaders who set bad examples for others by cultivating the communal lands (Springsguth 2013). Furthermore, the administrative council has intentionally ignored encroachment of wetland recession believing that encroachment could solve the shortage of farmlands. Similarly, land use and administration committees, which have established at kebele and sub-kebele level to administer land, are weak in discharging their responsibilities and exercising their power. Committee members have tried to hide cases of illegally occupied communal lands, and illegally occupied lands are found to be registered by the committee. The involvement of some committee members and their relatives in illegal land cultivation, and the heavy pressure of those who illegally encroaching make the committee members ineffective in administering the communal lands based on the land use and administration proclamation of the regional state. Other common leadership weaknesses observed include corruption and favouritism where influential farmers and relatives are rarely penalized; inadequate administrative and management capacities, lack of work continuity and inadequate penalties (Springsguth 2013). Fraudulent or unlawful conducts are not (adequately) prosecuted by kebele administrations or woreda authorities and courts. Police support to settle conflicts is inadequate. Legal decisions of court cases are pending. Wetland management is thus nearly impossible considering all the unresolved cases of conflicts. Cases of corruption (bribery, intimidation, drawing on beneficial relations) impede conflict resolution processes and complicate wetland management. Finally, *kebele* administrations are often too large to be governed efficiently and the kebele administration office is too far away.

Frequent restructuring of government institutions involved in natural resources management is also blamed for the failure of conservation programs in Ethiopia. According to Haileslassie et al. (2008), the organizations involved in land and water resource management were marked by frequent restructuring and re-organization over the last few years and the process seems to be going on. Though adjusting institutional responsibilities and redesigning organizational structures may be called for in the light of the changes and development needs of the country, the frequent restructuring process has certainly produced uncertainties, made capacity building difficult and affects the political will to push for change. The effect of restructuring of government organizations on natural resource management has investigated by Gete et al. (2006). Based on the findings of his study, the restructuring of government organizations at different tiers of government structures has resulted for lack of continuity of natural resource management activities by making natural resource management experts unstable; undermining a sense of ownership of program staff; resulting the high staff turnover and losing institutional memory (EEA/EEPRI 2006).

27.5.3 Technical Failures

Since Ethiopia has highly diverse physical and socio-economic environments as well as the spatial variations in the severity of resource degradation, the types of natural resource conservation efforts should be site specific (Bewket 2003). However, governments have tried to transfer conservation technologies with a blanket recommendation approach regardless of physical and socio-economic environments (Jonathan 2007). The SWC structures that were heavily promoted across the country were terracing and bunding, drainage channels, water ways and check dams occasionally accompanied by afforestation and reforestation. These technologies, which were disseminated with blanket recommendations, are not compatible with all local conditions. Moreover, the technical requirements of these measures are often overlooked (Jonathan 2007; Gete et al. 2006). They are poorly designed and constructed. The issue of integration among different technologies to make soil and water conservation measures more effective is not considered as well. Physical structures have often been applied without the integration of soil fertility and moisture management practices (EEA/EEPRI 2006; Shimelis 2012). These poorly designed, constructed and integrated conservation structures, therefore, were immediately collapsed and resulted for the formation of gullies.

27.5.4 Lack of Appropriate Policy

Underestimating the role of socio-cultural factors in the adoption of resource conservation practice by policy makers is another policy constraint in relation to conservation programs of the country. As argued by Alemneh (2003), the failures of conservation programs are partly emerging from the fact that policy makers ignore or fail to consider socio-cultural factors as the key determinants of the success or the failure of conservation programs. The policy makers viewed the land degradation and conservation problems as function of only physical and technical factors as explained by Alemneh (2003).

Land tenure insecurity is also another policy problem claimed by many scholars for its large contribution for the failure of conservation efforts. Several researchers have documented that tenure insecurity has been serious challenge for conservation efforts in Ethiopia and potentially hampering farmers' investments on natural resource management (Jonathan 2007; Gebremedhin 2004). The frequent redistribution of land by peasant associations all over the country especially during the Derge regime created a strong feeling of tenure insecurity among land users. As argued by Shibru and Kifle (1998), the frequent redistribution of land negatively affects the implementation and sustainability of land management in the country. Farmers have developed negative perceptions on the redistribution of land especially in areas where land redistribution were frequent. Hence, they have not made much effort toward improving their holdings due to lack of sense of ownership. An effort has been made recently by the current government to secure the land tenure of small farm holders and encourage sustainable land use practices.

The contradiction of the proclamation of environmental impact assessment and the Ethiopian investment proclamation are also the other area of problem in relation to the formulation of laws and policies in the country. The Environmental Impact Assessment Proclamation declares that an EIA both at the level of project implementation of activities, as well as for strategic assessments is very crucial (FDRE 2002a, b, c). Thus, the proclamation obliges the development projects not to start implementation without first getting the approval of the relevant government organization for its Environmental Impact Assessment. The proclamation also states that licensing institutions should ensure that a certain project must obtain an approval of the EIA document and a permit for the implementation from the authorized government organization before issuing investment or operation licenses (FDRE 2002a, b, c). The government organs are also requested by this proclamation to ensure that their policies, strategies and laws pass through an EIA process before their submission for approval (FDRE 2002a, b, c). However, the Investment Proclamation does not make EIA a requirement for obtaining an investment permit, and in practice the Ethiopian Investment Authority grants investment permits without EIA as a requirement (FDRE 2002a, b, c). This provision of investment proclamation is directly contradicted with the EIA proclamation that obliges any government body to ensure the approval of the authorized government body for implementation permits prior to issuance of operation licenses.

27.5.5 Lack of Enforcement of Policies and/or Laws

In order to effectively use, manage and govern the fishery resources of the country, FDRE enacted the Fishery Development and Utilizations Proclamation No. 315/2003. Following the federal government, the council of Amhara National Regional State enacted its own fisheries development, prevention and utilization proclamation No. 92/2003. However, in spite of initiative of production of a comprehensive, the proclamation often been left without implementation. Ketema et al. (2013) found that there is no formal control or prevention on the number, type of gill nets and fishing technologies in Lake Tana 10 years after the fisheries development and utilization proclamation was enacted. Neither the Bureau of Agriculture, which is responsible for controlling and monitoring fishery resource management, nor other concerned government bodies were discharging their responsibilities to resolve the fishery governance problems (Ketema et al. 2013). As a result, too much fishing pressure and illegal fishing activities were believed to be causing overexploitation of endemic fish stocks, threatening sustainable fish production, marketing and the livelihood of local fishermen (Ketema et al. 2013).

27.5.5.1 Overlapping and Conflicting Roles and Responsibilities

Organizational coordination and cooperation is vital for the achievement of the development and environmental resource management objectives. The coordination among different governmental organizations however is very weak or nonexistent. Rather, there is a high frequency of overlapping and conflicting roles and responsibilities between organizations (Ketema et al. 2013). In spite of various stakeholders' involvement in environmental resource related activities, there is no structural and coordinated linkage among them (Haileslassie et al. 2008; Ketema et al. 2013). Poor coordination is not only a problem among different organizations working on resource management but is also observed within organizations. The organization of ministries, bureaus and departments in different tiers of the government generally follow a 'disciplinary' orientation while problems in the sector call for an interdisciplinary and integrated approach (Haileslassie et al. 2008). Therefore, they cannot easily integrate their administrative efforts (Ketema et al. 2013). This could lead to inability to achieve the goals of socioeconomic development and natural resource management. Protective institutional responses, lack of effective coordinating mechanisms, conflicts of interests and jurisdictional confusion are accountable for the existence of weak coordination. Ineffective coordination is also exists between the government and other development partners. There is a lack of an effective approach to government and NGOs involvement in policy implementation (NBI 2006). Linked to this, the government does not consider informal grassroots institutions as potential partners for policy implementation (NBI 2006).

27.5.6 Conflicts of Interests

The dilemma of promoting irrigation agriculture on the one hand and protecting natural resources on the other hand is another factor that has constrained the sustainable management of natural resources. In response to increasing human population and concerns of food insecurity, the federal and regional governments have commenced the implementation of irrigation projects in Tana Sub Basin (Mulugeta 2013). The Koga Irrigation Project has been constructed and numerous other large-scale irrigation projects have been initiated around Lake Tana by the federal government (e.g., Rib, Gumara, Megech, Gilgel Abay and Jema) (Mulugeta 2013). The regional Bureau of Agriculture and its *woreda* office have also made efforts to promote small scale irrigation in the sub basin and consider wetlands as important places for promotion of irrigation. Evidence shows that the development of small scale irrigation around the lake is showing a steady increase in both yield and area coverage (Mulugeta 2013). Undoubtedly, the expansions of irrigation schemes in the sub basin have contributed for poverty alleviation and economic growth (Mekonnen and Sileshi 2007). They enhance agricultural production, open employment opportunities for the poor and support the national economy (Mekonnen and Sileshi 2007). Irrigation schemes, however, are often implemented at the expense of environmental considerations. According to SMEC (2008b), large scale irrigation projects which are under construction in the sub basin will have negative impact on hydrology of the lake and surrounding wetlands. The study conducted by Springsguth (2013) also provides the evidence for the conflicting opinions and interests about wetland use and management across sectors. According to this study, *Woreda* offices of Agriculture and BoA advocated wetland cultivation and recession farming. Similarly, local government administrations have an interest in distributing wetlands to landless, promoting drainage of wetlands and supporting youth using wetland resources as an alternative income source (e.g. collecting plants). By contrast, BoEPLAU pointed out the *woreda* government falsely associated the government's promotion of agricultural modernization and technological progress with the conversion of wetlands. As per BoEPLAU and its *woreda* office, recession agriculture is generally prohibited. The conflicting interests of wetland use between BoARD and BoEPLAU become most evident during the implementation of regulations at the *woreda* and *kebele* levels (Springsguth 2013). This indicates that the objectives of natural resource conservation have been compromised for the sake of addressing the socio-economic problems of the society.

27.6 Conclusion

The significance of managing and protecting natural resources of the country in general and Tana Sub-basin in particular for poverty alleviation and economic growth has been recognized by the Government of Ethiopia. It has articulated an overall comprehensive cross-sectoral environmental policy to guide the use and management of natural resources. Environmental management is also a cross-cutting issue for different national socio-economic policies and strategies as a key prerequisite for the success of socio-economic development. Multiple government organizations ranging from national to local governments have established and are involved in the accomplishment of the policies, strategies and laws related to natural resource management. Bilateral and multilateral organizations as well as NGOs have supported the endeavors of the government natural resource conservation programs in Lake Tana Sub Basin, although government agencies often do not consider religious organizations such as the EOTC and other informal community based organizations that have participated in the protection of forests as partners in natural resource management. The efforts exerted by numerous organizations from national to local governments and from formal to informal contributed for conservation of natural resources in Lake Tana Sub-Basin. Various sustainable land management technologies have been introduced and ten thousands of hectares of land in the sub basin are covered by them. The remnants of forests around the lake and in the upper sub basin have been preserved and protected. However, land degradation in various forms still continues to outweigh the rehabilitation of degraded lands and conservation of the natural resources in the sub basin. Considering the people as an instrument not as an actor of conservation programs, focusing on the quantity of conservation practices rather than their quality and effectiveness, organizational ineffectiveness, overlapping and conflicting of roles and responsibilities of government organizations and, conflicts of interest of the government are some of the reasons for the failures of institutions in achieving the objectives of environmental resources management. As this review indicates, making more effort to engage local-level participation and ownership of conservation programs and organizing multi-stakeholder platforms at various levels could go a long way to improving their effectiveness in environmental resource management.

References

Alemayehu W (2007) Ethiopian church forests: opportunities and challenges for restoration. PhD Thesis, Wageningon University, Wageningen, The Netherlands

Alemneh D (2003) Integrated natural resources management to enhance food security: the case for community-based approaches in Ethiopia: FAO working paper no. 1. Rome. Italy

Bandaragoda DJ (2000) A framework for institutional analysis for water resources management in a river basin context. Working Paper 5. International Water Management Institute, Colombo, Sri Lanka

Bewket W (2003) Towards integrated watershed management in the highland Ethiopia: the Chimoga watershed case study. PhD Thesis, Wageningen University and Research Center

Beyer S (2006) Environmental law and policy in people's Republic of China. Oxford University Press

BNWI (Blue Nile Water Institute, Bahir Dar University) (2014) Biophysical and socioeconomic characteristics of four wetlands around lake Tana. An Assessment done for Development of Integrated Wetland Management Plan by the Sponser of Nature and Biodiversity Union (NABU)

BoA (Bureau of Agriculture) (2014) Annual report on the implemented natural resource conservation practices in the region (2010/11-2013/14). Amhara Regional State, Bahir Dar. Unpublished

Demeke AB (2003) Factors influencing the adoption of soil and water conservation practices in Ethiopia. Discussion Paper No 37, Institute of Rural Development, University of Geottingen
Dessalegn R, Akalewold B, Yoseph E (2008) CSOs/NGOs in Ethiopia: partners in development and good governance. A Report Prepared for the Ad Hoc CSO/NGO Task Force, Addis Ababa, Ethiopia
EEA/EEPRI (Ethiopian Economic Association/Ethiopian Economic Policy Research Institute) (2006) Evaluation of the Ethiopian agricultural extension with particular emphasis on the participatory demonstration and training extension system (PADETES). EEA/EEPRI, Addis Ababa, Ethiopia
FAO (2005) Can trade work for the poor? The State of Food and Agriculture, Rome, Italy
Fasil G (1993) The subsistence crisis in Africa: the case of Ethiopia. Organization for Social Science Research in Southern and Eastern Africa. Nairobi, Kenya
FDRE (Federal Democratic Republic of Ethiopia) (1997a) Environmental policy of Ethiopia. Environmental Protection Authority In collaboration with the Ministry of Economic Development And Cooperation. Addis Ababa, Ethiopia. 28 p
FDRE (1997b) Rural land administration proclamation. NO. 89/1997. Addis Ababa
FDRE (2000) Ethiopian water resources management proclamation. No. 197/2000, Addis Ababa
FDRE (2002a) Food security strategy. Addis Ababa, Ethiopia
FDRE (2002b) Environmental pollution control proclamation. Proclamation No. 300/2002. Federal Negarit Gazeta, 9th year No. 12. Addis Ababa
FDRE (2002c) Environmental impact assessment proclamation. Proclamation No. 299/2002. Federal Negarit Gazeta, 9th year No. 11. Addis Ababa
FDRE (2003a) Rural development policies and strategies. Addis Ababa, Ethiopia
FDRE (2003b) Fisheries development and utilization proclamation. No. 315/2003, Federal Negarit Gazeta, 9th year No. 12. Addis Ababa
FDRE (2004) Institute of biodiversity conservation and research establishment. Proclamation No 381/2004, Federal Negarit Gazeta, Ethiopia, Addis Ababa
FDRE (2007a) Development conservation and utilization of wildlife proclamation. Proclamation No. 541/2007, Federal Negarit Gazeta, Addis Ababa
FDRE (2007b) National science, technology and innovation (STI) Policy of Ethiopia, Addis Ababa
FDRE (2007c) Solid waste management proclamation. Proclamation No. 513/2007, Federal Negarit Gazeta, Ethiopia, Addis Ababa
FDRE (2007d) Forest development, conservation and utilization proclamation. Proclamation No 542/2007, Federal Negarit Gazeta, Ethiopia, Addis Ababa
FDRE (2008) Ethiopian wild life development and conservation authority establishment. Proclamation No 575/2008, Federal Negarit Gazeta, Ethiopia, Addis Ababa
FDRE (2010a) The defnitions of powers and duties of excutive organs of the Federal Democratic Republic of Ethiopia. Proclamation No 691/2010, Negarit Gazeta, Ethiopia, Addis Ababa
FDRE (2010) Ethiopia's climate-resilient green economy: green economy strategy. Federal Democratic Republic of Ethiopia, Addis Ababa
FDRE Amhara Regional State (2013) Administration and use of watersheds rehabilitated and being rehabilitated with community participation proclamation. Proclamation No. 24/2013. Regional Zikre Hig, Year No. 13, Bahir Dar
FDRE Amhara Regional State (2014) Lake Tana and its environs biosphere reserve delination and adminstration detemination proclamation. Proclamation No. 125/2014, Regional State Council Zikre Hig, Bahir Dar
Gebremedhin B (2004) Economic incentives for soil conservation in the East African Countries. In: ISCO 2004—13th international soil conservation organisation conference—Brisbane, Conserving Soil and Water for Society: Sharing Solutions
Gete Z et al (2006) Stakeholder analysis for sustainable land management (SLM) in Ethiopia: an assessment of opportunities, strategic constraints, information needs, and knowledge gaps. Second draft report. Retrieved from http://www.efdinitiative.org/research/projects/

Gumindoga W (2010) Hydrologic impacts of land use change in the upper Gilgel Abay River Basin, Ethiopia. TOPMODEL APPLICATION. MA Thesis, Geo-Information Science and Earth Observation, The Netherlands
Haile EG, Assefa MM (2012) The impact of land use change on the hydrology of the angereb watershed, Ethiopia. Int J Water Sci 1:4. doi:10.5772/56266
Haileslassie A, Hagos F, Mapedza E, Sadoff C, Awulachew SB, Gebreselassie S, Peden D (2008) Institutional settings and livelihood strategies in the Blue Nile Basin: implications for upstream/downstream linkages. International Water Management Institute, Colombo, Sri Lanka, 81 p (IWMI Working Paper 132)
Hurni H (1988) Degradation and conservation of the resources in the Ethiopian highlands. Mt Res Dev 8(2/3):123–130
Jonathan M (2007) Ethiopia: country environmental profile. EC Delegation, Addis Abeba, available on www.un.org/esa/agenda21/natlinfo/wssd/ethiopia
Kassahun B et al (2004) Building capacity in Ethiopia to strengthen the participation of citizens'. Associations in Development: A Study of the Organizational Associations of Citizens. INTRAC for The World Bank
Kayambazinthu D et al (2003) Institutional arrangements governing natural resource management of the miombo woodland. In: Kowero G et al (eds) Policies and governance structures in Woodlands of South Africa. National Library of Indonesia Cataloging-in-Publication Data
Ketema DS, Chisholm N, Enright P (2013) Exploring the governance of Lake Tana fishery: interactive perspective on governance. Ethiop e-J Res Innovations Foresight 5(2):32–53
Leidreiter A (2010) Community participation in natural resource management in Lake Tana Watershed, Ethiopia. MA Thesis, Graduate Schools of Social Sciences. Universiteit van Amsterdam, The Netherlands
Markos E (1997) Demographic response to ecological degradation and food insecurity: drought prone areas in North Ethiopia. Doctoral Dissertation, University of Groningen, The Netherlands: PUDOC publication series
Mekonnen and Sileshi (2007) Impact of irrigation on livelihood and food security in the modern hare river irrigation scheme in Southern Ethiopia. Impact of Irrigation on Poverty and Environment in Ethiopia. Addis Ababa, Ethiopia available at eastafrica2.iwmi.org/Workshop/.../IIPE%20Proceeding_Final_N.pdf
Mehret A (2002) Decentralization in Ethiopia: two case studies on devolution of power and responsibilities to local authorities. In: Bahru Z, Pausewang S (eds) Ethiopia: the challenge of democracy from Below: Nordiska Afrikainstitutet, Uppsala and Forum for Social Studies, Addis Ababa
Moreaux R (2013) Remaining forests of the Lake Tana Region in Ethiopia: a participatory, social-ecological approach rehabilitation. Thesis, Landscape Ecology and Nature Conservation, University of Greifswald
Mukamuri et al (2003) Local organisations and natural resource management in the face of economic hardships: a case study from Zimbabwe. In: Kowero G et al (eds) Policies and governance structures in Woodlands of South Africa. National Library of Indonesia Cataloging-in-Publication Data
Mulugeta L (2004) Effects of land use changes on soil quality and native flora degradation and restoration in the highlands of Ethiopia: implications for sustainable land management. Doctoral Thesis, Swedish University of Agricultural Sciences, Uppsala
Mulugeta A (2013) Modeling and analysis of Lake Tana sub basin water resource systems, Ethiopia. PhD Dissertation, Universitate Rostock, Germany
NBI (Nile Basin Initiative) (2006) Baseline and needs assessment of national water polices of the Nile basin countries: a regional synthesis. Shared Vision Program. Water resources planning and management project
Oakley P (1995) People's participation in development projects: a critical review of current theory and practice

Samuel G (2006) Food aid and smallholder agriculture in Ethiopia: options and scenarios: a paper for the future agricultures consortium workshop. Institute of Development Studies, 20–22 March 2006, Addis Ababa, Ethiopia

Sands P (2003) Principles of international environmental law, 2nd edn. Cambridge University Press, New York

Sewnet AS (2013) Retrospective analysis of land cover and use dynamics in Gilgel Abbay Watershed by using GIS and remote sensing techniques, North Western Ethiopia. Int J Geosci. Available on http://www.scirp.org/journal/ijg

Shibru T, Kifle L (1998) Environmetal managment in Ethiopia: have the national conservation pland worked? environmetal publication series 1, organization for social science research in eastern and southern africa

Shimelis D (2012) Effectiveness of soil and water conservation measures for land restoration in the Wello area, northern Ethiopian highlands. PhD Dissertation, University of Rheinischen Friedrich-Wilhelms, Bonn, Germany

SMEC (2008b) Hydrological study of the Lake Tana sub-basin, main report. Ministry of Water Resources, Ethiopia

Springsguth M (2013) The impact of land administration and common pool resource management on Westland utilization along the Eastern Shore of Lake Tana, Ethiopia: alteration of Wetlands-risk or chance for rural livelihoods? Diploma Thesis, Earnest Moritize Arndt University of Greifswald

Swaminathan MS (1986) Sustainable nutrition security for Africa: lessons from India. The Hunger Project Paper No. 5 Rome

Thrupp LA (1998) Food security and environment in the greater horn of Africa. WRI (World Resources Institute), Washington, DC, USA

Uphoff N (1992) Local institutions and participation for sustainable development: international institute for environment and development, sustainable agriculture and rural livelihood program. Gatekeepers Series No. SA 31, New York, USA

World Bank (2000) World development report 1999/2000: entering the 21st century. Oxford University Press, New York

World Resource Institute (2005) Overview of the millennium ecosystem assessment. http://www.millenniumassessment.org/en/About.html

Chapter 28
Land Use and Watershed Management Practices in Lake Tana Basin

Wubneh Belete Abebe and Amare Sewnet Minale

Abstract Lake Tana Basin development corridor (LTSBDC) is a corridor in Amhara region which covers an area of about 1.5 million hectares. The area is known for the variety of annual and perennial crops grown, its breeds of livestock and diversified vegetation, and especially its forest resources. In the years between 1986 and 2013, both built-up areas and cultivated land cover increased. This indicates that, due to expansion of cultivated land and human settlements, the natural forest land cover declined. Similar to natural forest, grassland cover was also found to decrease. Land use and land cover change and consequent land degradation in the Lake Tana watershed limits the potential to develop a sustainable livelihood for its inhabitants. The effectiveness of the regional and local efforts to improve the environment and livelihoods of the residents in the Lake Tana watershed through natural resource conservation programs are limited due to capacity and financial barriers. Land use and water resource related barriers that were identified include weak policy implementation, low capacity (technical and financial), poor information management system, low or no incentives, and frequent restructuring of core and principal institutions at national/regional levels. Alleviation of these constraints will be important for better delivery of technical support and enabling local users to implement good land use related practices, to influence policies; reduce dependency on aids and implement environmentally friendly watershed management activities; terraces, gully rehabilitation check dams and alley cropping as biological measures.

Keywords Land use land cover · Land cover change · Watershed management · Remote sensing

W.B. Abebe (✉)
Amhara Design and Supervision Works Enterprise, Bahir Dar, Ethiopia
e-mail: wubnehb@yahoo.com

A.S. Minale
Department of Geography and Environmental Studies, Bahir Dar University, Bahir Dar, Ethiopia
e-mail: amare1974@gmail.com

K. Stave et al. (eds.), *Social and Ecological System Dynamics*,
AESS Interdisciplinary Environmental Studies and Sciences Series,
DOI 10.1007/978-3-319-45755-0_28

28.1 Introduction

The way people use land is the result of complex interactions between biophysical resources, socio-economic conditions, policy frameworks and issues related to various interested institutions in time and space. However, progressive land degradation and unplanned utilization of land resources coupled with climatic variability has aggravated the incidence of poverty and food insecurity situations over time. Hence, the need for optimum use of land has never been greater than at present from the immense agricultural and demographic pressures, where rapid population growth and turned the land into a relatively scarce commodity for various uses.

Land uses bring change to land cover and changes in land cover affect land uses. A change in either, however, is not necessarily the product of the other. Changes in land cover by land use do not necessarily imply a degradation of the land. However, many shifting land use patterns, driven by a variety of social causes, result in land cover changes that affect biodiversity, water and other processes that, cumulatively, affect global climate and biosphere (Riebsame et al. 1994 Cited in: ADSWE 2013).

This chapter analyzes current land use systems, farming practices, potentials, and constraints as a basis for understanding improvement requirements and extra importantly, as a platform for watershed management practices and the future watershed management objectives and strategies.

28.2 Land Use Characterization in Lake Tana Basin

28.2.1 Land Use Land Cover of Lake Tana Basin in 1986

The study by ADSWE (2013) indicated that Tana Basin land use land cover dynamics was planned to be studied based on the cover condition during the 1986 and then after about 25 years in 2013. In 1986, it is found that cultivated land has taken the larger share next to Water body; 19.2 and 19.3% respectively (Table 28.1, Fig. 28.1). The Lake Tana share about 18.8% of the Basin which is the largest water source to the Basin and also to the region. This Basin was extensively exploited even before 30 years (around 1986) where human intervened area coverage takes more proportion.

28.2.2 Land Use Land Cover of Lake Tana Basin in 2013

The cover size of cultivated land has almost doubled in about 30 years; from 19% in 1986 to 32% in 2013. Built up areas especially farm villages have also increased significantly; from 0.3% in 1986 to 19.0% in 2013 (Table 28.2, Fig. 28.2).

Table 28.1 Land Use and Land Cover of Lake Tana Basin in 1986

S.N.	LULC type	Area, ha	Percent
1	Cultivated land	309383.8	19.2
2	Farm villages with CL&FL	4953.7	0.3
3	Grassland	170742.9	10.6
4	Mixed forest	72139.3	4.5
5	Natural forest	162707.9	10.1
6	Open shrub land	182195.4	11.3
7	Shrub land	135681.8	8.4
8	Water body	311122.9	19.3
9	Wooded grassland	106695.7	6.6
10	Woodland	155505.8	9.7
	Sum	1611129.2	100

Note The area of the basin is 15,114 km^2; but 16,111 km^2 is put including 997 km^2 area from Angereb sub-basin being considered part of the dev't corridor

Wetland and water body have shown increase in 2013 where the community witnessed that this increase is obvious. This could be produced as a result of regulated flow incurred by the Chara-Chara weir at the outlet of Lake Tana Basin. On the contrary, farm village has shown decrease which could be because of the signature shifted to vegetation and cultivated land rather than to farm villages or the 1986 cover could compose higher proportion of cultivated land or trees with in farm villages.

28.3 Land Cover Dynamics

28.3.1 Land Cover Change Between 1986 and 2013

Changes in land cover driven by land use can be categorized into two types: modification and conversion. Modification is a change of condition within a cover type; for example, unmanaged forest modified to a forest managed by selective cutting. Conversion is a change from one cover type to another, such as deforestation to create cropland or pasture.

The Land Use and Land Cover Change of 1986–2013 for Tana basin is presented in (Table 28.3). The statistical land use change for the basin was computed from the loss and gain for each Major land use category between 1986 and 2013 and is further described in (Table 28.3).

The statistics table lists the initial state classes in the columns, the final state classes in the rows and the unchanged areas in the highlighted cells. For each initial state class (i.e., each column), the table indicates how these pixels were classified in the final state image. In another words the columns describe at what expense the

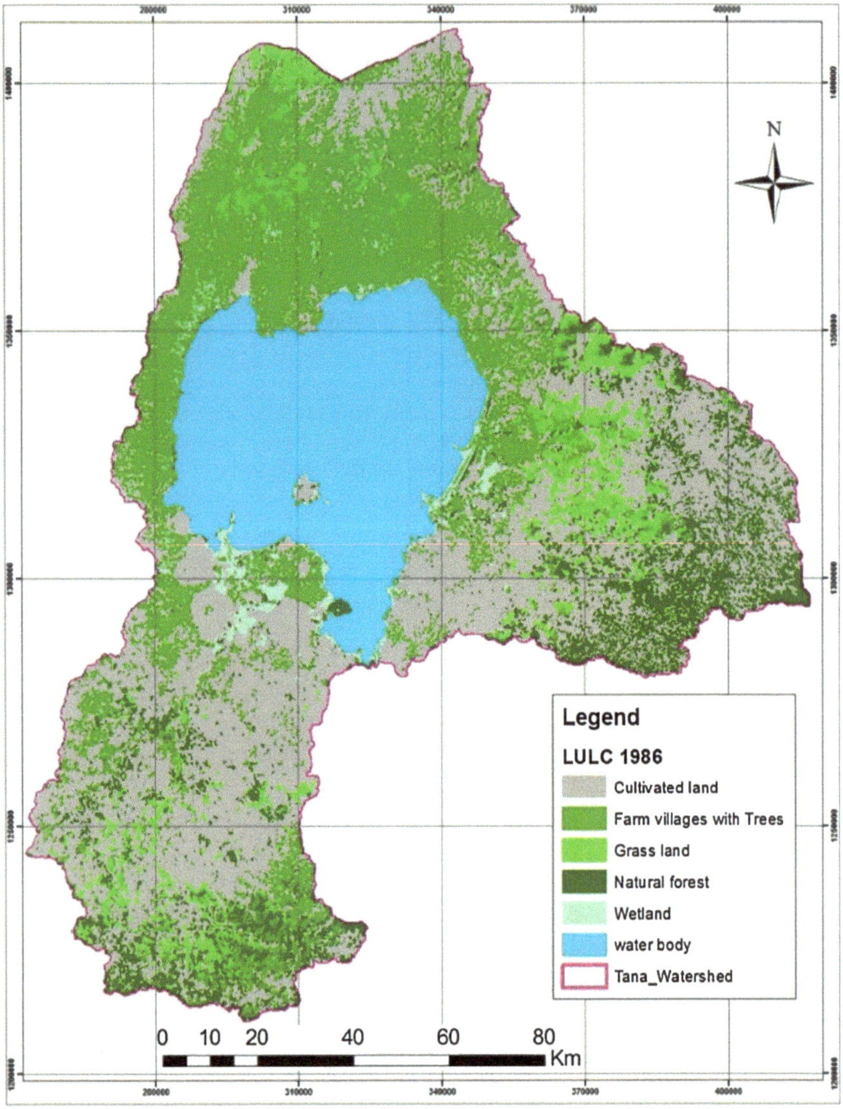

Fig. 28.1 Land use land cover map of Lake Tana Basin in 1986

new cover classes are established. For example much of the forest cover has been changed to farm villages and cultivated land as shown in Fig. 28.3.

In the initial state (1986) image 234,907 ha of land was classified as forest, but only 118548.5 ha of land was classified as forest in the final state image. This means a total of 116358.5 ha of forest was lost or converted to another land cover during the period 1986 to 2013 (Table 28.3). It appears that, referring to Table 28.3 and Fig. 28.3, the present forest land is resulted from losses as Woodland,

Table 28.2 Land use and land cover of Lake Tana Basin in 2013

S.N.	Land use land cover	Area, ha	Percent
1	Cultivated land	519417.1	32.239318
2	Built-up area	299685.9	18.600983
3	Grassland	182178.1	11.307482
4	Forest land	118548.5	7.3580998
5	Shrub land	59119.6	3.6694498
6	Woodland	33863.8	2.101868
7	Water body	315421.7	19.57768
8	Plantation forest with natural trees	14120.5	0.8764374
9	Wetland	68774.0	4.268682
	Sum	1611129.2	100

Grassland, Cultivated land, built up area, shrub land and others. As mentioned earlier, about 234,907 ha of land was classified as Forest land at initial state image (1986), while 2123.4 ha of land were classified as plantation forest; 25437.3 ha classified as Grassland; 7315.4 ha classified as woodland; 4560.6 ha as water body, 18964.3 ha as wetland, 11524.8 ha as shrub land, 47882.8 ha classified as farmers village and also 78525.8 ha classified as cultivated land in the final state image (2013) as losses from forest land in the initial state image (1986). Hence, forest conversion is too high for cultivated land and as farm village.

Apparently, referring to Table 28.3 and Fig. 28.4, most of the current cultivated land is established from forest land, shrub land and grass land. About 312887.6 ha of land were classified as cultivated land at initial state image (1986), while 549417.1 ha of land were classified as cultivated land in the final state image (2013). There is increase of 236529.5 ha cultivated land. It was observed that cultivated land increased from 19.2% in 1986 to 32.2% in 2013; this could be resulted greatly from deforestation.

28.4 Existing Land Use Characterization

28.4.1 General

Field site description using land use survey format were an input to characterize existing land use system, identify the constraints and opportunities and to propose wetland management plan in the Lake watershed.

The Current land use in the Lake watershed is the complex interaction between biophysical and socio-economic condition. As such, land use can be seen as the ultimate expression of everything else that is going on in the area. It provides a starting point and interactive framework for the study of both present and future land use. The existing land use map was produced by using standard cover legend

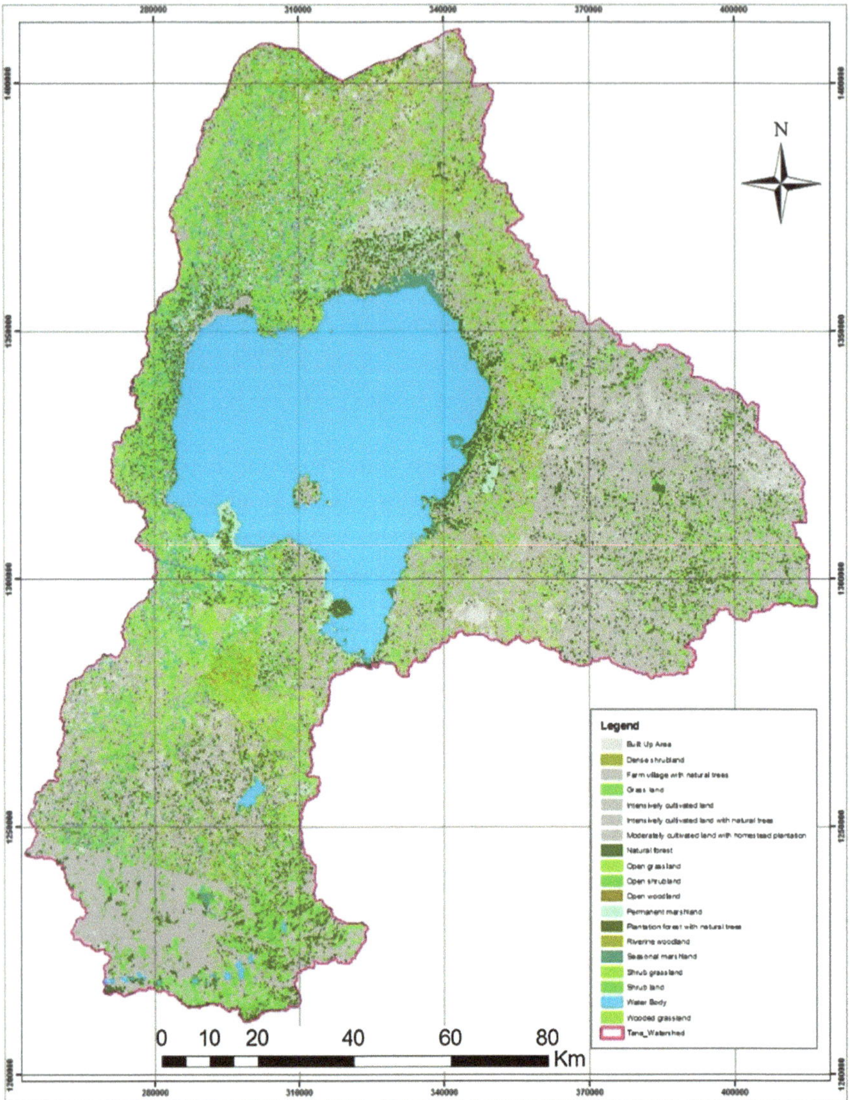

Fig. 28.2 Land use and land cover of Lake Tana Basin in 2013

and existing land cover and land use in the Lake watershed. The legend has been built up in an interactive manner, mainly from two sets of activities: observations and assessments in the field, and publications and project reports in the Lake watersheds. It gives information to define existing land use characterization. During reconnaissance survey the legend was used as a guide to link existing land cover with existing land use. Some description are presented as follows:

Table 28.3 Land use and land cover change area statistics in hectare (1986–2013)

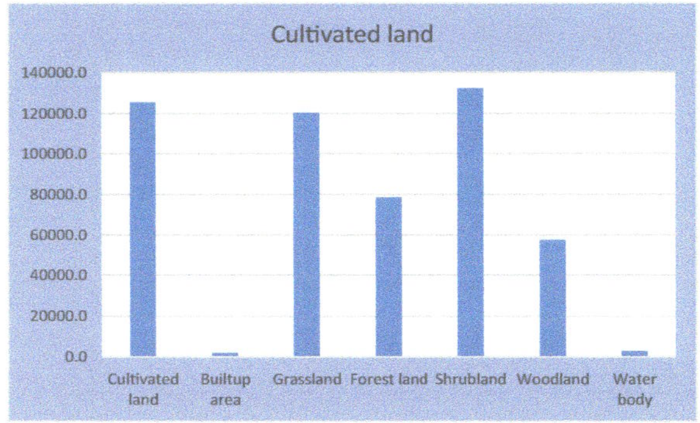

S. N.	LULC_type	LULC_major	Area	Percentage
1	Cultivated land	Built up area	242831051.3	8.002628
2	Cultivated land	Dense shrubland	53061419.7	1.748668
3	Cultivated land	Farm village with natural trees	540569650.4	17.814764
4	Cultivated land	Grass land	172552746.3	5.686569
5	Cultivated land	Intensively cultivated land	737956573.8	24.319757
6	Cultivated land	Intensively cultivated land with natural trees	499131329.2	16.449142
7	Cultivated land	Moderately cultivated land with homestead plantation	18848522.53	0.621163
8	Cultivated land	Natural forest	198624443	6.545776
9	Cultivated land	Open grassland	128985023.2	4.250771
10	Cultivated land	Open shrubland	52050338.41	1.715347
11	Cultivated land	Open woodland	21515063.08	0.709041
12	Cultivated land	Permanent marshland	62757602.75	2.068211
13	Cultivated land	Plantation forest with natural trees	36978700.37	1.218653
14	Cultivated land	Riverine woodland	37774261.38	1.244871
15	Cultivated land	Seasonal marshland	32641730.65	1.075726
16	Cultivated land	Shrub grassland	142276653.7	4.688804
17	Cultivated land	Shrub land	18367391.53	0.605307

(continued)

Table 28.3 (continued)

S. N.	LULC_type	LULC_major	Area	Percentage
18	Cultivated land	Water body	56029817.38	1.846493
19	Cultivated land	Wooded grassland	75923829.53	2.502111
20	Farm villages with CL&FL	Built up area	3764045.069	7.747273
21	Farm villages with CL&FL	Dense shrubland	45312.5	0.093264
22	Farm villages with CL&FL	Farm village with natural trees	12952786.44	26.659822
23	Farm villages with CL&FL	Grass land	2968101.496	6.109038
24	Farm villages with CL&FL	Intensively cultivated land	10948559.49	22.534661
25	Farm villages with CL&FL	Intensively cultivated land with natural trees	8749929.956	18.009374
26	Farm villages with CL&FL	Moderately cultivated land with homestead plantation	2039.0625	0.004197
27	Farm villages with CL&FL	Natural forest	6280727.95	12.927187
28	Farm villages with CL&FL	Open grassland	27796.875	0.057212
29	Farm villages with CL&FL	Open woodland	510978.8968	1.051712
30	Farm villages with CL&FL	Plantation forest with natural trees	4687.5	0.009648
31	Farm villages with CL&FL	Seasonal marshland	62500	0.128639
32	Farm villages with CL&FL	Water body	27187.5	0.055958
33	Farm villages with CL&FL	Wooded grassland	2021543.272	4.160802
34	Grassland	Built up area	129789378.1	7.750373
35	Grassland	Dense shrubland	28032575.73	1.673965
36	Grassland	Farm village with natural trees	269767274.2	16.109154
37	Grassland	Grass land	110850848.1	6.619459
38	Grassland	Intensively cultivated land	488816905.2	29.189704
39	Grassland	Intensively cultivated land with natural trees	260880739.8	15.578495
40	Grassland	Moderately cultivated land with homestead plantation	10697214.46	0.638784
41	Grassland	Natural forest	109505554.7	6.539125
42	Grassland	Open grassland	38433049.01	2.29503
43	Grassland	Open shrubland	27546483.07	1.644938
44	Grassland	Open woodland	10491106.88	0.626477
45	Grassland	Permanent marshland	37782899.92	2.256206
46	Grassland	Plantation forest with natural trees	19714382.41	1.177244
47	Grassland	Riverine woodland	22360111.64	1.335234
48	Grassland	Seasonal marshland	18196966.85	1.086632
49	Grassland	Shrub grassland	55767881.02	3.330179

(continued)

Table 28.3 (continued)

S. N.	LULC_type	LULC_major	Area	Percentage
50	Grassland	Shrub land	14792262.85	0.88332
51	Grassland	Water body	29316797.98	1.750653
52	Grassland	Wooded grassland	40537035.76	2.420669
53	Mixed forest	Built up area	52968838.49	7.486429
54	Mixed forest	Dense shrubland	7462001.446	1.054653
55	Mixed forest	Farm village with natural trees	106084711.8	14.993639
56	Mixed forest	Grass land	50917521.65	7.196503
57	Mixed forest	Intensively cultivated land	150359036.8	21.251216
58	Mixed forest	Intensively cultivated land with natural trees	100282808.7	14.173618
59	Mixed forest	Moderately cultivated land with homestead plantation	6961504.454	0.983914
60	Mixed forest	Natural forest	116817246.8	16.510537
61	Mixed forest	Open grassland	3969995.075	0.561105
62	Mixed forest	Open shrubland	6078779.391	0.859153
63	Mixed forest	Open woodland	3138499.086	0.443584
64	Mixed forest	Permanent marshland	47149375.66	6.663926
65	Mixed forest	Plantation forest with natural trees	4665318.515	0.65938
66	Mixed forest	Riverine woodland	9674185.844	1.367315
67	Mixed forest	Seasonal marshland	13976876.85	1.975442
68	Mixed forest	Shrub grassland	7175036.353	1.014094
69	Mixed forest	Shrub land	19146711.46	2.706129
70	Mixed forest	Water body	12264187.18	1.733377
71	Mixed forest	Wooded grassland	13019010.67	1.840061
72	Natural forest	Built up area	119745061.5	7.503694
73	Natural forest	Dense shrubland	20417029.76	1.279411
74	Natural forest	Farm village with natural trees	200029559.6	12.534634
75	Natural forest	Grass land	80520464.72	5.045727
76	Natural forest	Intensively cultivated land	324104686.3	20.309666
77	Natural forest	Intensively cultivated land with natural trees	188833045.3	11.833017
78	Natural forest	Moderately cultivated land with homestead plantation	14716950.78	0.922222
79	Natural forest	Natural forest	268907910.5	16.850821
80	Natural forest	Open grassland	15971166.23	1.000816
81	Natural forest	Open shrubland	24791792.33	1.553551

(continued)

Table 28.3 (continued)

S. N.	LULC_type	LULC_major	Area	Percentage
82	Natural forest	Open woodland	12950024.17	0.811499
83	Natural forest	Permanent marshland	108821392.2	6.819174
84	Natural forest	Plantation forest with natural trees	16569126.39	1.038286
85	Natural forest	Riverine woodland	47391011.51	2.969706
86	Natural forest	Seasonal marshland	19695272.46	1.234183
87	Natural forest	Shrub grassland	34761243.12	2.178275
88	Natural forest	Shrub land	37352106.66	2.340629
89	Natural forest	Water body	33341837.94	2.08933
90	Natural forest	Wooded grassland	48038188.41	3.010261
91	Open shrubland	Built up area	126040265.8	7.05339
92	Open shrubland	Dense shrubland	30166946.85	1.688185
93	Open shrubland	Farm village with natural trees	292172050	16.350357
94	Open shrubland	Grass land	100601362.9	5.629793
95	Open shrubland	Intensively cultivated land	483882798.1	27.078759
96	Open shrubland	Intensively cultivated land with natural trees	224348115.5	12.554835
97	Open shrubland	Moderately cultivated land with homestead plantation	13608344.4	0.761542
98	Open shrubland	Natural forest	105525632.8	5.905362
99	Open shrubland	Open grassland	52551224.29	2.94084
100	Open shrubland	Open shrubland	44274552.58	2.477666
101	Open shrubland	Open woodland	8546798.699	0.478291
102	Open shrubland	Permanent marshland	48340389.28	2.705196
103	Open shrubland	Plantation forest with natural trees	26596764.41	1.488392
104	Open shrubland	Riverine woodland	43510671.13	2.434918
105	Open shrubland	Seasonal marshland	21873488.43	1.224071
106	Open shrubland	Shrub grassland	100040889.2	5.598428
107	Open shrubland	Shrub land	18201177.72	1.018563
108	Open shrubland	Water body	37117015.86	2.07712
109	Open shrubland	Wooded grassland	36393283.78	2.036619
110	Shrubland	Built up area	126667054.3	9.518492
111	Shrubland	Dense shrubland	19395013.85	1.457453
112	Shrubland	Farm village with natural trees	200401392.4	15.059315
113	Shrubland	Grass land	85604580.29	6.432821

(continued)

Table 28.3 (continued)

S. N.	LULC_type	LULC_major	Area	Percentage
114	Shrubland	Intensively cultivated land	377315584.7	28.353666
115	Shrubland	Intensively cultivated land with natural trees	211990528.5	15.930189
116	Shrubland	Moderately cultivated land with homestead plantation	11608905.89	0.87236
117	Shrubland	Natural forest	111171706.6	8.354082
118	Shrubland	Open grassland	18092729.05	1.359592
119	Shrubland	Open shrubland	17495290.8	1.314697
120	Shrubland	Open woodland	8199555.481	0.616162
121	Shrubland	Permanent marshland	48022728.01	3.608704
122	Shrubland	Plantation forest with natural trees	10705820.48	0.804497
123	Shrubland	Riverine woodland	18766702.66	1.410238
124	Shrubland	Seasonal marshland	12448998.3	0.935489
125	Shrubland	Shrub grassland	26446346.35	1.987331
126	Shrubland	Shrub land	14890807.31	1.118981
127	Shrubland	Water body	21087947.35	1.58467
128	Shrubland	Wooded grassland	32555972.06	2.446443
129	Water body	Built up area	1486140.34	0.048703
130	Water body	Dense shrubland	151625.9616	0.004969
131	Water body	Farm village with natural trees	2966967.53	0.097231
132	Water body	Grass land	442611.2466	0.014505
133	Water body	Intensively cultivated land	2248900.097	0.073699
134	Water body	Intensively cultivated land with natural trees	25894047.26	0.848582
135	Water body	Moderately cultivated land with homestead plantation	673709.5673	0.022078
136	Water body	Natural forest	37529087.17	1.229878
137	Water body	Open grassland	59434.375	0.001948
138	Water body	Open shrubland	167568.6187	0.005491
139	Water body	Open woodland	50759.34956	0.001663
140	Water body	Permanent marshland	33941729.63	1.112316
141	Water body	Plantation forest with natural trees	695447.4027	0.022791
142	Water body	Riverine woodland	367260.7484	0.012036
143	Water body	Seasonal marshland	29312342.89	0.960604
144	Water body	Shrub grassland	608633.5613	0.019946
145	Water body	Shrub land	21284.34894	0.000698
146	Water body	Water body	2914711745	95.518971

(continued)

Table 28.3 (continued)

S. N.	LULC_type	LULC_major	Area	Percentage
147	Water body	Wooded grassland	276691.2789	0.009068
148	Wooded grassland	Built up area	87858626.81	8.395828
149	Wooded grassland	Dense shrubland	14334225.29	1.369788
150	Wooded grassland	Farm village with natural trees	148790487.8	14.218516
151	Wooded grassland	Grass land	66549593.79	6.359523
152	Wooded grassland	Intensively cultivated land	299622258.6	28.632099
153	Wooded grassland	Intensively cultivated land with natural trees	135645997.7	12.96242
154	Wooded grassland	Moderately cultivated land with homestead plantation	9228644.482	0.881895
155	Wooded grassland	Natural forest	105698786.4	10.100645
156	Wooded grassland	Open grassland	10920819.46	1.043601
157	Wooded grassland	Open shrubland	15441697.91	1.475619
158	Wooded grassland	Open woodland	5817410.526	0.555916
159	Wooded grassland	Permanent marshland	54561164	5.213901
160	Wooded grassland	Plantation forest with natural trees	9527325.821	0.910437
161	Wooded grassland	Riverine woodland	20697991.26	1.977914
162	Wooded grassland	Seasonal marshland	13350176.62	1.275752
163	Wooded grassland	Shrub grassland	16472416.12	1.574115
164	Wooded grassland	Shrub land	23135269.09	2.210821
165	Wooded grassland	Water body	18332585.11	1.751874
166	Wooded grassland	Wooded grassland	21751816.93	2.078618
167	Woodland	Built up area	107599027.6	7.05485
168	Woodland	Dense shrubland	22031675.89	1.444531
169	Woodland	Farm village with natural trees	224374428	14.711358
170	Woodland	Grass land	82434113.45	5.404884
171	Woodland	Intensively cultivated land	349843027.8	22.937846
172	Woodland	Intensively cultivated land with natural trees	214015055.1	14.032134
173	Woodland	Moderately cultivated land with homestead plantation	12954914.96	0.849403
174	Woodland	Natural forest	125423869.5	8.223555
175	Woodland	Open grassland	29755965.43	1.950983
176	Woodland	Open shrubland	41194986.28	2.700995
177	Woodland	Open woodland	6179659.101	0.405176
178	Woodland	Permanent marshland	69814040.07	4.577435
179	Woodland	Plantation forest with natural trees	15747813.63	1.032523

(continued)

Table 28.3 (continued)

S. N.	LULC_type	LULC_major	Area	Percentage
180	Woodland	Riverine woodland	60696039.55	3.979603
181	Woodland	Seasonal marshland	14990155.94	0.982846
182	Woodland	Shrub grassland	82299918.53	5.396085
183	Woodland	Shrub land	21149447.89	1.386687
184	Woodland	Water body	31988142.09	2.097338
185	Woodland	Wooded grassland	33205960.58	2.177186

	Land use land cover	Initial state image 1986										
		Cultivated land	Builtup area	Grassland	Mixed forest	Natural forest	Open shrubland	Shrubland	Water body	Wooded grassland	Woodland	Row total
Final state image 2013	Built up area	242831051.3	3764045.1	129789378.1	52968838.5	119745061.5	126040265.8	126667054.3	1486140.3	87858626.81	107599027.6	998749489.3
	Dense shrubland	53061419.7	45312.5	28032575.7	7462001.4	20417029.76	30166946.85	19395013.9	151626.0	14334225.29	22031675.89	195097827.0
	Farm village with natural trees	540569650.4	12952786.4	269767274.2	106084711.8	200029559.6	292172050	200401392.4	2966967.5	148790487.8	224374428	1998109308.3
	Grass land	172552746.3	2968101.5	110850848.1	50917521.7	80520464.72	100601362.9	85604580.3	442611.2	66549593.79	82434113.45	753441944.0
	Intensively cultivated land	737956573.8	10948559.5	488816905.2	150359036.8	324104686.3	483882798.1	377315584.7	2248900.1	299622258.6	349843027.8	3225098330.8
	Intensively cultivated land with natural trees	499131329.2	8749930.0	260880739.8	100282808.7	188833045.3	224348115.5	211990528.5	25894047.3	135645997.7	214015055.1	1869771597.0
	Moderately cultivated land with homestead plantation	18848522.5	2039.1	10697214.5	6961504.5	14716950.78	13608344.4	11608905.9	673709.6	9228644.482	12954914.96	99300750.6
	Natural forest	198624443.0	6280727.9	109505554.7	116817246.8	268907910.5	105525632.8	111171706.6	37529087.2	105698786.4	125423869.5	1185484965.3
	Open grassland	128985023.2	27796.9	38433049.0	3969995.1	15971166.23	52551224.29	18092729.1	59434.4	10920819.46	29755965.43	298767203.0
	Open shrubland	52050338.4	0.0	27546483.1	6078779.4	24791792.33	44274552.58	17495290.8	167568.6	15441697.91	41194986.28	229041489.4
	Open woodland	21515063.1	510978.9	10491106.9	3138499.1	12950024.17	8546798.699	8199555.5	50759.3	5817410.526	6179659.101	77399855.3
	Permanent marshland	62757602.8	0.0	37782899.9	47149375.7	108821392.2	48340389.28	48022728.0	33941729.6	54561164	69814040.07	511191321.5
	Plantation forest with natural trees	36978700.4	4687.5	19714382.4	4665318.5	16569126.39	26596764.41	10705820.5	695447.4	9527325.821	15747813.63	141205386.9
	Riverine woodland	37774261.4	0.0	22360111.6	9674185.8	47391011.51	43510671.13	18766702.7	367260.7	20697991.26	60696039.55	261238235.7
	Seasonal marshland	32641730.6	62500.0	18196966.9	13976876.8	19695272.46	21873488.43	12448998.3	29312342.9	13350176.62	14990155.94	176548509.0
	Shrub grassland	142276653.7	0.0	55767881.0	7175036.4	34761243.12	100040889.2	26446346.4	608633.6	16472416.12	82299918.53	465849018.0
	Shrub land	18367391.5	0.0	14792262.9	19146711.5	37352106.66	18201177.72	14890807.3	21284.3	23135269.09	21149447.89	167056458.9
	Water body	56029817.4	27187.5	29316798.0	12264187.2	33341837.94	37117015.86	21087947.3	2914711745.5	18332585.11	31988142.09	3154217263.9
	Wooded grassland	75923829.5	2021543.3	40537035.8	13019010.7	48038188.41	36393283.78	32555972.1	276691.3	21751816.93	33205960.58	303723332.3
	Class total	3128876148.2	48366196.0	1723279467.6	732111646.2	1616957870.0	1813791771.6	1372867664.4	3051605986.8	1077737293.6	1545698241.4	16111292285.9

(continued)

Table 28.3 (continued)

	Land use land cover	Initial state image 1986							
		Cultivated land	Builtup area	Grassland	Forest land	Shrubland	Woodland	Water body	Row total
	Land use land cover	Initial state image 1986							
		Cultivated land	Builtup area	Grassland	Forest land	Shrubland	Woodland	Water body	Row total
Final state image 2013	Built up area	242831051.3	3764045.1	217648004.9	172713900.0	252707320.1	107599027.6	1486140.3	998749489.3
	Dense shrubland	53061419.7	45312.5	42366801.0	27879031.2	49561960.7	22031675.89	151626.0	195097827.0
	Farm village with natural trees	540569650.4	12952786.4	418557762.0	306114271.5	492573442.4	224374428	2966967.5	1998109308.3
	Grass land	172552746.3	2968101.5	177400441.9	131437986.4	186205943.2	82434113.45	442611.2	753441944.0
	Intensively cultivated land	737956573.8	10948559.5	788439163.7	474463723.1	861198382.7	349843027.8	2248900.1	3225098330.8
	Intensively cultivated land with natural trees	499131329.2	8749930.0	396526737.5	289115854.0	436338644.0	214015055.1	25894047.3	1869771597.0
	Moderately cultivated land with homestead plantation	18848522.5	2039.1	19925858.9	21678455.2	25217250.3	12954914.96	673709.6	99300750.6
	Natural forest	198624443.0	6280727.9	215204341.1	385725157.3	216697339.4	125423869.5	37529087.2	1185484965.3
	Open grassland	128985023.2	27796.9	49353868.5	19941161.3	70643953.3	29755965.43	59434.4	298767203.0
	Open shrubland	52050338.4	0.0	42988181.0	30870571.7	61769843.4	41194986.28	167568.6	229041489.4
	Open woodland	21515063.1	510978.9	16308517.4	16088523.3	16746354.2	6179659.101	50759.3	77399855.3
	Permanent marshland	62757602.8	0.0	92344063.9	155970767.8	96363117.3	69814040.07	33941729.6	511191321.5
	Plantation forest with natural trees	36978700.4	4687.5	29241708.2	21234444.9	37302584.9	15747813.63	695447.4	141205386.9
	Riverine woodland	37774261.4	0.0	43058102.9	57065197.4	62277373.8	60696039.55	367260.7	261238235.7
	Seasonal marshland	32641730.6	62500.0	31547143.5	33672149.3	34322486.7	14990155.94	29312342.9	176548509.0
	Shrub grassland	142276653.7	0.0	72240297.1	41936279.5	126487235.6	82299918.53	608633.6	465849018.0
	Shrub land	18367391.5	0.0	37927531.9	56498818.1	33091985.0	21149447.89	21284.3	167056458.9
	Water body	56029817.4	27187.5	47649383.1	45606025.1	58204963.2	31988142.09	2914711745.5	3154217263.9
	Wooded grassland	75923829.5	2021543.3	62288852.7	61057199.1	68949255.8	33205960.58	276691.3	303723332.3
	Class total	3128876148.2	48366196.0	2801016761.3	2349069516.2	3186659436.0	1545698241.4	3051605986.8	16111292285.9

	Land use land cover	Initial state image 1986							
		Cultivated land	Builtup area	Grassland	Forest land	Shrubland	Woodland	Water body	Row total
Final stage image 2013	Cultivated land	1255936425.5	19700528.5	1204891760.2	785258032.4	1322754277.0	576812997.8	28816656.9	5194170678.3
	Builtup area	783400701.7	16716831.5	636205766.8	478828171.5	745280762.5	331973455.6	4453107.9	2996858797.6
	Grassland	519738252.7	5017441.6	361283460.2	254372626.2	452286387.9	227695958.0	1387370.5	1821781497.2
	Forest land	198624443.0	6280727.9	215204341.1	385725157.3	216697339.4	125423869.5	37529087.2	1185484965.3
	Shrubland	123479149.6	45312.5	123282514.0	115248421.0	144423789.1	84376110.1	340478.9	591195775.2
	Woodland	59289324.5	510978.9	59366620.3	73153720.6	79023728.0	66875698.7	418020.1	338638091.0
	Water body	56029817.4	27187.5	47649383.1	45606025.1	58204963.2	31988142.1	2914711745.5	3154217263.9
	plantation forest with natural trees	36978700.4	4687.5	29241708.2	21234444.9	37302584.9	15747813.6	695447.4	141205386.9
	Wetland	95399333.4	62500.0	123891207.4	189642917.2	130685604.0	84804196.0	63254072.5	687739830.5
	Class total	3128876148.2	48366196.0	2801016761.3	2349069516.2	3186659436.0	1545698241.4	3051605986.8	16111292285.9

(continued)

Table 28.3 (continued)

	Land use land cover	Initial state image 1986							
		Cultivated land	Builtup area	Grassland	Forest land	Shrubland	Woodland	Water body	Row Total
	Land use land cover	Initial state image 1986							
		Cultivated land	Builtup area	Grassland	Forest land	Shrubland	Woodland	Water body	Row Total
Final stage image 2013	Cultivated land	**125593.6**	1970.1	120489.2	78525.8	132275.4	57681.3	2881.7	519417.1
	Builtup area	78340.1	**1671.7**	63620.6	47882.8	74528.1	33197.3	445.3	299685.9
	Grassland	51973.8	501.7	**36128.3**	25437.3	45228.6	22769.6	138.7	182178.1
	Forest land	19862.4	628.1	21520.4	**38572.5**	21669.7	12542.4	3752.9	118548.5
	Shrubland	12347.9	4.5	12328.3	11524.8	**14442.4**	8437.6	34.0	59119.6
	Woodland	5928.9	51.1	5936.7	7315.4	7902.4	**6687.6**	41.8	33863.8
	Water body	5603.0	2.7	4764.9	4560.6	5820.5	3198.8	**291471.2**	315421.7
	plantation forest with natural trees	3697.9	0.5	2924.2	2123.4	3730.3	1574.8	69.5	14120.5
	Wetland	9539.9	6.3	12389.1	18964.3	13068.6	8480.4	6325.4	68774.0
	Class total	**312887.6**	**4836.6**	**280101.7**	**234907.0**	**318665.9**	**154569.8**	**305160.6**	**1611129.2**

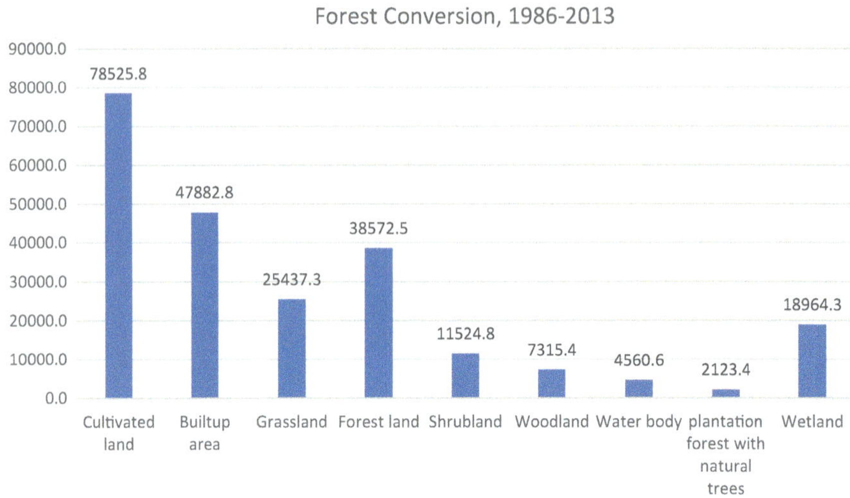

Fig. 28.3 Contribution of other land covers to the forest loss (1986–2013)

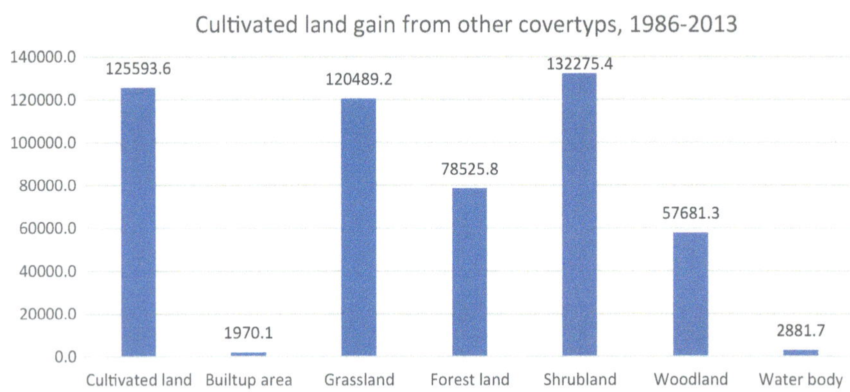

Fig. 28.4 Contribution of other land covers to the current expansion of cultivated land

Farm Villages with Trees and Cultivation: This legend included towns and farm villages which were recognized by patterns of constructions, trees grown around homesteads and along roadsides and grasslands left for livestock. Farm villages were also recognized being scattered in cultivated areas of various intensity classes. In the area main land-use activities include: residential, commercial, industrial, recreational and associated uses in urban built up; and, settlement, crop production, wood collection, grazing and browsing in rural farm villages.

Intensively Cultivated land: With main land-use activities include; small holder rain fed crop cultivation of cereals: Teff, wheat, barley, maize and sorghum; pulses, stimulants, spices and oilseeds; irrigated crop cultivation, cash/horticultural crops and sedentary small holder livestock grazing on unimproved pasture and fallow, and on aftermath of cattle, sheep, goats, horses and donkeys. This land is opened up for cultivation and natural grazing. 80% of the land is under annual crops during cropping season, and completely bare during the dry season.

Moderately Cultivated Land: Main land-use activities include as of intensively cultivated land but the portion of the land under natural shrub or bush often grazed and browsed by not cleared for crop cultivation. The ground is normally only partially bare in the dry season. 50% of ground is under annual crops during the cropping season, 10% under perennials. Land covers class observed in the area but not mapped was **perennial crop cultivation** with major perennial crops, coffee, chat, mango, and other tree crops with intercropping of various annual crops. Essentially no bare ground most of the year as this land use system occurs mostly on irrigated lands and moist surplus areas. 25% of the ground is estimated under annual crops, 30% perennial crops, and 30% under unimproved pasture for sedentary livestock grazing and browsing during the cropping season.

Forest Land: As major land cover with main land-use activities; forest harvesting, principally of fuel and construction wood, mostly of eucalyptus and rain fed small holder mixed agriculture, annual and perennial crops. It mainly occupies hills, mountains, escarpments, plateau and gorge sides where the rainfall is adequate and the terrain is rugged and not suited for crop cultivation. During this survey three forest subclasses were identify, includes natural, plantation and a mixture of plantation and natural forest. Eucalyptus woodland with main land use activities as wood harvesting for domestic use and seasonal grazing; **Note**: planted woodlots for domestic wood supply, mainly restricted to areas surrounding towns and farm villages are mapped as Farm villages with trees and cultivation but not as forest land or plantation forest.

Woodland: Main land use activities include livestock browsing and grazing, rain fed smallholder mixed crop production in patches and wood harvesting. Scattered trees mostly deciduous or succulent scattered on otherwise extensive grasslands known property of this cover. It occupies plain, mountainous and hilly areas.

Shrub Land: Cover has similar main land use activities as of bush land and recognized by woody plants that have multi-stem branches of short vegetation species. This mainly occurs on hills, escarpments, mountains and gorge slopes where the soils are very shallow and the slopes are very steep.

Grass land Covers: Grassland cover legend, main land use activity of livestock grazing of communal kinds, where woody species are very rare, shrubs and occasional trees often along watercourses. Livestock predominantly are of grazers however, in bushed and scrubbed grass land is with more frequency of browsers and some harvesting of wood. It is mainly recognized along rivers in valley bottoms, between crop fields, around villages and homesteads, on mountain and hill tops and side slopes and on plateaus having stony and shallow soils. Grasslands are characterized by the presence or absence of woody vegetation such as trees, bushes and shrubs. In most cases, grass lands in the Geldaw watershed have rock outcrops and stony areas.

Marshland/Wetland Covers: Which are perennial or seasonal types with main land use activity including year round grazing on the outer periphery in swamps have scatterings of woody species while marshes are predominantly grasses and sedges. These areas also serve as birds sanctuaries in permanent marsh while seasonal marsh area the main land use activities are crop production and dry season grazing. The marshland recognized and mapped occupies low-lying areas along riverbanks and plain areas around Lake Tana.

Water Body: As land cover ledged: this legend includes Lake Tana, with main land use activities of fishing, irrigation of crop land, and water supply for human and animal and transportation.

28.4.2 Major Existing Land Use Systems

Existing land use systems in the Lake watershed are dependent on the land cover, the agro climate and soil terrain characteristics, and of course, it can be modified further by the socio cultural and economic diversity in the Lake watershed.

28.4.2.1 Smallholder Rain-Fed Crop Production

This production conducted at the level of land holdings of local farmers, i.e., less than one hectare land. The land use is usually carrying out with low capital intensity, high family labour input and oxen, horse power used for land preparation. Rain will be the main source of moisture. In order to improve the current production level appropriate chemical fertilizers, manure, improved seeds and pesticides will be used as input of production. Family labour is used for land preparation, sowing, weeding, harvesting, threshing and winnowing. Choice of crops, planting dates, fertilization and all other management components are based on the season and typical commencement of rain for productive cropping. This system is based on the traditional farming practices; however, improved seeds, fertilizers and chemical usage are expected to improve performance.

In most productive zones (1500–3200 m a.s.l); wide range of crops [which include small holder cereals (Teff, corn, wheat, barley, sorghum, millet, oats, etc.); pulses (horse beans, chick-peas, haricot beans, field peas, lentils, soybean, and vetch); oil-seeds (linseed, Niger seed, fenugreek, rapeseed, sunflower, castor bean, groundnuts, etc.); stimulants (coffee, tea, chat, tobacco, etc.); fibres (cotton, flax, etc.); fruits (banana, orange, papaya, lemon, mandarin, mango, avocado, etc.); vegetables (onion, tomato, carrot, cabbage, etc.); root and tuber (potato, sweet-potatoes, etc.) and sugarcane] are grown and many species of livestock kept for different ends.

In the Lake watershed, rain fed crop production can be classified into two categories: Low and medium input rain fed crop production system. 'Input' here refers to material input such as seeds, fertilizer, pesticides, etc.

Low input rain fed crop production is not as extensive as in medium input production. Intercropping is very common. Bund for soil fertility and moisture retention is less extensive than in the medium input category. In low input cases, the use of local seeds is widely practiced but in medium input improved seed supplied from seed suppliers has been used. No use of fertilizers application. Farm fields are smaller than 1 ha.

Medium input rain fed crop production is more extensive than low input. Intercropping is not popular. Seed is mainly improved, fields are larger than 1 ha in average size. The use of fertilizer and manure is exercised but not at rate of recommendation in this category. In view of the above, it is very difficult to spatially distinguish between low input and medium input categories of rain fed crop production in the area. A general characterisation is therefore adopted for the purposes of this study. The following is a general characterisation of rain fed crop production in the area observation:

Land Management: Land management consists of soil water harvesting on bund. Even though the area is not deficit in total annual water; moisture distribution is very limited for production. Land degradation is a common phenomenon in the area.

Input Levels: The input levels ranges from low to medium. However, inputs are limited in this land use class with local seed being widely used. Mainly maize, rice, wheat and teff variety are the commonly supply to farmers.

Mechanization: Mechanization is rare, tractor as power for ploughing is not being used. Mostly oxen are used power for ploughing and crushing.

Farm Management: The most common form of farm management is soil bund to control soil erosion by water. Other forms of farm management include fencing around fields to protect crops against animals. Fences are cut from live trees, which together with fuel wood demand cause an increase in deforestation. Intercropping is common in the low input category.

Purposes of Crop Production: The Purposes of crop production are produced for food, market. In most cases, crop residues are stored as animal feed and fed to animals during the dry season.

Most limitations to production in the area include drought (moisture distributions), soil fertility degradation, crop failures attributed to the harsh climatic conditions hazards, pests, weeds and diseases, late planting due to lack of investment capital and poor farming techniques attributed to lack of extension. Yields of most crop production fall far below potential levels, which may be regarded as a crop failure. Low yields are attributed to low input levels that characterise this land use class. Other causes of low yields include moisture stress, poor farming techniques that are a consequence of limited farm training, pests and diseases and inappropriate seeds.

It could be concluded that the majority farming system of the Lake watershed is categorized under small holder agriculture with low capital intensity, low input, high labour intensity, hoe and animal plough cultivation, small holder farming with low management level, growing primarily subsistence annual and perennial crops with some sale of surplus.

28.4.2.2 Small to Medium Scale Crop-Livestock Production

In the Lake watershed, production systems are, in fact, of a mixed crop-livestock type with livestock for draught power being important. Rainfall is generally not limiting and growing seasons are often very long, with two crops per year in some parts of the Lake watersheds. Due to high population, farming is dominated by smallholdings. Medium scale dairying and fattening is found around rural towns. Marginal free grazing and browsing communal land (highly degraded shrub lands because of overgrazing and cutting of trees) around seasonal Lake Inundation area is a most common exercise of livestock feeding. In the dry season fodder becomes scarce, leading to poor animal health and starvation. Livestock products include milk, meat, hides and butter, for both domestic and commercial use. Most common constraints associated with livestock production include disease, poor quality pasture, pasture shortage, soil erosion, and low market prices for livestock products.

28.4.2.3 Smallholder Irrigated Crop Production

This is mainly practiced along the peripheries of the lake and in suitable areas near the river mouth in the wetlands using pumping irrigation by a few farmers around the lake and some rivers, differing from rainfed in terms of crops produced and management initiatives employed. Irrigated/recession production involves the growing of fruit trees, cereals and vegetables. Crops are produced mostly for sale at the market, but some produce is consumed at home.

Recession Farming: It is a typical seasonal wetlands and lakeshore cropping system using the seasonal residual moisture accumulated after the recession of Lake Tana and other seasonal marsh areas/wetlands. The area under recession agriculture could not easily be estimated but it is estimated more than a hundred hectares. On the wetlands, farmers are producing different early maturing crops by traditional agricultural practice on receded area. Teff, green pepper, cabbage, maize, and banana are the common crops grown on this area (Fig. 28.5).

Other characterizations include water harvesting (soil bund and shallow wells) as associated improvement systems. Other improvements associated with irrigated orchards include diversion furrows which, in some cases, are cement lined.

Fig. 28.5 Examples of recession crop productions

Produced mostly for sale at market, but some produce is consumed at home. However, long distances to markets coupled with poor roads puts farmers in a difficult position when it comes to selling farm produce. Constraints to production include low tillage capacity (labour shortage), market inaccessibility (demoralises farmers as they are unable to sell their produce), water scarcity, loss of irrigation infrastructure, poor roads, pests, lack of capital, lack of technical knowledge, and herbivore damage.

28.4.2.4 Smallholders' Perennial Crops and Forest Production

This mixed type of land use is targeted for areas with slopes less than 60% where land is available for forming by local farmers. Perennial crops that provide economic benefit from sales of fruits, leaves and stems could be planted and additionally serve as biological soil and water conservation measures. Together with this, farmers also grow trees that produce poles, timbers, firewood and products are being used for local construction purposes and making farm tools to be sold in markets. The system is also integrated with controlled livestock production. The livestock is still fed in order to avoid destruction of woody plants established by farmers. The land use system requires low to medium intensity of capital and labour. It is being practiced mainly on private land holdings of farmers. Inputs needed for the production system include improved multipurpose fruit crops and tree species seedlings and appropriate techniques of farming. Products of the land use system are sold in markets. It is important to note that sells can be delivered to investors who entered contractual agreement with local farmers for use in agro-industries. Appropriate fruit tree crops can be introduced to the appropriate environmental conditions through the government's extension system.

28.4.2.5 Small Scale Agro-Forestry Production

Agro-forestry is a system whereby the land use system integrates production of trees and annual crops (Fig. 28.6). The system focuses on conservation and production activities that sustain production and improve the conditions of the environment. In this system multipurpose trees are planted along field boundaries and terraces to support physical features and thereby protect the movement of soils down slope. In this system again woody plants planted to supply fodder, fuel wood and fruits as additional output. Leguminous woody species also add organic matter, nutrients and fix nitrogen into the soil to improve its fertility level. Improving the fertility status and water holding capacity of the soil improves the production level of annual crops grown in the system. Thus, farmers and investors can produce cereal crops, fodder and fruits that would be sold in markets for various uses by different interest groups and for home consumption.

Fig. 28.6 Farm forestry practices

28.4.2.6 Nature Conservation and Tourist Attraction

This type of land use is being implemented on areas having unique ecosystems such as wetland areas, lakes Tana water body and surrounding. This land use system is important to attract biodiversity (important bird breeding site), tourists and to generate income from the smokeless industry and conserve the unique biodiversity of the Lake watershed. It can properly be done with high capital intensity and low labour intensity. Construction of roads and integrating the existing land use system with nature conservation and tourist attraction is an important activity that needs to be agreed with the local land users and the government. Benefit sharing and integrating systematic livestock grazing with the intended investment could be best strategy to be developed with the local land resource users.

28.4.3 Constraints to Existing Land Use Systems in the Lake Watershed

Major constraints are lack of access to improve productivity through enhancing inputs, soil degradation, infertility and rugged topography, shortage of draught power, lack of technical knowledge, insect pest, disease and weed, natural hazards (frost, flood,), lack of capital which is attributed to lack of credit facilities and, poor market, road and veterinary service accessibility.

28.5 Opportunities for Future Wetland Management

- Policy and strategies in agricultural production, investment and natural resource management, nature conservation,
- Future opportunities for farmers include provision of farm power, soil and water conservation measures, extension training, pest and weed control,

- Introduction of early-maturing varieties and improved water harvesting techniques in the area,
- Market outlets and internal agro-industrial promotion.

28.6 Watershed Management Practices in Lake Tana Basin

28.6.1 History and Practice of Rainwater Harvesting

In Africa, the potential of rainwater harvesting for improved crop production was observed in the 1970s and 1980s and a number of rainwater harvesting projects were constructed in sub-Saharan Africa during the past decades with considerable amount of money. Rainwater harvesting in Ethiopia has a long history with strong attachment to the ancient Orthodox churches (Habtamu 1999).

Habtamu (1999) indicated that the history of rainwater harvesting in Ethiopia date back as early as the pre Axumite period (560 BC). During this period, rainwater was harvested and stored in ponds and tanks for agriculture and water supply purposes. Similarly, the Konso people have had a long and well-established tradition of building level terraces to harvest rainwater that is used to produce sorghum successfully under extremely harsh environment; low, erratic, and unreliable rainfall conditions (Habtamu 1999).

Most of the monasteries in the Amhara region and some churches own different types of rainwater harvesting structures. Some of these structures exist in Gondar Castles (15–16 centuries) and Lalibela Rock hewn churches (over 800 years ago). The water collected in the structures was used both for drinking and religious ceremonies. Rainwater harvesting structures in Monasteries like Mahbre Selassie in Gondar, Debrekerbe in Shoa, and in Kobo-Girana valley can be mentioned as an example.

In the region, production and productivity are very low mainly due to dependency on rainfall, which is uncertain and poorly distributed. Land degradation is critical in the region in general and in LTW in particular. Hence, Irrigation development using rainwater is paramount to alleviate problems of agricultural production. It may small scale but certainly has advantage over large scale schemes due to its low external input, their being labor intensive and are easily manageable and participatory in nature.

Since 2002/03, the Ministry of Agriculture has been involved in massive rainwater harvesting exercises in order to attain food self-sufficiency. The Ministry adopted the existing design and construction methods of storage tanks/cisterns from China. Initially, different types of underground storage tanks were constructed in Nazareth by Chinese expert in 1997. Within the framework of the National Agricultural Extension Program; the Amhara Regional State has started implementation of rainwater harvesting technologies to address household food security issue.

Since then, huge numbers of structures were constructed by governmental, non-governmental organizations (NGO) and the community. The highest numbers of structures were constructed in South Wollo and the minimum in Awi zone, which is one of zones of Lake Tana Watershed. The reason for the least in the watershed could be the availability of other alternative irrigation water resources and good traditional experience in diverting rivers for irrigation development.

The poor construction quality and the high loss of water through seepage resulted in the low acceptance of the technology by the farmers in some areas of the region. Absence of pilot schemes for awareness creation and technology adoption before starting region wide implementation have an impact on effectiveness and efficiency of the technology. Involving the communities at all stages of the development, from the inception to planning, and implementation could enhance sense of ownership and improves sustainability (Wubneh 2007). The Participation ranges from sharing idea to contribution of locally available materials and labor, particularly in the moisture sufficient Woredas of the region.

The planning and design of the structures was done at regional level by Bureau of Agriculture and Rural Development. In most of the areas site selection and construction was done not as per the design and construction guideline. At site level, the dimensions of the structures were interpreted differently and this resulted in construction variation across the region.

28.6.2 History and Experience of Watershed Based Development

28.6.2.1 History of Watershed Based Development

In Ethiopia, Different extension systems have been exercised for 54 years. These include the land grant extension system provided by the Imperial Ethiopian College of Agriculture and Mechanical Arts (IECMA), the Comprehensive Package Programs (CPPS), the Minimum Package Projects (MPPs), the Peasant Agricultural Preprogram (PADEP). However, the focus of extension system has been on crop production and to some extent on livestock, no attention was given to natural resource conservation and development. Watershed based natural resource management in general and soil and water conservation in particular has been commenced in the 1980s though the management was at large scale and top down. This has attributed largely to the unmanageable size of the target areas and the lack of community participation and limited sense of responsibility.

The average size of watersheds ranged from 30 to 40 thousands hectares and their number eventually grew from 19 (1980) to 116 (1990) with support of World Food Program (WFP), European Economic Commission (EEC) and Australian Government. The watershed conservation was strictly soil erosion control and afforestation. It did not include the socioeconomic transformation noticed in watershed development (Berhe 1996).

The lessons learned from this experience encouraged MoA and support agencies like FAO to initiate pilot watershed planning approaches on a bottom-up basis, using smaller units and following community-based approaches. A number of participatory planning tools and methodologies have been developed and tried out in Ethiopia. Most of these approaches emanate from combination of planning tools inspired by various methodologies and adaptations at the local level. However, the common ones are Minimum Planning, Participatory Rural Appraisal (PRA), Participatory Land Use Planning, Participatory Watershed Planning, and Local Level Participatory Planning Approach (LLPPA).

Recently, from several of these approaches Community-based Participatory Watershed Development guideline has been extracted to provide guidance to DAs and Woreda experts on how to engage and consult with communities to prepare a workable, socially acceptable, and technically sound community-based watershed plan.

Minimum planning at the initial stage involved shifting from larger watersheds to smaller sub-watersheds. The LLPPA (Local Level Participatory Planning Approach) developed within Managing Environmental Rehabilitation in Transition to Sustainable Livelihoods (MERET) project has gained national acceptance and ownership. It was developed for Development Agents, as a practical approach focusing mostly on integrated Natural Resource Management (NRM) interventions, productivity intensification measures, and small-scale community infrastructure such as water ponds and feeder roads.

LLPPA is Community based sub-watershed approaches originated from Watershed planning and the minimum planning. Historically, the project was conceived in 1992 and first training for trainers was conducted in 1993. First round of LLPPA plans were prepared for four regions, second round of training for trainers was conducted in 1994–1995. The Woreda level planning, which started in 1994, consisted of socio economic survey and planning, biophysical indicators (land use land cover) and Development plan prepared for 3 years.

Potential/Prospective of LLPPA

- Replaced the top down planning and initiated the bottom up planning
- Has built trust among beneficiaries
- Selection of appropriate measures by participation made possible
- Made focus on local priorities
- Raised awareness of beneficiaries
- Increased sense of ownership of land users.

Strengths and Constraints of LLPPA
Strengths

- Improved/encouraged participation
- Socio economic survey
- Farmers involved in mapping and area transects
- Merging of biophysical and socio economic parameters
- Gender equity

- Decentralization completed
- Participation encouraged
- Staff increased
- Land tenure issue addressed
- Land degradation, food security and drought issues addressed.

Constraints

- Lack of trained staff, confusion on approaches and MoA restructuring are some of identified constraints of the approach.

28.6.2.2 Experience in Watershed Management

Watershed management efforts in Amhara region have so far focused on the food insecure and degraded areas. The food secure areas, which are eroding at increasing rate, had given little attention. Existing efforts also focus on soil water conservation and water harvesting with heavy emphasis on physical measures, rather than on broader program of watershed development.

Watershed development has been problematic when applied without community participation and using only hydrological planning units. This resulted in various failures or serious shortcomings. For instance, large Borkena dam in South Wollo, all irrigation dams constructed by Co-SAERAR and Angered dam in North Gondar where constructed before sufficient conservation measures were in place. Similarly, about five large dams are going to be constructed in Lake Tana Watershed however none of these will offer sustainable benefit in the current situation of upstream degradation. Watershed management therefore, is important to the long–term effectiveness of the sustainable utilization of the dams and Lake Tana Reservoirs

Thus, the national government has designed and launched a Community Based Participatory Water shade Development (CBPWD) approach to lead the process of rural transformation, the generation of multiple and mutually reinforcing assets. Accordingly, Bureau of Agriculture and Rural Development is being heavily involved in watershed management and declared that all resources development projects should be watershed based. Four demonstration micro-catchments are being implemented directly through the bureau of which one is found in the Lake Tana Watershed, in Fogera Woreda. About 894 and 340 sub watersheds with recommended area of 200–500 ha are currently under study and implementation in entire areas draining into Lake Tana, respectively.

In the past decades, participatory watershed development approach has been taken on board by different organizations including NGOs and bilateral organizations. For instance, the MERET project started in 2002 has developed LLPPA and treated vast areas though it focused on food insecure areas and limited to soil and water conservation measures. Moreover, Organization for Rehabilitation and Development in Amhara (ORDA), German Technical Cooperation (GTZ), and The Amhara micro enterprise development, agricultural research, extension and watershed management (AMAREW) are some of organization working in Lake Tana watershed in participatory watershed development approach. GTZ has adopted Participatory Land Use-Planning (PLUP) approach in implementation of soil and water conservation and farming practices. South Gondar is one of zones were the organization has succeeded in mainstreaming the participatory element into the land use-planning and natural resources management approach.

The Amhara micro enterprise development, agricultural research, extension and watershed management (AMAREW) project, since its beginning in July 2002, has been conducting multi-faced and integrated rural development activities in targeted Woredas of selected pilot watersheds of the region in agricultural research, extension,watershed management and micro enterprise development. The organization has commenced operation in Sekota and Gubalafto since 2003. Besides, the pilot in Sekela Woreda, which is part of Lake Tana watershed, has been initiated since August 2005. It gained experience in participatory watershed management on these pilot watersheds.

A lot have been learnt from the report and work on ground of the aforementioned organization. These include participatory planning and implementation, demonstrating at small scale and scaling up of activities. These all reach experiences of organizations working in the watershed will be very helpful during implementation of the intended project.

28.6.2.3 Watershed Management Best Practices

Implementation of Sustainable Land Management Practices

The Tana Basin is under high pressure of human and livestock population, the forest cover is being depleted and some of the available pastures are over grazed through transhumance and improper use. Consequently, it is evident from the analysis of data that the heavy erosion of top fertile soils in the Basin is the major constraint in agriculture development.

The objective of the land management practices is to restore the ecology in the course of undertaking agricultural activities to ensure sustainable production. The land management shall be prepared and planned to be implemented through participatory approach and therefore, the report has been prepared keeping the aspiration of the farmers into consideration and based on the experience of the ongoing watershed project. Hence, the following mitigation measures are proposed referring the states of the arts in sustainable land management (SLM) practices.

28.6.2.4 Biological and Physical SWC Measures

Although no wide spread severe threat of soil erosion is revealed in the Basin; about 75% of the area is from moderate to severe soil erosion status. Therefore, the following interventions, according to climate variations, have been proposed in the Basin (MoA 2001, 2005).

(a) **SWC measures on moist-tepid (moist weyna-dega) and moist-cool (moist dega) climatic zones of the basin**

These climatic zones are of high agricultural activity with barley, wheat and pulses as main crop. Degradation wide spread, gully frequent. Old history of land use with high erosion damages. Soil is variable as the soil survey indicated; on slopes brown colored loams and many places are without soil. Grassland is overgrazed, forests are almost none existing, and most dung is used as fuel for cooking.

Measures on Cultivated land
Looking at the soil of the area; for sticky clay, generally apply **graded structures** (**bunds**) and look for **water ways** to be constructed. For soils which are sandy to silty having good infiltration, apply **level bunds** and no need of **water ways** but construct **cut off drains** to remove excess run off.

- For slopes less than 5% gradients, **grass strips** or **alley cropping** for soils with good infiltration (sandy to silty texture) combined with cut off drains shall be constructed. It is better to stop grazing until it establishes well and use **cut and carry** afterwards. Otherwise, **level bund** or **level fanya juu** shall be constructed. On clay soil, put graded structures but make water way in the first year. In the second year, apply graded bund. Careful construction and permanent maintenance is necessary, especially during heavy storms.
- For slopes 5–15%, use **graded terraces** on clay soils and soils with surface crusting. **Level terraces** can be used on light soils (sand, silt) having better infiltration capacity.
- For slopes between 15 and 50%, on soils with good infiltration, apply, **level bund** combined with **cut off drain** in between. For clay soils, make **water ways** in the first year and **graded bund** in the second year, cattle are excluded all the year. All structures need development into **bench terrace**. In case of high intensity rains apply graded bund.

In all structures, apply **Revegetation**. In gullies, make **check dam**. On degraded cultivated lands with shallow soil, apply area closure. Land above 50% slope must be changed into forest land (see impact analysis chapter). Agronomic practices useful in this agro-climatic zone include crop rotation, grass strips, strip cropping, ley cropping, green Manuring. Agro-forestry practices helpful in soil conservation include hedgerows, alley cropping and vegetation strips along graded contours.

Measures on Grassland: Normally, **controlled grazing** is sufficient. Apply **grassland improvement** where necessary. On degraded grassland with shallow soil, **area closure** is needed. There, **cut and carry** can be used. Above gullies, make **cut off drains** and in gullies **revegetation**.

Measures on Forestland: Normally, **area closure** is sufficient for tree planting. Cut and carry can be used for grass management. On steeper slopes, make **micro basin**, and on very steep slopes, **hillside terrace**. Below and within degraded forestland, make **cut off drain** to protect cultivated lands.

(b) **SWC Measures on Moist-warm (Moist kola) Climatic Zones of the Basin**

This zone is suitable for sorghum, Teff, cotton and Finger Millet. The soil is deeply weathered red clay loam to red clay and on flat areas, black heavy clay. Insects and pests on agricultural crops are frequent. Bush burning to control the long grass can be very dangerous in causing degradation in this zone. Gully is a problem due to over grazing.

Conservation measures resemble to the previous section except **graded fanya juu** can be applied on clay soils after the first year's only water way. And, on forestlands of steeper slopes of drier parts, **trench** combined with **hillside terrace** can be constructed in order to protect erosion during heavy storm and harvest run off water in the end.

(c) **SWC Measures on Sub moist-hot (dry kola) Climatic Zones of the Basin**

In this zone, Teff is grown during the short rainy season, as well as sorghum and sesame in years of sufficient rainfall. Drought is a recurring problem. Over grazing, wind erosion and the destruction of trees and bushes for charcoal making are serious. Soil erosion varies according to rainfall and also serious on grassland.

Conservation measures specially recommended in this zone are inter-bund management measures such as **ridge and furrow, mulching** etc.

(d) **Generally Recommended Bio-physical Measures**

The following measures with rough estimation of quantities are recommended:

(i) **Contour Cultivation**: The contour cultivation facilitate the retention of rainfall in situ in the small furrows across the slope. The contour cultivation is proposed to be popularized among the farmers through the Development Agents and extension workers. The cultivators shall be motivated to adopt this practice.

(ii) **Cut off Drain/Diversion Drain**: This diversion drain shall be laid across the slope to protect the agricultural land from erosion taking place due to high velocity of runoff from the non-arable sloping land. In due course, it is expected that some vegetative cover shall be developed which may be used for production purpose. This activity is very useful in the high rainfall area.

(iii) **Soil Bund**: This is one of the most important activities which will retain water all along the length of the soil bund. The rainfall of the project area is about 1100 mm; as such a uniform gradient is required in the soil bund.

(iv) **Stone Bund**: It is proposed to construct stone bund in cultivated area where stones are available in plenty. The removal of stones from the field shall improve the rooting condition for the crop. It is proposed to construct the stone bunds in the sloping terrain which in long run will retain the soil in the upstream side acting as barrier to retard the velocity of runoff. It is proposed to construct these bunds on the field boundary as such 100 m length needs to be covered per ha.

(v) **Fanya Juu**: The Fanya Juu is same as the soil bund but only the borrow pit area is at the downstream side of the bund. The extra efforts are required to uplift the soil and bund slowly acts a barrier to collect the displaced soil from the field and field become slowly levelled during the period as per the rate of aggradations the bund height is increased. Areas with steep sloping land above 10% is suggested to be covered under the activity. At a spacing of 50 m the total length of Fanya Juu shall be 200 m per ha.

28.6.2.5 Production Measures

It will consist of following measures;

Afforestation and Agro-Forestry
The afforestation can be popularized and implemented as community forestry, homestead gardens and agro forestry. The afforestation is shall be planned to in the highland areas of the Basin. Some of the problems which could encounter are shortage of manpower during the planting season as it coincides with cropping season. Free grazing of the cattle in watershed areas also causes high mortality of seedlings. Low consciousness of the society towards environment is also a factor for the weak response.

Presently looking to the scanty vegetation in the highland watersheds, the proposed strategy is to plant the seedlings at all such places where seedlings can be protected. As such the following places are proposed under afforestation;

- Communal lands with area enclosure and enrichment for increasing the plant population.
- Near gullies and streams
- Agro forestry on farm boundaries at soil bunds
- Along Roads
- Backyards or homestead Garden
- Different Institutions Such as religious places, Govt. institutions and public places.

It is proposed to distribute fodder and or multipurpose utility tree and seedlings at the rate of 20 per ha to the farmers. The various species recommended for the purpose of plantation are given in the Guidelines and Manual for Watershed Management (MoA 2001).

a. **Organic Farming/Compost Pits/Farm Yard Manure**

This will include the following items:

1. Compost pit—This pit is to be filled with 15 cm thick layer of waste material of crops followed with 5 cm thick layer of cow dung. Proper care is to be taken to maintain the moisture content and keeping the pit air tight after the completion of filling with mud.
2. Green Manuring
3. Integrated nutrient management system

Land Use Improvement Requirements

In improving the land use system in the Basin, it is critical to see the hard and soft issues complementarily; Policy and strategic issues and thereby implementation of study based sustainable land management practices are critical. The following alternatives are proposed (EPLAUA 2007)

- Strengthening capacity for integrated land use planning and implementation, at the regional and local levels and across sectors
- Development and Strengthening implementation of policies, regulations, and incentive structures,
- Strengthening and designing information management systems to support decision-making at all levels on integrated land use planning and management.
- Formulation of bylaws to govern wetland resource use as well as avert degradation of wetlands as a result of mismanagement of streams and rivers feeding the wetland.
- Creating awareness amongst communities on the useful functions of wetlands and the sustainable utilization of land and water resources
- Promoting integrated land-use planning with environmental conservation activities at watershed, micro watershed level
- Supporting and developing both extension staff and community capacity to develop/introduce sustainable land use systems.
- Improving the availability (quality and quantity) of potable water and irrigation water through rainwater harvesting, diversion and spring development
- Communication of Water Users along river resources at various spots is limited. Hence, the water users' complaint from different users can be resolved by involving representatives from different sites and setting equitable water use rights.

These alternatives can be accomplished in different ways; using state of the art extension systems and institutions. The detail could be seen as mentioned in the following sub-sections.

Conservation of Biodiversity and Climate Change

In association with land use/cover change, the consequent impacts to the environment at higher order impacts are biodiversity loss and aggravation of climate change. This Basin is known for its rich vegetation and wildlife. Although it is difficult to leave areas at its pristine state to bring about development and feed the human population, there is a need for sustainable use.

It is recommended that dense forest areas delineated in the land use land cover study shall be maintained. Conversion of the woodland into agricultural landscape should not be done without leaving some trees within cultivated lands which will be important to maintain the biodiversity from complete collapse or genetic erosion; especially, wildlife of larger mammals. In addition, the following conservation measures are recommended (IBC 2005):

In-situ Conservation

The Convention on Biological Diversity recognizes in situ conservation as the primary approach to biodiversity conservation (Article 8). Of particular importance is the balance to be struck between conservation measures within Protected Areas (Pas) and measures for sustainable use of natural areas outside of Pas in the wider countryside. It is generally recognized that activities, which occur in areas adjacent to Pas, may be critical to the viability of the Pas themselves. Adjacent communities ultimately control the PA to the extent that if the PA negatively affects the local population, then this area may be destined to fail. However, if local people are involved in the management of Pas and other forms of development compatible with the goals of the protected area are promoted in adjacent areas, then the protected area's long-term viability is likely to be enhanced.

Local community knowledge of natural ecosystems and wild taxa is widespread and significant, but the existing systems of research and information gathering have not adequately documented, exploited or supported such knowledge. There is an urgent need to facilitate the continuation, systematization, and, where appropriate, recording such knowledge and information. In addition communities do not have access to relevant information and data from outside resources, and it is important to find ways and means to access such information.

In Ethiopia, institutions involved in in situ conservation of biodiversity include the Institute of Biodiversity Conservation (IBC), the Ethiopian Agricultural Research Organization (EARO), Regional Agricultural departments, Higher Learning Institutions, etc. However, the impact of their work to conserve biological resources on the ground is very limited. Therefore, public participation in biodiversity conservation is critical and NRS Bureau of Agriculture in collaboration with ARARI shall take this responsibility.

Ex-situ Conservation

The Convention on Biological Diversity specifically recommends that ex situ measures be adopted as necessary in situations where in situ conservation programs do not prove to be adequate. These measures have most extensively been applied to conserve cultivated and domesticated agro-biodiversity, employing techniques such as seed banks, field gene banks, in vitro storage, and captive breeding measures. Other groups in need of ex situ conservation measures include: threatened species, wild relatives of cultivated plants and domesticated animals; medicinal plants; plant crops of local and regional importance; pasture and forage species; ornamental plant species; tree species; and micro-organisms. Ex situ conservation is complementary to the rehabilitation and restoration of degraded ecosystems, and the recovery of threatened species.

Ex situ conservation facilities provide excellent opportunities for researchers to study plants, animals, and micro-organisms in controlled conditions, and to improve collection, storage and regeneration techniques. Ex situ facilities can also be used for germplasm evaluation, as centers for documentation and information systems and for providing information on genetic resources on a commercial basis. Captive breeding of wild animals can be used to restore endangered species populations. It is important to increase populations as quickly as possible and reintroduce the animals back to their original habitat to minimize genetic erosion. Plants can also be re-introduced to their natural areas of occurrence. Such re-introductions should, however, be carried out in such a way that other indigenous species are not harmed or adversely affected. Here, some biodiversity conservation projects are being run by ORDA, in Bahir Dar. Therefore, this effort shall be strengthened by the regional government. The establishment of a regional Microbial Culture Collection would be essential for the preservation and use of the rich microbial diversity present in this Basin. In Ethiopia, institutions involved in ex situ conservation of biodiversity are very limited and include primarily the Institute of Biodiversity Conservation (IBC) and the Ethiopian Agricultural Research Organization (EARO). The IBC is maintaining crop genetic diversity in collaboration with farmers.

Similarly, care must be taken while collecting material/animals for ex situ conservation not to endanger other native species and genetic resources. The regulation and management of such transactions requires accurate information to determine the impact of collection on populations and ecosystems.

Management of Water Pollution

The previous section has its own positive effect on preventing water pollution. Beyond that it is critical to understand the pathways of water pollutants in a watershed. Sources could be agro-chemicals from cultivated lands or eroded soil from watersheds. The establishment of water and waste treatment plants in all pollution vulnerable areas of the Basin shall be well planned and implemented. In addition, most importantly, the application of watershed management through soil and water conservation and agro-forestry need to be planned and implemented in the study Basin in order to control water pollution.

Soil and water conservation works include:

- Physical measures: bunds, terracing, check dam, cut off drain, and water ways
- Biological measures:
  - Appropriate soil management (mulching, compost application, contour ploughing etc.)
  - Agro-forestry or Improve vegetation cover of the Basin (alley cropping, hedge rows, scattered tree etc.)
  - Agronomic practices (fallowing, relay cropping, inter cropping, ley cropping, crop rotation etc.)

Sustaining of Riverine Ecosystem and Fisheries

Release of Minimum Flow

It is essential for the development authorities to maintain the minimum flow for the sustenance of aquatic ecology and survival and propagation of invertebrates and fish. In order to avoid the possible loss of aquatic life, minimum flow will always need to be released from water abstractions.

Sustenance of Fisheries

The fish species reported in the Basin Rivers includes the *Barbus* spp., *Tilapia* spp. and Cat fish. However, the rivers do not support good fisheries and therefore commercial fishing is not in vogue in the Basin. Since the study envisages the creation of irrigation projects; a reservoir, and naturally the rivers have pondages during the dry season, it is proposed that the newly created reservoir could be used for the development of fisheries. This could boost the economy of the Basin. However, people need to be persuaded to practice aquaculture in the reservoir. The fish produced in the reservoir could be transported and marketed at various consumption centers, mainly Bahir Dar and/or Sudan.

The developments of reservoirs for subsistence and commercial fisheries have to be carefully planned. Since reservoir fisheries is principally based on Stock and Take system, introduction of stocking has to be based on carrying capacity of a water body. Considering the subsistence fishery in the proposed reservoir and river stretches upstream of the dam site, it is a justifiable and economical method to improve the availability of animal protein for local population. However, majority of the population inhabiting in the project Basin do not consume fish as it is not their food habit and fish contributes only as an occasional food item in the dietary requirements. But the stocking program in the proposed reservoir will help in increasing the potential impact, in terms of improving the existing catches leading to commercial fishing. Another aspect to be taken into consideration is the economic viability of stock-and-take program. Cost estimates vary according to the species recommended for stocking. For certain species, stocking may be relatively inexpensive, while others may require large capital inputs and thus would be more expensive. The cost of stocking also depends on the extent that the existing facility can be used to produce stocking material and possibility of joint utilization of any existing or proposed facilities between stocking and aquaculture practices in the region.

The anticipated time to establish fish stocks would depend on the species selected for introduction. In a stocking program, a rapid increase in abundance is expected in those portions of the reservoir, where stocking is introduced, as compared to other areas of the reservoir. Thus, it is expected that, in area where fish is stocked, significant stocks might develop in 2–3 years. The length of time required to achieve quicker results could be reduced considerably by multiple stocking of the same species over a wide range. Based on the abundance of fish species in the rivers, such as Barbus spp., and cat fish like Claris species, it seems to be the most appropriate for development of reservoir fisheries for commercial purposes.

The stocking program comprises of the following:

- Acclimatization stocking (a new fish species is introduced in a water course);
- Supplementary stocking (a species already living in a water body);
- Transfer stocking (transportation of mature fish from one water body to another);
- Repetitive stocking (species which do not propagate in natural conditions).

To carry out the stocking Programme on annual basis, suitable aquaculture facilities have to be created in the vicinity of the proposed project area to meet the requirements of fingerlings. Since the stock in the Rivers is very limited, the proposed aquaculture facilities have to meet the requirements of fingerlings of the reservoir.

In the absence of the separate Fisheries Department, the Agriculture and Rural Development Office is responsible for the management and development of fishery resources. So far, the Agriculture and Rural Development Office has not taken any reservoir fisheries development project in the Basin. Therefore, the technical expertise required for the development of hatchery and management of reservoir needs to be developed. Alternatively, the works for the development of a hatchery may be undertaken with the help of the Lake Tana Fisheries Research Centre and the management of the reservoir may be done by the office of Agriculture and Rural Development Office, with technical support from the Lake Tana Fisheries Research Centre.

Management of Water-logging and Soil Salinization

Water-logging problems are expected in various parts of the Basin, specially wide area is being flooded in Bahir Dar Zuria, Dembia and Fogera Woreda. Various measures are effective in preventing water-logging conditions, such as:

- Provision of surface or sub-surface drainage systems. The surface drainage system will be useful in the Basin with clayey soils having higher surface runoff and lower infiltration rate. On the other hand, sub-surface drainage system is more suitable for sandy or coarse textured soils with higher infiltration rate and lower surface runoff. Since the soil is silty clay loam to silty clay in texture which becomes heavier with depth, a surface drainage system is more suitable;
- Lining of canals to control seepage losses is an important control measure. Otherwise uncontrolled seepage from unlined canals, particularly in filling reaches can spoil a vast stretch of land on either side of the canal owing to the heavy seepage flow from them;
- Conjunctive use of surface and groundwater for irrigation in Basin of excessive recharge of groundwater can be very useful in prevention of water-logging.

The balance water available can be used to bring additional area under cultivation. Alternatively, water requirements of the Basin under silvo-pasture, community forestry, pasture, gully plantation, etc. can be met from this source. Thus, a comprehensive plan needs to be developed to utilize both surface as well as ground water resources in the Basin.

The requirements of a drainage system in the Basin need to be studied in greater detail through a separate study. Based on the findings of the study, provision of suitable measures to control water-logging need to be put.

Drainage

Drainage is necessary for successful rainfed or irrigated agriculture. The primary objective of drainage is to effectively increase the productivity of cultivated land, while irrigation is to supply water for crop growth. Thus, at the time of heavy rainfall, flooding, and irrigation, it is customary to plan for removal of excess water.

Storm Water Runoff

Storm water runoff is the major component of designed flow of the drains. Surface drains are normally designed to handle flows of 5-15 year return period. It has been found that the Soil Cover Complex Method of United States Soil Conservation Services (USSCS) gives best result for small watersheds. The details of USSCS method are given as below:

$$Q = \frac{(P - I_a)^2}{(P - I_a) + S}$$

Q depth of runoff (mm)
P depth of precipitation (mm)
I_a initial abstraction for the period between the start of rainfall and the runoff in equivalent depth over the catchment (mm)
S potential maximum retention of water in equivalent depth over the catchment (mm).

Period of Disposal

Prolonged submergence of fields is likely to saturate the root zone of the crops and cutoff oxygen supply. The period of disposal of rainfall depends upon type of crops and stage of their growth. Thus it is necessary to dispose excess water within allowable time.

Field Drains

Field drains need to be provided to drain excess runoff from rainfall in fields to avoid damage to crops due to stagnation of water. The size of the out let command determines the size of field drains. Field drains are constructed along the lowest part of the outlet command and lead to collector drains. For an area between 5 and 8 ha, the actual design requirements would be too small to warrant any design. A standard section 0.3 m wide at base width, 1:1 side slopes may be constructed for a depth of 0.6 m. These are constructed and maintained by the farmers. Accordingly, field drains have to be constructed to accommodate 5–12 l/s per sub outlet. Command collector drains are to be designed for 50–100 l/s. These drains are designed for non-silting, non-scouring velocities determined by the Manning's or Lacey's formulae.

Preventing Public Health Problems

The increase of water availability in the Basin provides suitable habitat for the growth of vectors of various diseases. Malaria has been found to be the major vector-borne diseases in the Basin. The main breeding season of the malaria vector is the rainy season. The preferred habitat is stagnant or slow moving water open to the sunshine or moderate shade. Malaria can be controlled by mosquito control and mosquito proofing measures. Mosquito control measures aim at destroying the habitat and interrupting the life cycle by mechanical or biological or chemical means. Special attention shall be given to the water resources projects which consist of various components and each requires a set of specific management measures. The suggested measures in irrigation schemes are explained in the following paragraphs:

Reservoirs (Lagoons or Ponds or Dams)

- Provision of marginal drainage for the zone between the maximum and minimum water levels of the reservoir;
- The marginal zone should be filled above the maximum water level and deepened in other Basin to a depth below the lower limits of marginal growth invasion;

Canal and Surface Drains

Mosquito breeding is generally associated with poor canal conditions. Some of the major measures to reduce the risk can be incorporated in the planning and design stages, and the other measures may be implemented during the operation and maintenance phase. These could be:

- Lining of canals to reduce seepage of water. This controls the creation of habitats suitable for the breeding of mosquitoes;
- Good alignment of canals and drains, effective canal and drains maintenance to ensure that they are in good shape and free from vegetation and silting at all times;
- Rotational policy for distribution of canal water to ensure weekly flushing of canals;
- Self-cleansing velocity of 0.6 m/s must preferably be maintained;
- Canal construction borrow pits, trenches, etc. must be drained if possible;
- Lining of Night storage reservoirs.

Settlement Sites

- The settlement sites should be located beyond the flight range of the vector mosquito (say 1 or 2 km) and should preferably on the upstream side of the reservoirs or lagoons or ponds so that with the air currents mosquito are carried away from the residential Basin;
- Site selected for habitation of workers should not be in the path of natural drainage.

Other Measures

- Adequate drainage system to dispose storm water drainage from the settlement Basin should be provided;

- Adequate vaccination and immunization facilities and prophylactic medication should be provided for the public at the site;
- Distribution of mosquito net,
- The water collections which are for no specific use to the community must be eliminated.

Introduction of carnivorous fish species in the water bodies can effectively combat malaria. In addition, there should be greater emphasis on providing adequate sanitary facilities.

Vector Control
Various aspects covered under vector control are:

- Reduce exposure to mosquitoes;
- Control of schistosomiasis;

a. **Reduce exposure to mosquitoes**

Encourage prevention of mosquito-borne disease by helping people in reducing their exposure to mosquitoes. Work with the malaria control programme in wereda to:

- Popularize the use of mosquito net;
- Conduct community education on the proper use of bed nets and how to avoid dawn-to dusk mosquito bites;
- Regular spray of insecticides;
- Implementation of various management measures for vector control (drainage, filling of breeding sites) as outlined in the earlier section.

b. **Control of Schistosomiasis**

Engineering interventions include the design of elements of the irrigation system to discourage proliferation of the host snails and, where relevant, to exclude human contamination of irrigation water. The host snails prefer slow moving or stagnant water (which encourages weed growth) and cannot survive long periods of desiccation.

Management of Agro-Chemicals: Chemical Fertilizer and Pesticides
A great deal of attention has been focused on agro-chemicals especially on pesticides as these are supposed to be carcinogenic, mutagenic, teratogenic and allergenic. The following measures are recommended to prevent pollution and associated adverse impacts due to over-use of agro-chemicals:

- Soil should be tested to analyze the nutrient status before each cropping season and fertilizer dose shall be fixed considering the concentration of soil nutrients and the type of crops to be grown;
- Organic manures should be used in place of chemical fertilizers. Agronomic practices such as crop rotation and green cover mulching, which reduce dependency on chemical fertilizers, should be incorporated in the cropping pattern itself;

- Soil conservation measures suggested as a part of watershed management are effective in controlling erosion by surface runoff. Agro-chemicals joining the eroded soil material is prevented from reaching the surface waters through soil conservation;
- Weeds should be controlled as far as possible using cultural practice, Chemicals that are only allowed to be used will be used;
- Obsolete chemicals shall be controlled and shall be reported to the concerned office.

The controlling of weeds in the growing crops with herbicides increases their yields and ensures the efficient use of irrigation, fertilizers and plant protection measures. The spraying of following herbicides and insecticides, which are in use in Ethiopia, are recommended by Ethiopian pesticide guideline: Malathion, Diazinone; Zinc phosphate; Acetylic 2%; and 2–4D Amine.

Promoting Sustainable Livelihood Activities

As mentioned in the impact analysis, there are people to change their farming system on the existing land and there are also people to be disengaged from their land being they were practicing cultivation on lands with slopes greater than 60%. Therefore, the implementing agency, the government, shall consider maintaining the lives of these people.

Changes in farming system, from cultivating field crops to agro-forestry or fruit trees cropping or conservation agriculture so to say, should be implemented in a sustainable manner. Prior to implementation of these activities it is proposed that awareness creation, training and support in inputs in terms of credit shall be done curiously. The following agro-forestry practices are proposed (MoA 2001):

- **Scattered tree**: trees are usually permanent and full sized and they may be dispersed either singly or in clumps. In some cases farmers plant or maintain trees in their cropland primarily to obtain valuable tree products. In other cases, the trees seem to increase the production of the surrounding crops and improve the soil and water condition for crop growth
- **Contour vegetation strips**: with multi-purpose trees and tree crops provide another agro-forestry practice that fits well into current farming system. Contour vegetation strips are usually introduced in order to prevent soil erosion on sloping croplands, while at the same time providing useful products such as food, fodder or wood.
- **Multipurpose trees, grass and other herbaceous plants**: are combined along the edges and uncultivated spaces of soil and water conservation structures. These plant combinations can produce useful items for home use or sale, while helping to stabilize and protect conservation structures.
- **Alley cropping or hedgerow intercropping**: is an agro-forestry practice used on crop land. While there are many variations, alley cropping most often consists of dense hedges of multipurpose trees planted in rows between wider strips of annual crops. The hedges are lopped to produce mulch, which is applied to the cropped areas to fertilize and cover the soil.

- The last but not least is **multi-story tree crop system** which consists of closely spaced trees intercropped with annual plants. In contrast to dispersed trees in cropland, this arrangement is often based on shade tolerant under story crops and on a greater diversity of tree and hedgerow species.

Besides, people who lose their land shall be provided with **urban land and credit** to participate in different newly established economic activities which changes their livelihood significantly where there is a need to work on livelihood adaptation. Studies revealed that rural people construct their livelihoods via three main strategies: agricultural intensification; livelihood diversification; and migration (Karim and John 1998). Hence, the following **non-farm** activities are proposed to diversify sustainable livelihoods (Karim and John 1998 and Gordon and Craig 2001):

- Rural-Non-Farm-Sector (RNFS) includes household and non-household manufacturing, handicrafts, processing, repairs, construction, mining and quarrying, transport, trade, communication, community and personal services, Carpentry, fish marketing and laboring etc.
- Whilst the women predominantly can be engaged in laboring, the production and sale of handicrafts and trading farm produce.

Training and Extension for Farmers

The change in development in the Basin needs to be managed by the government and the users of the land. The change for example from rainfed to irrigated cropping requires extension, training and demonstration programs for farmers. Considering these aspects, it is proposed that the Tana Basin authorities need to provide adequate training to farmers.

The training must include the following aspects of environmental protection:

- Prevention of spread of water-related diseases;
- Safe use of agro-chemicals; and
- Soil and water conservation programs.

Measures under prevention of spread of water related diseases must deal with:

- Hygiene and personal health care;
- Control of water spills, pudding etc.;
- Prevention and prophylactic measures for control of vectors;
- Disposal of human waste; and
- Disposal of drainage water.

Aspects to be covered under safe use of agro-chemicals include:

Control of Weeds in Channels

- Methods of cleaning and disposal of weeds;
- Detrimental environmental effects of agro-chemicals; and
- Information on biological weed control.

Pest Control

- Specific uses of pesticides and its optimization;
- Residual degradability;
- Effect on untargeted species;
- Safety procedures during application;
- Rate and frequency of application;
- Disposal of packing material and surpluses;
- Storage of chemicals;
- Information on cultural practices such as weeding, rotational cropping, cleaning of bunds, etc.

Fertilizer use

- Type, dosage, application techniques, timing and frequency of application and its relationship with type of soils;
- Disposal or storage of packing materials and surpluses.

Topics to be covered under soil and water conservation programs are:

- Establishment and protection of forest or trees;
- Control of tree felling for fuel wood and timber;
- Advice on establishment of village woodlot; and
- Soil conservation measures.
- Such training courses can be organized by Agriculture and Rural Development Offices of Woredas in consultation with Agriculture Development Bureau of Amhara Region and Tana Basin Organization.

28.7 Conclusion and Recommendation

28.7.1 Conclusion

Land use and land cover change and its consequent land degradation in the Lake Tana Watershed limits the potential to develop a sustainable livelihood for its inhabitants. Deforestation, sedimentation, and loss of fertility contribute to global warming, biodiversity loss, and restrict the availability of fresh water while altering the structure and integrity of local ecosystems. Those phenomena are exacerbated by inappropriate land use and damaging agriculture and grazing practices. The effectiveness of the regional and local efforts to improve the environment and livelihoods of the residents in the Lake Tana Watershed through natural resource conservation programs are limited due to capacity, and financial barriers.

It should be aimed to overcome barriers to integration of participatory land use system and SLM at large. Land use and water resource related barriers that were identified include weak policy implementation, low capacity (technical, and financial), poor information management system, low or no incentives, and frequent restructuring of core and principal institutions at national/regional levels. Alleviation of these constraints will be important for better delivery of technical support and enabling local users to implement good land use related practices, to influence policies; reduce dependency on aids and implement environmentally friendly activities.

28.7.2 *Recommendation*

Strategies to be adopted for removing the barriers to achieving Sustainable Land Management (SLM) in Lake Tana Watershed are (EPLAUA 2007):

1. **Integrated watershed management approach**: Sustainable land use can only be achieved through an integrated watershed management approach, which deals with the whole range of social, environmental, and agricultural problems. The natural resources conservation, land use intensification and the social aspects are major aspects to be included and addressed.
2. **Use of Participation as a Guiding Principle**:Participation of target groups/ local communities is the most important precondition for sustainable land use planning. Through participation, the involvement of the target groups will be high in the process of problem identification, problem analysis, constraints, opportunities and development priorities and interventions. Participatory approach increases community's awareness and is a key to promote new development strategies on sustainable basis.
3. **The integration of a gender perspective into participatory land use planning**: women are more affected by natural resources degradation. They walk long distances to fetch water, to collect firewood, and to collect feed for their animals. They are also the first victims of mal-nutrition and under nourishment because of food shortage. Therefore, natural resources degradation problems related with water and land will be solved with the participation of women. With this fact, the role of women in sustainable land use planning is not questionable.
4. **Small Area Based Approaches**: The general understanding should be that the SLM eventually, would cover whole areas of watershed. However, one should also understand that it is very difficult and challenging to introduce and practice new ideas to rural society. Thus, commencing in small area, such as micro-catchments, in some parts of a watershed would be of a great advantage. New land-use approaches and practices can be possible on private plots or communal areas, such as communal forest, communal grazing, etc. and then scaled up gradually.
5. **Supports to Livelihoods of local Population**: In order to alleviate the economic constraints of the rural population thereby problem of land degradation, different measures should be taken. To intensifying land-use and diversifying cropping systems based on certain key interventions, through linkage of

participatory land use planning and implementation initiatives are some of measures to alleviate the problems.

6. **Use of local resource centered approach**: As a strategy for sustainable land use/land management, the project cycle management has to be local resource centered approach. In doing so, the project will involve the concerned local government and community institutions in the management of projects.
7. **Provision of information and incentives on Sustainable Land Management**: To address the problems of natural resources degradation, community based natural resources conservation strategies must be launched. As experience elsewhere in the world shows, use of natural resources on sustainable basis can be achieved through good land husbandry and management practices. Incentives should be provided (in cash or in kind) to the farmers whose land management is better.

References

ADSWE (2013) Guideline for land use land cover and change detection study. Amhara Design and Supervision Works Enterprise, Bahir Dar, Ethiopia

Berhe (1996) Twenty years of soil conservation in Ethiopia; personal overview. Regional Soil Conservation Unit, Nairobi, Kenya

EPLAUA (2007) Community-based integrated natural resources management in Lake Tana Watershed; water resource, Watershed. Water Harvesting and Land Use, IFAD/EPLAUA GEF project, June 2007, Bahir Dar, Ethiopia

Gordon A, Craig C (2001) Rural non-farm activities and poverty alleviation in Sub-Saharan Africa. Policy series 14. Natural Resources Institute, Chatham, UK

Habtamu G (1999) Rainwater harvesting concepts and issues. Paper presented at founding conference of the Ethiopian rainwater harvesting association (ERHA). 17 December 1999. Addis Ababa, Ethiopia

IBC (2005) National biodiversity strategy and action plan. Institute of Biodiversity Conservation, Addis Ababa, Ethiopia

Karim H, John N (1998) Sustainable livelihoods and livelihood diversification. IDS Working Paper 69, Sussex university, UK, in collaboration with the International Institute for Environment and Development

MoA (2001) Guidelines and manual for community based participatory watershed development. Ministry of Agriculture, Addis Ababa, Ethiopia

MoA (2005) Soil and water conservation practices in Ethiopia. Ministry of Agriculture, Addis Ababa, Ethiopia

Wubneh B (2007) EIA implementation and follow-up: a case study of Koga irrigation and watershed management project. MSc thesis, UNESCO-IHE

Chapter 29
Establishment of the Lake Tana Biosphere Reserve within the UNESCO World Network of Biosphere Reserves

Ellen Kalmbach

Abstract In June 2015, the Lake Tana Biosphere Reserve was officially nominated as a new addition to the World Network of Biosphere Reserves under the UNESCO Man and the Biosphere program (UNESCO 1971). This official recognition of Lake Tana and its immediate surroundings as a UNESCO Biosphere Reserve was the culmination to date of a process which had started many years previously. This chapter describes why a Lake Tana Biosphere Reserve was proposed, the UNESCO Biosphere Reserve approach, the geographic description of the Lake Tana Biosphere Reserve, and why a biosphere reserve is a good solution for the Lake Tana Region. It explains the process followed to prepare the Lake Tana region to become a UNESCO Biosphere Reserve, including stakeholder contributions to the application/nomination process, vision and goals of the Lake Tana Biosphere Reserve, and its zonation. Finally, it discusses what will be needed for the successful management of the Lake Tana Biosphere Reserve, including a description of the Reserve management structure and the challenges faced in fulfilling the BR mandate of integration and networking.

Keywords Ethiopian biosphere reserve · Multi-stakeholder governance structure · Participatory biosphere reserve development · Multi-scale management framework · Biosphere reserve zonation

---

E. Kalmbach (✉)
The Nature and Biodiversity Conservation Union (NABU), Project Office Bahir Dar, Bahir Dar, Ethiopia
e-mail: ellen.kalmbach@gmx.net

E. Kalmbach
Michael Succow Foundation, Ellernholzstr. 1/3, 17489 Greifswald, Germany

K. Stave et al. (eds.), *Social and Ecological System Dynamics*,
AESS Interdisciplinary Environmental Studies and Sciences Series,
DOI 10.1007/978-3-319-45755-0_29

29.1 Introduction

In June 2015, the Lake Tana Biosphere Reserve was officially nominated as a new addition to the World Network of Biosphere Reserves under the UNESCO Man and the Biosphere program (UNESCO 1971). This official recognition of Lake Tana and its immediate surroundings as a UNESCO Biosphere Reserve was the culmination to date of a process which had started many years previously. Concerned individuals, public institutions, government representatives, scientists and NGOs joined forces over their concern about the evident environmental decline and deterioration of the basis for local livelihoods and future prosperity of one of Ethiopia's most important cultural and natural landscapes—the Lake Tana region.[1]

The idea to make the Lake Tana region a biosphere reserve (BR) was already born in 2005, when a group of regional experts from various disciplines (environmental protection, land management, fisheries and tourism) concluded in a regional workshop that a BR would be the best tool to address the already then existing severe environmental threats (zur Heide 2012). In 2006, the Regional Environmental Protection, Land Administration and Use Authority (later BoEPLAU) delivered a study about "Lake Tana and its Environs: Conservation, Utilization, Development and Threats". In it, a zonation model, including seasonal closures around river mouths and wetlands, was sketched out. The Bahir Dar Abay (Blue Nile) River Millennium Park was established in 2008 by the ANRS government as a category IV IUCN protected area. It stretches from the outlet of the Blue Nile to the Tis Abay waterfalls, covering an area of 4729 ha. In 2009 the Bureau of Culture, Tourism and Parks Development of ANRS finalized a study on the Protected Area Potential Assessment in and around Lake Tana. In this study, the possibility of establishing a Biosphere Reserve was again considered a good way forward as a framework to integrate human development with conservation for the Lake Tana region.

Concurrent with these regional governmental initiatives, the first two Ethiopian biosphere reserves were being prepared for nomination. The successful nomination of Kafa and Yayu forest biosphere reserves in 2010 was followed by negotiations between the Ministry of Science and Technology, UNESCO and conservation partners. The common understanding that emanated from these negotiations, laid out in an MOU, was to establish the biosphere reserve approach in other suitable locations in Ethiopia to tackle some of the most pressing problems of ecological degradation and development. Lake Tana was subsequently identified as the most suitable and urgent site for a next Ethiopian biosphere reserve.

[1]"Lake Tana Region" is a term differently cited in literature. It has no fixed boundaries. Lake Tana Region is a part of Lake Tana watershed and includes the lake as well as areas with strong ecological, cultural and socio-economic linkages to the lake. This is the sense in which I use the term "Lake Tana Region" in this chapter. Data provided in studies sometimes refer to the watershed, sometimes to the region. Only little data specifically for the area of the Lake Tana Biosphere Reserve exist to date.

In 2012, the Nature and Biodiversity Conservation Union (NABU), a German-based NGO, started the implementation of the project "For People and Nature—Establishment of a UNESCO Biosphere Reserve at Lake Tana, Ethiopia" through its Ethiopian sister organization NABU Ethiopia. The project was funded by the Ministry for Technical Cooperation (BMZ), Germany, and NABU, and gets scientific support from the Michael Succow Foundation. The ANRS government, represented by five signatory Bureaus including BoCTPD and BoEPLAU, adopted the project and process as part of its strategy for a sustainable future of the region and facilitated the submission of the BR application to UNESCO in September 2014 through the federal Ministry of Science and Technology.

In this chapter, I describe the process that led to this nomination. I will trace how the different stakeholders came together, why a Biosphere Reserve in line with the UNESCO criteria is considered an appropriate management approach for the protection and development of Lake Tana and its environs, and what the requirements are for a sustainable management and development of the Lake Tana Biosphere Reserve.

29.2 Why Was a Lake Tana Biosphere Reserve Proposed?

29.2.1 The UNESCO Biosphere Reserves Approach

Biosphere reserves are sites to demonstrate innovative approaches to living and working in harmony with nature. One of the primary objectives of BRs is to achieve a sustainable balance between conserving biological diversity, promoting economic development, and maintaining associated cultural values. BRs are also seen as learning sites for regional scale sustainable development, and therefore science, learning and networking play a key role in biosphere reserves. Correspondingly, the three main functions to be fulfilled by BRs are conservation, development and logistic support function (UNESCO MAB Programme). In June 2015, 20 new sites, including the Lake Tana Biosphere Reserve, were recognized and added to this World Network, making it a total of 651 sites in 120 countries named Biosphere Reserves by UNESCO's Man and the Biosphere (MAB) Programme in 2015.

Contrary to still prevailing misconceptions, biosphere reserves are thus not exclusive nature "reserves", as the name may suggest. They are regional models for sustainable (human) development in which biodiversity and ecosystems conservation is a cross-cutting and cross-sectorial issue. It is based on the notion that for any sustainable social and economic development, ecosystem integrity and its capacity to provide ecosystem services are a prerequisite (zur Heide 2012).

In the light of the current debate on the Green Economy to harmonize economic activities with the ecosystem′s long-term capacity to provide resources, biosphere reserves can also be considered to be *Model Regions for a Green Economy* (GIZ 2011). The Green Economy concept also aims at harnessing the potential to generate economic opportunities from biosphere reserves in the primary, secondary and tertiary sector, through e.g. ecotourism and economizing various ecosystem services that derive from biodiversity conservation.

Provided that an area fulfils some baseline biological criteria which qualifies it for inclusion in the World Network of Biosphere Reserves, the cornerstones of biosphere reserve establishment and management are participation, zonation and designated management mechanisms. Detailed criteria for the designation of UNESCO Biosphere Reserves are laid out in the Statutory Framework of the World Network of Biosphere Reserves (UNESCO 1996). They include (1) *Ecological and cultural criteria*, such that the area must be of significance for biological diversity conservation and include a mosaic of ecological systems and human interventions. (2) *Appropriate size and zonation*, whereby legally protected core zones are devoted to long-term protection, buffer zones are surrounding or contiguous to core areas, and the remaining transition areas, where sustainable resource management methods are developed and promoted. And (3) *Participatory and specifically designated management mechanisms*, which secure the involvement and participation of a suitable range of stakeholders in the design and carrying out of the functions of a biosphere reserve. In addition, provisions should be made to manage human use and activities in the buffer zone, a management policy and plan exists as well as a designated management authority. Programs for research, monitoring, education and training are established.

29.2.2 Geographic Description of the Lake Tana Biosphere Reserve

Lake Tana is the largest lake of Ethiopia and belongs to the ten largest lakes in Africa. It lies in the Amhara National Regional State (ANRS) in the Northern Ethiopian highlands and is one of the highest situated lakes in Africa. Lake Tana is the source of the Blue Nile River which contributes around 85% to the river Nile. Lake Tana and its surroundings host unique ecosystems with a high biodiversity especially of bird and plant species and is a bird wintering ground of global significance. A number of plant, bird, and fish species are endemic to the region, rendering Lake Tana region part of the Eastern Afromontane Biodiversity Hotspot. Beside its natural history, the Lake Tana Region also harbours a centuries-old agricultural and cultural history.

The Lake Tana Biosphere Reserve (LTBR) incorporates the lake itself and its immediate environs, including the whole mosaic of landscapes and habitats, ranging from wetlands to forests, from near-pristine to highly degraded, including a major urban center and small rural settlements (Fig. 29.1). Its total area coverage is 6 972 km^2, of which 3 042 km^2 represent aquatic surface and 3 930 km^2 terrestrial. Administratively, it covers 137 Kebeles (communities), which belong to 10 Woredas (districts).

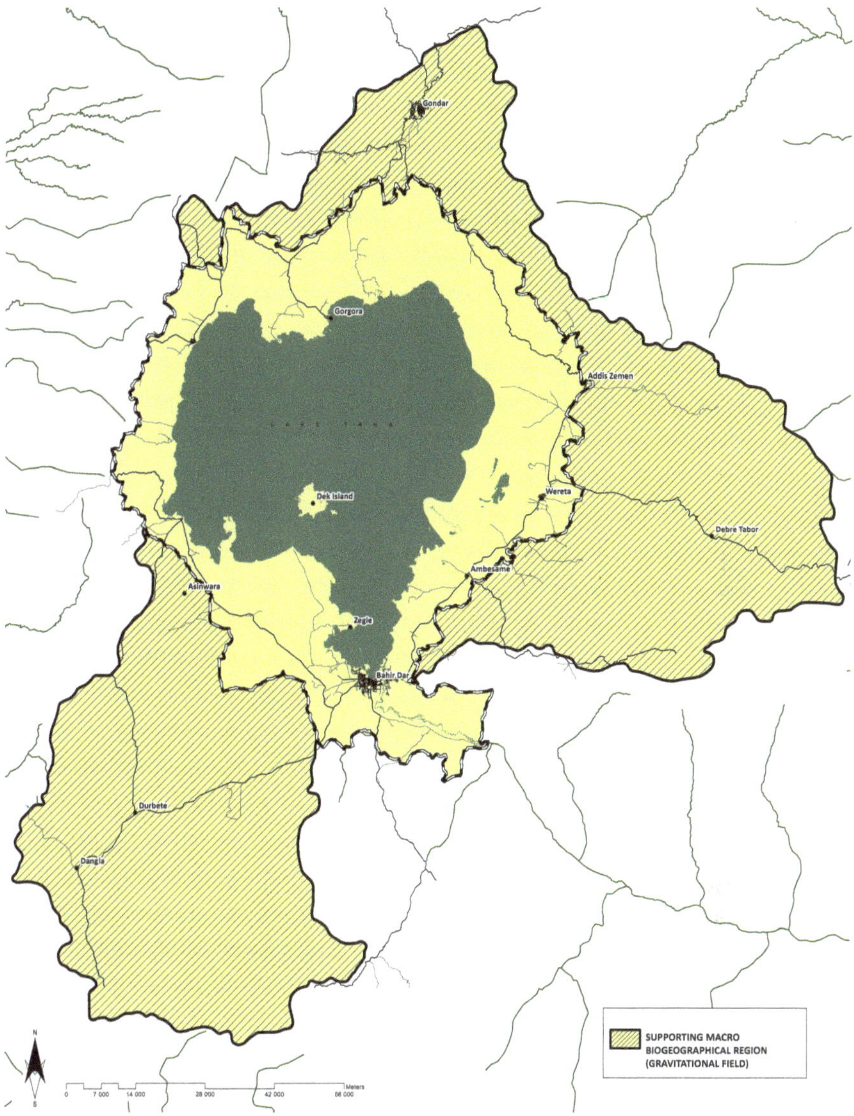

Fig. 29.1 Map showing the outline of Lake Tana Biosphere Reserve (*stippled line*) within the Lake Tana Watershed/Lake Tana Sub-Basin (adapted from FDRE application to UNESCO 2014)

Due to the heterogeneity of the area, the LTBR includes a great number of core zones and consequently buffer zones (Fig. 29.2). The 78 core zones of the LTBR mainly consist of aquatic and riparian habitats with little human disturbance, including remnants of papyrus reed beds, as well as patches of remaining natural forest. These are often situated around churches on islands and the mainland.

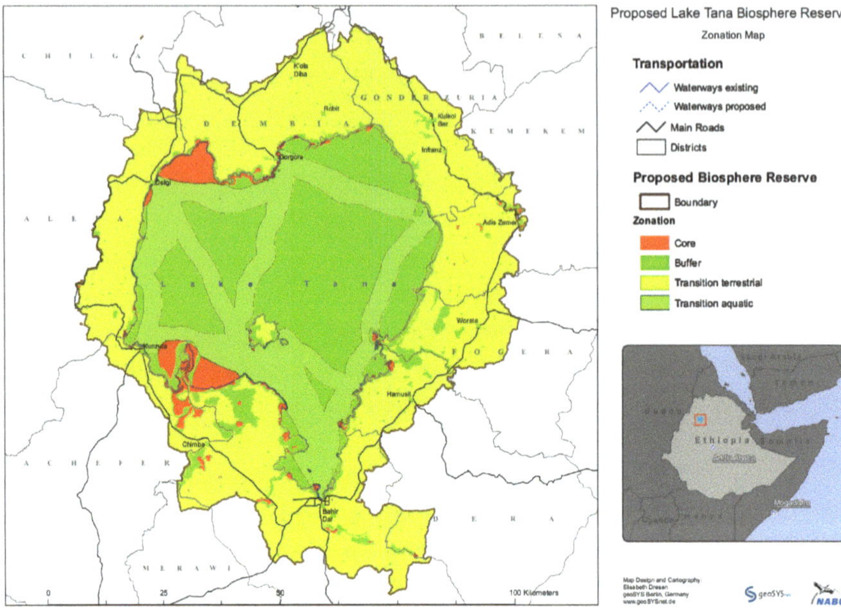

Fig. 29.2 Map showing the zonation of the Lake Tana Biosphere Reserve representing Core, Buffer and Transition zones as presented in the application to UNESCO; adapted from FDRE 2014

The buffer zones often surround core zones, but are also found independently in areas which play key roles for conservation, ecological restoration and functioning as ecological corridors, but where human activity is prevalent. These areas are focus points for community conservation schemes in tandem with restoration and regeneration activities in the management of the BR.

The transition zone of the LTBR comprises some parts of the water body of the lake, as well as extensive agricultural land and settlements, including Bahir Dar, the capital of the Amhara National Regional State. The Core Zone comprises 228 km^2 (3.3%), Buffer zone 1882 km^2 (27.0%), and the Transition zone 4681 km^2 (67.1%) (FDRE 2014).

29.2.3 Why Is a Biosphere Reserve a Good Solution for the Lake Tana Region?

Lake Tana Region is part of the Ethiopian Highlands and includes one of the highest situated lakes in Africa. In the context of the East-African Highlands, the proposed biosphere reserve will be the first of its kind in the biosphere reserve network. Until June 2015, there were three biosphere reserves in Ethiopia: Yayu (2010), Kafa (2010) and Sheka (2012). All three previously existing Ethiopian biosphere reserves are situated in the south-west of the country and dominated by forest ecosystems. Main objectives in these biosphere reserves are the conservation of the rain forests and genetic diversity of the coffee plant as well as the sustainable use of coffee species. The Lake Tana region, in contrast, is characterized by historical agricultural land use systems, a dense population, in parts highly degraded landscapes, but also remnants of unique aquatic, wetland and forest ecosystems. In this sense, the region takes a special position within the network of Ethiopian biosphere reserves (Fähser et al. 2015).

The Lake Tana region represents a network of complex, overlapping and sometimes contradicting interests, mostly centered around the lake ecosystem and the ecosystem services it provides. It combines internationally and locally important cultural history (e.g. world heritage sites), good conditions for agriculture (fertile soils and water availability), is part of a globally important biodiversity hotspot, a major Ethiopian and African water source with a need for local human development and poverty reduction. The biosphere reserve approach is uniquely suitable to tackle this situation by providing an integrative management approach which relies on the existing strong administrative structures and scientific capacity in the region.

The Lake Tana region has a centuries old cultural and agricultural tradition. Due to its fertile soils, year-round water supply form the surrounding highlands and tributary streams to the Lake, and the overall moderate climatic conditions, cereal crop production and small scale livestock rearing have been practiced in the region for centuries. The dominant agricultural system still practiced today is the smallholder agriculture.

At the same time, the lake and its many islands have provided refuge for religious life of the Ethiopian Orthodox Tewahido Church. Many orthodox churches and monasteries were established on islands, peninsulas and along the lake shore. At that time, much of the Lake Tana region was still covered with indigenous forest. The forest immediately surrounding the orthodox churches play an important part in the orthodox faith and the religious life of the church community and are therefore protected from most destructive activities, such as logging. Instead, the church communities have traditionally used forest resources in a non-destructive way, such as coffee production. Nowadays, the remaining indigenous forest in the area can be found almost exclusively surrounding churches. The average size of church forests near Lake Tana is several hectares, for example 5.2 ha for church forests in the South Gondar administrative zone northwest of Lake Tana (Cardelús et al. 2013).

The fragmented character of the Lake Tana region visible today, with its many small fields, dotted human settlements and remnant forest patches, is a legacy of these past and current human activities. However, the landscape and ecosystems of Lake Tana and its environs have not only played an important role for human activities, but they also provide habitats for animal and plant diversity of global importance.

In recent years, the intensified use of the resources of the Lake itself—above all water—have increased the pressure on the natural life support system of the region. Water, the basic defining element of the character and importance of the Lake Tana Region, has become increasingly valuable for electricity production and large scale irrigation at national and even international level. Fish, which used to be moderately exploited through traditional fisheries, has increased in market value and is being exported to neighbouring Sudan. Fishing methods have become more efficient and without strong regulations are likely to cause the complete collapse of fish stocks within the next 5–10 years (e.g. de Graaf et al. 2006; Tewabe 2013).

It is clear that additional protection for the natural life support system of the region has become necessary to safeguard the basis for future economic activity and health for the population as well as to preserve global biodiversity. At the same time, prevailing poverty, a lack of infrastructure and public services in the region, must be addressed.

Beside the natural and cultural treasures, the Lake Tana region also owns high academic and administrative capacities. It lies in the Amhara National Regional State (ANRS), one of ten states that together constitute the Federal Democratic Republic of Ethiopia (FDRE). The two largest cities of ANRS are on and close to the lake shore: Bahir Dar and Gondar. The ANRS is governed by a regional president, its cabinet of technical Bureaus, and a state assembly. Beside their administrative tasks, the Bureaus are also involved in research through attached research institutes, such as the Amhara Regional Agricultural Research Institute (ARARI). Additionally, three universities are found in or nearby the LTBR: Bahir Dar University, Gondar University and Debre Tabor University.

These are ideal conditions to have stakeholder contribution and involvement in the BR nomination and management process.

29.3 Preparation of the Lake Tana Region to Become a UNESCO Biosphere Reserve

Following the UNESCO principles, the successful establishment, management and development of a biosphere reserve is closely linked to active stakeholder participation (e.g. Seville Stragey, UNESCO 1996, and Madrid Action Plan, UNESCO 2008). The process that culminated in the nomination of the Lake Tana Biosphere Reserve by UNESCO in June 2015, was driven by local, national and international stakeholders from the beginning. At various stages throughout the process, many other stakeholders contributed to the design of the Lake Tana Biosphere Reserve.

29.3.1 Stakeholder Contributions to the Application Process

The key administrative government structures promoting the establishment of a Lake Tana Biosphere Reserve were the national MAB Committee, headed by the Ministry for Science and Technology (MoST) and the Amhara National Regional State government, mainly through its Bureau of Culture, Tourism and Parks Development

(BoCTPD) and Bureau of Environmental Protection, Land Administration and Use (BoEPLAU). From the beginning of the process, they were supported by two German-based conservation NGOs, the Nature Protection and Biodiversity Conservation Union (NABU) and the Michael Succow Foundation (MSF).

In 2007, NABU started its activities in Ethiopia with a project working towards nature conservation, regional development and the set-up and effective implementation of protected areas. Following a delegation travel to potential biosphere reserve sites as part of this project, NABU, MoST and UNESCO signed a trilateral Memorandum of Understanding (Umbrella MoU) in November 2009. The aim of the agreement was to work towards the protection of biodiversity and the use of natural resources from the perspective of sustainable development and conservation of cultural heritage. The contracting parties agreed to work towards the establishment of a national network of biosphere reserves in Ethiopia (zur Heide 2012).

With the support of NABU, Ethiopia applied for BR designation for Kafa in 2009, and received the official UNESCO nomination as Kafa Biosphere Reserve in 2010. In order to further facilitate and support the establishment of biosphere reserves in Ethiopia NABU, in accordance with the Umbrella MoU, invited international experts to Ethiopia, including the Michael Succow Foundation, for an in-depth evaluation of areas previously identified as potential biosphere reserves (zur Heide 2012). Consequently, the Michael Succow Foundation facilitated the realization of a feasibility study for the Lake Tana Region analysing opportunities and threats to becoming a UNESCO Biosphere Reserve. The German Federal Agency for Nature Conservation financed the study.

While conducting the feasibility study, zur Heide (2012) sought active involvement of a wide range of local stakeholders. This included several expert and community workshops, which prepared the ground for shaping the future Lake Tana Biosphere Reserve. First proposals on future zonation, on overarching objectives of a Lake Tana Biosphere Reserve, on development approaches, and on the most immediately needed nature protection measures were gathered in these workshops and documented in the feasibility study (zur Heide 2012). In this way, the feasibility study also bundled the voices of individual champions for a Lake Tana Biosphere Reserve, including for example academic researchers from the Bahir Dar University.

Backed by the results of the feasibility study, NABU and Michael Succow Foundation secured funding from the German Ministry for Technical Cooperation (BMZ) in order to support the ANRS and federal governments to prepare the Lake Tana region for the application as biosphere reserve. This NABU-led project, called "For People and Nature—Establishment of a UNESCO Biosphere Reserve at Lake Tana, Ethiopia", started in February 2012 and ended in December 2015. The project was implemented by NABU in cooperation with MSF.

The implementation of the project "For People and Nature" was closely coordinated with the ANRS government through the involvement of the technical bureaus, their associated research institutes and their structures down to Kebele (community) level. This cooperation was sketched out in a project agreement which was signed at the start of the project in 2012 between NABU and five technical Bureaus, referred to as the partner bureaus for the project. These five bureaus are the Bureau of Culture, Tourism and Parks Development (BoCTPD), Bureau of Environmental Protection, Land Administration and Use (BoEPLAU), Bureau of Agriculture (BoA), Bureau of Water Resource Development (BoWRD) and the Bureau of Financial and Economic Development (BoFED). Beside the facilitation by regional administrative structures for the implementation of the project activities, this working modality also aimed at providing institutional sustainability of the process.

The project "For People and Nature" consisted of four main components. The components were designed in order to prepare the Lake Tana region administratively for the application as UNESCO Biosphere Reserve, to pilot local projects on sustainable development, and to continue publicity, networking and scientific support activities for the region (Table 1). The implementation of these components was pursued through sub-contracting and partnering with a whole range of institutions and experts, including the partner bureaus, national and international consultants, academic institutions, non-governmental organizations and private companies.

Beside the direct involvement as sub-contractors for project activities, private, non-governmental, civil society and academic stakeholders have been integrated in the process towards the establishment of the Lake Tana Biosphere Reserve in the form of general and expert workshops and through the establishment of "focal persons" within institutions, such as universities. One of the main channels for the participation of communities in the design of the future biosphere reserve has been the participative zoning of the Biosphere Reserve, according to the UNESCO BR zoning scheme supporting core, buffer and transition zones.

Table 1 The four main components and corresponding sub-components of the project "For People and Nature—Establishment of a UNESCO biosphere reserve at Lake Tana, Ethiopia"

Component	Sub-components
1. BR development	1.1 Developing the BR administration 1.2 Participative zoning of the BR 1.3 Development of a BR Management and Business Plan 1.4 Implementation of the Management Plan 1.5 Proposal to UNESCO for official recognition of Lake Tana region as a Biosphere Reserve
2. Development of ecotourism and of regional products for alternative income generation	2.1 Development and promotion of ecotourism 2.2 Public-private partnership for the development and marketing of regional natural products
3. Conservation and use-concepts / sustainable resource management	3.1 Integrated wetland management 3.2 Reforestation and sustainable management of church forests 3.3 Development and testing of soil-friendly agriculture (conservation agriculture)
4. Communication and public relations (regional, national, and international) and interlinking with research projects	

29.3.2 Lake Tana Biosphere Reserve Zonation

The concept of zonation with strictly protected core zones, restricted-use buffer zones and transition zones for development activities lies at the heart of the UNESCO Biosphere Reserve approach. The zonation includes also the determination of the outer boundaries of the Biosphere Reserve. As mentioned above, first expert proposals for a suitable zonation for the Lake Tana Biosphere Reserve were forwarded during stakeholder workshops conducted as part of the feasibility study. These varied considerably, some including for example the whole Lake Tana watershed, some including the whole Lake as core zone, and all the riversides as buffer zones. Most of these proposals were made from scientific viewpoints. Subsequently, under the project "For People and Nature", an extensive and iterative process was conducted with expert input, community input, administration input to arrive at the final zonation proposal presented in the UNESCO application (Fig. 29.2).

The participative zonation process was carried out by the Bureau of Environmental Protection, Land Administration and Use (BoEPLAU), and facilitated by the NABU project staff. The various expert proposals and additional considerations of realities on the ground, such as management efficiency and potentially conflicting development projects, were combined into one proposed zonation map. This map provided the basis for the discussions with communities led by BoEPLAU experts.

In the Kebeles which were likely to contain core or buffer zones of the Lake Tana Biosphere Reserve, Kebele Zonation Committees with up to 16 members were formed and trained. The members of the Zonation Committees included Woreda experts, Development Agents, Kebele managers, community and religious representatives. The training was supported by a specifically developed training manual, outlining the UNESCO principles of biosphere reserve zoning and the reasons why Lake Tana is suitable to become a biosphere reserve.

After the training, the Zonation Committees were then presented with the expert proposals for core and buffer zones referring to their specific Kebele. At the same time, committee members contributed additional local information, for example on existing area closures. Through discussions and site visits, terrestrial core and buffer zones were locally agreed upon. Due to logistic difficulties, the aquatic zonation proposal is based on expert input only. For a reasonable zonation of the water body, it was suggested to assign boat transport routes, harbours and the most populated shorelines as the transition zone. The remaining area of the lake is designated buffer or core zone (see Fig. 29.2).

The reality of the participative zonation process as described above unrolled quite differently in the different Kebeles. In general, designation of an area as core zone met with the highest doubts from the side of the community. In several cases, proposed core zones were changed to buffer zones. In some Kebeles, some long-standing land use disputes within the community or between community interests and the administration were brought to the surface again through the zonation process. Many site visits and conflict resolution approaches were necessary to finally reach an agreement for some of the disputed areas.

The GPS data for the core and buffer zones were collected in the field by members of the Zonation Committees with Woreda expert support. This approach to the participative zonation process yielded a great many small core and buffer zones. Finally, some of the very small and neighbouring core zones were given a common name. Through this process the final number of core zones as listed in the legal document granting strict protection to these areas (Lake Tana Biosphere Reserve Regulation, ANRS 2014) numbers 78.

29.3.3 Vision and Goals of the Lake Tana Biosphere Reserve

The vision for the Lake Tana Biosphere Reserve that emerged during several stakeholder workshops and as stated in the Management Plan (Fähser et al. 2015) is:

For our sustainable future – Lake Tana, People and Nature

The detailed objectives and goals of the Lake Tana Biosphere Reserve are elaborated in the BR nomination application (FDRE 2014) and the Management Plan (Fähser et al. 2015). They are aligned with the three functions of biosphere reserves: conservation, development and logistic support.

At Lake Tana, the conservation foci lie on the regeneration of the lake itself by reducing sedimentation and eutrophication, on the protection of the remaining wetlands, and on the preservation and extension of natural forests. Furthermore, erosion in agricultural areas shall be reduced through soil friendly agriculture with a focus on soil protection and regeneration of the humus layer. This is the basis for a vast number of ecosystem services. The protection of rare and threatened species like the endemic *Labeobarbus* fish species, bird species, mammals, tree and other plant species is of high significance in the area of the biosphere reserve (Fähser et al. 2015).

Reduction of poverty, the creation of alternative and more diverse options for livelihoods, and ensuring food-security are the most urgent development goals in the region (Fähser et al. 2015). Soil- and environmentally friendly agriculture, regional product development and promotion of eco-tourism together with community-implemented natural resource management are the main focus areas to achieve these development goals sustainably.

In the region of this new biosphere reserve, a lot of educational and research activities have been established already through government programmes, regional as well as national and international universities. Coordination of these activities and alignment with biosphere reserve requirements, where applicable, should be supported and enhanced. This applies, for instance, to research and monitoring on erosion, nutrient transfer into the lake, and eutrophication of the lake (Fähser et al.

2015). Such coordination of research efforts in and around the Lake Tana Biosphere Reserve can greatly enhance the impact and relevance of focussed scientific enquiry for society and policy development.

29.4 Requirements for Successful Management of the Lake Tana Biosphere Reserve

Biosphere Reserves are seen by UNESCO to fulfil a key role as study sites and examples for sustainable development realized on a regional scale. This makes an integrated planning framework across disciplines and scales essential in order to guide development, land and natural resource use within the biosphere reserve. As BRs are instruments for integrated management of socio-ecological systems, their management needs to address regions comprehensively: abiotic and biotic factors (climate, water, soil, and landscape in its entirety, etc.), the local communities (cultures, traditions, knowledge, heritage, etc.), their practices (fishing, forestry, agriculture, livestock breeding, tourism, etc.) and the institutional and legal settings within which they act. Consequently, BR management involves many different interventions at many different levels at the same time, for example: protecting individual species, improving the water cycle, supporting the marketing of agricultural products, training local communities and monitoring.

In the Lake Tana region, a great number of policies and projects which aim at safeguarding the natural resources for current and future development and development initiatives have already been formulated or instigated. Examples are the Tana and Beles Integrated Water Resources Development Project, the Amhara Forest Proclamation, or the Lake Tana Tourism Development Strategy. On the other hand, there are activities and projects which threaten to undermine the biosphere reserve objectives and ecosystem integrity and are contra-productive to other interventions. For example, some envisaged irrigation schemes in or close to the wetlands bear the risk of undermining the IFAD project objective of rehabilitating the wetlands. Some plans for the intensification of agriculture would help to improve agricultural productivity but may undermine fish productivity and lake water quality (zur Heide 2012). Given this background, the need and potential for creating synergies, avoiding duplications and providing pro-active conflict resolution (between stakeholders and between policies) is high in the Lake Tana region. In order to be called “successful”, the Lake Tana Biosphere Reserve management must become the central player in the process of integration of policies and activities affecting the natural resource base and human development in the Biosphere Reserve area and its gravitational field.

Clearly, the management elements to achieve this goal must be evidence-based, participatory and transparent (e.g. UNESCO 1996). Since each biosphere reserve has its unique setting of social, political, environmental and legal factors, the detailed management structure of a given BR has to be suitable within that specific context. Several key elements of the Lake Tana Biosphere Reserve management have already been established, and others are proposed for future establishment. Since the Lake Tana Biosphere Reserve is the first BR in the Amhara National Regional State, the distribution of responsibilities and working modalities amongst the management elements will be part of an adaptive management and learning process, which has to be monitored and evaluated after an initial period.

29.4.1 Lake Tana Biosphere Reserve Management Structure

UNESCO clearly states that a dedicated management unit is central to successful biosphere reserve management (UNESCO 1996). Equal emphasis is put on the establishment of a local consultative framework, in order to promote the participation of the population. For the Lake Tana Biosphere Reserve, structures carrying out the tasks of a dedicated management unit as well as elements of a consultative framework have been established and started their activities (Fig. 29.3). However, the network of management elements is not yet complete and the working modalities and detailed responsibilities are still to be elaborated further.

During the process of LTBR development, a *Biosphere Reserve Management Unit (BRMU)* has been established in 2012 within the Bureau of Culture, Tourism and Parks Development (BoCTPD). Through the Lake Tana Biosphere Reserve Regulation (ANRS 2014), passed by the ANRS government in November 2014, the BoCTPD has received the legal mandate to manage and develop the LTBR. By end 2015, the BRMU had two full-time staff members and two part-time staff members. The core task of the BRMU is the facilitation of all processes and projects concerning LTBR development through, for example, networking, lobbying and mobilization activities, as well as publicity for the BR. For example, the BRMU facilitated the zonation process, the submission of the UNESCO application via the national MAB committee, and the drafting of the Lake Tana Regulation.

A *Biosphere Reserve Council (BR Council)* has been established in 2013. The BR Council is the highest stakeholder representation structure foreseen in the LTBR governance structure. It currently has 43 members, the majority of which are representatives of the state administration from regional to Woreda level, but some academic, NGO, community and religious representatives are also included. It is foreseen that the BR council meets on a half-yearly basis. The main role of the Council as stated the guideline for its establishment (BoCTPD 2013) is to "encourage the participation of major stakeholders in addressing challenges of Lake Tana and facilitate properly different developmental activities in/around Lake Tana". Additionally, the BR Council shall "advise the project For People and Nature" and "give direction or decision on required issues that facilitate the implementation of the project". Mobilization and public awareness creation is another objective of the council.

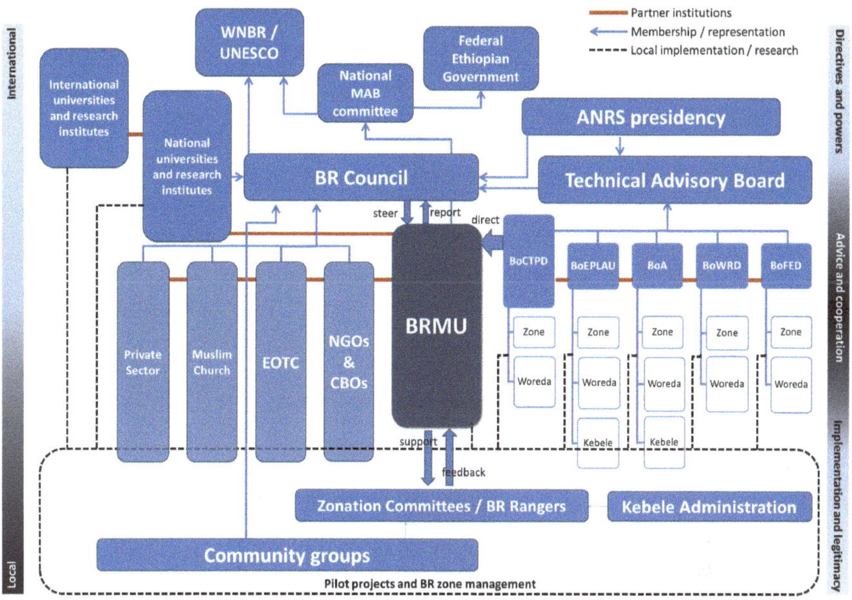

Fig. 29.3 Organigram of the foreseen governance structure for the Lake Tana Biosphere Reserve. According to the LTBR Regulation, the BoCTPD has the mandate to administer the LTBR. In 2012, a BRMU was established under the BoCTPD, which is expected to be strengthened with additional human and financial resources. The LTBR Technical Advisory Board is not yet established but proposed to replace the project steering committee. At the time of the LTBR inauguration, the BR Council membership was slightly more weighted towards bureau/ administration representation than suggested by this figure. After establishment of the Technical Advisory Board, the memberships and mandates of BR Council, BoCTPD and the Technical Advisory Board have to be adjusted to ensure complementarity and avoid conflicting mandates. This adjustment remains one of the challenges for successful LTBR management (see Sect. 29.4.1)

In relation to the project "For People and Nature", *BR project steering committees* have been established at regional, zonal and woreda level. In these steering committees, BR focal persons from the partner bureaus are represented as well as the central administration (presidency at regional level, Woreda Manager at Woreda level). In particular the woreda steering committees have played a key role in the participatory zonation process, and have facilitated some pilot project activities. Due to their active role, it is foreseen to maintain the Woreda steering committees for long-term, locally imbedded BR management. The regional level project steering committee as such might cease to exist after the end of a BR establishment follow-up project.

In place of the regional level steering committee, a new structure, the Lake Tana Biosphere Reserve *Technical Advisory Board*, is expected to be established as a permanent element of the LTBR management. Like the project steering committee, the Technical Advisory Board will encompass representatives of the five partner bureaus as well as the concerned zonal and city administrators. The exact mandate of the Technical Advisory Board is still to be determined. Its existence is foreseen to strengthen integration of BR activities across the concerned sectors, their policies and workplans. The Technical Advisory Board might also be mandated to steer and monitor the management of the BR on behalf of the ANRS presidency. In that case, the establishment of yet another governance element, a secretariat of the Technical Advisory Board, would become necessary.

At community level, *Kebele Zonation Committees* were established during the participatory zonation process (see Sect. 29.3.2). Through the training and active participation in the local zonation activities, the members of the committees represent a valuable source of knowledge and capacity for community-level BR management. An extension of the Zonation Committee mandate to community-level BR awareness and management activities would be an important step towards the establishment of a local consultative framework, as foreseen in the Statutory Framework for the World Network of Biosphere Reserves (UNESCO 1996).

Further Challenges for the Establishment of the BR Management Structure
A main challenge for establishing an effective BR management system will be to adapt existing and future management elements to the broader governance network. The aim is to create an effective structure, which nonetheless operationalizes the principles of participation and consultation, and also considers equity of the benefits that might accrue via the BR—tangible as well as non-tangible ones.

In this respect, the Lake Tana Biosphere Regulation gives a clear mandate to the BoCTPD to set up adequate structures and mechanisms: "The Bureau shall [...] put in place a mechanism in which the community inhabiting the biosphere reserve would amply benefit in terms of social and economic aspects as a result of conservation of development of same as well as follow up and ensure its implementation thereof [sic]" (LTBR Regulation Part III, Article 8.2.B). And "The Bureau may issue specific directives necessary for the full implementation of the provisions of this regulation" (Part IV, Article 12).

On the one hand, this means to further fine-tune the responsibilities and mandates of the various existing and future management elements in order to make them complementary and avoid duplications or gaps. On the other hand, management processes which are effective and reflect the BR management principles need to be established and implemented. In particular decision making rules and communication procedures need to be clear, understood, and simple. In line with this, an adaptation of membership to the various committees might have to take place. For example, if the Kebele Zonation committees are to be maintained and transformed into community BR management committees, they can form a key part of the local consultative framework. However, currently the only institutionalized routes which exist to carry community-level opinions and decisions towards the higher BR decision bodies, i.e. the BR Council or the future Technical Advisory Board, is via the government administration structures. To have a participatory community-led management system, decision-making structures and processes outside the governmental bodies need to be established or integrated. This could, for example, be achieved through creating Woreda-level stakeholder platforms and respectively adding 10 Woreda community representative members to the BR Council.

Similarly, the decision powers and responsibilities of the BR Council with respect to the envisaged Technical Advisory Board have to be clearly defined. As the highest stakeholder representation body, the BR council should have a significant weight in taking strategic decisions. This can be achieved by describing and following clear processes of decision making between the three highest BR management elements: the BR Council, the Technical Advisory Board and the BRMU. Feedback loops need to be included in the decision making processes.

In order to achieve the actual goal of sustainable development for the Lake Tana Biosphere Reserve, integration and efficiency needs to be established not only within and between the BR management elements, but on a much broader scale with regional, national and international partners and stakeholders.

29.4.2 Challenges Towards Fulfilling the BR Mandate of Integration and Networking

A biosphere reserve is most of all a management framework to promote innovative and locally appropriate solutions to sustainable development. A main task is the creation of synergies across different organisational and technical sectors. The working modalities through which the dedicated BR management team must achieve this is mostly by functioning as a broker and moderator: a broker of knowledge, access to resources, and ideas and a moderator between policies and stakeholders (see also Table 2).

The establishment of a representative management structure (Ref. Sect. 29.4.1) provides a good basis for the BR management team to be successful in their role as broker and moderator. Further than that, the BR management staff need to have the relevant technical knowledge and communication skills, but also the opportunity or even right to access relevant information. Therefore, the BR management would be strengthened considerably by the formalization of information exchange processes with key stakeholders. For example, in the planning process for infrastructural developments within or close to the BR, the consideration of effects on BR core zone integrity should be mandatory, similar to or as part of an Environmental Impact Assessment. This information exchange between planning agencies or technical bureaus and the BR management can be secured through its formalization.

In the case of the Lake Tana Biosphere Reserve, an important step towards successful integration of policies and activities will be the establishment of the Technical Advisory Board, where the most relevant regional bureaus are represented. However, the integration of initiatives from the private sector and civil

Table 2 Proposals to strengthen LTBR management

Proposal	Desired effect	Challenges
The BR establishes a central platform for Lake Tana Region knowledge management	The BR becomes the main point of contact for interested parties requesting information (scientific, xstatistical, narrative) about the Lake Tana region. The BR has an overview of all documented information and directs requests for detailed information to the relevant institution.	- Financial, human and technical resources - Pro-active support from relevant stakeholders
Provide financial support and flexibility to the dedicated BR management unit	The basic BR management is financially secured in the long-term and allows for long-term BR planning. The BR team can be flexible and react immediately to external funding opportunities or can quickly provide funds to BR initiatives	- Draw up an appropriate institutional setting to allow for financial independence and flexibility with partial state funding and third-party funding - Political will to secure long-term financial support
Ease regulations for business establishment with BR backing	More small and innovative businesses with sustainable business plans will start up, increase economic activity in the BR and improve local livelihoods	- Different regulations need to be adapted according to type of business - Long political and bureaucratic process
Transparency of finances and decision taking	Yearly plans, financial reports as well as minutes of Task Force and BR Council meetings are made publicly available via the BR website and upon request. This will strengthen trust, participation and a feeling of ownership by the stakeholders	- Adopting an organizational culture of transparency - Maintaining and updating an appropriately designed BR website
Give veto-right on developments within and close to the BR to the BR management	The BR management has an effective tool for pro-active management towards the BR goals.	- Overcome the perception that "development is good at any price" - Devolution of power to the BR management
Establish a national BR network	Active exchange and learning between Ethiopian BR experiences and common lobbying for national and BR support improves understanding and support for BRs	- Funding - Clear goals of an Ethiopian BR network

society is not yet given significant weight. For long-term economic and social sustainability the opportunities for cooperation with these sectors must be strengthened.

A smaller but equally important role of the BR management team will be the protection of core zone and ecosystem integrity through application of laws. On the one hand, this protection refers to unofficial intrusions of core zones. Those are likely to be incidental local events. Such cases are followed up retrospectively through community procedures or, in grave cases, legal procedures. More important, however, is the prospective protection of core zones and ecosystems in relation to planned, official activities. In the strongest case, this would require the possibility of the BR team to stop developments and bindingly request suitable adaptations of the plans to comply with BR regulations. This kind of "veto-law" for developments within the BR area exists for some other BRs in the world. It provides a very powerful tool for BR management, but needs strong political will to be introduced.

It lies at the hands of the BoCTPD to put in place appropriate internal organisational arrangements to make the BRMU functional, for example through the number of staff dedicated to it and powers given to them. Political will is necessary, however, to ensure the long-term viability of the BR through the commitment to provide the basic financing for its management. In this respect, the prospects for the Lake Tana Biosphere Reserve are encouraging. In the Lake Tana Biosphere Reserve regulation, the mandate for the LTBR management is clearly given to a government body, the BoCTPD. This implies that funds will be made available for the BoCTPD by the regional government to carry out its mandate. As a minimum, the BoCTPD's internal budget has to be adapted so as to apportion funds for LTBR management.

29.5 Conclusions

The early involvement of stakeholders and the prominent role of government institutions in the development of the Lake Tana Biosphere Reserve have led to a very positive starting situation for the future BR management. The dedicated LTBR management unit, the envisaged Technical Advisory Board, is anchored within existing administrative structures, elements of a local consultative framework exist, and a lot of BR capacity, in- and outside the administration, has been built up during the process. Management challenges remain mainly in filling management structure gaps, in setting up effective processes and in providing the necessary finances and powers to the management elements in order to secure long-term viability of the Lake Tana Biosphere Reserve.

An efficient governance structure with active stakeholder involvement at all levels will provide the best basis to address the challenges of ensuring ecological integrity and human development within the LTBR, as described above, in the respective BR documents (see Box 1) and throughout this book.

Box 1. Core Documents of the proposed Lake Tana Biosphere Reserve

The Lake Tana Biosphere Reserve Regulation

This regional-level law called "The Lake Tana and its Environs Biosphere Reserve Delineation and Administration Determination, Council of Regional Government Regulation No. 125/2014" was passed by the ANRS cabinet in November 2014 and gazetted in March 2015. It is the core legal document of the Lake Tana BR. It consists of four main parts and three annexes:

Part I General (includes mainly definitions of terms);
Part II Designation, Delineation and Boundaries of the Core Zones along with the prohibited activities therein;
Part III Objective and Administrative Condition of the Biosphere [Reserve];
Part IV Miscellaneous Provisions (includes penalties and additional powers of the BoCTPD);
Annex 1 Description of List and Geographical Direction of Core Zones (provides names and GPS coordinates for the 78 core zones)
Annex 2 Width [size in ha] of core zones and list of Kebeles in which they are found;
Annex 3 List of Kebeles incorporated in the Biosphere [Reserve]

Through this Regulation, the LTBR core zones have obtained a legally protected status, the Bureau of Culture, Tourism and Parks Development has been given the mandate to administer the LTBR as well as the power to issue further directives deemed necessary for its successful implementation.

The UNESCO application document for the Lake Tana Biosphere Reserve

A document of over 200 pages, following the UNESCO nomination format, was submitted by the Ethiopian national MAB committee to the MAB secretariat at the end of September 2014. The document contains detailed descriptions of all aspects of the future BR - abiotic and biotic factors, the local communities, institutional and legal settings, as well as goals and strategies for the BR. A core aspect of the application is the BR Plan, showing the proposed zonation. The application document had been signed by all implied Woreda administrations,as well as the Regional Presidency and the partner bureaus.

The LTBR Management Plan

A management plan for the initial 5-year period of the Lake Tana BR is being developed with stakeholder participation (workshops, interviews, commenting). The plan describes the main management goals for the future Lake Tana Biosphere Reserves. It contains concise descriptions of the currently most pressing problems to be addressed terms of conservation, development and logistic support for the Lake Tana Region. It further provides eight clear 5-year management goals and the corresponding actions that will help to achieve these goals. The table of contents of the plan have been presented to and approved by the BR Council in July 2014.The exact description of the goals and corresponding activities, budgets, timeline and monitoring activities are to be worked out in early 2015 and presented to the BR Council for revision and approval.

The BR Council guideline

A guideline for the establishment, mandate and membership of the Lake Tana Biosphere Reserve Council was drawn up by the BRMU, discussed and passed by the BR Council in its constituting session in December 2013. The Name of the guideline is "Guideline to establish Council to the proposed Lake Tana Biosphere Reserve". It is a guideline that is valid under the BoCTPD's mandate to administer the LTBR.

The Lake Tana Biosphere Reserve is amongst the largest BRs in terms of its total size. Considering size, population numbers, logistic and political realities on the ground, it was not possible to include the entire macro biogeographical area, the Tana sub-Basin, into the biosphere reserve. However, it is imperative to understand that the well-being of the larger area is a key requirement for the long-term viability of the Lake Tana Biosphere Reserve (FDRE 2014). Accordingly, the efficient governance of the complete Tana sub-Basin towards environmental sustainability is a key interest of the Lake Tana Biosphere Reserve management.

Acknowledgements For the writing of this chapter I benefited greatly from fruitful discussions with my colleagues both at NABU and at Michael Succow Foundation in Germany, in particular with Svane Bender-Kaphengst, Renée Moreaux and my co-authors of the LTBR management plan. My understanding of Ethiopian administrative and social functioning was much enhanced by daily interaction and discussions with my Ethiopian NABU colleagues and our BoCTPD partners, in particular with Getinet Fetene, Negash Atnafu, Teklu Damtie, Alemayehu Zewdie and Tadesse Adgo.

References

Amhara National Regional State (2014) A regulation to determine the zoning and administration of bio-sphere reserves of Lake Tana and its Vicinity, Regulation No. 125/2014. Zikre Hig Gazette No. 6, Bahir Dar, 21 November 2014

Bureau of Culture Tourism and Parks Development (2013) Guideline to establish council to the proposed Lake Tana Biosphere Reserve. Internal Document, BoCTPD

Cardelús CL, Scull P, Hair J et al (2013) A preliminary assessment of Ethiopian sacred grove status at the landscape and ecosystem scale. Diversity 5(2):320–344

de Graaf M, van Zwieten PAM, Machiels MAM et al (2006) Vulnerability to a small-scale commercial fishery of Lake Tana's (Ethiopia) endemic *Labeobarbus* compared with African catfish and Nile tilapia: An example of recruitment-overfishing? Fish Res 82:304–318

Fähser L, Kretschmer H, Pascall M et al (2015) Management Plan for the Lake Tana Biosphere Reserve. Final Draft submitted to NABU May 2015

Federal Democratic Republic of Ethiopia (2014) Application for nomination of 'Lake Tana Biosphere Reserve' prepared and submitted to UNESCO in terms of Article 4 of the statutory framework of the world network of Biosphere Reserves. Prepared by Dennis Moss

Partnership. On behalf of the project 'For People and Nature: Establishment of a UNESCO Biosphere Reserve at Lake Tana in Ethiopia' implemented by The Nature and Biodiversity Conservation Union (NABU) in cooperation with Michael Succow Foundation, supported by the German Federal Ministry for Economic Cooperation and Development (BMZ)

Gesellschaft für Internationale Zusammenarbeit (2011) Programme Progress Report (final). Sustainable Land Management – SLM. Eschborn, Germany

Tewabe D (2013) Status of Lake Tana commercial fishery, Ethiopia. ABC Res Alert 1(3). Available online at http://abcreal.weebly.com/uploads/1/4/2/3/14237537/status_of_lake_tana_commercial_fishery.pdf

United Nations Educational, Scientific and Cultural Organization (1971) Man and the biosphere programm. http://www.unesco.org/new/en/natural-sciences/environment/ecological-sciences/man-and-biosphere-programme/. Accessed 06 Jul 2015

United Nations Educational, Scientific and Cultural Organization (1996) Biosphere Reserves: the seville strategy and the statutory framework of the world network. UNESCO, Paris. http://unesdoc.unesco.org/images/0010/001038/103849Eb.pdf. Accessed 13 Jul 2015

United Nations Educational, Scientific and Cultural Organization (2008) Madrid action plan for Biosphere Reserves (2008–2013). UNESCO, Paris. http://unesdoc.unesco.org/images/0016/001633/163301e.pdf. Accessed 13 Jul 2015

zur Heide F (2012) Feasibility study for a Lake Tana Biosphere Reserve, Ethiopia. BfN Skripten, Bonn. http://www.bfn.de/fileadmin/MDB/documents/service/Skript_317.pdf. Accessed 13 Jul 2015

Chapter 30
Water-Induced Shift of Farming Systems and Value Addition in Lake Tana Sub-basin: The Case of Rice Production and Marketing in Fogera District, Northwestern Ethiopia

Berihun Tefera

Abstract Fogera district is one of the districts adjacent to Lake Tana in South Gondar zone in Amhara Regional State, Ethiopia. Floods and related water-borne diseases were serious problems in the plains of the Lake Tana Basin and Fogera district in particular for centuries. But with the introduction of rice as a crop, this environmental characteristic has become an opportunity instead of a challenge. To address the lack of documentation about this water utilization-induced change in farming systems in this area, and the resulting rice production and consumption value chain, this chapter presents three survey studies conducted from 2010/11to 2014 by structured interviews of farming households and marketing agents, field observations and group discussions supplemented by review of literatures. Survey data was collected from 200 farming households of high and medium level rice producing kebeles of Fogera district selected randomly, key informant experts and a variety of value chain actors. Participation in the rice industry in this area has increased from 30 participant households in 2 kebeles and area coverage of 6 ha in 1993/94 to 34,249 households in 24 kebeles covering an area of 20,230 ha of land in 2014/15. Fogera district jumps from no contribution in rice production before two decades to a share of 77% of the rice produced in Amhara and 60% produced in all of Ethiopia. Rice is now the primary crop in the district. In addition to rice production, farming households have also adapted to rice consumption by innovative value addition practices to local food items. About 82.5% of the respondents were rice producers, with non-producing households having no access to irrigation; rice production is significantly related to irrigation water utilization. These findings clarify the contribution of water resource utilization in shifting farming systems and

B. Tefera (✉)
Agricultural Economics Department, College of Agriculture and Environmental Sciences and Blue Nile Water Institute, Bahir Dar University, Bahir Dar, Ethiopia
e-mail: berihunt2@gmail.com

B. Tefera
Department of Agricultural & Resource Economics, University of Connecticut, Storrs, USA

K. Stave et al. (eds.), *Social and Ecological System Dynamics*,
AESS Interdisciplinary Environmental Studies and Sciences Series,
DOI 10.1007/978-3-319-45755-0_30

the value addition practices in the area. The study has implications for how to promote similar innovations where there are no cultural practices related to commodities targeted by extension interventions exemplify the impacts of proper water resource use.

Keywords Rice · Farming systems · Fogera · Irrigation · Tana Sub-basin · Agricultural extension · Value chain

30.1 Introduction

Until recently, the productivity of subsistence farming in the Lake Tana Basin was below 50% of its potential, mainly because of inappropriate rainwater management. Hence, poor rainwater productivity is one of the root causes of food-insecurity in the region and proper water management at grassroots level is important for improving food security (Mastewal 2012). In addition, floods and related water-borne diseases have been serious problems in the plains near the shores of Lake Tana and the Fogera district in particular (Sewemehon 2012). The introduction of rice as a crop in Fogera starting in the early 1970s changed the water-rich ecosystem of this district from an environmental problem to an economic and lifestyle opportunity. Rice production in the Fogera district has grown from almost nothing before the 1970s to a centrally important crop for the local community. It has altered not only the farming practices in the area, but also the economic engagement of local people in other aspects of the value addition and value chain of rice, and even changed local diets. Little is documented about the rice production system and its related effects on the economy and community, however. This chapter describes the introduction of rice as a crop and the impact of this on shift of farming system and value addition practices including adaptations in production, processing and consumption of the crop.

This chapter is written based on results of three surveys done by the author and existing literatures related to the district, water utilization and the commodity. These three surveys are rice farming system and value chain analysis of the Fogera district (selecting about 200 households from selected sample kebles[1] of the district), baseline survey of Machurit River irrigation system (around Woreta town, the capital of Fogera district), both conducted with the support of Japan International Center for Agricultural Science (JIRCAS) in 2010/11 and grass pea production and value chain analysis in South Gonder zone with the support of Bill and Melinda Gates Foundation conducted in 2014. The last is very important related to interaction of rice and grass pea in the mixed production system which is another interesting new adaptation in the production or farming system as a result of rice introduction. Farming households plant rice following the sowing of grass pea

[1]The lowest strata in the formal administration of Ethiopia.

before harvesting. They first harvest rice and the grass pea regenerates with new sprouts and finally harvest the grass pea grown by using the residual moisture.

The rice value chain analysis is done using Structure-Conduct-Performance (SCP) framework of value chain analysis and the result is described mainly in section eight of this chapter. This survey involves a selection of two purposively selected major rice producing kebeles and 200 randomly selected rice producer farmers. The data collected is analyzed descriptively using SPSS statistical software.

The remainder of the chapter is structured as follows: after an overview of the environmental characteristics and land use patterns in the Lake Tana Sub-basin and the Fogera district; the context, history and trends of rice production in Fogera is described. The last sections and sub sections present the rice value chain, including rice processing, markets and consumption.

30.2 Lake Tana Sub-basin

Lake Tana Sub-Basin or watershed is the headwater basin of the Upper Blue Nile River. It includes the largest freshwater lake of the country, Lake Tana, which is the source of the Blue Nile (Abbay) river and is located to the north part of the basin (Yared et al. 2012).

The Upper Blue Nile as a whole contributes the largest share of runoff (48.5 BMC per annum) of the Nile River flow. The Lake Tana basin is generally less prone to drought because of large amount of annual rainfall it receives and its location in the highlands of Ethiopia. The Upper Blue Nile basin is the largest river basin in-terms of volume of discharge and second largest in terms of area in Ethiopia. It is also the largest tributary of the main Nile and covering a total drainage area of 176,000 km^2 (Conway 2000).

The average annual rainfall of the Lake Tana watershed is about 1350 mm and falls primarily between June and October. The average annual temperature is about 28 °C. The watershed is dominated topographically by plains (about 48.4%), rugged land (36.1%) and valleys (15.5%). Undulating slopes and erodible soils lead to natural erosion and landslide problems in the North and South highlands. The watershed consists of three traditional Agro-Ecological Zones: Kola (Lowland) (15% of the watershed area), Dega (Highland) (25.5%) and Woina Dega (Midland) (59.5%). The topography is generally flat on the east and west side of the Lake Tana but is sloped varies greatly on the south and north edges of the watershed. In areas with flat topography, stream flow is relatively slow, and floodwaters tend to spread out into adjacent lands such the Fogera and Denbia floodplains around Lake Tana. The land use pattern consists of 32% farmland, 19.5% forest and bushes, 12.7% grazing, 7.1% construction, 6% water body and 22.7% others. The major crops grown are cereals (77%), pulses (17%), oilseeds (6%), and vegetables, root crops and fruits (<1%). The Lake Tana watershed also has a high climate variability resulting in unpredictable flooding and drought. The 2003 drought, exacerbated by hydropower generation demands, dropped the lake depth to 2 m and reduced the lake area by 35 km^2. Navigation was suspended for 3 months, affecting tourism, fisheries, and livelihoods for associated communities around the lake (Miheret and Ernest 2012).

30.3 Fogera District

The Fogera district is one of the districts adjacent to Lake Tana in the South Gondar Administrative zone in Amhara Regional State of Ethiopia (Map 30.1). It is situated at 11°46–11°59 latitude North and 37°33–37°52 longitude East. The altitude ranges from 1774 to 2410 meters above sea level (masl) and is predominantly classified as *Woina-Dega* (midland) agro-ecological zone. Mean annual rainfall is 1216 mm (with a range of 1103–1336 mm) from both the short (March and April) and long rains (June–September). Farmers depend on the long rainy season (*Kremt*) for crop production (IPMS 2005). Fogera district, which recently became independent administratively, has a total area of 102, 807.65 ha, subdivided into 27 rural kebeles and one town, Alem Ber. The topography consists of plains (about 76% of the total area), rugged land (11%) and valleys (13%) (Miheret and Ernest 2012). Fogera district is one of the eight districts bordering Lake Tana and one of the surplus food producing districts in the region from past years to the present. The dominant soil type on the Fogera plains is black clay soil (ferric Vertisols), while the medium and

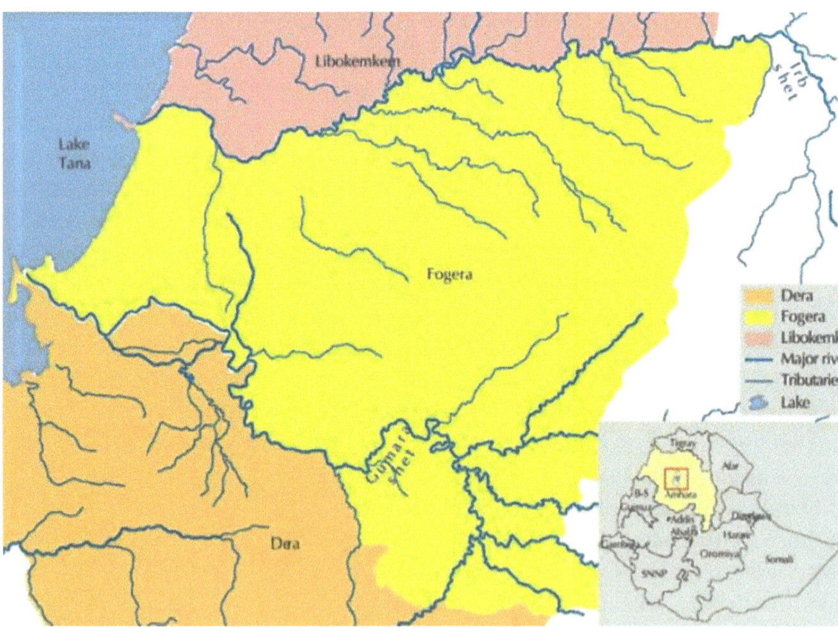

Map 30.1 Map of Fogera district. *Source* Tilahun et al. (2012)

high altitude areas contain primarily orthic Luvisols (IPMS 2005). Woreta town is the capital of Fogera district with 4 urban kebeles and 3 rural kebeles located 625 km far from Addis Ababa (the capital of Ethiopia) and 55 km from Bahir Dar (the capital of Amhara region). Fogera district is dominated by flat topography accounts for 76% of the total land area bordering Lake Tana. Its altitude ranges from 1774 to 2410 masl. Average annual rainfall is 1216 mm and average temperatures is 19 °C (Tilahun et al. 2012). The land use pattern during the survey period were about 44% cultivated land, 24% pasture land, 20% water bodies and the rest for others (Astewel 2010). The district had an estimated population of 269,350 in 2015 and nearly 85.4% of the population was lived in rural areas (CSA 2013). The larger rural population livelihood relies on a mixed farming system of crop and livestock as their main source of livelihood. Fogera district has an excess water supply that is underutilized (Sewmehon 2012).

30.4 Rice Production in Ethiopia and Amhara Region

Studies indicate that Ethiopia has huge potential for rice production. It is estimated that more than 20 million ha of land is suitable for rice production in 3 kinds of agro-ecological zones: rainfed lowland; rainfed upland and irrigated (FAO 2008; SG2000 2006). Rice production in Ethiopia was covered 8.5 thousand ha of land in 4 rice producing regions, and accounted for less than 0.05 % of the national potential, and 0.12 and 0.17% of total area and production in 2002/03 production year, respectively (CSA 2003). By 2014, rice production area and volume in Ethiopia was estimated to be 33,820 ha (nearly 0.17 % of the national potential) and 923,627 Quintal, respectively. The Amhara region contributed about 86.4% of the total area and 83.6% of the total volume of production of rice in Ethiopia. Amhara region production in 2014 was estimated to be 26,221 ha and 772,221 Quintal from 79,275 smallholders. Almost all of that rice was cultivated in the South Gonder administrative zone, which contains the Fogera district. In 2013/14, the Fogera district contributed 77% of the rice produced in Amhara and 60% produced in all of Ethiopia (CSA 2014; Fogera District Office of Agricultural 2015).

30.5 Use of Water Resources in the Fogera District

The Fogera district is one of the food surplus producing districts in the Amhara region because it has both rain fed and irrigated agriculture. For rice cultivation, the source of water for cultivation is rain. The use of irrigation was very low in the recent past. But, the study in 2010/11 showed that irrigation has increased dramatically in Fogera and almost all the rivers crossing the Fogera plain were used for irrigation. The district Agriculture and Rural Development Office estimates the land under irrigation to be about 80–83% of the potential area that can be irrigated by these water sources. About 55.8% of the 200 producers sampled in the 2010 survey use only rain but 45.5% of producers supplement rainfall with irrigation water from rivers and springs. Moreover, many farming households practice multiple cropping cycles per year using irrigation and residual moisture in swampy areas. For example two of the kebeles in Fogera irrigated 82.69 and 80.92% of their potentially irrigable land in the 2009/10 production season. Farming households irrigate their land by diverting or pumping rivers and springs sometimes using motor pumps.

30.6 Rice Introduction and Technology Adoption

Astewel (2010) explains that the basis for rice introduction in the district as well as in the Amhara region was the discovery of wild rice in the Fogera plain in the early 1970s. Rice production in Fogera was started by the Ethiopian Institute of Agricultural Research (EIAR) with two cooperatives and support of North Korean experts, in 1974. In the early 1980s through the technical support of North Korean experts', research on rice was initiated with Shaga cooperatives at the district and in a neighboring district of Dera and at a particular place of Jigna. This continued until farmers' cooperatives were dismantled in 1991 (Getachew 2000). Following this, Getachew Afework who was initially an expert in South Gondar Department of Agriculture started to collect seed locally from the previous introductions. Initial seeds were obtained from a farmer in Jigna kebele and that was why the major variety cultivated in the district and the surroundings was named as X-Jigna. After multiplication, seed was again distributed to other farmers for demonstration and research purposes under his supervision.

The Adet Agricultural Research Centre then released three other rice varieties called Gumera, Kokit and Tigabe. Adet Agricultural Research Centre has recently released the red colored Gumera variety (IAC 164) initially made it unacceptable by many producers and consumers who were not used to prepare red Injera, which led to low market demand within and outside the district. Rice attracted government attention in 1990s and became a focus of the extension service in Fogera and the region. During the early days of rice introduction, the extension system was geared towards promoting rice production to larger number of farmers. Farmers were provided with free inputs (seeds and fertilizer) and more farmers were also trained on the agronomic practices. Despite all these efforts, there was initial resistance due to an incorrect perception that rice causes infertility in humans. But, by collaborative and aggressive extension and research system the problem is resolved. The extension agency first promoted rice in the six kebeles in Fogera that usually flooded, using a lowland paddy rice variety. Next, extension started to focus to other areas of the district with additional rice varieties and agronomic practices (Tilahun et al. 2012).

The number of years that households in this survey have cultivated rice ranges from 1 to 15 years, with an average of 5.97 years (3.4 years standard deviation). About 82.5% of the 200 sample households were rice producers and, all but two sample households were planting the X-Jigna variety. Most of the sample households (77.5%) get their primary information about rice from other farmers. About a quarter also get their information from agricultural extension agents.

30.7 Trend of Rice Production in Fogera

Due to the importance of the rice crop and the abundance of water in Fogera, rice production has expanded significantly. In more than the last two decades from 1993/94 to 2014/15, the number of rice producing kebeles has increased from 2 to 24; the number of rice growing households increased from 30 to 34,249; and the area in rice cultivation grew from 6 to 20,230 ha. Rice became the second major crop of the district even though it had not been a major crop in the past. Figures 30.1 and 30.2 show the trend in adoption of rice production in this period.

Acceptance of rice technology in 1998/99 led to the considerable expansion of rice production at that time. By 2000/01 the production of rice could no longer expand into other kebeles because it was constrained by less favorable agro-ecology in the remaining kebeles which resulted in lower performance of the rice varieties being used. But, rice production and rice producing kebeles were significantly increased from 2007/08 onwards particularly due to introduction and better adoption of upland varieties in new kebeles and agro ecologies. After a jump of trend in 2007/08, it seems to reach at maximum in kebele coverage and consistently increasing in total production area and volume.

The government planned to continue this intensification and increase the national rice production to be produced by investors from 107,230 tons in 2010 to 280,627 tons after five years by 2015. In the same manner, the rice productivity is

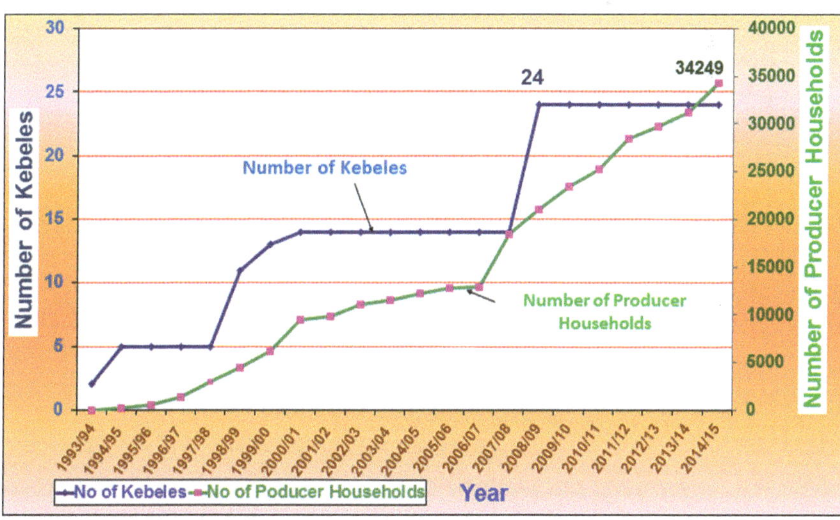

Fig. 30.1 Participant kebeles and households in rice production in Fogera district. *Source* Figure based on data from Fogera district office of Agricultural (2015)

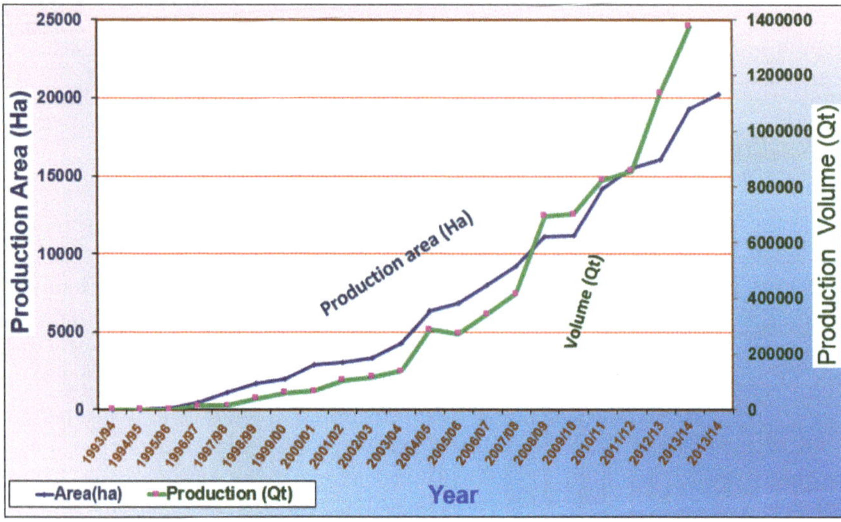

Fig. 30.2 Trend of rice production area and volume in Fogera district. *Source* Figure based on data from Fogera District Office of Agricultural (2015)

planned to increase from 0.8 Tons/ha in 2010 to 1.3 Tons/ha (Kejela 2011). In line with this, there will be expansion of rice production in Fogera district by smallholder farmers than investors with production on plots of land used to cultivate other crops and introduction and intensive extension of upland varieties in rice non-producing kebeles.

30.8 Structure, Conduct and Performance of Rice Value Chain

Rice has a variety of uses. The grain can be consumed by the producer or marketed. The grain straw and other by products can be used as animal feed, for house and roof construction, as a fuel source, and can also be marketed as a raw material for some factories. This sub section discusses the structure of the rice market, the conduct of market participants, and the efficiency and effectiveness of the market.

30.8.1 Rice Processing and Use

This section is primarily the result of the rice farming system and value chain analysis conducted in 2010/11 by interviewing about 200 sample rural households

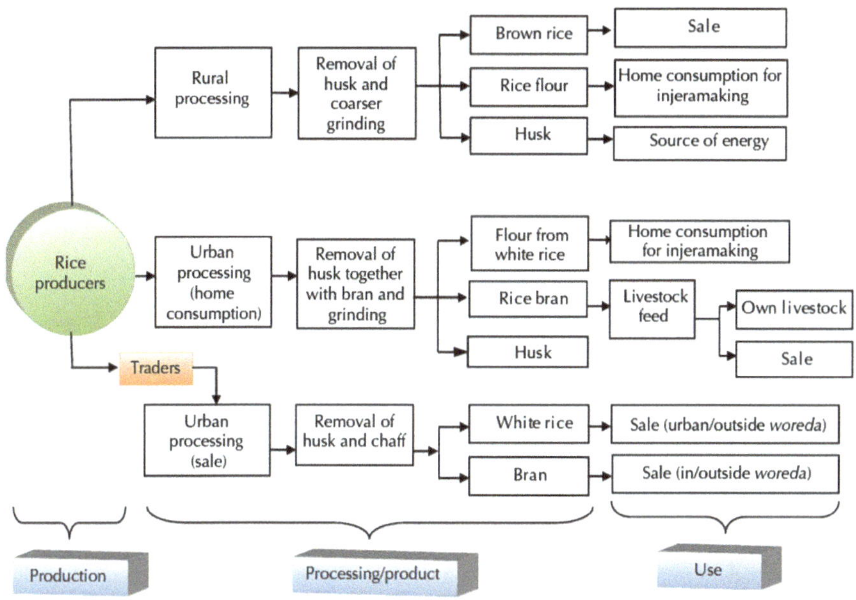

Fig. 30.3 Flow of rice processing in rural and urban areas in Fogera. *Source* Tilahun et al. (2012)

and key informant value chain actors and concerned experts and reviewing additional supportive related literatures.

Produced rice by farming households have three possible extension of processing, use and marketing based on "rural processing" for domestic consumption by rural household and "urban processing" for urban household consumption and selling. Figure 30.3 shows three channels of flow for rice processing, use and marketing (Tilahun et al. 2012).

i. "Rural processing (for local consumption and sale)": This channel shows rice processing, use and marketing when rural households target rice for household level consumption. First they remove husk from the grain by coarse grinding and winnowing followed by grinding of the rice grain for making Injera, sourdough-risen flatbread and Ethiopian staple food, and other household level processed food items, brown rice for sale and husk for household energy. In rural areas, paddy rice is taken to grinding mills where the hull is removed (dehulled) by coarse grinding.
ii. "Urban processing (for local consumption)": This is when rural households sell to processors/wholesalers or use grinding mills. It has three levels based on the agreement and interest, i.e. dehulling (removal of hull and converting to brown rice), polishing (removal of the brown inner seed coat and converting to white rice) and grinding to floor. The by-products, hull and bran, can be taken by farmers to home or can be sold to the processors. Processors market the rice husk as a livestock feed.

iii. "Urban processors (for sale in and outside of district)": This is relatively advanced way of processing the paddy rice particularly for marketing purpose to urban dwellers in and out of the district. The processing is similar with the second one above, but removal of hull and bran were done simultaneously. In addition, the by-product have had bran mixed with hull and some broken rice grain which is more nutritious and preferred to feeding animals and mainly to Bahir Dar and Gonder dairy and fattening farms.

30.8.2 Rice Market Structure

Market participants (actors)

The market participants can be classified into two main value chain actors and supportive value chain actors. Based on the survey, the main value chain actors were the following:

- *Shops and cooperatives*—Participated in delivering inputs like seed for rice products and also participate in the purchase of farmers product
- *Farmers*—are the main actors participating in the operation or production of rice
- *Polishing and grinding mills*—process the harvested rice and participate also in the buying and selling of rice product and act as assemblers, wholesalers and retailers.
- *Wholesalers*—are market agents at Woreta or other market centers that wholesale the product.
- *Retailers*—are market agents that buy from producers or wholesalers and retail to consumers and other retailers.
- *Brokers*—are mediators between two wholesalers and in some cases with producer farmers and wholesalers of different markets.
- *Factories*—buy rice bran from grinding mills and cooperatives and produce industrial commodities.
- *Transporters*—participate in the outbound logistic of the market value chain at different levels of supply.

Supportive Value chain actors: These were non-profit making market actors facilitating the proper activity of primary value chain actors and includes the following:

- *Agriculture office*—is mainly participating in the input delivery, technology introduction and innovation.
- *Non-governmental organizations*—like Improving Productivity and Marketing Success (IPMS) and Ethiopian Wetland and Natural Resource Association (EWNRA) that participated in research based extension system of rice technologies and crop land management.

- *Trade and Industry Office*—is a supportive value chain agent for legality of market middlemen in the value chain.
- *Research Centers*—are institutions participated in the development and pre-extension of rice technologies from past to the present. This includes the previous activities of the Present Ethiopian Institute of Agricultural Research and Adet Agricultural Research Center under Amhara National Regional State.

Rice Marketing Channels

The rice produced from the study areas was distributed in a number of channels. This can be treated from two points of view. These are flow of the commodity among number of market agents by diverse marketing functions and flow of the commodity in different locations during transfer of the product from producers to consumers by market transportation function. The flow of the commodity based on the survey conducted in 2010/11 in different market agents is explained in Fig. 30.4.

Rice producer farmers have an option to sell to polishing and grinding mills, cooperatives, local assemblers (collecting and selling to wholesalers), local retailers and consumers. Based on the response of sample households, about 85% of the product was sold to wholesalers and polishing and grinding mills. The other way of commodity flow other than agents is spatial flow of the rice product. Woreta town was the main distribution center of rice. Rice from Woreta town was distributed to all three transportation routes from Woreta in addition to the district level consumption and the surrounding. The major one was distribution extending to Addis Ababa including Bahir Dar and Debre Markos and the surroundings. The second was the distribution up to Mekele and Dessie towns and the surrounding including South Gonder. The third one was through Gonder to the surrounding and export to Sudan. Almost all farmers usually sold most of their rice product at Woreta. Only

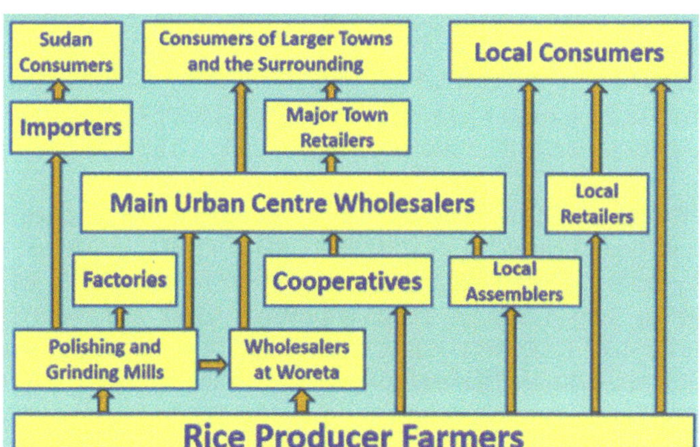

Fig. 30.4 Market value chain of rice from Fogera channeled through different agents. *Source* Figure based on field survey, 2010/11

24 and 45% of the farmers selling rice have had an experience to sell at local village level and nearby markets like Addis Zemen Town, respectively.

30.8.3 Marketing Conduct

Market conduct analysis of a market value chain is analysis of the behavior of all agents in the transfer of the commodity under consideration. However, this study focused primarily on the conduct of farming households as immediate and main producers and customers of rice to describe both the impact of water utilization in shifting production system and value addition and processing in consumption in the study area, Fogera district. Before the establishment of the cooperative and increased number of polishing and grinding mills at Woreta, farming households were forced to sell to the small number of wholesaling polishing and grinding mills and making them a price taker while selling their rice product. They had low bargaining power on the price of the product in these two cases. They could choose payment for the non-polished rice or payment after polishing the rice. During the survey period, farmers had an additional option, which was to pay only for the service of polishing. The payment to the farming households was done on cash basis. Cooperatives were other options for farming households to sell at an acceptable market price with additional profit repayment especially for members. Some passangers passing through Woreta were having a trend of purchasing significant volumes of rice for household consumption from polishing and grinding mills.

30.8.4 Rice Market Performance

The performance of the market is measured usually using price margin analysis. The average price of rice from 1991 to 1994 was 250 ETB (Ethiopian Birr)/ Quintal (Equivalent to 22.03 USD @ 1USD = 11.3487 ETB), and it was 950 ETB/Quintal (Equivalent to 83.71 USD @ 1USD = 11.3487 ETB) in 2009 due to the general price increase of all commodities nationally.[2] The producer price during the survey period in 2010 was 621 ETB/Quintal (Equivalent to 34.53 USD @ 1USD = 17.9841 ETB) and the wholesale price and retail price was 656 ETB/Quintal (Equivalent to 36.48 USD @ 1USD = 17.9841 ETB) and 695 ETB/ Quintal (Equivalent to 38.65 USD @ 1USD = 17.9841 ETB) with 35 ETB/Quintal (Equivalent to 1.95 USD @ 1USD = 17.9841 ETB) and 74 ETB/Quintal (Equivalent to 4.17 USD @ 1USD = 17.9841 ETB) price margin, respectively. The overall market share of farmers in percentage from the final retailers' price at

[2]Source of conversion rate from US Dollar (USD) to Ethiopian Birr (ETB) is from Exchange Rates.org.uk (http://www.exchangerates.org.uk/USD-ETB-exchange-rate-history.html—Accessed 06 July 2015).

Woreta market while including their production and marketing cost was ranged from 76.2 to 94.7% from the final price. But, this percentage was lower when the end prices of distant markets were considered.

30.9 Impacts of Water Resource Potential Utilization and Rice Production and Marketing

As described earlier, the major rivers in the district are Gumara and Rib Rivers. These rivers have resulted in serious flood in the district and they were damaging the farms and homes of the rural community. This was a serious challenge for the district for a number of years. But, due to the introduction of rice by using the excess water in the district, the challenge was shifted to an opportunity. The major impact of water potential utilization induced adoption of rice technology has resulted in a significant shift in farming systems and value addition practices in the following forms:

30.9.1 Change in Major Crop Types

In the previous years, rice was not known by the rural communities. It has been the first major crop of the district during the survey period followed by other major crops namely finger millet, teff, maize, sorghum, grass pea, chickpea, barley, oat and ground nut and niger seed from oil crops, and onion, peeper, papaya, orange, tomato, potato and cabbage from horticultural crops.

In the production system there was shift of crop enterprises due to the adoption of rice technology and irrigation practice. As a result, rice and horticultural crops like tomato and onion became dominant and replaced crops like teff, nigerseed, chickpea, grass pea and lentil. Sample households allocated 0.66 ha for rice, 0.65 ha for other cereals and 0.39 ha for non-cereal crops. The yields were 23.15 Quintal/ha rice, 8.56 Quintal/ha other cereals and 3.31 Quintal/ha non-cereals in 2009/10.

The other impact of rice in this regard was the results of farmers' innovation in integrating rice production with grass pea production by relay cropping. As a result, South Gonder was the second main producer zone next to North Wollo from all zones of the country based on the average production per year in the years considered (2006/07–2011/12) (CSA 2014).

The production of grass pea from Fogera district was by far higher than from the other districts of the South Gonder zone and it has been increasing over time. For example, the district's share of the total grass pea production of South Gonder zone in 2013/14 production year was about 28.6% (South Gonder Office of Agriculture 2014). The expansion of grass pea production in Fogera district was related to the cultivation of the crop by relay cropping with rice. Farming households plant grass pea seeds over the rice crop about two months before the rice is harvested. Farmers cut the grass pea along with harvesting the rice at maturity and it leads higher regeneration of the grasspea. By this practice, farmers conserve the residual moisture and efficiently use the land and water resource.

Land for grass pea was prepared from 4th week of August to October when it was produced as an independent crop and sowing was from October to November and spraying of insecticide was also done during this time to control pests. Harvesting and threshing of grass pea was done from December to January. Therefore, rice introduction results the two production systems of grass pea by independent production and mixing with rice.

30.9.2 Rice Consumption and Value Addition Practice

Farming households have started to consume this new crop by processing it in a number of forms. In this survey, it has been found that, all respondent farming households also adapt by innovative value addition practices to local food items including fermented pancake-like flatbread (*Injera*), porridge (*Genfo)*, various bread types and pizza (*Anebabro and Kita*), local Brewery (*Talla*), roasted rice grain (*Nifro*), distilled beverage (*Arekie*), rice souse (*Muke*), etc. From the total farming households, about 93.5% have consumed rice in the past. This indicated that, there were farming households that consume without producing rice. In addition, rice was consumed in combination with other crops especially when it was in *Injera* form. *Injera* from rice was preferred next to teff followed by millet, wheat and sorghum. The percentage of household consumption from the total production was considerably higher than Teff before a decade. Based on CSA report for 2001/02 production season, about 66.2% of the total farming households' rice production was consumed by the household level and the rest is sold and used as seed, in-kind wage payment, animal feed and others. While it was 62.8% of the total production consumed at household level for Teff in the same production year (CSA 2003). Therefore, adaptation in consuming of rice at household level was significantly higher than even Teff, the major staple food of the district. All these practices were adaptation mechanisms in processing and consumption as a result of water utilization induced shift in major crops of the area.

30.9.3 Alternative Animal Feed

Rice residue has 88.9% dry matter and gives better crude protein of 4.5% relative to other crop residues. Farming households feed their animals (Abebe 2007). According to Teshome (2009), it accounts for 60.7% of the feed resource in the study area. The rice bran was also given to livestock. Cattle, equines, small ruminants and poultry constitute about 91.4, 4.5, 3.8 and 0.3% of the total livestock population. The livestock ownership in TLU was 5.38 with a standard deviation of 3.42.

30.9.4 Marketing of Rice Grain and Residue and Income Generation

Rice is a multipurpose crop. In addition to consuming at household level, farming households use rice for income generation by marketing the grain and the crop residue. Woreta town is the main market center of rice for the rural community. In Fogera district, ownership of land, livestock and iron sheet-roofed houses are major indicators of wealth. Moreover, land which has potential for production of rice and access to irrigation and oxen from livestock has greater value. About 70–79% of the sample respondents has iron sheet-roofed houses. Farming households expressed that this was one of the main impacts water potential utilization and rice and horticultural crops production.

Farming households interviewed during the survey also believed that the low lying areas were fertile because of the silt deposited on the plots every year. As a result, fertilizer was not applied even on the fields which were repeatedly under rice. These farmers argue that whenever they apply fertilizer, there will be excessive vegetative growth and rice will lodge which will lead to poor seed setting. Organic fertilizer (compost) was a recent introduction to the district. Agricultural office of the district was working extensively with many farmers to adopt this technology. Farming households were also explain the extension service of the agricultural office to prepare their own compost from locally available materials including household wastes.

The Fogera rice in the market has had a reputation for poor quality because the grains are broken and the grain is sticky when boiled. Fractured grain is attributed to poor agronomic practices, post-harvest handling and low standard rice processing machines. In order to avoid breakage, farmers were advised not to harvest prematurely. In addition to grain production, straw and rice bran were by-products which have become the main sources of livestock feed but also generated household income through sale.

30.9.5 Enterprises Development

Another impact of use of the water resource and rice technology adoption was a diversified agribusinesses developed in rural and urban areas of the district and the surrounding areas. Yirgalem (2006) reports, there were nine polishers in Fogera district which were owned by six individuals in 2006. A study conducted in 2011–12 to quantify the number of processors and volume processed showed the number of processors at Woreta had increased dramatically and was estimated to be 120–150 with 82,782.6 tons of rice processed annually (Tilahun et al. 2012). This accounts for about 93% of the total production. The service charge, during the survey period in 2010, at rural kebele processors ranged from ETB 20/Quintal (Equivalent to 2.51 USD @ 1USD = 17.9841 ETB) during the dry season to ETB 60–70/Quintal in the rainy season. In addition to this processing, number of enterprises have emerged and involved in the post production transportation and marketing activities including wholesaling and retailing.

30.10 Conclusion

The introduction of rice technology in the study area and the use of untapped water potential have shifted the cropping pattern and agricultural enterprise choice or the farming system in general by replacing fallow land and grazing land and crops like Teff and millet by rice. Moreover, it has changed the threat of flood to opportunity. On top of its potential for rice production, proximity of the irrigation accessible and rice and horticulture producing kebeles to the town of Woreta and access to road has allowed the study area to derive economic benefit from production increase. The improvement of the farming households was not only in the production and increased income by selling the commodity rather it can also be depicted adaptability of the rice grain and residue utilization with the local previous experiences including the consumption of rice in different forms in combination with other crops and feeding animals. The water resource utilization and the introduction, diffusion and impact of rice technology in the study area have an important implication in the development and extension of agricultural or rural based technologies to areas of similar settings in Africa and the world.

Acknowledgements The author gratefully acknowledges the financial support of the Japan International Research Center for Agricultural Sciences (JIRCAS) in collecting the data on the general farming system of Fogera district and value chain of rice crop while implementing its "Development of Improved Infrastructure and Technologies for Rice Production in Africa" project.

References

Abebe H (2007) Supplementation of graded levels of concentrate mix on feed intake, digestibility, live weight change and carcass characteristics of Washera sheep fed urea treated rice straw. MSc Thesis, Haramaya University

Afework G (2000) Rice adaptation in Metema district North Gondar, Amhara Region, Ethiopia

Conway D (2000) The climate and hydrology of the upper blue Nile river. Geogr J 166: 49–62. 2000.tb00006.x/epdf. Accessed on 2 June 2015

CSA (Central Statistical Agency of Federal Democratic Republic of Ethiopia) (2014) Agricultural Sample Survey 2013/2014 (2006 E.C): Report on Area and Production of Major Crops (Private

Peasant Holdings, Meher Season), Statistical Bulletin Volume 1, No 532, Addis Ababa. http://www.csa.gov.et/images/general/news/land_use_2006_final. Accessed on 2 April 2015

CSA (Central Statistical Agency of Federal Democratic Republic of Ethiopia) (2013) Population Projection of Ethiopia for All Regions at Wereda Level from 2014–2017, Addis Ababa. http://www.csa.gov.et/images/general/news/pop_pro_wer_2014-2017_final. Accessed on 2 April 2015

CSA (Central Statistics Authority of Federal Democratic Republic of Ethiopia) (2003) Agricultural Sampling Survey. Results for Amhara Region, 2001\02 Bulletin Vol. III, Addis Ababa. http://csa.gov.et/. Accessed 10 May 2015

Exchange Rates.org.uk. http://www.exchangerates.org.uk/USD-ETB-exchange-rate-history.html. Accessed 06 July 2015

FAO (2008) Rice Production Trends in the World and Africa, Rome. Available on http://www.fao.org. Accessed on 2 April 2015

Fogera District Office of Agriculture (2015) Production area and volume of Fogera district from 1993/94 to 2014/15 production year, Woreta

Kejela G (2011) Coalition for African Rice Development (CARD): Mapping of Poverty Reduction Strategy Papers (PRSP), sector strategies and policies related to rice development in Ethiopia, Wabekbon Development Consultants Plc, Addis Ababa, Ethiopia. http://www.riceforafrica.org/. Accessed on 2 April 2015

Mastewal E (2012) Knowledge of rainfall partitioning for IWRM process: upper blue Nile basin, Ethiopia. In: Tefera B, Worie W, Wale M (eds) Proceedings of the second national workshop on challenges and opportunities of water resources management in Tana Basin, Upper Blue Nile Basin, Ethiopia, 26–27 Mar 2012. Blue Nile Water Institute—Bahir Dar University (BNWI-BDU), Bahir Dar, Ethiopia, pp 211–216. http://www.bdu.edu.et/bnwi/sites/bdu.edu.et.bnwi/files/Part%20II.pdf. Accessed on 2 April 2015

Miheret E, Ernest W (2012) Assessment of major threats of Lake Tana and strategies for integrated water use management. In: Tefera B, Worie W, Wale M (eds) Proceedings of the second national workshop on challenges and opportunities of water resources management in Tana Basin, Upper Blue Nile Basin, Ethiopia, 26–27 Mar 2012. Blue Nile Water Institute—Bahir Dar University (BNWI-BDU), Bahir Dar, Ethiopia, pp 281–292. http://www.bdu.edu.et/bnwi/sites/bdu.edu.et.bnwi/files/Part%20II.pdf. Accessed on 2 April 2015

SG2000 (Sasakawa Global Foundation 2000) (2006) Annual report

Sewmehon D (2012) Water and rural livelihoods in the crop-livestock system of Amhara Region, Ethiopia: Multiple use system (MUS) approach for water productivity improvement. In: Tefera B, Worie W, Wale M (eds) Proceedings of the second national workshop on challenges and opportunities of water resources management in Tana Basin, Upper Blue Nile Basin, Ethiopia, 26–27 Mar 2012. Blue Nile Water Institute—Bahir Dar University (BNWI-BDU), Bahir Dar, Ethiopia, pp 251–257. http://www.bdu.edu.et/bnwi/sites/bdu.edu.et.bnwi/files/Part%20II.pdf. Accessed on 2 April 2015

South Gonder Agricultural Office (2014) South Gonder Administrative zone production area and volume of production of 2013/04 production year, South Gonder Office of agriculture, Debre Tabor

Teshome D (2009) On-farm evaluation of urea treated rice straw and rice bran supplementation on feed intake, milk yield and composition of Fogera cows, North Western Ethiopia, A Thesis Submitted to the Department of Animal Science and Technology School of Graduate Studies of Bahir Dar University in Partial Fulfillment of The Requirements for the Degree of Master of Science In Agriculture (Animal Production), Bahir Dar University, Bahir Dar. http://www.ipms-ethiopia.org/content/files/Documents/publications/MscTheses/Final%20Thesis%20(Teshome%20Derso).pdf. Accessed on 2 April 2015

Yared A, Dimitri P, Moges SA, Schalk J. Van Andel, Shreedhar M, Yunqing X (2012) Spatio-temporal assessment of drought under the influence of varying record length: the case of Upper Blue Nile Basin, Ethiopia. In: Tefera B, Worie W, Wale M (eds) Proceedings of the second national workshop on challenges and opportunities of water resources management in Tana Basin, Upper Blue Nile Basin, Ethiopia, 26–27 Mar 2012. Blue Nile Water Institute—

Bahir Dar University (BNWI-BDU), Bahir Dar, Ethiopia, pp 113–130. http://www.bdu.edu.et/bnwi/sites/bdu.edu.et.bnwi/files/Part%20II.pdf. Accessed on 2 June 2015

Yirgalem A (2006) Progresses, on-going activities and the way forward based on two years' experience of IPMS Fogera, 2005 and 2006

Further Reading

Astewul T (2010) Analysis of rice profitability and marketing chain: the case of Fogera district, South Gondar, Amhara Region, Ethiopia. MSc thesis, Haramaya University. https://cgspace.cgiar.org/handle/10568/816. Accessed on 2 April 2015

Tilahun G, Kahsay B, Dirk H, Bogale A (2012) Rice value chain development in Fogera woreda based on the IPMS experience. Nairobi, Kenya: ILRI. https://www.google.com.et/url?sa=t&rct=j&q=&esrc=s&source=web&cd=1&cad=rja&uact=8&ved=0CBsQFjAA&url=https%3A%2F%2Fcgspace.cgiar.org%2Fbitstream%2Fhandle%2F10568%2F16850%2FIPMS_Rice_CaseStudy.pdf%3Fsequence%3D8&ei=r3ObVbtZytlTyNqAuAs&usg=AFQjCNFLamrHNW9lvPMQvbqkHcYM52CB0A&sig2=8mjopsIHan2cVBi7UHniDQ&bvm=bv.96952980,d.bGQ. Accessed 10 July 2013

IPMS (Improving Productivity and Marketing Success) (2005) Fogera District Pilot Learning Site Diagnosis and Program Design, Addis Ababa. https://www.google.com.et/url?sa=t&rct=j&q=&esrc=s&source=web&cd=1&cad=rja&uact=8&ved=0CBsQFjAA&url=https%3A%2F%2Fcgspace.cgiar.org%2Fbitstream%2Fhandle%2F10568%2F16747%2FFogera.pdf%3Fsequence%3D1&ei=-3ObVY26FYSAU-PYm4gL&usg=AFQjCNGBMgnRF3KuJ9jRJVIfRH_tAmAk1 g&sig2=cnp8zxARkbbTP3BjCErfBg&bvm=bv.96952980,d.bGQ. Accessed 10 July 2013

Chapter 31
Trends and Driving Forces of Eucalyptus Plantation by Smallholders in the Lake Tana Watershed of Ethiopia

Berihun Tefera and Habtemariam Kassa

Abstract Lake Tana watershed is one of the growth corridors of the country identified by the government of Ethiopia. The competition for land among the different alternative land uses is intense. One of these is eucalyptus plantation not only on marginal lands but also on fertile plots used for producing annual crops. The trends and driving forces of eucalyptus plantation in the Lake Tana watershed and in Amhara region of Ethiopia in general are not well documented. This chapter reports on findings of a study on the trends and driving forces of eucalyptus plantation in Lake Tana sub basin based on preliminary assessment of major eucalyptus production areas of Amhara region and a case study from Mecha district of West Gojjam, Amhara Regional State. The district represents the major eucalyptus plantation area of the watershed. The study was conducted in 2011 and 2012. In addition to the visit of the major eucalyptus production and marketing sites and key informants interview in Amhara region, formal survey was administered to 400 sample households randomly and proportionately selected from high, medium and low level eucalyptus planting kebeles of Mecha district. Results show that close to 77% of the sample households had planted eucalyptus. The trend of eucalyptus plantation is increasing and has been particularly high since 2002. Wald test has been used to check the presence of the structural break of trend in the year in the yearly cumulated number of farming households who started eucalyptus plantation. The result confirms the presence of significant structural break in trend for the year ($p < 0.01$). More than 60% of the eucalyptus growers plant eucalyptus in woodlots on plots with plain topography and with medium to high levels of soil fertility. This indicates that eucalyptus plantation is

B. Tefera (✉)
Agricultural Economics Department, College of Agriculture and Environmental Sciences and Blue Nile Water Institute, Bahir Dar University, Bahir Dar, Ethiopia
e-mail: berihunt2@gmail.com

H. Kassa
Center for International Forestry Research, Ethiopia Office, Addis Ababa, Ethiopia
e-mail: h.kassa@cgiar.org

B. Tefera
Department of Agricultural & Resource Economics, University of Connecticut, Storrs, USA

K. Stave et al. (eds.), *Social and Ecological System Dynamics*,
AESS Interdisciplinary Environmental Studies and Sciences Series,
DOI 10.1007/978-3-319-45755-0_31

expanding at the expense of crop production. The major drivers for the expansion raised by key informants are the increasing price of farm inputs (mainly fertilizer) for crop production, the Ethio-Sudan cross-border trade, the rising demand of wood for construction, industries and woodwork shops, and inadequacy of wood from natural forests due to deforestation. Besides these socioeconomic drivers, certain biological properties of eucalyptus make it especially appealing. Adaptability to wider agro ecological zones, coppicing ability, fast growth rate, straight poles, low labor demand, and resistance to diseases and pests are its commonly cited qualities. Given the growing demand for wood in the domestic and international market, and the expressed intention of most rural households to plant eucalyptus, eucalyptus plantation is likely to expand even more in the Lake Tana watershed. Therefore, proper land use development and organization of appropriate institutes for optimum level of eucalyptus plantation and balanced enterprise choice for the rural households and overall development of Lake Tana watershed and the region are suggested.

Keywords Amhara regional state · Land use changes · Markets · Mecha district · Smallholders · Woodlots · Wald test

31.1 Introduction

31.1.1 Lake Tana Watershed

The Lake Tana watershed is the headwater basin of the Upper Blue Nile River. The basin covers an area of approximately 15,354 km^2, of which Lake Tana covers 3150 km^2. The average annual rainfall of the watershed is about 1350 mm and much of the rainfall occurs between June and September. The average annual temperature is 28 °C. About 48.4% of the Lake Tana watershed has plain topography. Rugged lands and valleys cover 36.1 and 15.5% of the watershed, respectively. The soils of the watershed are considered as fertile and suitable for crop cultivation. Commonly the soils are classified as red, brown and black covering respectively 40.3, 32.8 and 26.8% of the watershed. Based on mainly altitude, three major agro-ecological zones are known to exist: midlands (*Woina dega*) share 59.5% whereas highlands (*Dega*) and lowlands (*Kolla*) cover 25.5 and 15% of the watershed, respectively (Miheret 2012).

31.1.2 Development Plans and Practices in Lake Tana Watershed

The Lake Tana watershed has relatively developed urban centers and dense settlements, good roads, and reasonable air connectivity to other regions. It has potential for growth in many sectors with strong multiplier effects, for instance in smallholder agriculture, agro-industry, tourism, fisheries, livestock and energy, that can stimulate further national and regional economic growth and improved livelihoods. Lake Tana is also the largest multipurpose lake of the country, which faces many environmental threats. There is little integrated baseline information on its ecological, social, cultural, biodiversity and economic values (Miheret 2012). Some trials, such as the six year (2010–2016) water resource based strategic plan of the Tana growth corridor of Amhara Regional state was not properly implemented. Planning that integrates different sectors and taken a holistic and long term objectives to develop the watershed is lacking.

As part of its effort to alleviate food security problems, the federal government of Ethiopia has started construction of small- and large-scale irrigation schemes on most of the rivers in the Lake Tana watershed. Also, the regional government of Amhara National Regional State has launched projects that will irrigate over 128,000 ha with submerged area of about 10,000 ha at different sites and from different rivers in the watershed (Sewmehon 2012). The submerged area is expected to be utilized for fishery that will contribute to increasing fish supply and reduce the pressure on fish resources of Lake Tana.

31.1.3 Eucalyptus in Ethiopia

Eucalyptus species were introduced to Ethiopia towards the end of the 19th century. About 55 species of eucalyptus have been introduced and only few of them are widely planted. The most widespread species include *E. camaldulensis, E. citriodora, E. globulus subsp, E. globulus, E. regnans, E. saligna* and *E. tereticornis* (Friis 1995; Dessie and Erkossa 2011). Certain unique characteristics of Eucalyptus make it highly desirable for growers. It is not browsed by livestock, has high survival rate when planted, grows well even on marginal soils, grows fast, produces straight poles, and has coppicing ability. These qualities have encouraged farmers to plant Eucalyptus on available sites (Dessie and Erkossa 2011). Between the two main species, *E.globulus* and *E.camaldulensis*, eucalyputs can be found in nearly all the major agro-ecological zones of the country, in wet and dry highlands, as well as in mid-altitudes. In Ethiopia *E.globulus* grows from about 1900 to 2800 m above sea level and above; while *E. camaldulensis* grows between 500 and 2000 (Azene 2007; Mekonnen 2013).

Trees provide significant socio-economic and ecological benefits. In many areas, there is tradition to plant trees in order to reduce dependence on natural forests and to reduce degradation of natural resource. Early maturing tree species are often chosen by tree growers over more environmentally friendly species, such as nutrient-replenishing leguminous trees (Garay et al. 2004).

Eucalyptus is a common species introduced into agricultural systems in areas where trees can be combined with the production of crops (Kidanu et al. 2005).

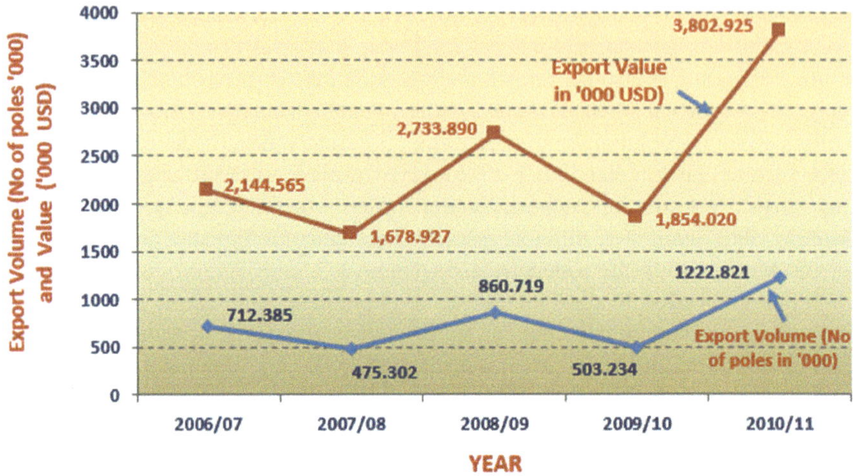

Fig. 31.1 Export volume (Number of poles in '000) and export value (in '000 USD) to Sudan through Metema. *Source* Compiled from data gathered from the Ethiopian Customs Authority, Metema Branch (2012)

Traditional agro-forestry practices in Ethiopia involve tree planting in various spatial patterns to meet the demand for fuel wood and construction. Single rows of Eucalyptus species planted along field borders have become a common feature of the central highlands of Ethiopia, including the study area. Eucalyptus is now being planted on farm plots that are even suitable for annual crops. This can bring a considerable return to the farmer from marketing poles and posts. Although the changes in eucalyptus planting can be seen in the landscape, the major drivers and the trends overtime in eucalyptus plantation in the watershed have not been systematically studied. This information is needed for proper planning of land use and land management in the region in the general and Lake Tana watershed in particular.

Amhara region has since recently been exporting eucalyptus poles. Records at Metema District Customs office (Metema Customs 2012) show that between 2006/07 and 2010/11, 3,774,461 poles were exported to Sudan through Metema with a total value of 12,214,327 USD (Fig. 31.1). Much of the exported pole is believed to have come from Merawi market that collects eucalyptus poles from smallholders in Mecha district and the surrounding areas. It became one of the exportable items to Sudan in this port Next to pulse crops and live animals.

31.2 Methodology

This chapter reports on a study conducted in 2011 and 2012 on eucalyptus plantation trends and driving forces among smallholder in the Lake Tana Basin. Survey data was mainly collected from Mecha district following preliminary eucalyptus production and marketing assessment in the region. The district was purposively selected for its widespread plantation of eucalyptus and its importance as a source of eucalyptus exports to neighboring regions and to Sudan. We used a structured questionnaire to collect information primarily from eucalyptus growers. We also interviewed key informants and made physical observations in Mecha district and other major eucalyptus plantation and marketing areas/districts of Amhara region, including Gondar Zuria and Metema from North Gondar zone, Farta and Gayent districts from South Gonder zone and Yilmana Densa and Achefer districts from West Gojjam zone. Major key informants were eucalyptus growers, experts, market middlemen including exporters and consumers who were selected purposively.

Three-stage random sampling technique was used. In the first stage, all 39 rural kebeles (the lowest administrative unit in Ethiopia) in the district were stratified into three categories based on the level of eucalyptus plantation in the kebele—low, medium and high. Each stratum was further divided into four categories based on access to roads. These four strata are kebeles having access to the main Addis Ababa—Bahir Dar asphalt road, kebeles with access to all weather non-asphalt roads, kebeles who have no access to any of the roads but with access to transport facilities in dry season after harvest by crossing fields to transport eucalyptus, and kebeles with no any access even during dry season in any of the above possibilities. Table 31.1 shows the number of sampled households in each category. The categorization of kebeles to different strata and determination of percentages were based on the secondary data and key informant interview with experts of the district agricultural office. One kebele from each of the three levels of plantation was selected. Within those kebeles, the proportion of households in each road access category was calculated and sample households in each kebele were selected proportionate to households' access to road.

Ultimately, data was collected from 165 households in Ambomesk kebele (high level plantation area), 150 households in Berakat kebele (medium level plantation area) and 85 households in Goragote kebele (low level plantation area) (a total of 400 households).

The questionnaire focused on six main categories: household characteristics; household asset, income and production costs; eucalyptus plantation and production; eucalyptus use and marketing; access to institution, infrastructures and information; and general challenges in eucalyptus production and marketing. In addition to the descriptive analysis, statistical tests like t-test, chi sqaure test and Wald test were used to analyze the data.

Table 31.1 Number and distribution of selected kebeles and sample households considering level of eucalyptus plantation and accessibility (n = 400)

Level of eucalyptus plantation	No. of kebeles and randomly selected kebele	Degree of road access	Number of HHs	% Share from total population	Sample share from 400 sample HHs	Actual sample HHs taken
– High	14 (Ambomesk Kebele)	– Accessible by asphalt road	8674	13.59	54.35	54
		– Accessible by all-weather non-asphalt road	15,282	23.94	95.76	96
		– No road but accessible during the dry season	1886	2.95	11.82	12
		– No accessible by transport year round	450	0.70	2.82	3
		Sub total	26,292	41.19	164.75	165
– Medium	14 (Berakat Kebele)	– Accessible by asphalt road	0	0	0	0
		– Accessible by all-weather non-asphalt road	10,876	17.04	68.15	68
		– No road but accessible during the dry season	11,778	18.45	73.80	74
		– No accessible by transport year round	1255	1.97	7.86	8
		Sub total	23,909	37.45	149.82	150
– Low	11 (Goragote Kebele)	– Accessible by asphalt road	0	0	0	0
		– Accessible by all-weather non-asphalt road	650	1.02	4.07	4
		– No road but accessible during the dry season	7510	11.76	47.06	47
		– No accessible by transport year round	5474	8.58	34.30	34
		Sub total	13,634	21.36	85.43	85
Total	39 kebeles		63,835	100.00	400.00	400

31.3 Results and Discussion

31.3.1 Household Characteristics and Resource Ownership

31.3.1.1 Household Socioeconomic Characteristics

Demographically, the heads of households in the watershed and the region in general are usually male, married and illiterate. This is also true for the sampled households of the case study. Table 31.2 summarizes the demographic characteristics of the heads of the sampled households in terms of sex, religion, ethnicity, marital status, academic status, and leadership role in the community of planter and non-planter households. The total number of households who have planted eucalyptus is 307, constituting 76.75% of the total samples. Thirteen percent of the sampled households are female-headed households. About 82% of the respondents are married and 55% of the heads of households were illiterate. Nearly 59.3% of the household heads have a leadership role in the community, which could be political, religious or other non-religious social role. Households headed by males, married, literate and having leadership role are significantly ($P < 0.01$) more likely to have planted eucalyptus more than others (Table 31.2).

Table 31.2 Socioeconomic characteristics of planter and non-planter households

Households' characteristics	Categories	Planters		Non-planters		Total		X^2-value
		Freq.	%	Freq.	%	Freq.	%	
Sex	– Male	277	90.2	71	76.3	348	87.0	12.165*
	– Female	30	9.8	22	23.7	52	13.0	
Marital status	– Married	265	86.3	64	68.8	329	82.2	15.121*
	– Never married	21	6.8	14	15.1	35	8.8	
	– Divorced	13	4.2	10	10.8	23	5.8	
	– Widowed	8	2.6	5	5.4	13	3.2	
Academic status	– Illiterate	149	48.5	70	75.3	219	54.8	22.214*
	– Read and write only	84	27.4	13	14.0	97	24.2	
	– Grade 1–6	51	16.6	6	6.5	57	14.2	
	– Grade 7–8	13	4.2	2	2.2	15	3.8	
	– Grades 9 and above	6	1.9	2	2.2	8	2.0	
	– Religious school	4	1.3	0	0.0	4	1.0	
Leadership role	– No role	169	55.0	67	73.6	236	59.3	13.574*
	– Religious	26	8.5	8	8.8	34	8.5	
	– Political	49	16.0	5	5.5	54	13.6	
	– Social	55	17.9	8	8.8	63	15.8	
	– Others	8	2.6	3	3.3	11	2.8	
Total		307	76.75	93	23.25	400	100.0	

Source Compiled from the survey data
Remark *significant at 0.01 level

Table 31.3 Mean and standard deviation of age and family size of planter and non-planter households

Socioeconomic characteristics	Planters		Non-planters		Total		t-Value
	Mean	SD	Mean	SD	Mean	SD	
Age of household head	46.25	12.82	43.73	14.09	45.66	13.15	1.622*
Family size (Absolute number)	6.45	2.38	4.67	2.20	6.04	2.45	6.458**
Family size (Adult equivalent)	3.87	1.74	2.91	1.41	3.65	1.72	5.425**

Source Compiled from the survey data
Remark *, **significant at 0.1 and 0.01 levels, respectively

The age of household heads ranged between 22 and 83. Average age of the sample household head is 45.7 years and planters are significantly ($P < 0.1$) older than non-planters. Family size of the sample households is 6.04 in absolute numbers and 3.65 in adult equivalent, and the difference between the planters and non-planters was highly significant ($P < 0.01$) (Table 31.3).

31.3.1.2 Household Resource Ownership

The average area of land owned per household is 0.93 ha (s.d. 0.59), divided into an average of 3.37 (s.d. 1.36) parcels. But, farmers also cultivate land they rent from others. When rented land is included, the total land households used in 2011 becomes 1.04 ha (s.d. 0.58). The average oxen and total livestock ownership are 1.56 oxen (s.d. 1.05) and 5.38 TLU (s.d. 3.33). Moreover, sample households who have planted eucalyptus are better-off in all the specified resources and the difference is significant ($P < 0.01$). The total number of households that rented in and rented out land is 64 and 31 respectively. The average size of rented in land is 0.47 ha (s.d. 0.33) and rented out amounts to 0.52 ha (s.d. 0.38).

The average area of land allocated for eucalyptus is 0.24 ha (s.d. 0.27) (Table 31.4). The high standard deviation shows the significance of variability of eucalyptus growers in the study area. About 8.14% of the growers were found to have planted all their land and rented in additional land from others.

31.3.2 History and Trend of Eucalyptus Plantation

As natural forests and woodlands are shrinking rapidly, smallholders establish their own woodlots using fast growing tree species. About 58% of the area under planted forests is covered with Eucalyptus, with an estimated area of over 0.5 million ha in 2010 (Gil et al. 2010). Respondent planter household heads were asked the year when they started planting eucalyptus for the first time. Only 269 households from the total 307 eucalyptus planters could recall the year when they first planted eucalyptus. The cumulative number of eucalyptus planters in the district in each

Table 31.4 Land, trees and livestock asset ownership of planter and non-planter households

Resource description	Planters		Non-planters		Total		t-Value
	Mean	SD	Mean	SD	Mean	SD	
Total land owned (ha)	1.00	0.60	0.66	0.45	0.93	0.59	5.836*
Total land used (ha)	1.13	0.59	0.77	0.45	1.04	0.58	6.085*
Land planted with eucalyptus	0.24	0.27	–	–	0.24	0.27	
Number of plots	3.53	1.36	2.82	1.20	3.37	1.36	4.797*
Oxen ownership	1.65	1.08	1.26	0.88	1.56	1.05	3.172*
Livestock ownership in TLU	5.73	3.31	4.14	3.09	5.38	3.33	4.005*

Source Compiled from the survey data
Remark *significant at 0.01 level

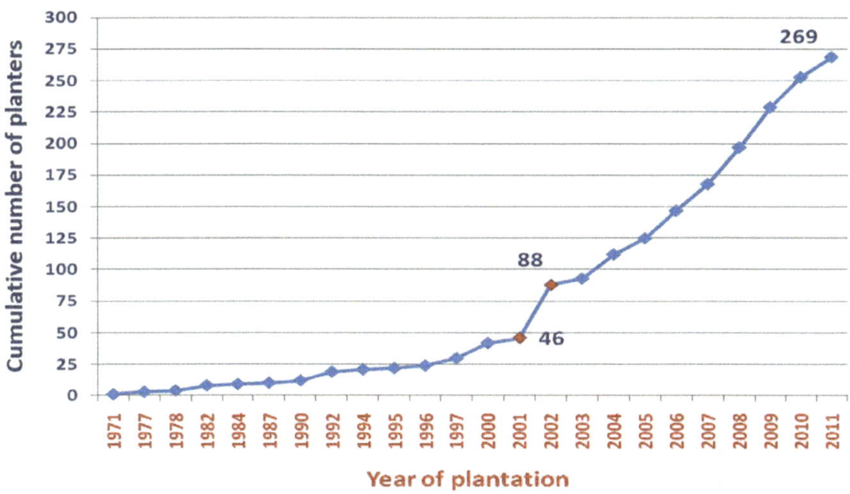

Fig. 31.2 Cumulative number of Eucalyptus planters in the selected kebelees of Mecha district between 1971 and 2011. *Source* Compiled from the survey data

year from the total sample households is presented in Fig. 31.2. The figure shows a rapid rise in plantations began from mid 1990s to early 2000s. Wald test has been used to check the presence of the structural break during the trend jump between 2001 and 2002 in the yearly cumulated number of farming households who started eucalyptus plantation. The test for the year (2002) based on the robust regression analysis of cumulative number of farmers on year has shown that there was a significant structural break in the trend of cumulative number of farmers choosing the enterprise ($p < 0.01$). The period around the structural break year are times when a number of policy and market interventions that can affect the farmers' enterprise choice were implemented. Some of these are removal of subsidies and credits for agricultural inputs and begining of issuance of agricultural land use certificate.

31.3.3 Eucalyptus Tree Plantation Practices

Most of the eucalyptus trees planted by survey respondents are on plots that have the following characteristics: woodlots (59%), plain topography (92%) and good soil fertility levels (88%). This is an indication that tree growing is expanding onto farm plots suitable for cereal crop production. Many (almost 93%) practice intercropping of eucalyptus with cereal crops like maize, teff and wheat during the first year of plantation (Table 31.5).

31.3.4 Drivers of Eucalyptus Plantation

Demand for wood in the country is increasing as a result of population and economic growth. At the same time, supply is declining sharply due to high deforestation rate of natural forest and woodlands and a low level of investment in plantation forests. Thus, the gap between supply and demand is widening and forcing the country to import industrial wood and paper products. As early as the 1980s, the Government of Ethiopia introduced fast growing trees, particularly Eucalyptus species. These re-greening and plantation programs grew out of increased recognition by policy makers of the importance of expanding forest cover to increase the supply of forest products, conserve biodiversity and reduce the decline in forest-based ecosystem services (Limeneh and Kassa 2014).

Table 31.5 Eucalyptus plantation practices of sample households

Plantation characteristics	Categories	No. of eucalyptus plantings or plots reported	% of all responses
Plantation niche	Woodlot	181	59.0
	Farm boundary	119	38.8
	Scattered in the farms	32	10.4
Topography of planted plots	Plain	284	92.5
	Sloped/rugged	108	35.2
Soil fertility level planted plots	High fertility	161	52.4
	Medium fertility	201	65.5
	Low fertility	38	12.4
Cropping pattern	Practice intercropping	286	93.2
	No intercropping	9	2.9
Total planters		307	100.0

Source Compiled from the survey data
Note that respondents identified more than one response

Farming households do react rationally to policy and market signals taking into account the socio-economic benefits and costs of their decisions. This is also true when farmers make decision to plant trees. The benefits and costs that farming households take into account for a decision are influenced by different factors internal or external to them. Studies in other areas of Ethiopia have found that farmers' willingness to grow trees on their farms is a function of their attitudes towards the advantages of growing trees, their perception on the opinions of formal and informal community influencers and other socioeconomic factors (Zubair and Garforth 2006).

In this study, three main categories of factors have been identified as major drivers of the expansion of eucalyptus plantation in the Lake Tana watershed of Ethiopia. These are socioeconomic factors, the ecosystem characteristics in the region and the biological characteristics of eucalyptus.

i. **Socioeconomic drivers**

Farmers who planted eucalyptus were asked to identify main socioeconomic drivers that are driving the expansion of eucalyptus plantations by smallholders in the study area. As indicated in Table 31.6, the respondents identified higher price of eucalyptus logs and hence better economic returns relative to annual crops due to expanding domestic and export markets following the opening and growing importance of Ethio-Sudan border trade, rising prices of agricultural inputs notably chemical fertilizers and improved seeds, ability to sale anytime when needs for cash arises and get more cash income in the future, the need to meet own wood demands for energy and construction, and its lower demand for labour that enables household members to be engaged in nonfarm activities after planting.

The responses of farmers are in line with information from literatures. The following findings from literature and personal observations of the authors can describe the major drivers of eucalyptus plantation expansion under this category.

Table 31.6 Main socioeconomic drivers of Eucalyptus plantation expansion identified by planter households ($N = 307$) specified by sample households to plant trees

Socioeconomic drivers	Frequency	%
Higher economic returns relative to growing annual crops	295	96.09
Rising price of agricultural inputs	265	86.32
Possibility to have more cash in the future	288	93.81
Meeting own/household wood demand	272	88.60
Time availability after plantation to participate in nonfarm activities	196	63.84
Others	28	9.12

Source Compiled from the survey data
Note that most respondents identified more than one response

- **Higher economic returns from growing trees and the rising prices of agricultural inputs**

The price of fertilizer has been steadily increasing, making cereal crop production more costly (Tomoya and Takashi 2010). Based on the information from the experts of Mecha district agricultural office, the price of chemical fertilizer in April 2015 was Birr 4628/quintal (approximately 230 USD). The current extension recommendation level of fertilizer use in the study area implies that a farmer needs to buy fertilizer worth 5700 Birr (approximately 281 USD) per hectare of crop cultivation. In addition, annual crops are more susceptible to rainfall variability than eucalyptus plantations. Thus, shortage of capital to buy fertilizers and seeds together with lesser climatic and biotic risks in growing trees motivates farming households to establish and expand eucalyptus plantation.

In addition to the generally increasing price of fertilizers and other inputs such as improved seeds, there was "the collapse of maize prices in 2001" owing also to under developed seed distribution system for all cereal crops in general and maize in particular (Spielman et al. 2015). The main crop produced in Mecha district is maize. This increasing cost of inputs and decreasing price of grains is making cereals production, particularly maize, less attractive for smallholders. In a graph of hybrid maize seed-to-grain price ratio from the 1991/2 to 2007/8 production season, Spielman et al. (2015) show that the seed-to-grain price ratio has increased sharply from nearly 6 in 1999/00 to about 10 in 2000/01, even though it is followed by a considerable decrease in the following years. Getaw and Atle (2011) also indicated that the real price of grains including maize in Ethiopia is far less than the nominal price and the disparity is increasing significantly as of 2000. At the same time, selling eucalyptus brings a high economic return and can also satisfy household construction and fuelwood demand. For example, a study by Tilashwork (2009) found that about 96% of the Eucalyptus tree growers in Mecha district grow trees for income. This is in line with the findings of other studies related to eucalyptus in other parts of Ethiopia that growing of eucalypts was financially more profitable, with a considerable positive net present value, compared to the alternative agricultural crops in different areas (Tesfaye 1997; Daba 1998; Asaye 2002; Demamu 2002; Zerihun 2002). For example, eucalyputs wood products contribute 78% of the firewood, 100% the construction poles and posts, and 20% of the charcoal supplied to buyers at Huruta town in Oromia region of Ethiopia (Mekonnen 2010). Therefore, one of the major incentives for this is the market opportunity and economic incentive for growing eucalyptus.

- **Ethio-Sudan cross-border trade**

The Ethio-Sudan Trade Agreement Ratification Proclamation was signed between both countries in November 2003 (Proclamation No. 221/2000) and strengthened by Ethio-Sudan Preferential Trade Agreement Ratification Proclamation (Proclamation No. 318/2003) (FDRE 2000, 2003). These agreements led to the emergence and significant growth of an Ethio-Sudan cross-border eucalyptus pole trade that has resulted in increasing the price of eucalyptus and hence income to tree growers. Although exact data are not available for the volume of trade, marketing middlemen we interviewed indicated that about half of the Eucalyptus produced in Mecha district goes to the Sudanese market through Bahir Dar, the regional capital and the major market for much of the wood produced in Lake Tana watershed.

- **Increasing domestic demand due to urbanization and infrastructure expansion**

Buildings in towns and infrastructures in rural areas are expanding. Bahir Dar city and other urban centers in the region are also expanding. These constructions consume wood, particularly eucalyptus poles for scaffolding.

- **Increased demand of industries and woodwork shops**

The number of graduates from Agricultural, Technical, Vocational and Educational Training Colleges (ATVETCs) has been increasing rapidly in the past few years. These graduates establish additional woodwork shops. Even though, it is not regionally and nationally economical compared to establishing a factory closer to the source area for better advantage of both the input suppliers and product consumers, Maychew cheap wood factory in Tigray is also one of the market niches for eucalyptus wood from the study area.

- **Declining supply of woods from natural forests**

The national forest law prohibits cutting timber trees from natural forests. As a result, the supply of indigenous timber tree logs notably *Cordia africana* has declined significantly. Woodwork shops and charcoal producers that relied on indigenous species in the past are trying to shift to use eucalyptus logs.

ii. **Ecological characteristics of Mecha District**

Although eucalyptus trees can grow in a wide range of altitude, farming households and experts explained that the following agro-ecological factors of the study area (Mastewal 2012; Mecha District Agriculture Office 2011; Tilashwork 2009) make it ideal for plantation of Eucalyptus:

- **Altitude**: The altitudinal range of the district ranges from 1500 to 2500 m.a.s.l. (meter above sea level). This provides year round ideal temperature for tree growth.
- **Topography**: The district is dominantly plain, as only about a quarter of its area is characterized by rugged topography, valleys and mountains.
- **Soil fertility**: The soil in the district is considered as generally fertile.
- **Rain fall**: The district annually receives between 1000 and 2000 mm of rainfall.

iii. **Properties of the of Eucalyptus species**

Eucalyptus, which has more than 500 species, has become the most planted genus of tree in the world (Teketay 2000). Debates about whether the costs of eucalyptus outweigh the benefits are common worldwide among experts and researchers in diverse disciplines. The main arguments against the eucalyptus are its contribution to higher intake of water and drying of water resources, promotion of soil erosion, suppression of understory vegetation, depletion of soil nutrients and allelopathic effects (Davidson 1985; FAO 1988; Teketay 2000; Amare 2002; Nduwamungu et al. 2007). Arguments in favor of eucalyptus plantation are that the trees grow quickly, require minimal care, grow in a wide variety of ecological zones and poor environments, has ability to coppice, resist environmental stress and diseases, and the seeds are easy to collect, store and do not require pre-sowing treatment (FAO 1979; Zerfu 2002; Mekonnen et al. 2007; Nduwamungu et al. 2007). Proponents emphasize its lower water use per unit of dry matter produced, its nursing ability to support emergence of indigenous species on deforested landscapes, and stress the need to use better site-species matching and proper management practices instead of blaming the species.

All sample households were asked to select among lists of identified merits and demerits. The coppicing ability, straight pole growth, fast growth and hence shorter maturity period, multiple uses of the wood, low level of labor required for its management, and its drought, disease and pest resistance. The disadvantages identified in their order of importance were allelopathic effect of eucalyptus on nearby crops, challenges to convert the land back to crop land after eucalyptus harvesting, non-palatability of eucalyptus leaves means it cannot be used to supplement animal feed, high consumption of water and high labour demand during planting (Table 31.7). Almost all (97%) of the respondent households believed that the advantages of eucalyptus outweigh its disadvantages. This is supported by the finding that nearly 76% of the respondents expressed their intention to plant eucalyptus in the future.

Table 31.7 Merits and demerits of Eucalyptus plantation identified by respondent households (n = 400)

Merits of eucalyptus	Freq.	%	Demerits	Freq.	%
Coppicing ability	362	90.50	Its allelopathic effects on nearby crops	336	84.00
Straight pole growth	360	90.00	Hard to convert eucalyptus plot to crop land	272	68.00
Fast growth	357	89.25	Leaves not used for animal feed	233	58.25
Multipurpose use of the wood	338	84.50	High consumption of water	207	51.75
Its management demands less labor	304	76.00	High labour demand during plantation	200	50.00
Drought resistance	287	71.75	Others	45	11.25
Diseases and pests resistance	282	70.50			
Others	32	8.00			
Total	400	100.00	Total	400	100.00

Source Compiled from the survey data
Remark Most respondents identified more than one response

31.4 Conclusions

Though few in Ethiopia, including in Amhara region, have long years of experience in growing Eucalyptus, its expansion is recent. Rapid plantation expansion has been observed following removal of subsidies for agricultural inputs in mid 1990s and the issuance of land use certificates in the early 2000s that improved security of tenure on agricultural lands. Mecha, Adet, Sekela, Durbete and Dangla districts of West Gojjam Administrative Zone; Gayent, Farta Dabat, Ambagiorgis, Debark, Gonder Zuria, Denbia and Lay Armacheho districts of North Gonder; and Meket and Wadla districts of North Wollo Zones are the major eucalyptus growing areas in Western Amhara, most of which are situated in the Lake Tana watershed. These are also the major source areas that supply fuelwood and construction wood for neighboring urban centers, other zones and the neighboring Tigray region besides supplying the export market to Sudan. Mecha district represents these districts in terms of rapidly expanding eucalyptus plantation onto agricultural lands.

Eucalyptus planting in Amhara region in general and the study area in particular is expanding at high rate. The sharp rise in the price of agricultural inputs (mainly fertilizers), growing demand and rising prices of poles and posts from eucalyptus owing to rising domestic and export market (notably the Ethio-Sudan cross-border trade) and the emerging possibilities for household heads to be engaged in non-farm activities have been cited as important drivers. These socioeconomic drivers are reinforced by the bioliogical characteristics of eucalyptus. The major factor constraining the expansion of eucalyptus plantation is the households' on-going food and income generation demand.

Eucalyptus growing and marketing has become among the major livelihood strategies of people in the study area. Nearly a quarter of the land owned by the households is devoted to growing eucalyptus and a significantly higher portion of the total annual household cash income is obtained from eucalyptus woodlots. These woodlots can be transformed into cash when need arises unlike income from crop production that mostly generates income after the harvesting season. Woodlots are also creating new jobs even for the landless poor, for e.g. by cutting, carrying, etc. Most farmers intend to continue to plant eucalyptus in the future. Thus, eucalyptus plantation is expected to continue to increase in the Lake Tana watershed. Therefore, proper land use plan and mechanism to enforce this is needed and appropriate institutional mechanisms and support systems should be developed to supporting farmers to make informed and balanced decisions when expanding eucalyptus plantations. Mechanisms need to be put in place to support farmers to produce more from small areas of plantations. Additional research is needed to generate more evidence and better inform the development of legal and fiscal instruments that reduce trade-offs and maximize synergy between short term economic benefits and long term environmental needs in the Lake Tana watershed of western Ethiopia.

Acknowledgements The authors thank Center for International Forestry Research (CIFOR) for covering the cost of the research that led to the write up of this chapter.

References

Amare G (2002) Eucalyptus farming in Ethiopia: the case for eucalyptus woodlots in the Amhara region. 2002 Bahir Dar Conference Proceedings, pp 137–153, Ethiopian Society of Soil Science

Asaye A (2002) Growth performance and economics of growing *Eucalyptus camaldulensis* by smallholder farmers of Amhara Region: the case of Gonder Zuria District, North Gonder, Ethiopia. MSc Thesis. SLU, Sweden

Azene B-T (2007) Useful trees and shrubs for Ethiopia. Identification, propagation and management for 17 Agroecolgy Zones. Technical Manual No.6. In: Bo Tengnas, Ensermu Kelbesa, Sebsibe Demissew, Patrick Mandu (eds). World Agroforestry Center Nairobi, Kenya. p 550

Daba W (1998) The economics of growing *Eucalyptus globulus (Labill.)* on the Highlands of Oromia, Ethiopia; with special reference to Intoto and Chancho areas. MSc. Thesis. SLU, Sweden

Davidson J (1985) Setting aside the idea that eucalyptus are always bad. FAO-UNDP Project BED/79/017, Working Paper no 10. http://www.ifpri.org/sites/default/files/publications/esspwp20.pdf. Accessed on 28 Mar 2015

Demamu M (2002) Economic analysis of *Eucalyptus globulus* plantation in the former Dessie Fuel wood Project, South Wollo, Ethiopia. MSc. Thesis. SLU, Sweden

Demel T (2000) Facts and experience on eucalyptus in Ethiopia and elsewhere: ground for making wise and informed decision. Workshop on Eucalyptus Dilemma, 15 Nov 2000

Dessie G, Erkossa T (2011) Eucalyptus in East Africa Socio-economic and environmental issues, Working Paper FP46/E, Planted Forests and Trees Working Papers, FAO-Forestry Department, Rome, Italy http://www.fao.org/3/a-am332e.pdf. Accessed on 28 Mar 2015

Ethiopian Customs Authority Metema Branch (2012) Commodity export volume and value through Metma border from 2006 to 2011, Metema

FAO (1979) Eucalyptus for planting. FAO forestry and forest products study, No.11. FAO, Rome

FAO (1988) The Eucalyptus Dilemma, FAO, Rome

FDRE (2000) Proclamation on Ethio-Sudan Preferential Trade Agreement Ratification, Federal Negarit Gazeta, Proclamation No. 221/2000, Addis Ababa, Ethiopia. http://chilot.me/2000-01-1993-g-c-proclamations/. Accessed on 1 Jun 2015

FDRE (2003) Proclamation on Ethio-Sudan Trade Agreement Ratification, Federal Negarit Gazeta, Proclamation No. 318/2003, Addis Ababa, Ethiopia. http://www.fsc.gov.et/resources/Negarit%20Gazeta/Gazeta-1995/Proc%20No.%20318-2003%20Ethio-Sudan%20Preferential%20Trade%20Agreement%20.pdf. Accessed on 28 Mar 2015

Friis I (1995) Myrtaceae. In: Edwards S, Mesfin T, Hedberg I (eds) Flora of Ethiopia and Eritrea, vol 2(2). Addis Ababa University, Ethiopia

Garay I, Pellens R, Kindel A (2004) Evaluation of soil conditions in fast-growing Plantations of *Eucalyptus grandis* and *Acacia mangium* in Brazil: a contribution to the study of sustainable land use, Brazil. Applied Soil Ecology. 27, pp 177–187. http://www.researchgate.net/profile/Irene_Garay/publication/262006199_Evaluation_of_soil_conditions_in_fast-growing_plantations_of_Eucalyptus_grandis_and_Acacia_mangium_in_Brazil_a_contribution_to_the_study_of_sustainable_land_use/links/0f3175364038937494000000.pdf. Accessed on 1 June 2015

Getaw T, Atle G (2011) The behavior of commodity prices in Ethiopia, Agricultural Economics 42 (2011). pp 87–97. https://www.google.com.et/url?sa=t&rct=j&q=&esrc=s&source=web&cd=1&cad=rja&uact=8&ved=0CBsQFjAA&url=http%3A%2F%2Fwww.researchgate.net%2Fpublication%2F227353740_The_behavior_of_commodity_prices_in_Ethiopia&ei=XfZ2VcXmF8aTsAHwyKW4BA&usg=AFQjCNFnzlw2EAu8_kfQZbkiO9wDWoTp2Q&sig2=cKeY5qZ-vSX0spgx4_DfGDQ&bvm=bv.95039771,d.bGg. Accessed on 28 Mar 2015

Gil L, Wubalem T, Eduardo T, Rosana L (eds) (2010) Proceeding of the conference on: *Eucaluptus species* management, History, Status and Trends in Ethiopia. 15–17 Sept 2910, Ethiopian Institute of Agricultural Research. Addis Ababa, Ethiopia. 409 p

Kidanu S, Mamo T, Stroosnijder L (2005) Biomass production of *Eucalyptus* boundary plantations and their effect on crop productivity on Ethiopian highlands vertisols, Ethiopia. 63, pp 281–290. https://www.wageningenur.nl/de/Publicatie-details.htm?publicationId=publication-way-333430313337. Accessed on 28 Mar 2015

Limeneh M, Kassa H (2014) Re-greening Ethiopia: history, challenges and lessons, Forests 5: 1896–1909. www.mdpi.com/journal/forests. Accessed on 1 June 2015

Mecha District Agriculture Office (2011) Baseline information of Mecha District, Merawi

Mastewal E (2012) Knowledge of rainfall partitioning for IWRM process: Upper Blue Nile Basin, Ethiopia. Spatio-temporal assessment of drought under the influence of varying record length: the case of Upper Blue Nile Basin, Ethiopia. In: Berihun T, Workiye W, Melaku W (eds) Proceedings of the second national workshop on challenges and opportunities of water resources management in Tana Basin, Upper Blue Nile Basin, Ethiopia, 26–27 Mar 2012. Blue Nile Water Institute—Bahir Dar University (BNWI-BDU), Bahir Dar, Ethiopia, pp 211–216. http://www.bdu.edu.et/bnwi/sites/bdu.edu.et.bnwi/files/Part%20II.pdf. Accessed on 28 Mar 2015

Mekonnen Z (2010) Community opinion, marketing and current debates on eucalyptus in Huruta District, Arsi Zone of Oromia Region, Ethiopia, In: Gil L, Tadesse W, Tolosana E, López R, (eds) Eucalyptus Species Management, History, Status and Trends in Ethiopia, 15–17 Sept 2910. Ethiopian Institute of Agricultural Research. Addis Ababa, Ethiopia, pp 131–144

Mekonnen Z (2013) Productivity of *Eucalyptus camaldulensis (Dehnh.)* in Goro Woreda of Bale zone, Ethiopia, Research Journal of Agriculture and Environmental Management. Apex J Int Full Length Res Paper 2(9): 252–260. http://www.apexjournal.org—ISSN 2315–8719 © 2013. Accessed on 10 July 2015

Mekonnen Z, Kassa H, Lemenh M, Campbell BM (2007) The role and management of Eucalyptus in Lode Hetosa district, central Ethiopia. For Trees Livelihoods 17: 309–323 http://www.cifor.org/library/2364/the-role-and-management-of-eucalyptus-in-lode-hetosa-district-central-ethiopia/. Accessed on 28 Mar 2015

Miheret E, Ernest W (2012) Assessment of major threats of Lake Tana and strategies for integrated water use management. In: Berihun T, Workiye W, Melaku W (eds). Proceedings of the second national workshop on challenges and opportunities of water resources management in Tana Basin, Upper Blue Nile Basin, Ethiopia, 26–27 Mar 2012. Blue Nile Water Institute—Bahir Dar University (BNWI-BDU), Bahir Dar, Ethiopia, pp 281–292. http://www.bdu.edu.et/bnwi/sites/bdu.edu.et.bnwi/files/Part%20II.pdf. Accessed on 2 Apr 2015

Nduwamungu J, Munyanziza E, Nduwamungu JD, Ntirugulirwa B, Gapusi RJ, Bambe JC, Ntabana D Ndizeye G (2007) Eucalyptus in Rwanda: are the blames true or false? Institute Des Sciences Agronomiques Du Rwanda (ISAR)

Sewmehon Demissie Tegegne (2012) Water and rural livelihoods in the crop-livestock system of Amhara Region, Ethiopia: Multiple use system (MUS) approach for water productivity improvement. In: Berihun T, Workiye W, Melaku W (eds). Proceedings of the second national workshop on challenges and opportunities of water resources management in Tana Basin, Upper Blue Nile Basin, Ethiopia, 26–27 Mar 2012. Blue Nile Water Institute—Bahir Dar University (BNWI-BDU), Bahir Dar, Ethiopia, pp 251–257. http://www.bdu.edu.et/bnwi/sites/bdu.edu.et.bnwi/files/Part%20II.pdf. Accessed on 28 Mar 2015

Spielman D, Kelemwork D, Alemu D (2015) Seed, fertilizer, and agricultural extension in Ethiopia. Ethiopia Strategy Support Program (ESSP) II Working Paper 020. https://www.researchgate.net/publication/268372918_Seed_Fertilizer_and_Agricultural_Extension_in_Ethiopia. Accessed on 10 Jul 2015

Tesfaye T (1997) Problems and prospects of tree growing by smallholder farmers: a case study in Feleghe-Hiwot locality, eastern Tigray, Ethiopia. MSc. Thesis. SLU, Sweden

Tilashwork C (2009) The effect of Eucalyptus on crop productivity, and soil properties in the Koga watershed, western Amhara region, Ethiopia. Thesis, Cornell University. http://www.mu.edu.et/mejs/pdfs/V6N1/7-MEJS%2000151-58-69.pdf. Accessed on 28 Mar 2015

Tomoya M, Takashi Y (2010) The impacts of fertilizer credit on crop production and income in Ethiopia production and income in Ethiopia, National graduate institute for policy studies, Discussion Paper: 10–23, Tokyo, Japan

Zerfu H (2002) Ecological impact evaluation of eucalyptus plantation in comparison with agricultural and grazing land-use types in the highlands of Ethiopia. Ph.D. dissertation, Institute of Forest Ecology, Vienna University of Agricultural Science, Vienna

Zerihun K (2002) Profitability and household income contribution of growing *Eucalyptus globulus (Labill.)* to smallholder farmers: The case of the Central Highlands of Oromia, Ethiopia. MSC Thesis. SLU, Sweden. http://www.idosi.org/ejas/2(1)10/6.pdf. Accessed on 28 Mar 2015

Zubair M, Garforth C (2006) Farm level tree planting in Pakistan: the role of farmers' perceptions and attitudes. Agrofor Syst 66:217–229

Chapter 32
Waste Management in Lake Tana Basin—Case of Rapidly Urbanizing Bahir Dar City

Biruk Abate and Goraw Goshu

Abstract Urbanization is a global phenomenon with more pronounced consequences on waste management in developing countries where the rate of infrastructure development is mostly outpaced by the rate of waste generation. Bahir Dar, a rapidly urbanizing city at the southern tip of Lake Tana, is not an exception. This paper reviewed the current waste management system in Lake Tana basin taking Bahir Dar city as case. Bahir Dar city produces more wastes and will continue to produce more with the ever-increasing population of the city. The waste management practice is challenged by low prioritization of waste management and limited revenues for financing waste management. The review result also indicates that the municipality's mechanism of coordination of the public and private sectors has played a vital role in waste management. Effective involvement of both private and public sectors has made possible to improve waste management and provide door-to-door collection. However, the daily monitoring of waste management has not been sufficient. Therefore, an integrated solid and liquid waste management practice should be implemented for the City-Lake Tana basin and also for the surrounding environment. This has to include development plans for improving sustainable sanitation and disposal of the sewage system, and adopt the best practices of waste management for the City-Lake basin ecosystem.

Keywords Municipal wastes · Urbanization · Solid waste · Liquid waste · Waste prioritization

B. Abate (✉)
School of Chemical and Food Engineering, Biotechnology Institute,
Bahir Dar Institute of Technology, Bahir Dar University,
P.O. Box 26, Bahir Dar, Ethiopia
e-mail: abatebiruk@gmail.com; biruka@bdu.edu.et

G. Goshu
Blue Nile Water Institute and College of Agriculture and Environmental Sciences,
Bahir Dar University, P.O. Box 26, Bahir Dar, Ethiopia
e-mail: gorawha@gmail.com; gorawg@bdu.edu.et

K. Stave et al. (eds.), *Social and Ecological System Dynamics*,
AESS Interdisciplinary Environmental Studies and Sciences Series,
DOI 10.1007/978-3-319-45755-0_32

32.1 Introduction

The Federal Democratic Republic of Ethiopia Solid Waste Management Proclamation No. 513 (2007) defines Solid waste and solid waste management (SWM) as follows: 'Solid Waste' implies anything that is neither liquid nor gas and is discarded as unwanted. 'Solid Waste Management' means the collection, transportation, storage, recycling or disposal of solid waste, or the subsequent use of a disposal site that is no longer operational. The term 'Municipal Solid Waste' (MSW) covers solid wastes generated by households, by commercial and industrial premises like shops, hotels, garages and agriculture, by institutions such as schools, hospital care homes and prisons and from public spaces such as streets, bus stops, parks and gardens (Christian 2012).

Waste is classified differently in different contexts. In the context of subject of this review, the following classification is adapted from Suryawanshi et al. 2013.

Biodegradable waste: Typically, it originates from plant and animal sources, which may be broken down by microbes or other living organisms; While these wastes may appear physically different, they tend to be fairly homogeneous in biochemical composition (carbohydrates, fats and proteins) for anaerobic digestion for biogas production by virtue of their high methane potential.

Hazardous/toxic waste: It is a waste with properties that make it potentially dangerous or harmful to human health or the environment. They can be the by-products of manufacturing processes, discarded used materials, or discarded unused commercial products (cleaning fluids, pesticides).

Recyclable waste: It is the removal of items from the waste stream to be used as raw materials in the manufacture of new products (paper, glass bottles and ceramics).

Inert waste: It consists of construction and demolition waste, dirt, rocks, debris etc. with relatively lower environmental impact by virtue of its non-biodegradability.

Ethiopia is one of the many developing countries in sub-Saharan Africa where municipal authorities are struggling to provide adequate urban environmental services. Bahir Dar is one the fastest growing cities in Ethiopia with a current population of >290 thousand. If the current annual population growth rate of 6.6% continues, the city population will be doubled in just 11 years. Mekete et al. (2009) mentioned the need to have adequate SWM system in Bahir Dar city.

32.2 Solid Waste Management

The main stakeholders in the MSW system in Bahir Dar include Bahir Dar City Administration, Amhara Regional Bureau of Environmental Protection, Land Administration and Use BoEPLAU, Amhara Regional Health Bureau (BoH), Amhara Regional Government, United Nations Development Programme (UNDP), Federal Environmental Protection Authority (EPA) accordingly to the study of

United Nation Environmental program source document for the city waste management.

Dream Light (DL) is a private company responsible for collection, transportation, disposal and recycling of municipal solid waste in eight out of nine Kebeles of Bahir Dar. Green Dream (GD): This Community Based Organization (CBO) is comprised of 30 female workers and responsible for solid waste collection in a door-to-door manner in Shumabo kebele.

"*Koralews*" are informal waste material buyers and they also collect recyclable and reusable material. "*Lewaches*" exchange recyclable materials with clothes and utensils. Dumpsite pickers collect recyclables & reusable materials from the disposal site and sell it to "*Koralews*". Children and beggars ask for food leftovers, reusable and recyclable materials that they use themselves or sell to middlemen (Worku 2012).

Parts of the recyclable materials collected by "Koralews", "Lewaches", street persons and formal waste collectors are sold to middlemen, who in turn sell them to brokers of recycling companies in Addis Ababa. There are 55 middlemen collecting and selling metals, plastics and glasses and one middleman for textiles and shoes (Worku 2012). These middlemen are registered at the Bureau of Trade and Industry, thus have a license for trading materials. The waste generators in Bahir Dar are households, commercials and institutions. Bahir Dar has roughly 80,000 households (extrapolation based on CSA 2007).They are responsible for filling their solid waste in collection bags and for payment of the service fees. The business sector includes shops, hotels, restaurants, markets, garages etc. counts 7040 commercials (UNEP 2010a).They are responsible for filling their solid waste in collection bags and for payment of the service fees. Institutions such as governmental and non-governmental bureaus, schools and universities are responsible for filling their solid waste in collection bags and for paying the service fee. According to UNEP (2010a), there is no formal structure/platform within the municipality for communication among stakeholders. In some cases the stakeholders contact each other informally (Fenzie 2011; Getahun 2011).

The daily generation of MSW in Bahir Dar amounts to a total of 102.5 t/d (tons per day), commercial waste is 28 t/d, residential waste is 54 t/d, the institutional waste is 17 t/d and the street sweepings is 3.56 t/d, and 32% of the total MSW consists of ash and soil, 30% is food waste and 13% is made up of yard waste (UNEP 2010a). The large share of ash and soil component in residential waste (47%) is explained by the predominant use of firewood and charcoal in households. The ash residues are usually disposed on the ground, later put in the waste collection bag from where it is collected by Dream Light workers. The seasonal variation is expected to be minimal due to steady consumption behavior throughout the year (Alemnew 2011). Per capita generation of waste in Bahir Dar was assessed to be 0.25 kg/day for residential and 0.45 kg/day for all residential, commercial, institutional and street sweeping waste streams (UNEP 2010a). The waste projections by UNEP (2010a) show that the waste generation will increase similar to the

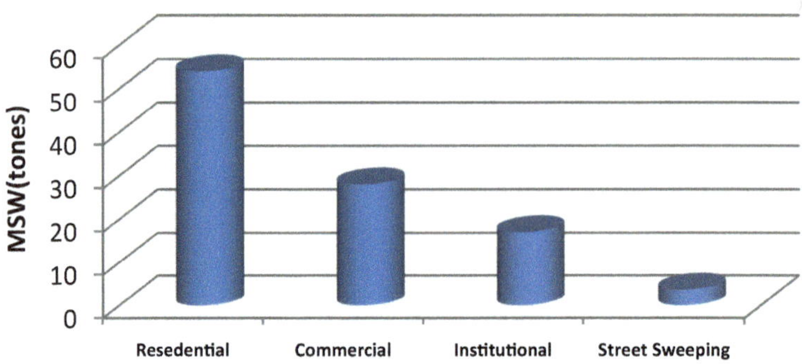

Fig. 32.1 MSW composition quantities in tones in Bahir Dar City. *Source* Modified based on data from UNEP (2010a)

population growth. Hence in 2021, when the population is doubled, waste generation will also be doubled (199 t/day).Based on the share of organic content (food and yard wastes) in the different waste sources (UNEP 2010a) the sum of food and yard wastes from the residential (16 t/d), commercial (16 t/d), institutional (12 t/d) and street sweeping waste streams (1 t/d) results in a total organic waste quantity of 45 tons per day, which equals 44% of the total amount of daily generated waste in Bahir Dar City.

The sources of hazardous waste are hospitals (biological wastes), contaminated containers (chemicals), institutions and schools (laboratory wastes), factories (expired drugs and printing inks) and also stores (pesticides), etc. Residential hazardous waste amounts to 156.6 kg/d (0.3% of total residential waste stream), commercial hazardous waste was recorded to be 124.8 kg/d (0.5% of commercial waste). Institutional hazardous waste was 120.7 kg/d (0.7%) and street sweeping hazardous waste was 0. This makes a total of 402.1 kg of hazardous waste generated per day, which is 0.4% of the total MSW in Bahir Dar according to the written document source (UNEP 2010a) (Figs. 32.1, 32.2 and 32.3).

32.2.1 Waste Collection and Transport

Since Dream Light's entry into the SWM system of Bahir Dar in 2008, it is the waste generators responsibility to put their mixed waste into any (non-standardized) bags and place them in a designated location on their compound or along the road. Some high-standard hotels require having their wastes collected up to twice per day. There are controllers who can organize and supervise their collection team to empty the generators waste bags into push-carts or into strong plastic bags and bring them to collection points There, the workers await the Dream Light collection truck to empty the bag contents. According to the UNEP study an overall collection rate of 71% is stated, which is 73 tons/day from a total of 102.5 t/day generated. The

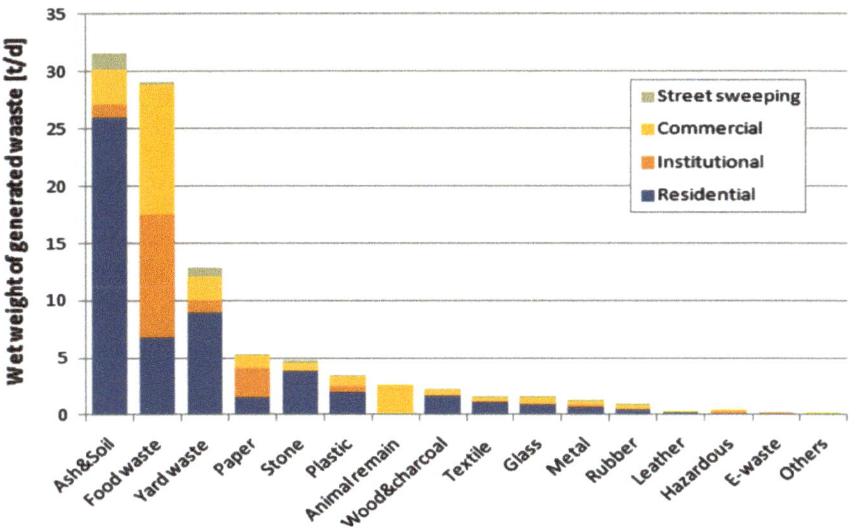

Fig. 32.2 MSW streams in Bahir Dar City and their compositions as wet weight of generated waste. *Source* Based on data from UNEP (2010a)

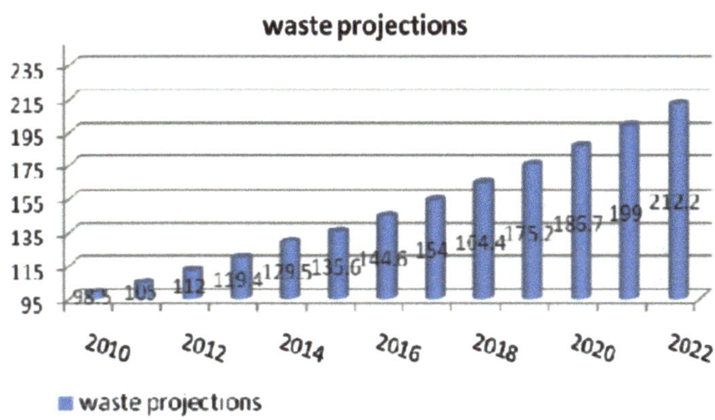

Fig. 32.3 Waste projections for future in Bahir Dar City. *Source* Based on data from UNEP (2010a)

remaining amount of 29.5 t/d is not being collected and according to UNEP (2010b) burned, buried or simply dumped on the lakesides or into rivers. Some parts of the street sweeping are outsourced to 'Million and Guadenutshu' (Million and Friends), the rest are swept by the city service itself. The street sweeping workers fill the street waste into plastic bags and leave them along the streets from where they are picked up by Dream Light or Green Dream workers.

There are no officially designated waste collection locations in Bahir Dar, but around 100 widely accepted collection points on the side of the road which were selected based on easy accessibility with pushcarts, workers, collection trucks and acceptable' distance to residents in order to avoid complaints due to odour and aesthetics.

Recycling of solid wastes in Bahir Dar seems insignificant (<1%) (UNEP 2010b). However, this is an underestimation because the UNEP study only accounted municipal composting as recycling activity and did not include informal recycling activities. Organic recycling is currently practiced as follows: the municipal composting site is located 3 km south of the city and the municipal workers transport the filled container by truck to the composting site about once per week. Thus 8 m^3 (3t) of fresh substrate arrives approximately 4 times per month, where it is manually sorted, turned and decreased in size. After 3–4 months, the finished compost is picked up by the municipality and brought to the city to be used for planting of flowers.

According to UNEP (2010b), no treatment facilities exist in Bahir Dar. All healthcare and industries and some governmental institutions follow their own way of removal. Most of them burn their waste; while some others dispose it to the nearby river Blue Nile or into Lake Tana (UNEP 2010c). There is no documented evidence that shows the criteria used for the selection of the area as dumpsite regarding prior study of hydrology, geology, socio-economic and environmental issues. The dumpsite is surrounded by land use activities such as informal settlement and agricultural activities. The liquid human waste (from emptying of septic tanks in Bahir Dar) is also dumped in close proximity of Gordma landfill site.

Similarly, there are problems of solid waste disposal sites selection. There are no standard transfer stations in the city. Institutions and industries follow their way of removal of waste and the available dumping sites are not well planned. Applying and integrating GIS and remote sensing techniques to select the best possible solid wastes dumping is one way of solving the problem. The study has shown land use, slope, water sources, settlement and transport facilities as determining factor in order to find appropriate site for solid waste dumping. Considering this study analysis output into account the most suitable sites were located in southern and south east of the town and are bare and grass lands (Fig. 32.4).

32.2.2 Sustainability Aspects

In case of sustainability aspects this sub-section provides an overview of the enabling environment for the municipal solid waste system in Bahir Dar and hence the Lake Tana Basin. It combines the results from the document 'Assessment of the SWM system in Bahir Dar City-Lake Tana basin and the gaps identified for the development of an ISWM plan' (UNEP 2010b) with field observations and information gathered through interviews with SWM stakeholders.

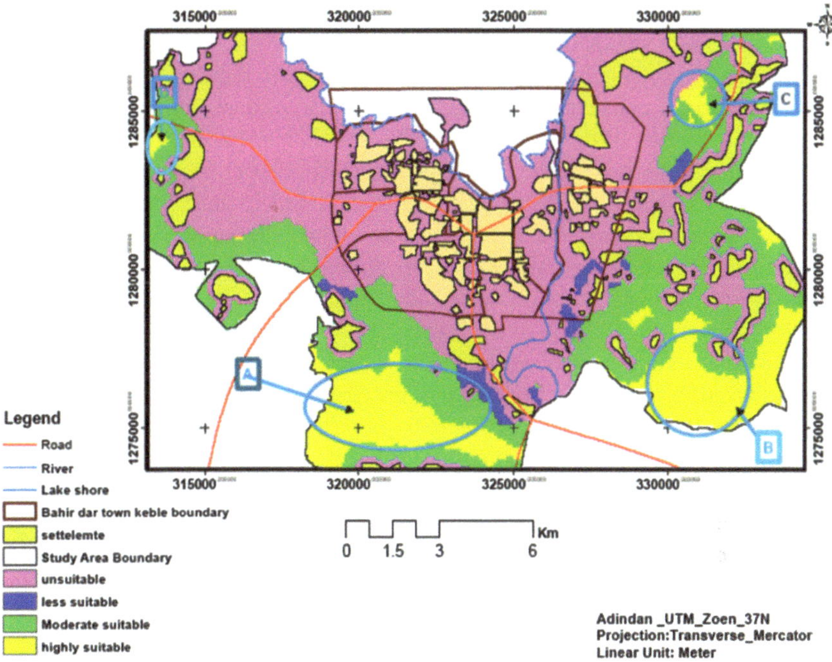

Fig. 32.4 Suitability map for solid waste dumping. *Source* Tirusew and Amare (2013)

32.2.3 Technical Aspects

With regarding to collection and transport no waste segregation is currently practiced at source, apart from the small-scale segregation of recyclables for informal recycling. There are no standard solid waste transfer stations in Bahir Dar City. It is difficult to designate unofficial collection points for temporary storage of waste during collection. Solid waste collection (71%) does not cover all solid waste generators. Transportation trucks are not standardized for solid waste transportation purposes, spare parts are not locally available—they have to be imported from the capital Addis Ababa.

Regarding to resource recovery there is no financial or technical support, no public recognition for waste recyclers only a small amount of organic solid waste (0.5 t/d) is used to produce compost in the municipal composting plant. There are no treatment facilities in Bahir Dar for MSW; and there is an open dumpsite (no sanitary landfill) and there are no sound operation practices performed at disposal site and about 71% (73 t/d) of generated MSW is collected and disposed of at disposal site. The environmental aspects of *Gordema* dumping siteis close to rural settlements and leads to pollution of groundwater sources through leachate (Worku 2012) even though this influence is not that much accounted for the Lake Tana basin waste management study but the problem may contribute indirectly on City and the Lake basin sanitation effects.

Concerning to socio-cultural aspects the pparticipation required at the source of waste generation is considered to be good, particularly considering the short time since the introduction of this new system. However, payment rate of collection fee is low (around 50%) thus considered as poor (UNEP 2010b). Awareness of public towards SWM is still low. When Dream Light started its business, it organized awareness raising campaigns to teach people how to handle waste (Alemnew 2011). Regarding to financial-economic aspects the payment rate for the solid waste collection service has been low until now. Only about 50% of the households pay the collection fee. About 90% of the commercials and institutions receiving waste collection services pay the fee regularly. Integration and coordination of the current institutional arrangement of the SWM is unsatisfactory. Each sector is working independently (UNEP 2010b) and there is no clear bridge between the federal (national) institutions and the regional (or local) institutions; and the region has not yet enacted any law on environmental issues. It is rather using the federal laws but facing difficulty in implementation, enforcement and monitor. Bahir Dar City does not have its own detailed rules and regulations specific to SWM that clearly indicates the responsibilities of the actors involved in SWM (Kassa 2009).

32.3 Liquid Waste Management

In Bahir Dar 20% of the households do not have access to latrines, using open fields and disposing waste water into the available open spaces. Because of the lack of latrines, waste disposal sites and poor collection practices, only 35% of the city's liquid waste was being collected (Fesseha 2012). In the absence of a city sewerage system, the overall drainage in the town is a problem aggravated by the plain geographical terrain and rapid urban growth. Waste water generation potential increases with increasing use of flush toilets, which in turn requires increased wastewater treatment plants (Fesseha 2012). This implies the ever-growing Bahir Dar City with modern buildings using flush toilets will produce more waste water that should be treated.

The waste water management practices of the major institutions of Bahir Dar City-Lake Tana basin are not also environment friendly. These institutions do not have their own wastewater treatment and management systems. They simply discharge waste water into Lake Tana through tubes and open ditches. Such discharges pollute the environment and create offensive smell and aggravate the conditions for the spread of communicable diseases. There was no well-organized coordination among the concerned bodies regarding waste water management. Eutrophication, waterborne diseases, shortage of water and adequate sanitation are still a major challenge for Bahir Dar-Lake Tana basin.

This review showed that only 28% of the housing units used septic tanks to collect and treat the generated wastewater. Among the total respondents, who had no space in their compound for waste water discharge, 64.3% discharged the waste water they generated into streets. In this study, 80% of the respondents had access to latrines, while 20% used open fields (Tables 32.1 - 32.3) (Fesseha 2012). The fecal and chemical pollution levels have been significantly increased and clearly discernible in the Bahir Dar Lake Tana basin (Goraw 2013).

Table 32.1 Type of waste sources and % by weight in Bahir Dar City

Type of waste	% by weight
Food	86.6
Paper	3.3
Plastic, leather and plastic	2.2
Glass	0.6
Textile	2.2
Metals	0.3
Others	4.8

Source Based on data from UNEP (2010a)

Table 32.2 Distribution of latrines availability by education, monthly income (in Birr) and space availability in Bahir Dar City

	Latrine availability		Total	X^2	P-value
	Yes	No			
	No. (%)	No. (%)	No. (%)		
Educational status (n = 270)					
Illiterate	78(29.0)	15 (5.5)	93 (34.4)	0.84	$P < 0.001$
Literate	139(51.5)	38 (14.0)	177 (65.6)		
*Monthly income** (n = 270)					
Less than 500	139(51.5)	49 (18.1)	188 (69.6)	10.3	$P < 0.001$
501–750	39 (14.4)	3 (1.1)	42 (15.6)		
751–1000	19 (7.0)	1 (0.4)	20 (7.4)		
Above 1000	18 (6.7)	2 (0.8)	20 (7.4)		
Space availability (n = 270)					
Have space	66 (24.5)	10 (3.7)	76 (28.1)	3.9	$P < 0.001$
Have no space	148 (55.8)	46 (17.0)	194 (71.9)		

Source Fesseha (2012)

Table 32.3 Logistic regression predicting likelihood of reporting for availability of latrine, March 2008, Bahir Dar City

	B	S.E.	Wald	df	Sig.	Exp. (B)
Education						
Illiterate (Ref.)						
Literate	0.526	0.350	2.258	1	0.133	1.693
Space availability						
No	(Ref.)					
Yes	0.751	0.407	3.398	1	0.065	2.119
Income						
Income group			9.024		0.029	
Less than 500 Birr (Ref.)						
500–750	−1.466	0.629	5.435	1	0.020	0.231
751–1000	−1.730	1.051	2.709	1	0.100	0.177
More than 1000	−1.016	0.775	1.718	1	0.190	0.362
Constant	−2.00	10.462	18.733	1	0.000	0.135

Source Fesseha (2012)
S.E. Standard error, *Wald* Wald statistic, *df* Difference of the chi-squared statistic, *Sig.* Significant, *B* Defendant, *Exp. (B)* Exponential of the defendant

32.4 Conclusion

This review shows that even though various studies and programs are undertaken to curve the problem of solid waste, the service still falls short of the required level. Technical measures, including waste sorting, recycling and composting, and infrastructural measures for leachate collection and gas venting should be upgraded to standardize the city waste management system. Concern for financial viability and long-term planning for waste management are important indicators of sustain the strengthening of SWM planning. Similarly, the formation of the Dream Light and others with their motivating role in waste management at local level has been driving this practice towards sustainability. Therefore, it is good practice, leading towards the fulfillment of the municipal vision of a safe and clean municipal area enable SWM practice of the Bahir Dar city-Lake Tana basin.

The municipality's mechanism for coordinating the public and private sectors has played a vital role in waste management. However, the daily monitoring of waste management by the Community Development Section has not been sufficient. Effective involvement of both private and public sectors has made it possible to improve waste management and provide door-to-door collection. The role of the private sector in recycling is important and it can contribute to sustainable waste management by reducing the quantities of plastics and sound financial management and regular and reliable payment of the contractor is important for satisfactory private sector participation. The Municipality of the city can achieve its target of solid waste management with active involvement of the private sector, minimizing municipal expenditure by means of effective management practice.

According to the review nearly two-thirds of all households in Bahir Dar discharge waste water into streets and flood water drainages. There is a poor level of awareness about existing regulations on sanitation among the experts as well as the public. The City Service Administration Office authorities did not seem to give due attention to waste water management in the City. There is weak implementation of the regional hygiene regulations. In addition, space availability is an important factor affecting waste water management at household level in the city.

32.5 Recommendations

The following measures need to be taken to improve the Bahir Dar City-Lake Tana basin waste management practice from now:

- In order to increase the quality and efficiency of services in the local operational context, more research and studies in waste management should be undertaken;
  - Characterization of solid and liquid wastes in the basin
  - Assess solid and liquid waste management practices of the residents
  - Evaluation of technical and socioeconomic competiveness of solid and liquid waste management technologies such as waste minimization, recycling, waste conversion to energy, bio-fuels, chemicals or other useful products, and this should include but is not limited to the following technologies of waste-to-energy, anaerobic digestion, composting and bio-fertilizer production, other thermal or biological conversion technologies and also strategies to promote diversion to higher and better uses (e.g. organics diversion, market analysis, optimized material management, logistics, etc.) and land filling.
  - Perform economic or cost/benefit analyses, feasibility studies for untested technologies or management strategies, life cycle analysis or inventory, and analyses of policies that relate to the above (e.g. extended producer responsibility, recycling goals, carbon legislation, bottle bills, etc.).
- Waste management training and resource centers should be established. This will improve the standards of waste management in the city and hence the city administrator could formulate a clearer role for its sustainable waste management and resource mobilization center and so the necessary instructions should be given to waste management.
- In order to implement effective waste management in the city, the administration could formulate short-term and long-term plans for waste management and the municipality could employ urban planner for urban development planning.
- Increasing public awareness about waste management; controlling the offenders through strict enforcement of regulations.

- Assigning qualified environmental health workers to each kebele and sub-city to enforce the sanitation regulations and coordinating the efforts at the grassroots level.
- The urban health extension program should also take measures to mitigate the problems.
- Therefore, an integrated solid and liquid waste management should be implemented for the City-Lake Tana basin and also for the surrounding environment. This has to include development plans for improving sustainable sanitation and disposal of the sewage system, and adopt the best practices of waste management system for the City-Lake basin areas.
- All stakeholders in the city and around it should participate in striving towards sustainable construction in order to embark upon the environmental impact issues. Since waste perceived as major obstacle particularly in the construction infrastructures, the city administrator is obligated to develop tools or model to enhance the quality of waste management during the infrastructure construction life cycle.
- The City Administrator should have a well-designed and carefully implemented waste management policy structures to all three "pillars" of sustainable development (environmental, economic and social) in the city by:

  - Improving economic efficiency, especially in resource extraction and use (e.g. through waste prevention, reuse, recovery or recycling);
  - Reducing the budget needed for waste collection services;
  - reducing or eliminating adverse impacts on health and the local and general environment;
  - Delivering more attractive and pleasant human settlements and social amenity; and
  - Creating sources of employment and potentially a route out of poverty for some of the poorest members of the community.

- Waste management delivers benefits to subsequent generations, by providing them with a more robust economy, a fairer and more inclusive society and a cleaner environment, thereby facilitating intergenerational equity and the result of this type of sustainable waste management can provide opportunities to the poor, for example, by enabling waste pickers to earn a sustainable income. As the waste management activities are typically performed by the poor, often poor and vulnerable women, and so these activities can deliver significant economic and social benefits by improving autonomy and recognition for the people concerned and hence implementing waste management policy will facilitate and promote the principle of intra-generational equity for the city.
- Policies of the city administrator should be designed and practiced for incorporating a careful, balanced and integrated give effect to other principles of sustainable development, such as the precautionary principle.
- Therefore, in the absence of policies to produce a different result in the city, the rate of waste generation typically increases with economic growth, advances in technology and the appearance of new products incorporating technological

advances and changes in the range of products on the market may lead to increased waste through the use of more disposable products or greater amounts of packaging and at the same time the hazardous nature of the waste may increase as product composition changes.

References

Alemnew G (2011) Dream Light PLC, Head of Marketing and Sales Department, 2011

Christian R.L (2012) Feasibility assessment tool for urban anaerobic digestion in developing countries, a participatory multi-criteria assessment from a sustainability perspective applied in Bahir Dar, Ethiopia, MSc Thesis Environmental Science(Major in Environmental Technology),Wageningen University Netherlands,Final 24 January 2012

CSA (2007) Central statistical agency of Ethiopia: population and housing census of Ethiopia. Amhara, pp 1–4

Fenzie (2011) Regional Bo EPLAU, Head of ensuring sustainable environmental protection authority core process

Fesseha H (2012) Liquid waste management: the case of Bahir Dar, Ethiopia

Getahun (2011) City administration, head of sanitation, beautification and park development core process

Goraw G, Byamukama D, Manafi et al (2010) A pilot study on anthropogenic fecal pollution impact in Bahir Dar Gulf of Lake Tana, Northern Ethiopia. Ecohydrol Hydrobiol 10(2–4): 271–280

Goraw G (2013) Anthropological faecal and chemical pollution impact on ecosystem and public health, Bahir Dar gulf of Lake Tana , Northern Ethiopia.Bahir Dar University Research and Community Service Site Visting and Research Papers Presented on Waste Management, November14, 2013

Kassa G (2009) Management of domestic solid waste in Ethiopia. VDM Verlag Dr, Müller, Saarbrücken, Germany

Mekete D, Atikilt A, Hana T (2009) Solid waste management in Bahir Dar City. School of Civil and Water Resources Engineering, Bahir Dar University

Suryawanshi PC et al (2013) Solid and liquid wastes: avenues of collection and disposal. Int Res J Environ Sci 2(3):74–77

Tirusew A, Amare S (2013) Waste management day. Bahir Dar University Research and community service. Solid waste management by using geographic information system (GIS) and remote sensing: The case of Bahir Dar Town, Bahir Dar, 14 Nov 2013

UNEP (2010a) Assessment of the Solid Waste management system in Bahir Dar town and the gaps identified for the development of an ISWM plan. Forum for Environment, June 2010

UNEP (2010b) Solid waste characterization and quantification of Bahir Dar City for the development of an ISWM plan. Forum for Environment, June 2010

UNEP (2010c) Target setting for ISWM of Bahir Dar City, Ethiopia. Forum for Environment, August 2010 workshop in the Amhara National Regional state. 23 Jan, Bahir Dar, Amhara

Worku D (2012) Recycling practices and potentials in Bahir Dar City and the influence of landfill leachate on groundwater quality, MSc thesis, Cornell University

Part V
Synthesis

Chapter 33
Problems, Efforts and Future Directions of Natural Resources Management in Western Amhara Region of the Blue Nile Basin, Ethiopia: Review

Yihenew G. Selassie

Abstract Natural resources degradation is a serious problem in Western Amhara Region of the Blue Nile basin. To generate information on problems, extension efforts and future directions in natural resources management of the Amhara Region of the Nile basin, literatures from various sources were consulted and compiled. The prevalence of huge natural resources degradation are manifested in the form of soil loss, soil fertility deterioration, organic matter depletion, water resources degradation, siltation of dams, reservoirs and agricultural lands, forest resources exhaustion, environmental pollution and climate change were main once. About half of the highland's land area (about 27 million ha) is significantly eroded and over one-fourth (14 million ha) are seriously eroded. Over 2 million ha of land are permanently degraded that the land is no longer able to support cultivation. Some estimates indicate that the average annual soil loss from arable land was 100 tons ha^{-1} and the average productivity loss on cropland was 1.8%. Others provided estimates of soil loss rates of 42, 8, 5, 70, and 5 tons ha^{-1} yr^{-1} from cropland land under perennial crops, grazing and browsing land, productive land, currently uncultivable land, and wood and bush land, respectively. The main causes of these problems were identified to be natural factors (rugged nature of the topography and high and erratic rainfall); political and economic history of the highland areas; livelihood, backward farming practices, socio-economic problems & poor land use policy enforcement. Moreover, it was identified that there are limited but encouraging efforts in research, technology dissemination and policy formulation in the region to solve the problems. It is possible to conclude that even though there are tremendous amounts of natural resources degradation in the Nile basin, the efforts being done in the region to mitigate the problems are encouraging. Therefore, scaling up successful practices can enhance the endeavor towards sustainable natural resources management.

Y.G. Selassie (✉)
College of Agriculture and Environmental Sciences, Blue Nile Water Institute, Bahir Dar University, Bahir Dar, Ethiopia
e-mail: yihenewgs@gmail.com

K. Stave et al. (eds.), *Social and Ecological System Dynamics*,
AESS Interdisciplinary Environmental Studies and Sciences Series,
DOI 10.1007/978-3-319-45755-0_33

Keywords Blue nile basin · Land degradation · Policy support · Scaling up · Western Amhara

33.1 Introduction

The principal environmental problem in the Nile basin is natural resources degradation manifested mainly in the form of soil erosion, gully formation, soil fertility loss, deforestation, water resources degradation, environmental pollution, climate change and crop yield reduction. Natural resources degradation is a concern for farmers, development agencies, and governments throughout the world since it is affecting soil, land, forest and water resources upon which man depends for his sustenance. Today, this problem is universally recognized as a serious threat to man's well being and the threats are very severe mainly in developing countries. Soil degradation can be described as a reduction of resource potential or a combination of processes acting on the land, such as soil erosion by water and wind, bringing about deterioration of the physical, chemical and biological or economic properties of soil (Maitima and Olson 2001). The excessive dependence of the Ethiopian rural population on natural resources, particularly land, as a means of livelihood is the major underlying cause for land and other natural resources degradation (EPA 1998).

Some forms of land degradation are the result of normal natural processes of physical shaping of the landscape and high intensity of rainfall. The scale of the problem, however, dramatically increased due to the increase in deforestation, overgrazing, over-cultivation, inappropriate farming practices, and increasing human population. Removing vegetative cover on steep slopes (slopes ranging between 15 and 50%) for agricultural expansion, firewood and other wood requirements as well as for grazing space has paved the way to massive soil erosion. It is reported that Amhara Region, which is the sources of the Blue Nile, covers 58% of the country's soil loss (Tesfahun and Osman 2003).

The variations in level of erosion in different places could be due the several factors that control the spatial pattern and amount of soil loss in an area. Sediment degrades water quality, and carries soil-adsorbed polluting chemicals. Sediment deposition in irrigation canals, stream channels, reservoirs, water conveyance structures, reduces their capacity and would require costly operation for removal (Foster 1982). In terms of economic value, Pitmentel et al. (1995) estimated that worldwide costs of soil erosion to be about four hundred billion dollars per year, more than 70 dollars per person per year.

Forest and water resources degradation due to mismanagement, over utilization and environmental pollution are other forms of natural resources degradation. Forest cover in the Ethiopian highlands as a whole decreased from 46 to 2.7% of the land area between the 1950s and the late 1980s (USAID 2004). Moreover, environmental pollution and the consequences in climate change are the other threats for development manifested in the region.

Due to the aforementioned problems, it is reported that agricultural productivity in Amhara Region declines by 2–3% per year (Tesfahun and Osman 2003). It is, however, worthwhile to note that total production has increased due to extensive agriculture. Reviewing the amount of natural resources degradation, determining the causes of this degradation and suggesting preventing mechanisms are useful to envisage development activities and policy decisions. In this paper review of literature on intensity of natural resources degradation, causes of land degradation, current efforts on natural resources management and suggestions on natural resources management are presented.

33.2 Nile degradation in the Nile Basin

Natural resources degradation can be defined as the deterioration of physical, chemical and biological properties of objects of nature like forests, water and soils. It is the decline in soil, water, and vegetation quality—the very things we depend on for life. Ethiopia is considered to have one of the most serious soil erosion problems in the world. In the high lands of Ethiopia soil losses are extremely high with an estimated average of 20 tons ha^{-1} yr^{-1} and it can radically exceed this value on steep slopes (Mulatie 2009). Other report also indicated that Ethiopia is considered to have one of the most serious soil degradation problems in the world. The average annual rate of soil loss in Ethiopia is estimated to be 12 tons ha^{-1} yr^{-1}, and it can drastically exceed on steep slopes with soil loss rates greater than 300 tons ha^{-1} yr^{-1} (USAID 2000).

The maximum soil loss tolerance for tropical regions is 25 tons ha^{-1} yr^{-1} (Ringo 1999). A commonly used soil loss tolerance rate is 5–12 tons ha^{-1} yr^{-1} for shallow to deep soils (Lal 1984). The tolerance value for tropical soils has not yet been formulated at international level, but Hudson (1986) established annual soil loss tolerance limits that vary between 0.2 and 11 tons ha^{-1} yr^{-1} (cited in Ringo 1999). From this, one can collude that soil loss in the Amhara Region is above the tolerance level. It is estimated that over 1.9 billion tons of soil are lost from high lands of Ethiopia annually (EHRS 1986). The Ethiopian highlands, which are the center of major agricultural and economic activities, have been the victim of soil erosion for many years. It is concluded that about half of the highland's land area (about 27 million ha) is significantly eroded and over one-fourth (14 million ha) are seriously eroded. Moreover, 2 million ha of land are permanently degraded that the land is no longer able to support cultivation (EHRS 1986).

Measured values range from 0 to 300 tons ha^{-1} yr^{-1} on specific plots (Hurni 1985; Nyssen et al. 2006 Impact of land use and land cover on sheet and rill erosion rates in the Tigray Highlands, Ethiopia, unpublished). As reported by the Ethiopian Highland Reclamation study (EHRS 1986), 27% (over 14 million ha) of the high land area of Ethiopia were seriously eroded and some 6 million ha completely withdrawn from agricultural use. A further 13 million ha were moderately eroded. Of the remaining 28 million ha, about 54% is susceptible to erosion, requiring

Table 33.1 Annual soil loss under various land use systems and topographic features

Land use type	Land area (million ha)	Documented range of annual soil loss				
Cultivated land	18	288	900	1800	3600	5400
Pasture and range land	60	960	3000	6000	12,000	18,000
Total	78	1248	3900	7800	15,600	23,400

Source Tamrie (1995)

some form of soil conservation measures. Tamrie (1995) also estimated that the amount of annual soil movement (loss) by erosion ranges from 1248–23,400 million tons per year from 78 million ha of pasture and rangelands and cultivated fields in Ethiopia (Table 33.1).

Soil erosion is considered to be a major agricultural problem in Ethiopia, particularly in the high lands (above 1500 m above sea level) which constitute 43% of the total area of the country. The Ethiopian highlands including the Amhara Region contain 88% of the countries population, 67% of its livestock and over 90% of its permanently cultivated area. Because of population pressure, all possible arable lands including steep slopes up to 60% have been cultivated. The Ethiopian Highland Reclamation Study (EHRS) estimated that the average annual soil loss from arable land was 100 tons/ha and the average productivity loss on cropland was 1.8% (Constable 1985 Ethiopian Highlands Reclamation Study (EHRS): Summary (EHRS) FAO/MoA Joint Project, Addis Ababa, unpublished). Hurni (1986) has provided estimates of soil loss rates of 42, 8, 5, 70, 5 and 5 t/h/yr from cropland land under perennial crops, grazing and browsing land, productive land, currently uncultivable land, and wood and bush land, respectively in Ethiopia.

The amount of eroded soil and sediment load reaching rivers and lakes in the country is estimated differently by different investigators ranging from 16–300 tons ha^{-1} yr^{-1} (Tamrie 1995). According to the assessment of land degradation in Ethiopia by FAO (EHRS 1984) some 1900 million tons of soil are annually eroded from the highlands. This is equivalent to an average net soil loss of 100 tons/ha and an annual loss of 8 mm of soil depth. The highest rate of erosion s found in Wollo, Gondar, Gojam and Shewa. The estimated soil loss due to sheet and rill erosion ranges between 201 to over 300 tons ha^{-1} yr^{-1}. This is on average, equivalent to 2.5 cm depth of soil from a hectare of land. Teshome (1995) used the AGNPS model to evaluate water erosion processes in Tikurso on the highlands of central Ethiopia. The results of his study showed that erosion rate was in the range of 51–100 tons ha^{-1} yr^{-1}.

In Amhara Region, the soil loss due to water erosion is estimated to be 58% of the total soil loss in the country (Tesfahun and Osman 2003). This has already resulted in a reduction of agricultural productivity by 2–3% per year, taking a considerable area of arable land out of production. The situation is becoming catastrophic because increasingly marginal lands are being cultivated, even on very steep slopes (Tesfahun and Osman 2003). Another study by Woody Biomass Inventory and Strategic Planning Project (WBISPP 2000) indicated that 82% of the

Table 33.2 Classification of slope classes and soil loss in ANRS

Slope class (degrees)	Share of slope class (%)	Soil loss class (tons ha^{-1} yr^{-1})	Share of soil loss class (%)
0–2	60.2	≤3.125	63.31
2–8	34.05	3.125–6.25	9.43
8–16	5.4	6.25–12.5	9.01
16–32	0.34	12.5–25	8.26
>32	0.01	25–50	5.84
		50–100	2.63
		100–200	1.42
		200–400	0.11
		400–800	0

Source ANRS-BoFED (2009)

Amhara Region has a soil loss rate of less than 12.5 tons ha^{-1} yr^{-1} while 18% suffers a soil loss of 12.5–200 ha^{-1} yr^{-1}.

The report of the ANRS-BoFED (2009) also indicated that all ranges of soil loss classes are available in the region due the differences in slope, plant cover, rainfall intensity and other climatic variations existing in the region (Table 33. 2).

Chemical degradation is the other form of land degradation manifested in the region. It is the deterioration in the chemical properties of the soil due the effect of accumulated chemicals or chemical processes going on in the soil. The most important aspects in soil chemical degradation in Amhara Region are depletion of nutrients and organic matter, acidity and salinity. Soil fertility depletion associated with low input agriculture is big threat to boost agricultural production. Most of the lands in Amhara Region have been cultivated for 3000 years with minimum return of nutrients both as inorganic and organic fertilizers. The problem of acidity in the highlands of Gojjam and Gondar are limiting crop yield. Many lime trials conducted in Awi, West Gojjam and East Gojjam zones has exhibited improved crop yield with lime application. Salinity is a problem in the lowlands of North Shewa Zone around Shewa Robit, in Awash basin part of the region, where the rate of precipitation is lower than the rate of evapo-transpiration, and in ground waters of Lake Tana area.

33.3 Major Causes of Land Degradation in Amhara Region

The ANRS is the most severely eroded and eroding region in the country. Even if the problem severely existed in Gonder and Wollo areas for the many decades back, by now it is more serious and active in West and East Gojam, Awi and the surrounding areas of Lake Tana, which were considered as the surplus producing areas of the region as well as the nation. The main causes of natural resources degradation in ANRS can be summarized as follows.

33.3.1 Natural Factors

Topography: the area is characterized by rugged and steeply sloped topography which causes the intensity of erosion to be very high. Mountains cover 52% of Asia, 36% of North America, 25% of Europe, 22% of South America, 17% of Australia and 3% of Africa. As a whole, 24% of the Earth's land mass is mountainous. Nevertheless, of the 3% of African mountains, Ethiopia takes 60% (Wikipedia 2006). In Ethiopia, about 44% of the land area (> 1500 m asl) has typical highland characteristics. Contrary to the country's average, the highland part in the Amhara Region is about 70%. The annual report of ANRS-BoFED (2009) indicated that most of the land area in Amhara Region have land capability classes that are not suitable for annual crop production (Table 33.3). However, most of the lands have been put under cultivated land uses which aggravated the erosion rate.

High and erratic rainfall pattern: the western highlands of the Amhara National Regional State receive heavy rainfall that in some places reaches well above 2000 mm. All the rainfall falls in few months time that does not usually exceed 4 months. This situation causes excessive land slide and erosion besides to development of acidity problems.

33.3.2 Political and Economic History of the Highland Areas

Coincidentally, the impacts of droughts and land degradation have been scrolled along with the route of the center of gravity to political and economic activities. For instance, the following places northern highlands (Axumite Empire), northeast (Wag Empire), central highlands (Shewa Dynasty), northern highlands (Gondar and

Table 33.3 Land use/Land cover type and land capability classes in Amhara National Regional State

Land use/Land cover	Share (%)	Land capability class	Area of each class (ha)	Share of each class (%)
Afro-alpine	0.4	I	104,535.70	0.68
Cultivated land	36.3	II	2,669,337.6	17.31
Grass land	17.1	III	1,352,783.9	8.77
Natural forest	0.6	IV	2,600,018.8	16.86
Plantation forest	1.2	V	4,510.8	0.03
Wood land	6.6	VI	196,215.8	1.27
Shrub land	27.6	VI	5,238,631.0	33.98
Water	2.0	VII	3,251,381.8	21.09
Wet land	0.8	Water bodies	315,521	–
Others	7.5			

ANRS-BoFED (2009)

Tigray Kingdoms), central highlands (the present Ethiopia) which were sits of the political powers attracted more people. Unlike most of the world's hotspots of civilization, situated around river deltas and downstream places, centre of economic and political activities in Ethiopia is peculiarly situated in the mountain system where land resources are sensitive to degradation agents and where the resources support immense economic and ecological functions to both upstream and downstream riparian countries. Compared with the average population pattern of settlement of the world (out of 10 people lives in mountain areas), it is estimated in Ethiopia that 9 out of 10 people or about 90% of the total population live in the highlands (Biru 2007).This caused tremendous amount of land degradation.

33.3.2.1 Livelihood

Livelihood is the financial means whereby one lives. Wild animals and fruits could be sources of livelihoods in some areas, agricultural products are sources of livelihoods for farmers; monthly salaries for government employees and profits for traders. Natural resources have been misused or destroyed for livelihood purposes. Many of the forest resources have been destroyed mainly because the trees have been used for fuel, construction material and building different implements. Many farmers bring damage to the land they plough due to continue cultivation, removal of crop residues from their farmlands for fuel and construction, clearing trees in their farm, etc. Activities directly associated to livelihood that are damaging the natural resources are the following.

33.3.2.2 Agricultural Expansion

Subsistence farmers in developing countries practice low input agriculture. Such practices are based on horizontal expansion of agricultural lands to increase total harvestable yield instead of improving productivity of the land under cultivation. This practice is called extensive agriculture. Extensive agriculture usually brings about loss of forest cover and eventual land degradation.

33.3.2.3 Complete Removal of Crop Residues from the Field

Farmers remove the residues for animal feed, plastering of their houses and fuel energy source. Such practices have negative effects on soil organic matter content, soil nutrient levels, soil structure and moisture holding capacity of the soil. Moreover, it aggravates soil erosion.

33.3.2.4 Collection of Animal Dung from the Field

Animal dung is collected from farm lands as the source of fire energy and plastering of houses. Poor farmers are involved in collecting dried animal dung which is locally called *Kubet*. Such practices have similar effects as that of collected crop residues from the field on soil physical and chemical properties (organic matter content, CEC, soil nutrient levels, soil structure and moisture holding capacity of the soil).

33.3.2.5 Overstocking

Ethiopia has the largest livestock population in Africa. The farmers consider that livestock is their source of cash during drought seasons or when crop fails; they are sources of drought power for plowing, trampling and threshing. Therefore, they keep tremendous amount of livestock on a limited piece of land. This livestock population grazes in uncontrolled grazing system on mainly communal lands. Eventually, livestock productivity is one of the lowest in Sub-Saharan Africa. In such situation, neither the land nor the livestock are benefited. This overgrazing and over browsing by livestock diminishes the capacity of the grass land to restore exposing the land for erosion. Over grazing and over browsing also cause soil compaction leading to herbivore damage of seedlings and hampered natural regeneration.

33.3.2.6 Clearing Forest Resources

About 35% of the land area of Ethiopia was believed to be once covered with high forest and it is now estimated that less than 2.7% of the land is covered with high forests (EFAP 1994). The vast stretches of land which formerly had a dense vegetative cover, are today a landscape where man cannot protect himself from the scorching heat of the sun. The Ethiopian Forestry Action Program (EFAP) study estimated that between 80,000 and 200,000 ha of forests are destroyed each year. Due to this reason, 1.9–3.5 billion tons of fertile topsoil are being washed away each year mainly as the ultimate result of deforestation; and 20,000–30,000 ha of crop land in the country's highlands is being abandoned each year because of this soil degradation. At the present rate of land degradation, the farmlands of some 10 million highland farmlands will be destroyed by the year 2010 (EFAP 1994). Though the forest resource is very limited, the demand remains high. According to the Ethiopian Forestry Action Program (EFAP), the national woody biomass demand is estimated at 50 million m^3 of which 45 million m^3 is in the form of fuel wood. The supply was estimated at only 14.4 million m^3 indicating a deficit of some 35.6 million m^3 (African Development Fund 2001) or 71% of the demand.

Destruction, clearing or incineration of forest and woodland resources (deforestation*)* is practiced for various reasons. Farmers expand crop cultivation spurred by the ever-increasing human population coupled with several constraints preventing/narrowing possibilities of options for economic diversification. Charcoal and wood production meant for a domestic requirements of energy is consuming a lot of wood. Construction material and carpentry/household utensils as well as for sale to generate income and support household livelihood are also demanding forest materials. Some urgently required socio-economic/infrastructural development, namely re-settlement, mining, and road construction consume a lot of forest cover. Commercial logging is also causing a serious treat for the environment and agricultural development. Moreover, unsustainable utilization involving improper and unplanned harvesting practices of forest trees is causing wastage of wood (because of very low recovery rates) and damage to the residuals trees/plants and stands.

33.3.2.7 Infrastructure and Industrial Development

Development of infrastructures like roads and buildings are useful to ensure development of any country. Nevertheless, the wrong way of doing the development efforts have been damaging the natural resources of Ethiopia. For example, huge amount of lands are washed away by road side drainage canals, many road construction comps were installed near river bodies and lakes; stone extraction quarries have left untreated after stone mining. What is exiting is that there is no accountability for the damage made on the land in the country.

33.3.3 Backward Farming Practices

Continues cultivation and complete removal of crop residues from the field is one of the problems of soil degradation. Due to the small land holding per household (<1 ha in Northwestern Ethiopia), there is a practice of continues cultivation without replenishing the land with appropriate amount of input. Improper farming practices are also the other one. Farmers use a 3000 years old farming practices that does not protect the land yet. Among others, frequent plowing of the land destroying the soil structure; plowing across the contour, leaving the land uncovered during the main rainy season (tef field), cultivation of annual crops on steep slopes, and trampling etc. contributes a lot for soil degradation. According to Tadele et al. (2006) indicated that trampling increases both run-off and soil loss (Table 33.4).

Table 33.4 The effect of land preparation techniques on runoff and soil loss

Treatment	Total soil loss (kg/ha)	Total runoff (m^3/ha)
Level	1968	431.072
Not trampled	1518	447.399
Trampled	3549	898.049
P	0.0010	0.0000

Source Tadele et al. (2006)

33.3.4 Socio-Economic Problems

33.3.4.1 Over Population

As population grows, there will be fragmentation of land holdings. This situation will make the natural resources overexploited and deteriorated. This problem usually happens because of the fact that natural resources are limited in space and time. Hence as population grows, there will be distribution of the limited resource among the people which avoids fallowing, soil and water conservation and afforestation practices. The Amhara National Regional State (ANRS) with the area of 157,076.52 km^2 has a population of about 19,122,515 and the population density is 121.7 persons km^{-2}. From the total population, 85% is dependent on agriculture and the average land holding per household is less that a hectare.

33.3.4.2 Poverty

Defining poverty is very difficult; because one who is poor in one country could be categorized under the rich category in another country. In 1997, a Kenyan "poor man" was asked what poverty is. And he said: "*Don't ask me what poverty is because you have met it outside my house. Look at the house and count the number of holes. Look at my utensils and the clothes that I am wearing. Look at everything and write what you see. What you see is poverty.*" Similarly a Latvian "poor man" defined poverty in 1997 as "*Poverty is humiliation, the sense of being dependent on them, and of being forced to accept rudeness, insults, and indifference when we seek help.*" (http://siteresources.worldbank.org/INTPOVERTY/Resources/335642-1124115102975/1555199-1124115187705/ch2.pdf.) From the two definitions alone, we can see that in Kenya, poverty is a matter of life or death; while in Latvia, it is a matter of honor and prestige. In West Gojjam Zone of the Amhara National Regional State, a farmer with one pair of oxen and 2 ha of land could be considered as a "rich man" while such a man can be categorized as extremely "poor man" in Borena Zone of Oromia National Regional State, where most farmers have cattle counted in dozens and large tract of grazing area.

It is, however, already well recognized that poverty and environmental degradation in underdeveloped countries are interlinked and cause unsustainable production and consumption patterns. The protection of the environment and of natural resources is an essential part of development: without adequate environmental capital, development is undermined and this in turn may reduce the resources available for investing in combating environmental damage, and hence poverty alleviation is not only a moral imperative but also a prerequisite for environmental sustainability and sustainable development.

Poor people have little access to agricultural inputs and alternative energy sources. Moreover, poor people who are not living properly today cannot think of the future generation. They think of how to pass the day only. In such situations poverty can be cause of land degradation. Poor people live at the expense of the future generation. Due to such poor economic status, the rural poor will be forced to depend on forests for subsistence income.

33.3.4.3 Lack of Alternative Sources of Livelihood

Ethiopian farmers lack alternative livelihoods except agriculture. The declining standards of livelihood of the farming communities and their close dependence on forests and woodlands have lead to clearing/burning for subsistence farming and cutting of trees/shrubs for cash. These problems are linked with lack of other alternatives.

33.3.4.4 Lack of Awareness on the Effect of Erosion

The farmers lack of awareness on effects of erosion and they do not realize that erosion is the main cause of yield reduction year after year.

33.3.5 Poor Land Use Policy Enforcement

The ANRS has land use policy (Proclamation No. 46/2000). Nevertheless, the policy never been properly implemented. Due to this reason, nobody takes accountability for the damage they make on their lands due to violating land use policies.

33.4 Research, Extension and Policy Efforts in Natural Resources Management

33.4.1 Research Efforts

The regional natural resources research efforts and outputs generated by the research system have been limited compared to the intensity of the problem in the

Region. Nevertheless, the research efforts have generated some basic and applied information. Among others, the following are the major ones.

- Bio-fertilizers
  - Adaptations of green manuring plant species;
  - Evaluation of green manure as cover crop;
  - Introduction of N fixing microorganisms like symbiotic blue green algae (*Azolla*) and Rhizobium;
- Mineral fertilizer recommendations
  - Blanket N and P fertilizer recommendation;
  - Soil test based n and p fertilizer rate recommendations of maize for West Amhara;
  - Integrated nutrient management including the use of farmyard manure and mineral fertilizer integrated application rate;
- Land management
  - Quantifying the impacts of livestock trampling on runoff, soil loss and crop yield under traditional teff cultivation system;
- The role of camber bed, broad bed and furrow for moisture drainage;
- Determination of crop water requirement
  - Crop water requirement of selected crops;
  - Determination of optimum irrigation scheduling for selected crops;
- Agro-meteorology
  - Agro metrological and cropping pattern analysis;
  - Rainfall variability and farmers' copping strategies;
- Forestry
  - Study on indigenous tree and shrub species of churches and monasteries;
  - Adaptations of multi-purpose trees and shrubs;
- Water harvesting
  - Evaluation of different water harvesting techniques in improving the survival rate of tree seedlings in drought affected woredas;
  - Evaluation of different water harvesting and conservation structures (clay pond, geo-membrane, tie-ridge);
- Resource characterization
  - Natural resources limitations and potentials in Amhara National Regional State;
  - Inventory and characterization of potentials and management of wetlands in Eastern Amhara;
  - Selected chemical and physical properties of selected areas;

- Soil and water conservation
  - On farm demonstration of physical and biological soil conservation measures for gully stabilization and biomass production.
  - Evaluation of different soil and water conservation structures
  - Watershed management at Anjeni, May ber, Andit Tid and farmers' fields.

33.4.2 Extension Efforts

The extension effort, led by the Amhara National Regional State Bureau of Agriculture and Rural Development (ANRS-BOARD), on natural resources conservation in ANRS is tremendous regardless of the limited technologies presented from the research system.

Based on the unpublished information obtained from the Forest Resources Development Case Team of ANRS-BOARD, huge accomplishments have been carried out from 2005–2009. As indicated in the report, major accomplishments in areas of watershed study and development, soil and moisture conservation on crop lands, mountain development, gulley management and forest development have been done. It is clearly seen than there was a huge growth in accomplished activities across the years.

The same report, however, indicates that from the total 3,416,213 ha of soil and water conservation works, only 1,808,868 ha (53%) sustained. Similarly, USAID (2000) reported that at the national level the government launched a massive soil conservation program beginning in the mid-1970s. Between 1976 and 1990, 71,000 ha of soil and stone bunds, 233,000 ha of hillside terraces for afforestation, 12,000 km of check dams in gullied lands, 390,000 ha of closed areas for natural regeneration, 448,000 ha of land planted with different tree species, and 526,425 ha of bench terrace interventions were completed. However, by 1990, only 30% of soil bunds, 25% of the stone bunds, 60% of the hillside terraces, 22% of land planted in trees, and 7% of the reserve areas still survived. Moreover, the number of seedlings that became trees have never been reported. If works done cannot be sustained, it is quite difficult to move forward. Therefore, such massive efforts need to be strengthened in order to cover fast the yet uncovered areas of the region.

33.4.3 Policy Efforts

It is now well recognized that the most serious NRM problems stem from institutional rather than technical factors. The success of NRM policy depends largely upon institutional organization, stability and capacity.

33.4.3.1 Land Administration and Use Policy of ANRS

It is apparent and natural that lands owned by private enterprises are more protected than communal grazing lands and forests. Even though there is a need for thorough study, the land redistribution activities that took place in the Amhara Region in 1996 might have negatively affected the soil and water conservation and land development endeavors. Cognizant of the extent of the problem, in 2000, the ANRS presented the Rural Land Administration and Use proclamation (Proclamation No. 46/2000) in the absence of such a policy at the federal level (ANRS 2000). The springboard for the policy development was the Amhara Regional Conservation Strategy (RCS) document. This document, in which important recommendations related to rural land were made, can be taken as the significant step forward.

The most important legal basis taken from previous efforts were the Regional Land Administration Proclamation, detailed directive formulated for the implementation of the proclamation; and the Regional Environmental Protection, Land Administration and Use Authority Proclamation. The proclamation No. 46/2000 developed following the RCS was issued to determine the Administration and Use of the rural land in ANRS. The proclamation defined the right of possession as well as the right to use, rent and inheritance of land use rights. In this land administration and use law, Article 6(3) states that "As long as the land users utilize the land according to the established rules, this proclamation assures and secures their holding and use rights". The policies stated objectives are to ensure the long-term land use rights of the landholders and to encourage productivity and sustainable development. It also aims to initiate a sense of security of rights among land users to encourage them to safeguard the soil and thereby sustain its productivity.

33.4.3.2 Other National Policies Adopted by ANRS

The Constitution of FDRE: the constitution of the Federal Democratic Republic of Ethiopia (FDRE) put the following points under Article 92 to protect the environment.

- Government shall endeavor to ensure that all Ethiopians live in a clean and healthy environment.
- The design and implementation of programmes and projects of development shall not damage or destroy the environment.
- People have the right to full consultation and to the expression of views in the planning and implementations of environmental policies and projects that affect them directly.
- Government and citizens shall have the duty to protect the environment.

Environmental Policy of Ethiopia: The environmental policy of Ethiopia (EPE) was approved on 2 April 1997. The draft document was prepared by the Environmental Protection Authority in collaboration with the Ministry of Economic

Development and Cooperation (EPA and MEDC 1997). The overall policy goal was to improve and enhance the health and quality of life of all Ethiopians and to promote sustainable social and economic development through the sound management and use natural, human-made and cultural resources and the environment as a whole so as to meet the needs of the present generation without compromising the ability of future generations to meet their own needs.

Environmental Pollution Control Proclamation: The environmental pollution control proclamation (Proclamation No. 300/2002) was declared on December 3rd, 2002 based on the following premises.

- Social and economic development endeavors may inflict environmental harm that could make the endeavors counter-productive;
- The protection of the environment, in general, and the safeguarding of human health and well being, as well as the maintaining of the biota and the aesthetic value of nature, in particular, are the duty and responsibility of all; and
- It is appropriate to eliminate or, when not possible to mitigate pollution as an undesirable consequences or social and economic development activities.

Environmental Impact assessment (EIA) proclamation: environmental impact assessment means the methodology of identifying and evaluating in advance any effect, be it positive or negative, which results from the implementation of a proposed project or public instrument (FDRE House of Peoples Representatives 2002). Environmental impact assessment is used to predict and manage the environmental effects which a proposed development activity, as a result of its design sitting, construction, operation or an ongoing one as a result of its modification or termination, entails and thus helps to going about intended development (Proclamation No. 299/2002).

33.5 Conclusions and Recommendations

The Nile basin area is having tremendous levels of natural resources degradation. However, the efforts being done in the region to mitigate the problems are encouraging. Therefore, scaling up successful practices can enhance the endeavor towards sustainable natural resources management. For successful outcomes, there are enabling conditions in the region. Among others, agriculture is the focus of the government; strategies to eliminate poverty are formulated; decentralization of power and empowerment of the local people to plan on own problems and resources are put in place; and the MDG stimulates the effort.

From the results of the study the following recommendations to mitigate the NR degradation can be drawn.

(a) Strengthen soil and water conservation efforts through proper implementation of soil and water conservation practices with proper designs and integration of physical and biological measures; practicing area closure and tree plantation schemes on sloppy lands and watersheds above dams; and delivering alternative fuel energy sources to minimize dependence on fire wood
(b) Improve land productivity and reduce sloppy areas under cultivation through the use of improved seed, integrated nutrient management; amelioration of problematic soils (acidic, saline and waterlogged soils); integrated pest management; and improved agronomic and land management practices.
(c) Practice stall feeding (better feed supply and quick economic return)
(d) Reduce pressure on the farm lands by providing alternatives to the rural community other than agriculture
(e) Controlling population growth, with particular emphasis in the rural areas
(f) Improve awareness of the farmers on effects of erosion
(g) Expand education and health *services*
(h) Improving the physical and social infrastructures (Irrigation schemes, roads, credit services)
(i) Follow the watershed approach in NRM and scale up watershed by watershed
(j) Involve farmers and relevant stakeholders at all stages of implementing NRM activities
(k) Implement the land use policy.

References

African Development Fund (2001) Koga irrigation and watershed Management Project (Eth/Paai/2001/01) appraisal report, February 2001. Available vial DIALOG http://www.afdb.org/fileadmin/uploads/afdb/Documents/Project-and-Operations/ET-2001-025-EN-ADF-BD-WP-COMPLET-KOGA-IRRIGATION-AND-WATERSHED-PROJECT.PDF

ANRS (Amhara National Regional State (2000) Proclamation No 46/2000. Rural land administration and use proclamation, Bahir Dar

ANRS-BoFED (Amhara National Regional State-Bureau of Agriculture and Economic Development) (2009) 2007/2008 Budget Year Annual Statistical Bulletin, Bahir Dar, Ethiopia

Biru Y (2007) Land degradation and options for sustainable land management in the Lake Tana Basin (LTB), Amhara Region, Ethiopia. Dissertation, University of Bern

EFAP (1994) Synopsis report. Ethiopia, Ethiopian Forestry Action Plan (EFAP) Secretariat, Ministry of Natural Resouces and Environmental Protection, Addis Ababa, Ethiopia

EHRS (Ethiopian Highlands Reclamation Study) (1986) Ethiopian highland reclamation study, final report, Vol. I, FAO, Rome

EPA (Environmental Protection Authority) (1998) National action programme to combat desertification. Federal Democratic Republic of Ethiopia Environmental Protection Authority, Addis Ababa

EPA (Environmental Protection Authority) and MEDC (Ministry of Economic Development and Cooperation) (1997) Environmental Policy of Ethiopia. Addis Ababa

FDRE House of Peoples Representatives (2002) Environmental impact assessment. Proclamation No. 299/2002. 9th Year No. 11, 3rd December 2002, Addis Ababa

Ethiopian Highlands Reclamation Study (EHRS) (1984) Annual Research Report. (1983–1984), Addis Ababa

Foster GR (1982) Modeling the Erosion Process. In: Haan CT, Johnson HP and Brakensiek DL (eds) Hydrologic modeling of small watersheds. ASAE Monograph No. 5, American Society of Agricultural Engineers, St. Joseph, Michigan pp 297–380

Hudson NW (1986) Soil conservation. Bats ford Ltd., London

Hurni H (1985) Erosion-Productivity-Conservation systems in Ethiopia. In: Proceedings 4th international conference on soil conservation, Maracay, Venezuela, p 654–674

Hurni H (1986) Degradation and conservation of the soil resource in the Ethiopian highlands. In: A paper presented at the first international workshop on African mountains and highlands, Ethiopia. October 19–27, 1986. Agriculture, CADU Publication No. 74

Lal RE (1984) Soil erosion: research methods. St. Lucie Press. Soil and water conservation society. Delray Beach. Ankeny

Maitima JM, Olson JM (2001) Guide to field methods for comparative site analysis for the land use change, impacts and dynamics (LUCID): Project Working Paper Series Number: 15. International livestock Research Institute, Nairobi

Mulatie M (2009) Soil erosion assessment, runoff estimation and water harvesting site selection using gis and remote sensing at Debre Mewi Watershed (West Gojam). MSc thesis, Bahri Dar University, Bahir Dar

Nyssen J, Poesen J, Veyret-Picot M, Moeyersons J, Mitiku Haile, Deckers J, Dewit J, Naudts J, Kassa Teka, Govers G (2006) Assessment of gully erosion rates through interviews and measurements: a case study from Northern Ethiopia. Earth Surf Proc Land 31(2):167–185

Pitmenlel D, Harvey C, Resosudarmo P et al (1995) Environmental and Economic Costs of Soil Erosion and Conservation. Science 267 (5201):1117–1123. doi:10.1126/science.267.5201.1117

Ringo DE (1999) Assessment of erosion in the Turasha catchment in the lake Naivasha area, Kenya. MSc thesis

Tadele A, Gete Z, Tesfaye et al (2006) Quantifying the impacts of livestock trampling on runoff, soil loss and crop yield under traditional tefe cultivation system. In Yihenew GS, Enyew A, Zewudu A et al (eds) Proceedings of the 1st annual regional conference on completed research activities on natural resources management. 14–17 August 2006, ARARI, Bahir Dar

Tamrie H (1995) The survey of the soil and water resources of Ethiopia, UNU/Tokyo

Tesfahun F, Osman A (2003) Challenge and prospects of Food security in Ethiopia. In: Proceeding of the food security conference. UNNCC, August 13–15, 2003. Addis Ababa

Teshome A (1995) Modeling of water erosion processes by agricultural non-point source pollution model in Tikurso Watershed, North Shewa. M.Sc. Thesis Research. Wageningen Agricultural University, The Netherlands

USAID (2000) Amhara National Regional State food security research assessment report USAID Collaborative research support programs Team, May 2000

USAID (United States Agency for International Development) (2004) Ethiopia land policy and administration assessment. Final report with Appendices. USAID Contract No. LAG-00-98-00031-00, Task Order No. 4

Wikipedia (2006) Mountain. The free encyclopedia.htm, https://en.wikipedia.org/wiki/Mountain. Last modified 21:01, 17 October, 2006

WBISPP (2000) Manual for woody biomass inventory and strategic planning project. Addis Ababa, Ethiopia. http://www.energycommunity.org/reCOMMEND/reCommend4.pdf

Chapter 34
Participatory System Dynamics Mapping for Collaboration and Socioecological Integration in the Lake Tana Region

Krystyna Stave, Birgit Kopainsky, Mesfin Anteneh, Abebe Ameha Mengistu, Mesenbet Yibeltal Sebhat, Shimelis Aynalem, Berihun Tefera, Alemayehu Wassie, Chalachew Aragaw, Belachew Getenet, Biazenlegn S. Beyene, Azanaw Abebe, Goraw Goshu, Aklilu Tilahun Tadesse and Melak Mesfin Ayenew

K. Stave (✉)
School of Public Policy and Leadership, University of Nevada Las Vegas, Las Vegas, NV, USA,
e-mail: krystyna.stave@unlv.edu

B. Kopainsky · A.T. Tadesse
System Dynamics Group, University of Bergen, Bergen, Norway

M. Anteneh · B. Getenet
Faculty of Social Sciences, Bahir Dar University, Bahir Dar, Ethiopia

A.A. Mengistu
Department of Biology, Adama University, Adama, Ethiopia

M.Y. Sebhat
Department of Water Resources and Irrigation Engineering, Bahir Dar University, Bahir Dar, Ethiopia

S. Aynalem · B. Tefera · A. Wassie · A. Abebe · G. Goshu
College of Agriculture and Environmental Sciences, Bahir Dar University, Bahir Dar, Ethiopia

S. Aynalem
College of Natural Science, Addis Ababa University, Addis Ababa, Ethiopia

S. Aynalem
Research Associate of International Crane Foundation, Baraboo, USA

C. Aragaw
Amhara Region Livestock Resources Development Promotion Agency, Bahir Dar, Ethiopia

B.S. Beyene
College of Science, Bahir Dar University, Bahir Dar, Ethiopia

M.M. Ayenew
Division of Mathematics, Addis Ababa Science and Technology University, Addis Ababa, Ethiopia

K. Stave et al. (eds.), *Social and Ecological System Dynamics*,
AESS Interdisciplinary Environmental Studies and Sciences Series,
DOI 10.1007/978-3-319-45755-0_34

Abstract This chapter presents a system map of causal links and feedback loops among social and ecological components in the Lake Tana region. The map was collaboratively developed by twenty-seven regional researchers, managers, and development experts who participated in a one-day systems mapping workshop entitled "Strengthening Links between Policy and Research for Sustainable Development in the Lake Tana Basin" held in Bahir Dar, Ethiopia in November 2014. The chapter describes the map development process, content of the map, participant perspectives on the value of the process and the workshop outcomes, and steps for building on the workshop.

Keywords Participatory system dynamics · Group model building · SES · System integration · System mapping · Causal loop diagram

34.1 Introduction

The first part of this book introduced the interrelated social and ecological issues in the Lake Tana region and proposed that an integrated view be developed using a system dynamics approach to better understand the drivers of change. Parts II and III described the details of the region's ecosystem, socioeconomic system and management characteristics and trends. This last synthesis section turns our attention forward. This chapter presents a system map of causal links and feedback loops among social and ecological components in the Lake Tana region that was collaboratively developed by regional researchers, decision-makers, and development practitioners. The initial map was created in a one-day systems mapping workshop entitled "Strengthening Links between Policy and Research for Sustainable Development in the Lake Tana Basin" that was held in Bahir Dar, Ethiopia in November 2014. The workshop was coordinated by the Blue Nile Water Institute (BNWI) at Bahir Dar University.

Twenty-one regional researchers and six managers, directors and senior staff of regional resource management agencies participated in the problem definition and mapping activities of the workshop. Participants brought diverse disciplinary backgrounds to the discussion and represented both research and decision-making perspectives. Approximately 70% of the participants were grounded in biophysical disciplines and 30% in management and social science disciplines. They had no previous systems training. While they had experience with the domain being addressed in the workshop, they had not previously worked together to determine a common systems perspective on the region's issues or a collaborative framework to address the system as a whole.

The intent of the workshop was two-fold: to begin a structured dialogue among scientists and decision-makers in the Lake Tana basin and to begin to develop a high-level, integrative framework for understanding the causes of complex social and ecological issues, research needs and potential policy solutions. This chapter describes the workshop process, products, and participant experiences, and discusses how to build on this work.

34.2 Need for an Integrative Framework

As the headwater catchment for the Blue Nile River, one of two primary tributaries of the transnational Nile River system, the Lake Tana basin is the focus of widespread attention at regional, national, and international levels. In spite of the importance of the basin and the issues summarized in Chaps. 1 and 2 and detailed in the following sections of the book, there is no comprehensive, integrated, system-wide description of its characteristics and dynamics that can serve as a basis for its sustainable development. Policy questions and research needs are dynamic and often involve tradeoffs that are not easy to resolve. Although many researchers and decision-makers are focusing on the region, and there are already some avenues for the use of scientific research in regional decisions such as the work that has been done by the Blue Nile Water Institute since its establishment in 2012, research efforts are still quite fragmented and are often not visible or useful to decision-makers. The utility of research is hampered by limited applied research, lack of access to good system-wide data, and sometimes inefficient knowledge transfer to the grassroots level. Much more can be done.

This workshop was undertaken as a step toward connecting the baseline research done by scientists at Bahir Dar University with the environmental and development questions being asked by regional decision-makers. The purpose of including both decision-makers and researchers in the workshop was to strengthen the connections between regional researchers and decision-makers, promote ongoing dialogue between researchers and policy-makers and identify specific directions for policy-relevant research. The workshop followed a system dynamics approach to develop a system map, or conceptual model, of causal relationships that underlie problematic trends in the system. Further work is expected to build on the outcomes of this initiative, including the potential creation of a simulation model for policy evaluation.

34.3 Approach

The workshop followed the first stages of a system dynamics group modeling process. System dynamics group model building, also called collaborative or participatory modeling, is a process of engaging a group of stakeholders in structured problem-solving following the steps of the system dynamics modeling process (e.g., Vennix 1996; Andersen et al. 1997; Hovmand 2014). In a set of facilitated activities, stakeholders develop a collective understanding of the key problematic trends that characterize their issues and create maps of the causal relationships in the system that give rise to the trends. In later stages of group modeling, these structural diagrams can be translated into sets of mathematical relationships that can be simulated on a computer. The group uses the computer simulation model to identify possible policies or management interventions and test their effects on the problem. Andersen and Richardson (1997) and Hovmand (2014) provide guidance for conducting group modeling activities.

The degree of stakeholder engagement in model development can vary, depending on a number of factors such as the amount of time available, purpose of the group collaboration, or nature of the problem. Antunes et al. (2015) describe three possible modes of system dynamics group modeling along the continuum of stakeholder participation in model development. The first mode is the use of an existing simulation model for collaborative learning about and analysis of a problem. The second mode is participatory systems mapping, in which participants develop diagrams of the causal relationships in the system using causal loop or stock-and-flow techniques. In the third mode, participants collaborate in the development of a working simulation model and use it for analysis.

The system dynamics group model building approach is particularly useful for building shared understanding of a dynamic problem, the system connections and feedback underlying the problem, and the concerns of other stakeholders in the system. System maps can serve as boundary objects to show stakeholders how they are connected across disciplinary or other boundaries (Black 2013). When a simulation model is built, stakeholder engagement in the development process and use of the model for evaluating policy alternatives also builds shared ownership of the resulting policy recommendations (e.g. Stave 2002).

Since the purpose of the Lake Tana workshop described here was to develop a collective understanding of the system and a platform for further collaborative research, the workshop was structured as Antunes et al.'s (2015) second mode of group modeling, participatory systems mapping.

34.4 Workshop Activities

The workshop was opened with addresses by Bahir Dar University President Dr. Baylie Damtie and U.S. Ambassador to Ethiopia Patricia M. Haslach. Participants then spent the morning in facilitated problem-definition and systems mapping exercises. In the afternoon, participants reviewed the maps, suggested modifications, discussed observations from the maps and the mapping process, reflected on issues raised by the exercise and suggested next steps. Table 34.1 presents an overview of the activities.

The workshop focused on three key activities for developing the participant-generated systems map: identification of problematic trends, small group collective development of causal maps working backward from key problem trends, and full group discussion of issues raised by the system mapping. In the first activity, participants first generated individual sets of key problematic trends. These were then clustered in a full group session to develop a collective sense of the most pressing issues (Fig. 34.1).

The trends served as the starting point for the small group mapping. After reviewing the set of key problematic trends in the full group, participants divided

Table 34.1 Workshop activities

Timeframe	Agenda items	Actions and outcomes
Morning	• Workshop and participant Introductions	
	• Overview of regional sustainability issues	• Facilitators introduced causal loop analysis for Lake Tana region issues
	• First issue focusing activity	• Participants individually generated graphs of key problematic trends
	• *Refreshment break*	• Facilitators organized trend graphs in clusters
	• Second issue focusing activity	• Full group reviewed graphs and added missing trends
	• Small group mapping activity	• Participants divided into two groups to develop causal maps underlying problematic trends
	• *Lunch break*	• Facilitators transferred small group maps to computer for display and discussion
Afternoon	• Full group discussion of mapping	• Facilitators presented the small group maps to the full group
		• Participants reflected on issues that emerged from the maps and map development process
	• *Refreshment break*	
	• Full group discussion of further steps	• Participants discussed issues that were missed in the mapping and how the process could continue
	• Closing remarks, Goraw Goshu, BNWI director	

Fig. 34.1 Individual generation of problematic trend graphs, clustered graphs. Used with permission

into two subgroups to develop maps of the cause-and-effect relationships underlying the problematic trends. Each subgroup chose one or more problem indicators as starting points. They worked backward to describe what factors were causing the observed trends and forward to show how changes in one variable led to effects in other variables. Creating the causal diagram (Fig. 34.3) required discussion of which factors and causal links were relevant as well as the polarity, or direction of causality between each pair of variables. Facilitators then entered the variables and causal links into a software program during the lunch break to make them more legible for use in the afternoon discussion. These versions (Fig. 34.4) included all the variables from the hand-drawn maps as written, but rearranged to untangle the connections and highlight loops where possible. They were used as the basis for the group discussion (Note that Figs. 34.3 and 34.4 show only parts of the completed diagrams to illustrate the steps in the process. The variable names are not intended to be legible in these diagrams.). After the workshop, the facilitators created two more versions of the maps based on Fig. 34.4 and the group discussion. Figure 34.5 is a consolidated map reconciling the two participant-drawn maps. Figure 34.6 is a higher-level sector diagram that was abstracted from the consolidated causal loop diagram as suggested in the workshop discussion.

34.5 Workshop Outcomes

Problematic trends identified

Each participant identified the 2 or 3 top issues they considered most problematic from their professional perspective. They chose indicators for those issues and sketched a graph representing their perception of the problematic trend in that indicator. The graphs were clustered by issue into the indicators listed in Table 34.2. After reviewing the clustered graphs as a group, participants discussed any discrepancies in the graphs and added other problematic trends they felt were missing. Of the 80 top problematic trends identified, 40% were ecological (8 different indicators), 23% were socioeconomic (7 different indicators), and 37% were issues of human use of environmental resources (19 different indicators).

System map drafted and revised

Figure 34.2 shows the progression of the causal diagrams and the discussions in the two subgroups. Figure 34.3 shows sections of the resulting hand-drawn maps, which one participant called "bird's nests." In these diagrams, a link represents a hypothesized cause-and-effect relationship, that is, a hypothesis that a change in the variable at the tail of the arrow will cause a change in the variable at the head of the arrow. Participants identified the polarity of each link, or direction of the relationship, shown by a "+" or "−" sign at the head of the arrow. A "+" indicates that the connected variables change in the same direction if all other factors are held constant; a "−" indicates the variables change in the opposite direction.

Table 34.2 Key problematic trends identified by participants

Category	Indicator (trend)	Number of participants selecting as one of top three problem indicators
Ecological	Invasive species (increasing)	5
	Sedimentation (increasing)	5
	Soil erosion (increasing)	5
	Water quality (decreasing)	5
	Biodiversity (decreasing)	4
	Fish biomass (decreasing)	4
	Wetland area (decreasing)	3
	Flood frequency (increasing)	1
		32 (40%)
Socioeconomic	Population (increasing)	12
	Entrepreneurship (low, not increasing as fast as desired)	1
	Financial resources (remaining low)	1
	Health (low, not increasing as desired)	1
	Human capital, educated human power (low, not increasing as desired)	1
	Poverty (high, not decreasing as desired)	1
	Tourism (increasing faster than infrastructure)	1
		18 (23%)
Resource use and management	Land use intensity (increasing)	4
	Land degradation (increasing)	3
	Waste, sanitation capacity (increasing waste, higher than capacity)	3
	Agricultural land (decreasing)	2
	Environmental and nutrient pollution (increasing)	2
	Food demand (increasing)	2
	Institutional coherence of resource management organizations (low)	2
	Agricultural yield (decreasing)	1
	Community based development, participation and empowerment (low)	1
	Competition between conservation and development (increasing)	1
	Economic pressure on resource utilization, (increasing resource use)	1
	Energy (increasing fossil fuel, hydropower and solar remaining low)	1

(continued)

Table 34.2 (continued)

Category	Indicator (trend)	Number of participants selecting as one of top three problem indicators
	Greenhouse gas emissions (increasing)	1
	Importance of trans regional issues such as hydropower (increasing)	1
	Infrastructure: water-related, roads (low relative to increasing demand)	1
	Policy implementation and integration (low, not increasing)	1
	Research knowledge gap (increasing)	1
	Stakeholder conflict (increasing)	1
	Water use conflict (increasing)	1
		30 (37%)

Fig. 34.2 Two subgroups of small group causal mapping. Used with permission

Participant reflection on map development and content

Participants used the electronic versions of their maps (excerpted in Fig. 34.4) to discuss the map development process, content of the map, value of the mapping process to themselves, and how the map could be used to facilitate communication with policy-makers. Their comments and suggestions were incorporated into the consolidated map (Fig. 34.5).

During thc map development session, participants began by describing causal relationships in their direct area of expertise and then broadened their focus to

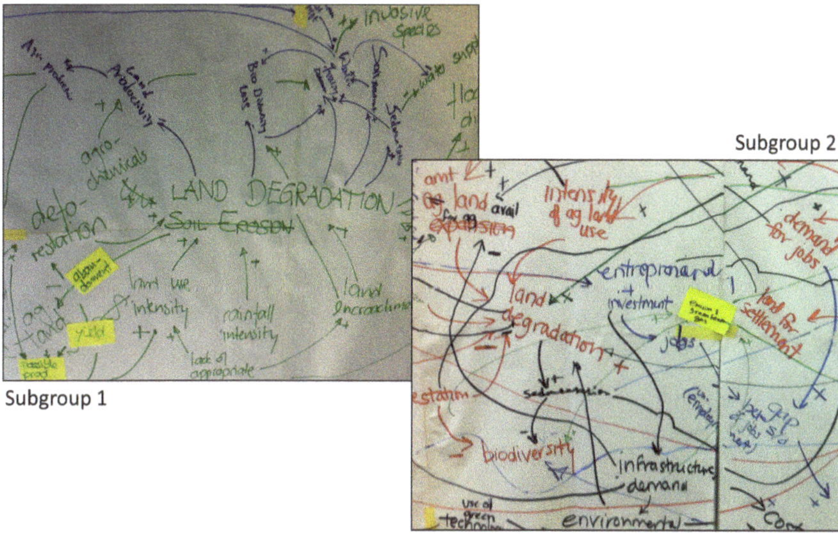

Fig. 34.3 Sections of the maps produced during small group causal mapping

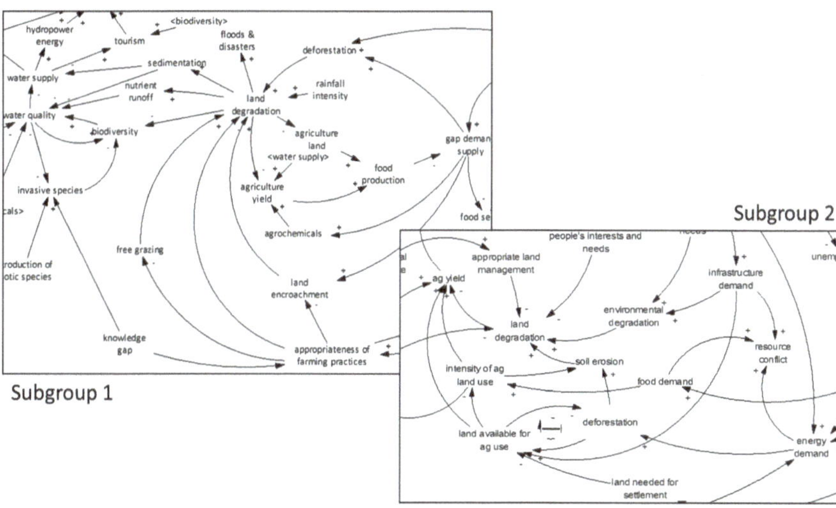

Fig. 34.4 Sections of the electronic versions of Fig. 34.3 maps used in workshop discussion. These are shown to illustrate the steps in the process. Variable names are not intended to be legible here

describe the connections they saw with other sectors. Biophysical scientists first focused on relationships between *deforestation* and *soil erosion*, for example, or reinforcing relationships between *biodiversity* and *land and water quality*, while socioeconomic experts described causal links between *education* and *economic*

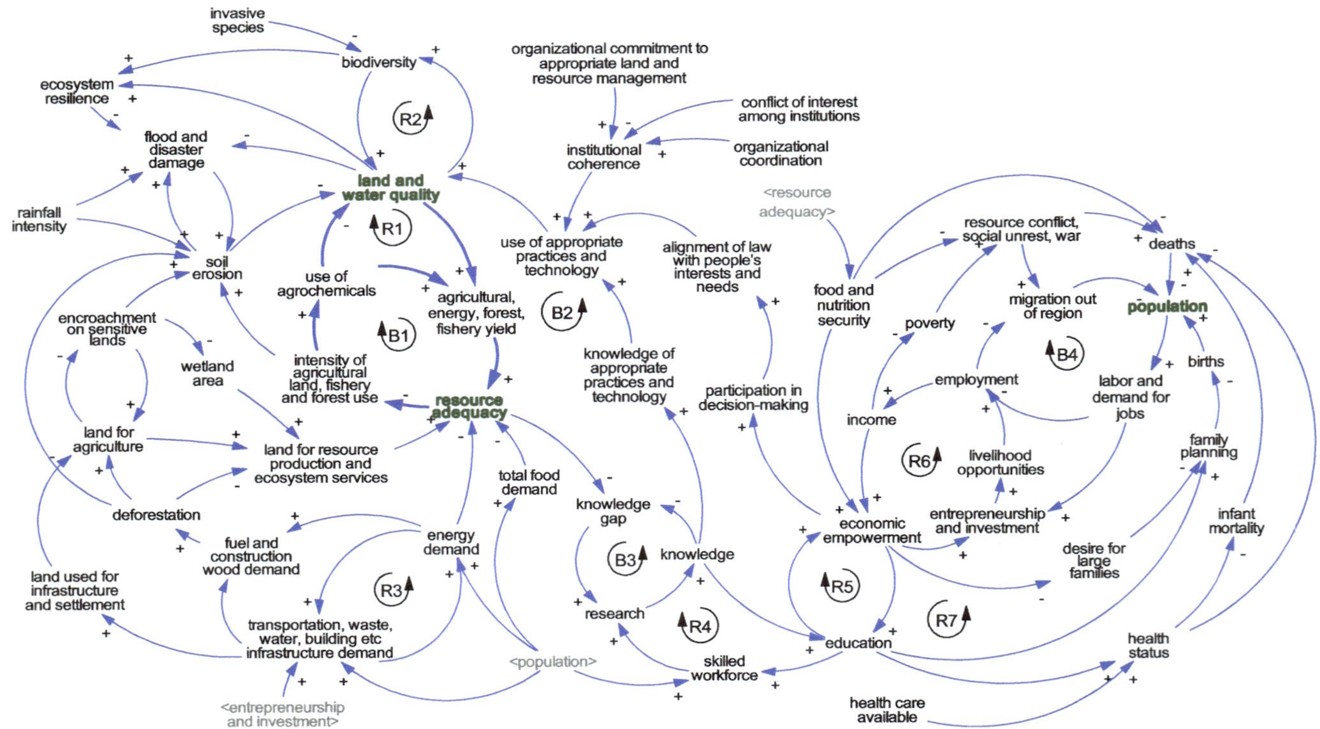

Fig. 34.5 Consolidated causal loop diagram produced by combining maps from Subgroup 1 and 2 shown in Fig. 34.3. Selected feedback loops are labeled, with "R" indicating a reinforcing feedback loop and "B" indicating balancing feedback. Bolded variable names indicate key map development starting points

empowerment, between *employment*, *economic empowerment* and *entrepreneurship*, and between *population*, *labor demand*, *employment*, and *migration*, among others. The starting points for different stakeholders in these subgroups are shown in bold green type in Fig. 34.5. The discussions quickly expanded to the connections between variables in different sectors.

In the full group debriefing, participants discussed whether causal loops in the diagrams were reinforcing or balancing, and what that might mean for the system's behavior. Reinforcing feedback loops continue, or amplify, the direction of an initial change in any of the variables in the loop. They are indicated by a "R" inside the loop in Fig. 34.5. Loops that balance, or counteract, an initial change are indicated by a "B" inside the loop. The loops described above thus became loops R2, R5, R6, and B4 in Fig. 34.5. In addition, two other cross-sector connections were highlighted in the discussion. One concerned the role of research (loops B3, R4, and R5). Loop B3 shows how a change in *resource adequacy* relative to what is known about producing adequate resources (*knowledge*) can drive *research*, which then increases *knowledge* and reduces the initial *knowledge gap* (B3). Better *knowledge of appropriate conservation practices and technologies* contributes to improvements in *land and water quality*, ultimately improving *resource adequacy* (B2). Increasing levels of *knowledge* contribute toward relevant *education* that increases the *skilled workforce* that conducts further *research* (R4). *Education* also reinforces *economic empowerment* (R5) which has many further positive changes that ripple through the socioeconomic sector. A potential leverage point for improving many system aspects is thus improving research. A second area of discussion concerned changes in *population*. Loop B6 describes the pressure for migration out of the region that stems from increased population, increased labor demand, and decreased employment. In the post-mapping discussion, participants also pointed out the potential benefits to entrepreneurship and livelihood opportunities that could arise from population growth. They noted that there could be other benefits as well that were not yet incorporated into the map.

One loop that was identified by both groups was what became the R1 loop in the upper left portion of Fig. 34.5 (indicated in bold type). This proposes that if *resource adequacy* decreases, the *intensity of agricultural land, fishery and forest use* would likely increase, leading to an increase in the *use of agrochemicals*, a decrease in *land and water quality*, a further decrease in *agricultural, energy, forest, fishery yield*, and a further decrease in *resource adequacy*. This long-term reinforcing loop is connected to a short-term balancing loop, B1 (shown in bold in Fig. 34.5), in which the increased *use of agrochemicals* leads to a short-term increase in *agricultural, energy, forest, fishery yield*, and an increase in *resource adequacy* that counteracts the initial decline. These kinds of linked balancing and reinforcing loops are often found in social—ecological systems and can generate counterintuitive long-term responses to policy and management interventions. One of the loops dominates the system's behavior in the near-term until other factors accumulate to a point that the other loop becomes more important in the system. In this case, land and water quality degradation from the use of agrochemicals may take time to build up to a level that it affects yield.

Value to participants

In addition to discussing how the diagrams helped them visualize system connections, participants felt the map and the mapping process helped them better understand how problems are created, where research is needed, and where policies need to be implemented. They also described other, less tangible benefits of the map and mapping process. Some thought the way the map put individual disciplines in a larger context would help for purposes such as writing grant proposals as well as illustrating the need for integrated management. "This comprehensive perspective," said one participant, "shows that no one stakeholder 'owns' the lake or the management problems. It opened up my view." Another appreciated the way the process was able to incorporate the ideas of all the participants. He said: "I think I have a voice on the map," and suggested this would be a way to help more stakeholders understand the system and contribute to sustainable management.

Communication with policy-makers

There was much discussion about how this process and the map could support communication between researchers and policy-makers. Participants felt the limited participation of decision-makers limited the development of the framework. A number of researchers said they thought decision-makers who did not participate in the mapping would say the main problems in the region are implementation problems rather than policy problems. They felt that if more decision-makers could be involved in map development, they could clear up any misunderstanding or mistakes, expand the representation of implementation issues, and show how and where those problems arise.

For communicating the messages from the integrated system map to policy-makers who were not involved in its development, participants felt the map should be simplified. "They do not need to see the full complexity but they need to know the results." Figure 34.6 is a simplified version of the consolidated map that leaves out the details of specific causal relationship but shows high-level connections among sectors. Participants thought this kind of diagram could be used to explain the concept of the map to policy-makers to encourage them to participate in future map development and communicate general messages from the more detailed map.

34.6 Reflection and Discussion

The workshop had two goals: to foster dialogue among scientists and decision-makers in the Lake Tana basin and to initiate development of a high-level, integrative framework for understanding the causes of complex social and ecological issues, research needs and potential policy solutions. The discussions and products generated in this workshop provide a substantial foundation for integrating social and ecological components of the Lake Tana region for sustainability.

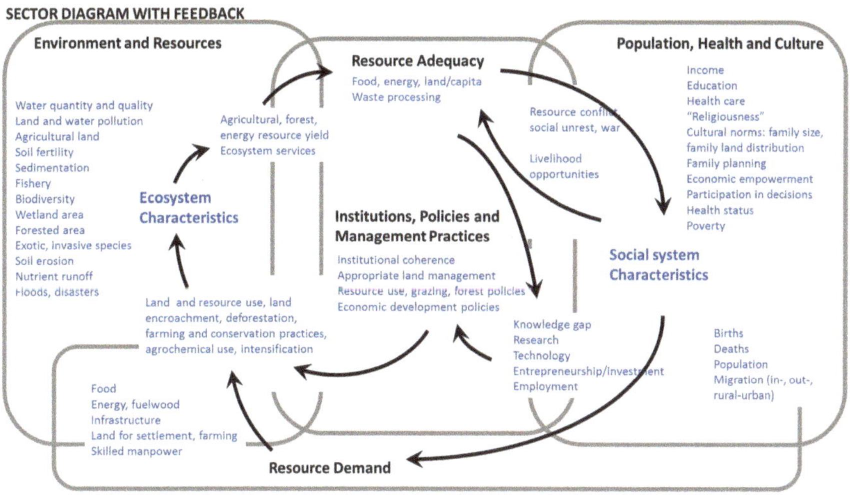

Fig. 34.6 Sector diagram showing high-level feedback among major sectors

Participants recognized that the short time available limited the extent to which they could develop and analyze the system map. They suggested several next steps that could improve the map and promote collaboration among Lake Tana Basin stakeholders. These included working more on the development of the causal diagram, holding additional workshops with more stakeholders, simplifying the diagram for communicating with policy-makers, spending more time analyzing the diagram, and developing a full simulation model for formal policy analysis.

In addition to expanding the diagram further, such as the suggestion to incorporate the potential benefits of population as well as the costs, some participants suggested indicating different strengths of relationships by varying the thickness of the causal link lines, for example. From a dynamic perspective, however, this would be difficult to implement as the strength of a relationship changes over time. Others suggested bringing a wider representation of stakeholders into future workshops, including farmers, non-governmental organizations, and other resource management agencies. Additional mapping workshops with a more diverse set of stakeholders would enhance the way the sustainability problem is defined. Other stakeholders might have concerns about different trends, or they may confirm that the set identified here is sufficient. Additional stakeholders would also enhance the contents of the causal map by adding missing variables and relationships.

Participants wanted to spend more time analyzing the diagram. At the end of the discussion, one person remarked: "Now we have to sort out what are the policy ideas and what are the research ideas. There are different policies. We have to differentiate. But we can't just show the map, we should summarize the ideas in terms of policy and research." Participants thought the ultimate outcome of this work should be the ability to tell policy-makers what they should do to address regional problems. "The research should tell policy-makers 'this is how to protect the Lake.'" This kind of policy analysis is a possible extension of the workshop. After further revisions, the causal map can be used as the basis for identifying potential policy interventions. Formal policy analysis would require development of a computer simulation model based on the system map, however. Such a model would allow potential policies to be evaluated using the model.

More can be done to build on the foundation established in this workshop. It is important to emphasize, however, the value of what was accomplished. In one day, participants collectively developed a list of key problematic trends, a conceptual model of how the system produces those problematic trends in the form of an integrated causal map, and a broader vision of how their own work fits into the context of the whole system. The workshop demonstrated how integration and collaboration can be done.

In addition to the direct output from the workshop, participants also reported longer-term effects in the way they viewed resource management issues. Approximately six months after the workshop, participants were sent a short, anonymous follow-up survey. They were reminded of the title, date, and agenda outline of the workshop, but not given any other information about the activities, and asked to reflect on the day and any thoughts they took away from the workshop. The survey asked what stood out about the content of the workshop and whether the workshop activities influenced their thinking or work since the workshop. About 40% (11/27) of the participants responded.

In response to the questions "What stands out in your mind about the content of the workshop? What were the main idea(s) about the Lake Tana Basin you took away from the workshop?" respondents identified aspects of cause-and-effect relationships, system integration, and stakeholder interactions as the main themes:

> The major idea I have taken from the work shop is that Lake Tana is one of the most resource full areas in Ethiopia. Although it is a resource rich area in the country, it is also highly susceptible to resource degradation. So during the workshop we have tried to identify the causes and consequences of resource degradation in around the lake and the measures to be taken tackle the problem for the future.

> The high level of complexity of the factors interacting with each other over time and space that determine the Lake Tana system and its dynamics.

> How soil erosion affects the basin, land use change effect on sedimentation to Lake Tana, wetland shrinkage around the Lake basin, food insecurity and low income.

> The presence and interaction of different stakeholders in Lake Tana Basin, visualizing the relationship of the stakeholders and their interaction by mapping exercise.

> Integration among various stakeholders has a paramount importance to solve Tana basin problems. The fact that a complex issue can be seen piece by piece and if each piece is addressed carefully a system model can be developed that ultimately shows the cause and effect.

> Identifying the causes and effects (i.e. the potential problems and their consequence) and the possible solutions that will be taken and strengthen links between policy and research for sustainable development in the Lake Tana ecosystem.

> Lake Tana and its catchment area must be protected and sustainably developed. To achieve this, all stakeholders participation is very important.
>
> I got a brief introduction to the mapping, particularly; how socio-economic sphere can be integrated with the well calibrated physical systems to reproduce more reliable and sustainable models.
>
> The problems we have observed in the Lake Tana basin are multifaceted and interlinked with each other.
>
> Need to look at Lake Tana as a whole. The problems and the solutions are complex to visualize.
>
> Stakeholder participation, system thinking approach, holistic approach to development.

In response to the question of how workshop activities influenced their thinking or work since the workshop, several respondents described using aspects of systems thinking in planning, implementation, and communication:

> Most of our experience in solving issues/problems is only from our own professional perspective. However, issues are always complex and beyond a single profession. Here I understood system thinking is a way out.
>
> System thinking is indispensable for research problem analysis.
>
> After that workshop I understand that there must be good communication between researcher and policy makers. Because if there is smooth relation or strong linking between policy makers and researcher we are fruitful and make the natural resource sustainable.
>
> I have understood that Lake Tana and its basin have multiple challenges other than fisheries. My understanding was only related to fisheries. Currently I am developing Lake Tana Fisheries Management Plan to be submitted for regional government officials for decision making. This workshop helped me to incorporate issues in the Fisheries Management Plan other than fisheries challenges.
>
> Before the workshop I was not able to see the problems of the basin in a concrete manner. Then I was able to see the parts as a whole and the whole as parts.

These responses indicate that such systems mapping workshops can also strengthen critical systems perspectives among participants. This is an important prerequisite for any follow-up activity to the workshop and particularly for further development of the system map into a computer simulation model for policy analysis and design. Without a social environment and institutional setting that allows the insights from a simulation model to be received, appreciated, understood and eventually implemented by stakeholders on the ground, formal simulation efforts will not lead to effective change. The workshop described here demonstrates one approach for supporting the development of collaborative social capital among stakeholders in addition to developing the map itself. For the Lake Tana Basin, it sets a foundation for further sustainability analysis.

References

Andersen DF, Richardson GP (1997) Scripts for group model building. Syst Dyn Rev 13(2):107–129

Andersen DF, Richardson GP, Vennix JAM (1997) Group model building: adding more science to the craft. Syst Dyn Rev 13(2):187–201

Antunes MP, Stave KA, Videira N, Santos R (2015) Using participatory system dynamics in environmental and sustainability dialogues. In: Ruth M (ed) Handbook of research methods and applications in environmental studies. Cheltenham, UK: Northampton, MA, USA: Edward Elgar Publishing

Black LJ (2013) When visuals are boundary objects in system dynamics work. Syst Dyn Rev 29 (2):70–86

Hovmand P (2014) Community based system dynamics. Springer, New York

Stave KA (2002) Using system dynamics to improve public participation in environmental decisions. Syst Dyn Rev 18(2):139–167

Vennix JAM (1996) Group model building: facilitating team learning using system dynamics. Wiley, Chichester

Chapter 35
Research Needs in the Lake Tana Basin Social-Ecological System

Goraw Goshu, Shimelis Aynalem, Baylie Damtie and Krystyna Stave

Abstract This chapter synthesizes the research needs presented in the book and proposes actions for addressing them. It highlights the socioeconomic and biophysical characteristics, development, production and consumption patterns, challenges and threats and management approaches that were discussed in the preceding chapters and discusses the priorities for future Lake Tana basin research that were raised. The analysis shows that the basin has many unique features, and great potential for sustainable development, especially in water resources. The lake and basin provide multiple benefits to the local community, the region and downstream riparian countries. Many of the problematic trends in the region result from an imbalance between resource production and consumption patterns, and limited or poorly implemented conservation measures. The many and complex challenges and threats in the basin are caused, in part, by the lack of an integrated basin plan and poor collaborative mechanisms to share data and find sustainable solutions. The available data and information about the basin is limited. It is not well-organized, accessible, or used well for the basin's resource management. The identification of major research

G. Goshu (✉) · S. Aynalem
Blue Nile Water Institute and College of Agriculture and Environmental Sciences, Bahir Dar University, Bahir Dar, Ethiopia
e-mail: gorawha@gmail.com

S. Aynalem
e-mail: shimelisay@yahoo.co.in; shimelis.aynalem@gmail.com

S. Aynalem
College of Natural Science, Addis Ababa University, Addis Ababa, Ethiopia

S. Aynalem
Research Associate of International Crane Foundation, Baraboo, USA

B. Damtie
Washera Geospace and Radar Science Laboratory, Department of Physics, Bahir Dar University, Bahir Dar, Ethiopia
e-mail: baylied@bdu.edu.et

K. Stave
School of Public Policy and Leadership, University of Nevada Las Vegas, Las Vegas, USA
e-mail: krystyna.stave@unlv.edu

K. Stave et al. (eds.), *Social and Ecological System Dynamics*,
AESS Interdisciplinary Environmental Studies and Sciences Series,
DOI 10.1007/978-3-319-45755-0_35

gaps and priorities for future research highlighted in this book, as well as the systems framework for integration and collaboration, provide a basis for further research and policy decisions to promote sustainability in the region.

Keywords Lake Tana basin · Upper Blue Nile River · Ethiopia · Sustainable development · Collaboration · Integration · Knowledge gaps

35.1 Discussion

Lake Tana and its catchment area host a rich array of ecological resources as well as over three million people who interact with and shape the environment in diverse ways. The basin provides multiple benefits to the local community, region, and downstream riparian countries. The lake is the source of the transboundary Blue Nile River. Its basin and the entire upper Blue Nile region have become geopolitically sensitive areas and receive considerable attention from the lower riparian countries. The construction of the Great Ethiopian Renaissance Dam across the Blue Nile river near the Ethio-Sudan border for hydropower generation is likely unique in Africa because of its production capacity and the regional economic and political integration opportunities it will create. This investment will bring many opportunities for all Nile riparian countries in terms of economic, social and political and technical cooperation. Lake Tana is also a natural source of water for the Tana Beles hydroelectric power plant which has already been operating for a number of years.

Lake Tana and its tributaries constitute the largest fresh water resource in the country. Because of its huge water resource potential, conducive agroecology and fertile soil, it has been identified as a major economic corridor. The aquatic, wetland and terrestrial ecosystems in the basin support a myriad of biodiversity which is not accounted for very well (microorganisms, animals and plants). Lake Tana is a home for a unique fish species flock of Labeobarbus and others (Nagelkerke and Sibbing 1996; Degraaf et al. 2004). The wetlands are important feeding and breeding ground for birds including migratory species (Shimelis et al. 2008). In addition, Lake Tana provides unique religious and social values to the riparian community. It is a home of centuries-old island monasteries dating back to the 13th century. These are important to the local communities for the many religion ceremonies that bring people together in celebration as well as to the local tourist economy.

Beyond its unique characteristics, the Lake Tana Basin is also an example of the sustainability challenges faced by many rapidly growing regions around the world that are making the transition from heavily resource-dependent, rurally-oriented societies to more urban-focused societies. Sustainable development in this context means considering the present needs of a primarily rural population and the future needs of a more urban population, as well as managing the transition smoothly. As population in the Lake Tana region grows, it puts pressure on the interconnected ecological and socioeconomic system. The resulting environmental degradation and social change raise questions about how to best to manage the system in order to meet the region's needs now and in the future. Critical socioeconomic development needs in the Lake Tana basin include reducing poverty, achieving food security, agricultural transformation, transforming rural livelihoods and income, reversing land degradation to avoid on-site and off-site effects and reducing sedimentation of the Lake and dams lakes. Many environmental conservation and related polices listed in the previous sections have been implemented for more than two decades. However, the reality on the ground is that environmental problems persist.

The chapters in this book survey the status and trends of a broad set of biophysical and socioeconomic characteristics that define the region, as well as the environmental policies and practices that govern human-environment interaction. They identify key problematic trends and challenges in the basin, describe a variety of research needs and questions, and suggest ways forward.

35.2 Research Needs Identified in the Book

35.2.1 The Lake Tana Basin Social-Ecological System as a Whole

In Chap. 2, Goraw Goshu and Shimelis Aynalem present a set of challenges faced by the region, which are explained in detail in the sections that follow. Their overview and the detailed chapters that follow describe the interconnections among ecosystem and social systems in the region, and build the case for an integrated approach to managing human-environment relationships. In Chap. 33, Yihenew G. Selassie sums up the most critical manifestation of natural resource degradation as soil loss, with related soil fertility deterioration, organic matter depletion, water resources degradation, and siltation of dams, reservoirs and agricultural lands. The problem is widespread in the basin, with about half of the highland's land area significantly eroded, over one-fourth seriously eroded, and over 2 Million ha of land permanently degraded to the point that the land can longer support cultivation. Soil and land degradation is due to natural factors (rugged and steeply sloped topography) as well as intense cultivation which aggravates the erosion rate. Anthropogenic factors include agricultural expansion, practices that remove crop residues and animal dung completely from fields, overstocking of livestock, forest clearing, and infrastructure and industrial development. These problematic practices are rooted in social issues including poverty, lack of alternative sources of livelihood, and lack of awareness of the effect of erosion. Policies for mitigating these factors exist, but are often not enforced. As Yihenew G. Selassie notes, "It is now well recognized that the most serious NRM problems stem from institutional rather than technical factors. The success of NRM policy depends largely upon institutional organization, stability and capacity." *In sum, despite the magnitude of the problems, there are promising efforts in research, technology dissemination and policy formulation in the region to address the problems. Even with the high degree of degradation in the region, mitigation efforts are encouraging, and could be scaled up.*

Problematic trends in soil and environmental degradation, changes in land cover and water quality, and changes in livelihoods and quality of life in the Lake Tana basin arise from the interconnections and feedback between the social and environmental characteristics of the region. In support of developing an integrative framework for policy-making and collaboration, Krystyna Stave and Birgit Kopainsky review social-ecological system approaches in Chap. 3 and propose system dynamics participatory modeling for the Lake Tana basin. This approach was applied in a systems mapping workshop held in Bahir Dar in 2014 by a group of researchers, decision-makers and practitioners in the region. The process and the system map that was produced are discussed in Chap. 34 by workshop facilitators and participants Krystyna Stave, Birgit Kopainsky, Mesfin Anteneh Wubie, Abebe Ameha Mengistu, Mesenbet Yibeltal Sebhat, Shimelis Aynalem, Berihun Tefera, Alemayehu Wassie, Chalachew Aragaw, Belachew Getnet, Biazenlegn S. Beyene, Azanaw Abebe, Goraw Goshu, Aklilu Tilahun and Melak Mesfin. The workshop demonstrated one approach for developing collaborative social capital and strengthening systems perspectives among participants in addition to developing the map itself. Participants recommended engaging a wider set of stakeholders in further development of the system map, with a possible extension to a computer simulation model for policy analysis and design. *Going forward, they stress the importance of collaboration among stakeholders from diverse biophysical and socioeconomic disciplines and including researchers, decision-makers, and practitioners.*

35.3 Ecosystem Characteristics

35.3.1 Climate Distribution and Variability

Chapters 5 and 6 examine the distribution and variability of climate characteristics across the basin. Wubneh Belete, Tesfahun G/Michael, Elias Sime, Biazenlegn S. Beyene, and Fenta Negate give an overview of climatic conditions in the basin in Chap. 5, describing a small seasonal change in air temperature around an annual average of 20 degrees C, a tropical highland monsoon with one rainy season between June and September, and rainfall distribution controlled by the movement of the inter-tropical convergence zone (ITCZ) to the northward and southward. *They note that the small number and uneven distribution of gauging stations limit what is known about the distribution of climatic variables and the effect of climate change on rainfall amount and distribution*. This further limits understanding of the potential effect of climate change on river flows in the basin.

In Chap. 6, Tadesse Terefe Zeleke and Baylie Damtie add a detailed analysis of climate variability. They review a number of studies conducted in Ethiopia that show a consistent warming trend in the maximum and minimum temperatures over the past few decades, with minimum temperatures increasing faster than maximum temperatures, but with no consistent pattern or trends in rainfall. Then they report the results of their study which investigated the variability and driving forces of extreme climate events in the Abay river basin. The study used three different observation datasets including mean monthly rainfall data from 430 unevenly distributed meteorological stations for the period 1979–2014, a blended gauge and satellite product and a station-based dataset from the Global Precipitation Climatology Project (GPCP). *Analysis shows that rainfall varies spatially across the basin, with southwestern parts of the region receiving more rainfall for longer periods than northeastern areas and variability is dominated by inter-decadal signals, likely tied to large-scale circulation.*

35.3.2 Hydrogeology, Soils, Hydrology, and Water Quality

In Chap. 7, Fenta Nigate, Tenalem Ayenew, Wubneh Belete, and Kristine Walraevens explain how the complex hydrogeology in the region affects groundwater occurrence. The center of the Lake Tana basin is underlain by multi-layered volcanic formations covered with thick alluvial and residual Quaternary sediments. Tectonic activity has led to a high degree of fracturing and weathering which makes the groundwater system productive, but difficult to characterize. *Information about the system is limited, and critical as development expands in the region.*

Mekonnen Getahun and Yihenew G. Selassie highlight the serious problem of soil degradation in Chap. 8. They identify the six major soil types in the Lake Tana basin as mainly low-fertility soils. Much is known about the characteristics and distribution of these soils. However, given the importance of soil conditions for agricultural productivity and other human requirements, there is a need to learn more about how best to manage the use of soils and improve soil degradation. *More information is needed for practices to build soil organic content, guide inorganic fertilizer use, minimize soil erosion, and irrigate properly for best crop productivity.*

In Chap. 9, Elias Sime and Biazenlegn Solomon Beyene explain the basin's hydrology. Precipitation in the basin comes from moist air originating in the Atlantic and Indian oceans that follows the north-south movement of the Inter Tropical Convergence Zone, falling on the basin mostly during one main rainy season between June and September. Rainfall flows in four main rivers to Lake Tana, a large and shallow water body. Despite the importance of surface water in the region for the many ecosystem services it provides as well as its uses for transportation, fishing, irrigation, and hydropower, and the problems of soil erosion, there are insufficient gauging stations in the catchment areas and the lake to fully understand the hydrological processes. *There is a need to better understand erosion processes, sediment transport and deposition mechanisms, the main factors driving soil erosion (soil, land use/cover, slope, climate), the effects of scale (watershed to basin) in soil erosion modeling, the effect of soil and water management efforts being implemented and the value of critical management areas for maintaining cultivated land productivity.*

Goraw Goshu, A.A. Koelmans and J.J.M de Klein identify similar data needs in Chap. 10 for water quality. Limited monitoring stations and the lack of a sustainable monitoring program in the Lake Tana basin mean existing data about water quality are fragmented and incomplete. Threats to Lake Tana include excessive nutrient loads, sedimentation, and fecal pollution and toxigenic cyanobacteria at the shore and river mouths. *Establishing a good base of knowledge about the hydrologic links between anthropogenic activities and water quality in and around Lake Tana is critical.*

35.3.3 Plankton and Fish in Lake Tana

Plankton is a vital link between water quality and fisheries since it is fundamental in the transformation of solar energy to higher trophic levels. Increased phytoplankton production generally results in increased zooplankton and fish production and is affected by water quality. In Chap. 11, Ayalew Wondie and Seyoum Mengistu present what is known about the composition, distribution, abundance, biomass and production of plankton communities in Lake Tana from a review of a number of research studies. *They recommend four areas for further plankton research: studies on zooplankton feeding behaviour, the effect of sediment on growth of zooplankton, the role of detritus in the feeding of zooplankton, and the role of protozoan and microbes in the planktonic food web.* As Abebe Ameha Mengistu, Chalachew Aragaw, Minwyelet Mengist Zerihun and Goraw Goshu note in Chap. 12, the fish of Lake Tana are impressively diverse and unique. "The lake harbors the only species flock of large cyprinids in the world, and of the 28 known species, approximately 68% are endemic to the lake. Most of the habitats harboring the ichthyofaunal diversity of this lake are relatively intact, and still naturally attractive. "The fish and fisheries are an important resource for the local population and are under threat from human-induced impacts. *Research is needed to understand the impacts of human activity on the fish and on how development and conservation activities can best preserve these fish and fishery resources.*

35.3.4 Birds, Herpetofauna, and Mammals

The Lake Tana basin is a particularly rich area for birds, as Shimelis Aynalem describes in Chap. 13. Wetlands around the lake and along the rivers in the basin are especially important habitats, providing year-round and winter roosting sites.

There is great potential for economic benefit through eco-tourism and bird watching, but birds are under threat in a variety of ways, including changes in their habitat through natural causes and human activities. Human-induced factors such as wetland conversion to agricultural lands, deforestation, sedimentation, pesticides and fertilizer application, and recent occurrence of water hyacinth are the main ones. *Comprehensive studies and investigation of the avifauna resource are still lacking; there is a need to know more about the way the ecological context is changing, as a result of human factors as well as ecological dynamics, in order to promote conservation and appropriate utilization of the resource.*

In Chap. 14, Shimelis Aynalem and Abebe Ameha Mengistu review what is known about mammals and herpetofauna. A total of 19 amphibian species, 35 reptile species and 28 mammalian species are listed in the Lake Tana basin. This diversity is likely associated with the presence of diverse habitats such as woodlands, wooded grasslands, cultivated land, natural forest, mixed forest area, water bodies and open wetland. However, the total number of species listed is also likely an underestimate. There is little information about Order-Rodentia (rodents), Lagomorphs (lagomorphs), Erinaceomorpha (hedgehogs and gymnures), Soricomorpha (shrews, moles, and soledons) and Chiroptera (bats), for example. *The listed species provide a starting point, but much more should be learned about these and other species in the area, including the diversity, richness, distribution and status of each species. We also need to know much more about their habitats, how those habitats are being degraded, and what management practices can be implemented to balance the integrity of the Lake Tana ecosystem and to sustainably use the resources. This requires capacity building in the field of Herpetology, as well as community awareness on the value and function of wildlife.*

35.3.5 Forest Resources, Wetlands, and Invasive Plants

Alemayehu Wassie explains the changes in forest resources in the Amhara region in Chap. 15. Forest composition and extent is changing, with an overall small increase in coverage in the last 15 years, but with a decrease in structural and species composition. The changes in forest cover and composition result from human activity that both increase forest area and decrease indigenous biodiversity. Eucalyptus plantations are increasing, as are efforts to rehabilitate degraded forests, but the size and quality of Afroalpine and Subafroalpine vegetation, the high Dry afromontane forests, Bamboo forests and the Woodlands have decreased significantly in the same period, resulting in a serious loss of indigenous biodiversity and opportunities for invasive species. The major threats devastating the region's forest resources are uncontrolled grazing, cropland expansion and massive clearance of wood land for investments and resettlements. *Sustainably managing forest resources will require a better understanding of the social and economic motivations for the way people use forest resources.*

The Lake Tana basin contains considerable wetland resources. In Chap. 16, Shimelis Aynalem, Goraw Goshu and Ayalew Wondie describe the different types, which extend from the headwaters of the basin's rivers to its floodplains. They include delta wetlands at the river mouths, lacustrine wetlands, riverine wetlands, man-made wetlands formed by dams and others formed by hot and cold springs. Wetland ecosystems provide habitat for diverse species, including globally threatened and endangered species. However, wetland loss is occurring as a result of the development of dams, irrigation schemes and conversion projects, as well as overgrazing and conversion to agricultural land. There is little information about the rates or details of the changes in the Lake Tana basin. *Research is needed about the dynamics of the resource itself—the changes in wetland characteristics—as well as the human activities driving the changes in wetlands, and what strategies would work for conservation and sustainable utilization.*

Land cover change and wetland destruction open the door for problems with exotic and invasive vegetation. In Chap. 17, Ali Seid and Banchiamlak Getenet review threats posed by one particularly problematic species: water hyacinth in Lake Tana. Water hyacinth infestation in the lake can be a symptom of broader watershed management problems, such as nutrient loading from excessive fertilizer runoff from upland agricultural fields and wastewater from human settlements. In turn, it causes problems in water bodies, where plants physically obstruct fishing, transport, and other activities. Other effects include degradation of water quality and increases in water-related diseases. Although water hyacinth has some beneficial uses, it incurs tremendous economic costs, as well as quality of life and health costs. *More needs to be studied about how it grows and spreads as well as how to control it. Control measures range from reducing nutrient inflows to physical removal and biological control.*

35.4 Socioeconomic System Characteristics and Land Use

35.4.1 Demography, Gender and Rural Livelihood, and Stakeholders

Chapters 18, 19, and 20 describe the characteristics of the people in the Lake Tana region. Mesfin Anteneh notes in Chap. 18 that the age group ranging from 0–14 years old is the largest population category and the number of women in the reproductive age group is also very high. This predicts continuing rapid population growth in the Lake Tana basin, with a high and growing dependency ratio. This indicates that the future population will put increasing pressure on resources in the region. *A better understanding of population distribution and demographic trends is needed to plan for future resource needs and manage population change.* Sewmehon Demissie and Azanaw Abebe focus specifically on the issues of women in Chap. 19. Women are a significant part of the labor force in the region, with responsibilities for subsistence agriculture and all domestic and reproductive roles such as taking care of the health and education of children, fetching water, collecting fuel wood, and cooking. Women are constrained in their productivity and self-enhancement by poor access to credit, technology, and information, however. *Since any effort for sustainability planning should address the disparities between male and female livelihoods, more should be done to understand the obstacles to closing gender gaps.* In Chap. 20, Dessalegn M. Ketema, Nicholas Chisholm, and Patrick Enright look more broadly at the characteristics of stakeholders in the Lake Tana region's resource use, management and governance. They describe the range of stakeholders in the system by their interests in and influence over resource decisions. *This information can assist policy makers in the development of sustainable natural resource management policy, taking into account the interest and influence of a wide range of stakeholders.*

35.4.2 Role of Natural Resources in the Lake Tana Region's Economy

Daregot Berihun examines the significant role of natural resources in the region's economy in Chap. 21. Tax data shows the contribution of activities such as agriculture and fisheries, tourism and navigation, trade and industry, and transport to the regional revenue is substantial. There is great potential for irrigation development and improvements to make agriculture more resource-efficient. Industry is also in an early stage of development. *These sectors need more attention to be developed with optimized use of the resources of the basin.*

35.4.3 Land Use Distribution and Change, Agriculture and Livestock

In Chap. 22, Amare Sewnet Minale and Wubneh Belete Abebe describe changes in different land cover units such as forest, wood and bush lands, grass, wetlands and water bodies, and farm and settlements. Expansion of agriculture and settlement lands, driven by poverty and population growth, has led to rapid deforestation and land degradation. But despite increased land under cultivation, farmer incomes have decreased and farmers have not produced enough food to meet demand, leading to further expansion of agricultural land and land degradation. *There needs to be better coordination between land distribution and family planning policies. A detailed inspection of land use and cover change is also crucial for those planners and decision makers to consider environmental and socioeconomic impacts in the region and design sustainable solutions.*

Merkuz Abera examines agriculture, the mainstay of the Ethiopian and Lake Tana region economy, in Chap. 23. Crop diversity is high, including cereals, legumes, root crops, oil crops, vegetables, fruit crops and other cash crops. Crop production includes rain-fed systems, irrigation, residual moisture and minor recession cropping. Agricultural inputs are rising, and insect pests and diseases are production constraints for farmers. Most farming systems are mixed crop and livestock production. In Chap. 24, Kefyalew Alemayehu and Asaminew Tassew explain livestock production in the region. Not only does livestock represent a significant portion of the local and national economy, it also serves many roles for smallholder farmers. It is a source of food, income from animal products, inputs to crop production (draught power and manure) and a major saving mechanism to help people buffer against crop failure. They also provide transportation, cultural value and fuel from dung. Cattle are the most important, followed by small ruminants (goats and sheep) and chickens. Donkeys, horses and mules provide power for crop production, transport to market. Then, crop residue feeds the livestock and manure puts nutrients back in the field. The many challenges to increasing farm animal productivity include limited access and low use of feed technology, inadequate veterinary service provision, genetic limitation of the indigenous livestock breeds, weak linkages among stakeholders, limited access to credit, and climate change. Increasing demand for livestock products also provides opportunities. *Research is needed to enhance the complementarities between crop and livestock production and integration with watershed management activities, remove barriers to livestock production, and capitalize on opportunities.*

35.4.4 Urban Areas

Ethiopia is currently one of the least urbanized countries with under 20% of its population living in urban areas, but it is expected to urbanize rapidly (around 4% per annum), over the next 40 years. Projections for the Lake Tana region are similar. In Chap. 25, Genet Gebre Eyesus reviews the distribution of urban areas and towns in the Amhara region, gives an overview of the type of infrastructure available, and explains the approach to planning for these urban areas. *With a growing population and high rate of urbanization, careful coordination of plans and resources will be critical in the region for sustainable development.*

35.5 Management

35.5.1 Policies, Institutions, and Practices for Environmental Resource Management

A number of legal frameworks for conservation exist in Ethiopia at the national and regional level for protecting resources in the Lake Tana basin. In Chap. 26, Solomon Addisu Legesse gives an overview of environmental protection measures in Ethiopia, including the 1997 Environmental Policy of Ethiopia, Environmental Protection Authority founded in 1994, National Policy on Conservation of Biodiversity, Environmental Impact Assessment requirements, and a variety of land use laws and water resources management proclamations. However, he points out there are major environmental challenges that stem from population pressure and related agricultural expansion, industrial activities, and energy demand. *The extent of environmental degradation, together with limited on-the-ground implementation of policy and limited stakeholder participation, strain existing policy framework and monitoring programs.* Solomon Addisu Legesse summarizes the problems of environmental protection in three categories. "First, the policies and strategies are too general and do not give adequate direction/guideline to those who are implementing them. Second, even when there are laws that could be implemented, there are cases where there are no institutions or capable institutions that implement them. Thirdly, there are overlapping and, sometimes contradicting objectives of different institutions and hence there is little cooperation and coordination among different institutions."

Belachew Getenet and Berihun Tefera examine institutions for environmental resource management further in Chap. 27. They ask whether the policies, laws, proclamations and organizations created to protect the environment fulfill their objectives and find that, on the whole, they do. The rate of land degradation is still greater than the land rehabilitation and conservation efforts, however. *They suggest that policy effectiveness can be further improved by genuinely engaging local people in conservation programs, coordinating government organizations better to reduce overlapping and conflicting roles and responsibilities, addressing conflicts of interest, organizing multi-stakeholder platforms, and focusing on the quality rather than quantity of conservation programs.*

In Chap. 28, Wubneh Belete Abebe and Amare Sewnet Minale discuss land use and watershed management practices in the Lake Tana basin. The major change in land use and land cover in the Lake Tana basin in the period 1986–2013 was a loss of forest land and an increase in cultivated land. Cultivation systems range from smallholder rain-fed crop production to larger-scale crop and livestock production to agroforestry. Some of the challenges facing existing land use systems in the basin include lack of access to inputs for improving productivity, lack of technical knowledge, natural hazards, and poor access to markets and veterinary services. Land use is also clearly connected with water management. The authors present a history of practices in rainwater harvesting, watershed-based development, and best practices for watershed management and the management of water pollution. They conclude that land degradation resulting from land use and land cover change limits the potential for sustainable livelihoods for people in the Lake Tana basin. *Despite the long history of land and watershed management efforts, there are barriers that should be addressed, including weak policy implementation, low technical and financial capacity, poor information management, minimal incentives, and frequent restructuring of core and principal institutions at national/regional levels. They recommend an integrated watershed management approach that addresses social and environmental as well as agricultural issues; a strong focus on of local target groups, especially women; and improvements to information and incentives.*

35.5.2 Lake Tana Biosphere Reserve

In Chap. 29, Ellen Kalmbach describes the history and process that led to the establishment in June 2015 of the Lake Tana Biosphere Reserve in the World Network of Biosphere Reserves under the UNESCO Man and the Biosphere program. This designation is the result of regional governmental initiatives dating back to 2005 that sought to integrate human development and conservation for the Lake Tana region. The biosphere reserve approach provides an integrative management approach which relies on the existing strong administrative structures and scientific capacity in the region and is well suited to its complex social and ecological characteristics. Stakeholders and government institutions have been an integral part of the development of the management structure that has been developed for the Reserve. *Some of the management challenges that remain include filling gaps in the management structure, setting up effective processes, and providing sufficient funding and authority to ensure its long-term viability. Coordination between the Lake Tana Biosphere Reserve and management of the larger Tana sub-basin will also be important.*

35.5.3 Farming Systems Innovation

In Chap. 30, Berihun Tefera describes an innovation in farming—the introduction of rice as a crop—that took advantage of an environmental challenge that was a long-standing problem for people in the Fogera district. The introduction of rice took advantage of the area's frequent flooding. With a change in resource use, this environmental characteristic has become an opportunity instead of a challenge. This research shows the potential for promoting such innovations where there are no cultural practices related to commodities targeted by extension interventions and *raises a question about what other opportunities exist for similar innovations.*

35.5.4 Trends and Driving Forces in Smallholder Eucalyptus Plantations

Berihun Tefera and Habtemariam Kassa describe the driving forces underlying a major landscape change in the Lake Tana basin—the establishment of large areas of eucalyptus plantations—in Chap. 31. The drivers include a variety of social and ecological factors, properties of the tree itself, as well as the economic and social context. This study indicates that eucalyptus plantation is expanding at the expense of crop production and will continue to do so, given the growing demand for wood and several characteristics that make it appealing for smallholder farmers to plant eucalyptus. *The authors recommend that mechanisms be developed to help rural households balance their choices and follow proper land use development.*

35.5.5 Waste Management

In Chap. 32, Biruk Abate and Goraw Goshu discuss waste management in the city of Bahir Dar. Rising population in the urban areas of the Lake Tana basin, especially in Bahir Dar city, brings with it increasing waste generation. Bahir Dar is like many cities, where the pace of waste generation exceeds the rate of infrastructure development. Limited financing and low prioritization of waste management are two challenges for waste management. Coordinated public and private sector efforts have improved waste management and are able to provide door-to-door collection of solid waste, but these efforts are not sufficient to address the problem fully. *The authors recommend that an integrated solid and liquid waste management practice be developed for the basin.*

35.6 Summary of Research Needs and Priorities

Two main themes have been repeated throughout this book: the need for an integrative approach and the need for better information about the resources and human activities in the region. In all areas, researchers point to limited information about the resources themselves, effects of human activity on environmental conditions, and practices that could avoid or remediate degradation as a barrier to sustainable resource management. Integration is needed between research and decision-making, across organizations responsible for making and enforcing rules about resource use, and across socioeconomic and ecological disciplines. Developing an integrated plan for sustainability that can support evidence-based decisions requires scientific research, technological tools and good policy practices. Scientific research provides data and knowledge that could be transferred to a user either to make an evidence-based decision or improve resource consumption or management practices.

35.7 Role of Bahir Dar University in the Way Forward

Bahir Dar University, with its motto of Wisdom at the Source of the Blue Nile, has a historical obligation to work towards the sustainability of the Tana Basin. Bahir Dar University's proposition of sustainability demands that human activities become guided by sound scientific analyses and recommendations. In order to carry out sound scientific and comprehensive analyses and give proper recommendations, comprehensive and holistic information about the region is integral. To date, a number of organizations have come to the region, established their own data acquisition infrastructure, analyzed that data, published some papers based on the data and closed the project. Often, unfortunately, research is repeated without knowing even a trace of the previous efforts. As a result it is hard to build a holistic understanding of the Tana Basin using continuous and reliable data. The realization of this unfortunate status quo makes decision and policy makers highly skeptical of the recommendations coming out of fragmented research efforts; as a result, policy makers often follow their own instincts and use economic dimensions to inform policies rather than working to find the delicate balance among all the three dimensions of sustainability.

Similarly, it is not easy for external or local researchers to find comprehensive and centralized information on the current challenges and researchable topics on the Tana Basin. Instead they need to investigate many publications by many different institutions around the world to find the missing gaps and challenges that need further investigation. External researchers also often overlook the research efforts of those grounded in the region.

Also, the general public does not have access to updated information on the Tana Basin and is therefore left without a proper educational input to shape their activities and to properly teach their children how to engage with their environment. This means we have a serious disconnection and fragmentation between our efforts in information gathering, scientific activities and actions by decision and policy makers and the general public. The way forward must create solid and dynamic connections among these players and establish a system that can provide holistic information about the Tana Basin, which in turn will be the basis of our scientific recommendations and actions. Bahir Dar University calls institutions, researchers, decision and policy makers and general public to join hands in its effort to establish an integrated and holistic system that can be used towards sustainability by carrying out three core projects mentioned below.

Project 1

The first project aims to establish a comprehensive and modern data acquisitions and management system on the region. The data infrastructure includes information about social aspects of the system (such as population, economic activities, and development projects) and natural resources (including soil, crop, water, forest, climate and other characteristics). The university has made good progress in terms of establishing infrastructure that can measure soil erosion (loading and unloading), precipitation, water quality, vegetation changes, evaporation and underground water, health, population and economic status and poverty levels. Also, advanced measurement tools including weather radar and phased array radar system have been initiated and installed by the university and in collaboration with other national and international institutions. This is significant progress and has allowed us to obtain a more holistic picture about the region.

However, still we have major gaps in measurement, which need to be filled in order to capture the spatial and temporal dynamics of the natural and social variables we are interested to monitor. Many measurement devices are needed to fill these gaps. Moreover, we have challenges in terms of acquiring the technology and the practical skills needed in collecting, transferring, storing, analyzing and distributing such a diversified and comprehensive data set. We welcome institutions and researchers to work with us in dealing with these challenges.

Project 2

The second project focuses on extracting holistic and useful scientific recommendations from the information gathered in the region. This project is basically a scientific project with major focus on establishing the system level science on the region. However, system level science is highly complex and the cause and effect analysis is extremely complicated and interwoven. It demands professionals who are traditionally divided and very often without communications among themselves. With volumes after volumes of publications and techniques available in each discipline and thereby each professionals being immersed in his/her perspectives, it is not an easy job to create effective communications among them. Bahir Dar University invites researchers, institutions, policy makers and general public to work with us to overcome these challenges and be able to extract system level science applicable for practical purpose.

Project 3

The third project scales up and extends the collaborative system dynamics analysis described in Chap. 34. The initial systems workshop produced a causal map and sector diagram showing connections among disciplines and highlighting points of leverage for research and decision-making. As noted by the participants in the first workshop, the analysis must be extended to include a wider diversity of stakeholders. Perspectives of stakeholders at all scales—from local farmers and direct users of other resources to urban stakeholders, to regional resource managers and national-level policy-makers—must be represented. A further step is to develop a fully operational simulation model from the extended causal map that could be used for identifying and evaluating potential policy levers to promote sustainability. Such a model should be developed using participatory methods and produced in a way that would suit ongoing local needs for decision-making and analysis. This third project aims to guide practical actions of the decision and policy makers, researchers and the general public using system level scientific recommendations. Scientific research should guide our actions by convincingly showing the cause and effect relationships. The human activities, strategies and policies and projects in the Tana Basin region can be analyzed in terms of their influence on social, environmental, political and economic issues. Thus if we have a system level analysis tool that can show such negative or positive influences in a convincing manner, the activities or the policies and strategies envisioned can be adjusted accordingly.

35.7.1 Conclusion

Bahir Dar University has a strong commitment to influence the general public and policy and decision makers to follow sound science in their actions. This means high level and policy relevant information needs to be synthesized and regularly disseminated in order to capture attention. The university needs to hold briefings regularly for all the relevant stakeholders about the status of the region. Regular publications like this book, magazines, and electronic information with frequent updates need to be established. Seminars, workshops and conferences with active participation of all stakeholders are also very important. Present technology allows us to provide real time usable, and relevant information using systems such as mobile phones, radio and TV. Bahir Dar University invites organizations, researchers and general public to join hands in our endeavor to create a dynamic scientific society. The way forward is to integrate all our efforts in data gathering, scientific activities and in creating a system level scientific descriptions for the Tana Basin, which can allow us to strike the delicate balance among all dimensions of sustainability.

References

Degraaf M, Machiels MAM, Wudneh et al. (2004) Declining stocks of Lake Tana's endemic Barbus species flock (Pisces; Cyprinidae): natural variation or human impact? BiolConserv 116:277–287

Nagelkerke LAJ, Sibbing FA (1996) Reproductive segregation among the large barbs (Barbus intermedius complex) of Lake Tana, Ethiopia. An example of intralacustrine speciation? J Fish Biol 49:1244–1266

Shimelis A, Afework B, Abebe G (2008) Species diversity, distribution, relative abundance and habitat association of the avian fauna of modified habitat of Bahir Dar and Debre Mariam Island, Lake Tana, Ethiopia. Int J of Ecol and Env Sci 34(3):259–267

Index

K. Stave et al. (eds.), *Social and Ecological System Dynamics*,
AESS Interdisciplinary Environmental Studies and Sciences Series,
DOI 10.1007/978-3-319-45755-0

FSC
www.fsc.org
MIX
Papier | Fördert gute Waldnutzung
FSC® C083411

Kolibri 360 GmbH
Ferdinand-Jühlke-Straße 7
99095 Erfurt, Deutschland
produktsicherheit@kolibri360.de